STRUCTURAL ANALYSIS

VOLUME II

STRUCTURAL ANALYSIS

VOLUME II

By

Dr. R. Vaidyanathan
B.E. (Hons), M.Sc.(Engg.), Ph. D., F.I.E., M.I.S.T.E.,
Formerly Principal, Government College
of Engineering, Tirunelveli
Tamil Nadu

Dr. P. Perumal
B.E., (Distn.), M.E. (Struct) , Ph. D., F.I.E., M.I.S.T.E.
(Professor and Head of Civil Engineering (Retd.)
Government College of Engineering Salem)
Professor, Dr. Mahalingam College of Engineering and
Technology, Pollachi
Tamil Nadu

LAXMI PUBLICATIONS (P) LTD
(An ISO 9001:2008 Company)

BENGALURU • CHENNAI • COCHIN • GUWAHATI • HYDERABAD
JALANDHAR • KOLKATA • LUCKNOW • MUMBAI • RANCHI • NEW DELHI
BOSTON (USA) • ACCRA (GHANA) • NAIROBI (KENYA)

Printed and bound in India

Second Edition: *2005;* Third Edition*: 2016, 2019*

New Edition : 2025

ISBN **:** 978-81-318-0782-8

Branches

✆	Bengaluru	080-26 75 69 30
✆	Chennai	044-24 34 47 26
✆	Guwahati	0361-254 36 69
✆	Hyderabad	040-27 55 53 83
✆	Kochi	0484-405 13 03
✆	Kolkata	033-40 04 77 79
✆	Lucknow	0522-430 36 13
✆	Ranchi	0651-224 24 64

Published in India by

LAXMI PUBLICATIONS (P) LTD

An ISO 9001:2015 Company

113, GOLDEN HOUSE, GURUDWARA ROAD, DARYAGANJ,

NEW DELHI - 110002, INDIA

Corporate Office : B-33, Sector-5, Noida,

Gautambuddha Nagar (U.P.) - 201 301, INDIA

Telephone : +91-120-44 46 900

www.laxmipublications.com

info@laxmipublications.com

CONTENTS

डॉ एन लक्ष्मणन
Dr. N. Lakshmanan
निदेशक
Director

स्ट्रक्चरल इंजीनियरिंग रिसर्च सेंटर-मद्रास
STRUCTURAL ENGINEERING RESEARCH CENTRE -MADRAS
वैज्ञानिक तथा औद्योगिक अनुसंधान परिषद्
(Council of Scientific & Industrial Research)
डा.सं/P.B. No. 8287, तरमणि, Taramani,
चेन्नै /Chennai - 600 113, भारत/INDIA

FOREWORD

The two volume publication on Structural Analysis by Dr. R.Vaidyanathan and his disciple Dr. P. Perumal is totally geared to meet the syllabus of Anna University.

The first volume deals comprehensively with analytical methods while volume two deals with moving loads and analysis of different structural configurations.

The speciality of the book is that it addresses the students as if a teacher would address them in a classroom.

Thanks to several decades of teaching Structural Engineering subjects in almost all the Government colleges in the state, the authors have had access to the students' minds and most of their doubts and difficulties and have used this book to answer them and clarify many a mystery relating to the behaviour of various structural shapes.

They have rendered the mathematical parts to painfree simplicity and concentrated on all practical applications of every structural shape like arches, suspension cable, rigid frames and curved beams.

I am sure students of Civil Engineering all over the world find these volumes immensely helpful in aiding their comprehension of structural behaviour and analysis.

(N. Lakshmanan)

Phone : (O) 2254 2139 (R) 2550 7550
Fax : 91-44-2254 1508 / 2254 1973
Grams : SERCENTER
E-mail : nlaxman@sercm.csir.res.in
director@sercm.csir.res.in
director@cssercm.ren.nic.in
http://www.sercm.org

PREFACE TO THE THIRD EDITION

We feel happy to note the continuous increase in the reception of the second edition of Structural Analysis – Volume II. The thoroughly revised book, now in its third edition includes improved/simplified calculations, addition of new problems on influence lines for trusses, influence lines for indeterminate structures, Arches, Cables and suspension bridges and space trusses. We have made the contents appropriate for a wide range of readers so that any student or professional would find the desired contents.

M/s Laxmi Publications have been very tolerant and patient with the Authors. Our thanks are due to M/s Laxmi Publications whose professional comments and suggestions have made value addition to this book. The authors owe the publishers a lot for their flexibility and accommodation.

The continued reinforcing feed backs and encouraging words from Professors and students all over, in various forms, acts as catalysts in our effort.

—Authors

PREFACE TO THE FIRST EDITION

The authors have great pleasure in presenting to the student community the second volume of their book on Structural Analysis. This book meets the syllabus for Structural Analysis II of Anna University, Tamil Nadu. However, this also will serve the students and teachers of all other Universities, AMIE students and practicing Engineers. The objective of this book is to present the material in a simple way. The authors have taken care to answer the doubts that they themselves had when they were students of Civil Engineering.

This volume contains eight chapters, namely Rolling loads, Influence lines, Influence Lines for Indeterminate Structures, Arches, Cables and Suspension bridges, Beams Curved in Plan, Space Trusses and Plastic Analysis of Structures. All the chapters are presented in a simple and lucid style with a lot of numerical examples and illustrative sketches for the benefit of the teachers and students.

A large number of problems are also available at the end of each chapter to enable the student to test his/her understanding of the subject. The authors have freely consulted many good books on the subject and the help is gratefully acknowledged.

The authors are aware that the book is not perfect. They would appreciate if errors discovered, if any, are brought to their attention. Suggestions from the readers for improvement of this book will be thankfully appreciated. The next edition is sure to take into account all such suggestions and would be more free from errors and omissions.

The authors are very confident that the staff and student community who are in need of a single, simple book covering the entire syllabus with simple presentation will get the fullest benefits.

—Authors

ACKNOWLEDGEMENTS

We are not sure we saw farther than others, but we had a lot of giants on whose slippery shoulders we could stand to peek into the intricacies of Structural Analysis. We thank them all.

First and foremost, we acknowledge the following authors whose books enlightened us in several aspects.

Punmia, Jain & Jain, Gupta Pandit & Gupta, Prakash Rao, Ramamrutham, Reddy, Khurmi, Negi and Jangid, Vazirani & Ratwani, Ramchandra, Junnarkar and Shah and Kinney.

We acknowledge help from the following students and research scholars in doing numerical verifications, compilation of problems and fair copying.

Ramalakshmi, Shanmugaraj, Kolanjinathan, Balamurugan, Prabhu, Manimegalai, Saravanan, Shirly, Mohanraj, Thiruvannamalai and Shaji.

The following were patient enough to unscramble our maze of numbers and sketches and gave them font and figures. We are really indebted to *L. Murugesan, DTP Designer, Mepco Schlenk Engineering College, Sivakasi and M.S. Shahul, M.S. Computers, Melappalayam, Tirunelveli.*

M/s Laxmi Publications have been really very quick, prompt, understanding and considerate when we were regularly passing deadlines and overhauling our own texts. We are thankful to them for their accommodative partnership and hope to deserve the same in future.

We both are greatly beholden to Seetha and Saraswathi, our wives and our family for putting up with our night owl habits and irregular mealtimes. They have suffered a lot and we are trying to prove they have not suffered in vain.

A lot of people gave us words of encouragement which sustained us in our grind. We say, "thank you folks; the words may not have cost you much, but they were invaluable to us".

Several sets of students have served as sounding boards for us, the brilliant, the slow, the inattentive and the indifferent. But all of them have helped us decide what kind of presentation would be effective.

We therefore dedicate this book to the student community.

—Authors

CHAPTER 1
ROLLING LOADS

1.1 INTRODUCTION

In Civil Engineering structures, loads can change their positions. In such a case, certain load positions can be critical to certain parts or components of the structure. Shifting of load positions is common enough in buildings. But they are more pronounced in bridges, over which vehicles keep rolling. Certain wheel positions can be again crucial. We should be able to identify such positions and their influence on structural members. In this chapter we will consider beams and try to evolve criteria for load positions causing maximum bending moments or maximum shears.

1.2 SIMPLY SUPPORTED BEAMS

1.2.1 Load Categories

We can consider 5 categories of loads on beams :

1. Single concentrated loads.
2. udl longer than the beam span
3. udl shorter than the beam span
4. Two wheel axles separated by a fixed distance
5. Multiple wheel axles (train of loads).

The objectives of this study on rolling loads are:

1. To find the load position and values of maximum shear force and bending moment at a given section due to a given system of rolling loads.
2. To find the location and values of the absolute maximum shear force and bending moment that may occur on the span due to the given system of rolling loads and
3. To find the equivalent udl due to a given system of rolling loads to make the designer's work simple.

1.2.2 Simply Supported Beam with a Single Concentrated Load W Moving from the Left End to the Right

As a load W moves from the left support A to the right we will try to get the maximum positive shear force, maximum negative shear force and the maximum bending moment at any typical section X, at a distance x from A. We will call the distance of the load from A as a. (see Fig. 1.1).

(*a*) **Maximum positive shear at X (Load in XB)**

For any load position (for any a),

$$R_A = \frac{W.(l-a)}{l} \qquad R_B = \frac{Wa}{l}$$

[when $a < x$, (when W is on AX), SF at X is equal to R_B and is negative ; when $a > x$ (when W is on XB), SF at X is equal to R_A and is positive.]

SF at X, $F_X = R_A = \frac{W(l-a)}{l}$, when $a \geq x$. ...(1.1)

The maximum value of positive F_x occurs when a is least, *i.e.*, $a = x$

and $F_{x\,max} = \frac{W(l-x)}{l}$ ...(1.2)

This is plotted in Fig. 1.1 (*c*).

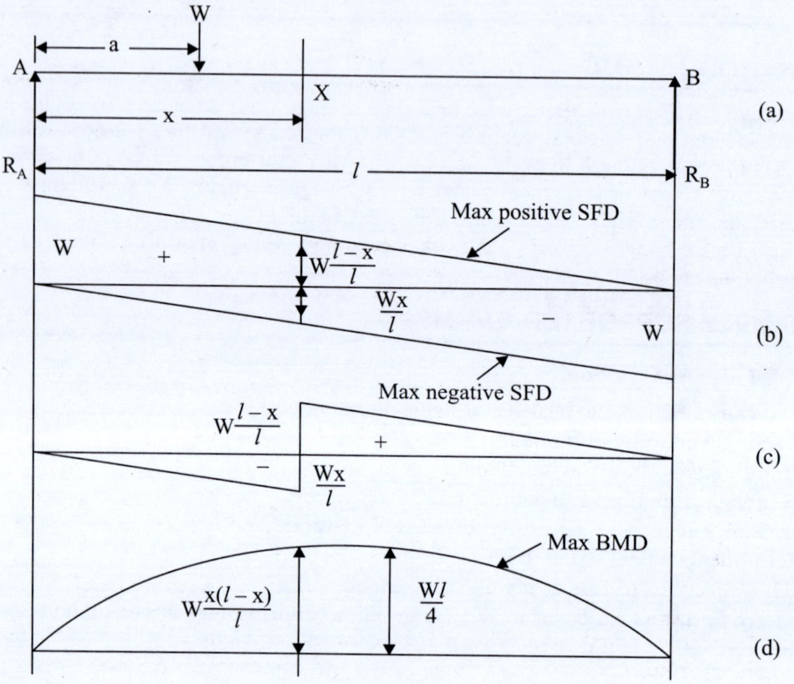

Fig. 1.1

(*b*) **Maximum negative shear at X**

Negative shear occurs at X when the load is on AX ($a \leq x$) and

$$F_x = -R_B = -\frac{Wa}{l} \quad \text{...(1.3)}$$

The maximum value of R_B occurs at X, when $a = x$. This is seen in Fig. 1.1 (*c*). Maximum negative shear, $F_{x\,max}$ is then $-\frac{Wx}{l}$

From the above 2 cases, we can conclude that the maximum positive shear occurs at a section X when the load is on the section and just to the right of it. Its value is $\frac{W(l-x)}{l}$. The maximum negative shear at X occurs when the load is at X (and just to the left of X). Its value is $\frac{Wx}{l}$.

(*c*) **Absolute maximum shear ($F_{max\ max}$)**

To get this, we can vary x to get $F_{max\ max}$.

$$F_{max\ max} + = \frac{W(l-0)}{l} = W, \qquad \text{when } x = 0 \qquad \ldots(1.4)$$

$$F_{max\ max} - = \frac{-W \cdot l}{l} = -W \qquad \text{when } x = l \qquad \ldots(1.5)$$

F_{max} + and F_{max} – are plotted in Fig. 1.1 (*b*)

(*d*) **Maximum bending moment diagram**

(*i*) Load on AX

$$M_x = R_B \cdot (l - x) = \frac{Wa}{l}(l - x) \qquad \ldots(1.6)$$

Here, the max value of a is x

$$M_{x\ max} = \frac{Wx}{l}(l - x) \qquad \ldots(1.7)$$

(*ii*) Load on XB

$$M_x = R_A\, x = \frac{W(l-a)}{l} \cdot x \qquad \ldots(1.8)$$

The maximum value of this occurs at $a = x$ and its value is

$$M_{x\ max} = \frac{W(l-x)}{l} \cdot x \qquad \ldots(1.9)$$

Obviously, (1.7) and (1.9) happen to be identical. It represents the bending moment at X when the load is at X ($a = x$)

When we plot (1.7), we get the diagram of maximum bending moment at X, when x varies from 0 to l. This is a parabola, [Fig. 1.1 (*d*)] and its maximum value occurs at $x = \frac{l}{2}$ and $M_{max\ max} = \frac{Wl}{4}$, this being the absolute maximum bending moment in a S.S. beam due to a moving load W. For $M_{max\ max}$ to occur, the load position is, $a = \frac{l}{2}$ and the section is at $x = \frac{l}{2}$.

1.2.3 S.S. Beams with a Moving udl Longer than the Span

This happens when long trains traverse small span bridges. The train wheels if they are close enough can be treated as transmitting a udl to the bridge girder.

Again we will consider the load front to be at a variable distance '*a*' from the L.H. support A. Let us consider a typical section X, at a distance x from A.

(*a*) **Maximum positive shear at X**

When the load head H is left of X, section X will have only negative shear, equal to R_B. For considering positive shear, the tail of the load T, has to be at a variable distance b from A. [Fig. 1.2 (*a*)]

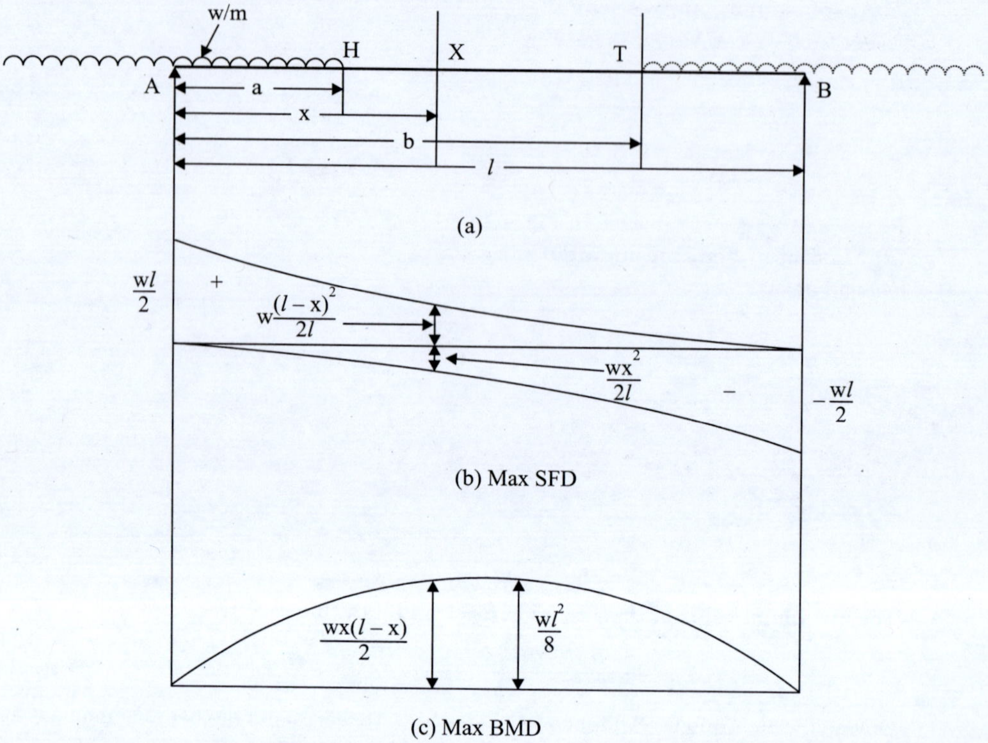

Fig. 1.2

(*i*) $b \le x$

when T is on AX, $R_A = w\dfrac{(l-b)(l-b)}{2l}$.

Then positive shear at X is

$$F_x = R_A - w(x-b) = w(l-b)\frac{(l-b)}{2l} - w(x-b)$$

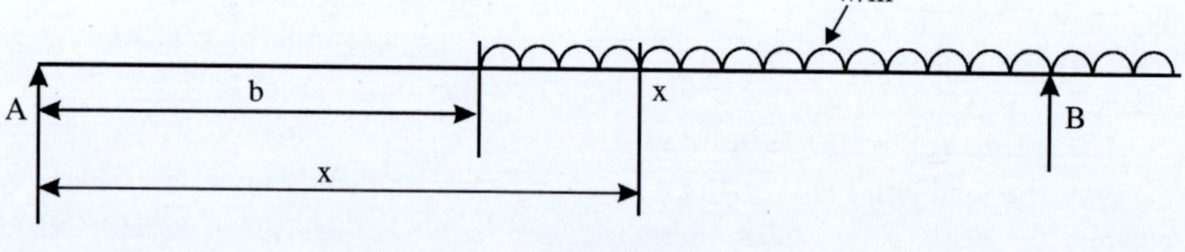

Fig. 1.3

when b is less than x, $(x - b)$ is positive. Hence F_x is a maximum when $b = x$

Hence, $$F_{x\,max} = \frac{w\,(l-b)^2}{2l} = \frac{w\,(l-x)^2}{2l} \quad ...(1.10)$$

We can plot $F_{x\,max}$ as x varies from 0 to l. This is a second degree curve and is shown in Fig. 1.2 (*b*). $F_{x\,max}$ is maximum when b is zero ; *i.e.,* when the load covers the entire span and occurs at A. Its value is $\frac{wl}{2}$.

(*ii*) $b > x$ need not be considered

(*b*) **Maximum negative shear**

By the reasoning made in section (*a*) above, the maximum negative shear at X occurs when the load head H is at X. Its value is

$$F_{x\,max} = -R_B = -\frac{wa^2}{2l} = -\frac{wx^2}{2l} \quad ...(1.11)$$

As 'a' varies from 0 to l, F_x varies parabolically from 0 to $-\frac{wl}{2}$. This is also plotted in Fig. 1.2 (*b*).

(*c*) **Maximum bending moment**

(*i*) Load head H on AX. ($a \le x$)

$$M_x = R_B\,.\,(l-x) = \frac{w\,.\,a^2\,.\,(l-x)}{2l}$$

This obviously is maximum when $a = x$ and

$$M_{x\,max} = \frac{w\,x^2\,(l-x)}{2l} \quad ...(1.12)$$

(*ii*) Load head H on XB ($a \ge x$)

$$M_x = R_B\,(l-x) - \frac{w\,(a-x)^2}{2}$$

$$M_x = \frac{w\,.\,a^2\,(l-x)}{2l} - \frac{w\,(a-x)^2}{2} \quad ...(1.13)$$

To maximize M_x, let us equate $\frac{\partial M_x}{\partial a}$ to zero.

$$+\frac{w\,a\,(l-x)}{l} - w(a-x) = 0$$

$$wa - \frac{w\,.\,a\,.\,x}{l} - wa + wx = 0$$

$$a = l, \text{ when } x \neq 0$$

Hence the load shall cover the entire span.

$$M_{x\,max} = \frac{wl}{2}(l-x) - \frac{w\,(l-x)^2}{2} = \frac{wx\,(l-x)}{2} \quad ...(1.14)$$

This will have a maximum value when $x = \frac{l}{2}$,

$$\left(\text{Since } \frac{\partial}{\partial x} M_{x\,max} = -\frac{wl}{2} + wx = 0 \quad \text{and} \quad x = \frac{l}{2}\right)$$

And $M_{max\ max} = \frac{wl^2}{8}$, and occurs at mid span when the load covers the entire span. It so happens that the max BM diagram is also the same as the BMD when the entire span is loaded uniformly.

1.2.4 S.S. Beams with a Moving udl Shorter than the Span

This situation occurs in long span railway bridges. This would give rise to more number of load positions than in the long udl case. Let us assume *a* udl of intensity w and length c. The load head H is at a distance 'a' from the left end A of the span. Let us consider the shears and bending moments at a typical section X, distance x from A. (Fig. 1.4)

(*a*) **Positive shear force F**

(*i*) $a < x$. (load head to the left of X)

Shear force at X = $-R_B$. This will not give any positive shear at X.

(*ii*) Let us shift the load such that the tail of the load is at a distance b from A (T to the right of X), as in Fig. 1.4 (*b*).

$$F_{x\ max} = R_A = \frac{wc}{l}\left(l - b - \frac{c}{2}\right) \quad ...(1.15)$$

F_x will have its maximum value when b is a minimum, *i.e.*, $b = x$. Then

$$F_x = \frac{wc}{l}\left(l - x - \frac{c}{2}\right) \quad ...(1.16)$$

Now let us explore if shifting the load train to the left of x would increase or decrease F_x. When b becomes $x - \Delta b$, equation (1.15) becomes

$$F_{x\ max} = \frac{wc}{l}\left(l - x + \Delta b - \frac{c}{2}\right) - w\Delta b, \quad \text{[see Fig. 1.4 (c)]}$$

$$F_x = \frac{wc}{l}\left(l - x - \frac{c}{2}\right) + \boxed{\frac{wc}{l}\Delta b - w\Delta b}$$

In the above equation, the quantity in the box is negative since $\frac{c}{l}$ is less than 1. So we conclude that if we move the load train to the left of X, the shear force does not increase. Hence equation (1.16) represents $F_{x\ max}$.

(*b*) **Negative shear force F_x**

(*i*) $a < x$

$$F_x = -R_B = -wc \cdot \frac{\left(a - \frac{c}{2}\right)}{l}$$

This will have the maximum value when $a = x$.

$$F_{x\ max} = -R_B = -\frac{wc}{l}\left(x - \frac{c}{2}\right) \quad ...(1.17)$$

Moving the load head H beyond X will not increase this value, by the same reasoning as for the positive shear force.

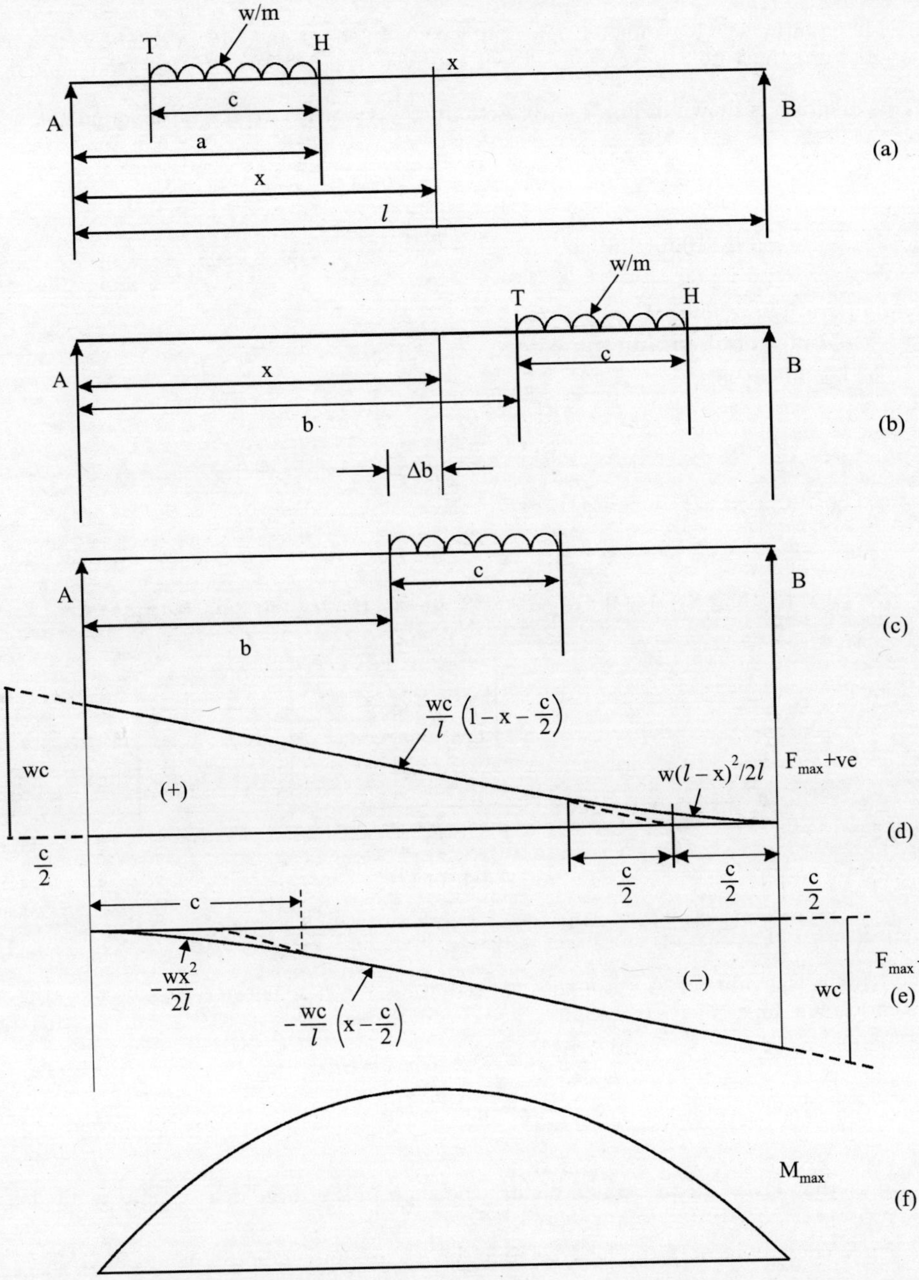
w/m
T
H
x
c
A
B
a
x
l
(a)
w/m
T
H
c
A
B
x
b
Δb
(b)
c
A
B
b
(c)
wc/l (l − x − c/2)
wc
(+)
w(l − x)²/2l
F max +ve
(d)
c/2
c/2
c/2
c/2
c
−wx²/2l
−wc/l (x − c/2)
(−)
F max −ve
wc
(e)
M max
(f)

Fig. 1.4

(*c*) **$F_{max\ max}$**

In equations (1.16) and (1.17), we can vary x. In equation (1.16), x can be varied from 0 to $(l - c)$. For values of $x > (l - c)$, the S.F. will be given by equation (1.10) (for loads longer than l). This diagram is shown in Fig. 1.4 (*d*). Actually F_{max} reaches wc at a notional point $x = -\frac{c}{2}$.

$$x = 0 \text{ to } (l - c) \qquad F_{x\,max} = \frac{wc}{l}\left(l - x - \frac{c}{2}\right)$$

$$x = (l - c) \text{ to } l \qquad F_{x\,max} = \frac{w\,(l - x)^2}{2l}$$

$F_{max\ max}$ –ve

Figure 1.4(*e*) shows $F_{max\ max}$ –ve.

(*d*) **Maximum bending moment**

(*i*) Load head on AX.

$$M_x = R_B \,.\, (l - x)$$

$$M_x = \frac{wc}{l}\cdot\left(a - \frac{c}{2}\right)(l - x)$$

(*ii*) Load tail on XB.

$$M_x = R_A\, x = \frac{wc}{l}\left(l - b - \frac{c}{2}\right)x$$

(*iii*) Load head on XB. Load tail on AX. X divides the load in 2 parts (Fig. 1.5).

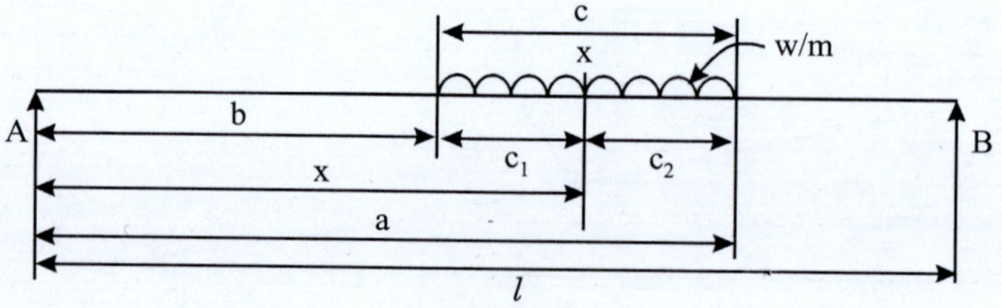

Fig. 1.5

$$M_x = R_A \,.\, x - \frac{w(x - b)^2}{2} = \frac{wc}{l}\left(l - b - \frac{c}{2}\right)x - \frac{w(x - b)^2}{2} \qquad \text{...(1.18)}$$

This is the value of M_x (at a chosen section X). We can choose the value of b (the load position) to maximize M_x.

$$\frac{\partial M_x}{\partial b} = 0 = -\frac{wc}{l}\,.\,x + w(x - b)$$

Hence,
$$\frac{x}{l} = \frac{x - b}{c} = \frac{c_1}{c} \qquad \text{...(1.19)}$$

Equation (1.19) means that X divides the span in the same ratio $\left(\frac{x}{l}\right)$ as it divides the load $\left(\frac{c_1}{c}\right)$. That represents the load position for maximum beading moment at X. $M_{x\,max}$ is obtained by substituting the value of b from equation (1.19) into equation (1.18).

Thus,
$$M_{x\ max} = \frac{wcx}{l}\left(l - x + \frac{cx}{l} - \frac{c}{2}\right) - \frac{w}{2}\frac{c^2x^2}{l^2}$$

$$M_{x\ max} = \frac{wcx}{l}(l - x)\left(1 - \frac{c}{2l}\right) \qquad \text{...(1.20)}$$

The absolute maximum bending moment is obtained by choosing x for maximum M_{max}. Differentiating $M_{x\ max}$ with x and equating to zero, we get $x = \frac{l}{2}$.

and
$$M_{max\ max} = \frac{wc}{4}\left(l - \frac{c}{2}\right) \qquad \text{...(1.21)}$$

Examples 1.1, through 1.5 demonstrate the application of the above formulae to numerical and practical problems involving rolling loads.

Example 1.1 *A single load of 100 kN rolls along a girder of 20 m span. Draw the diagrams of maximum bending moment and shear force (positive and negative). What will be the absolute maximum positive shear force and bending moment?*

Solution **Maximum bending moment and shear force:**

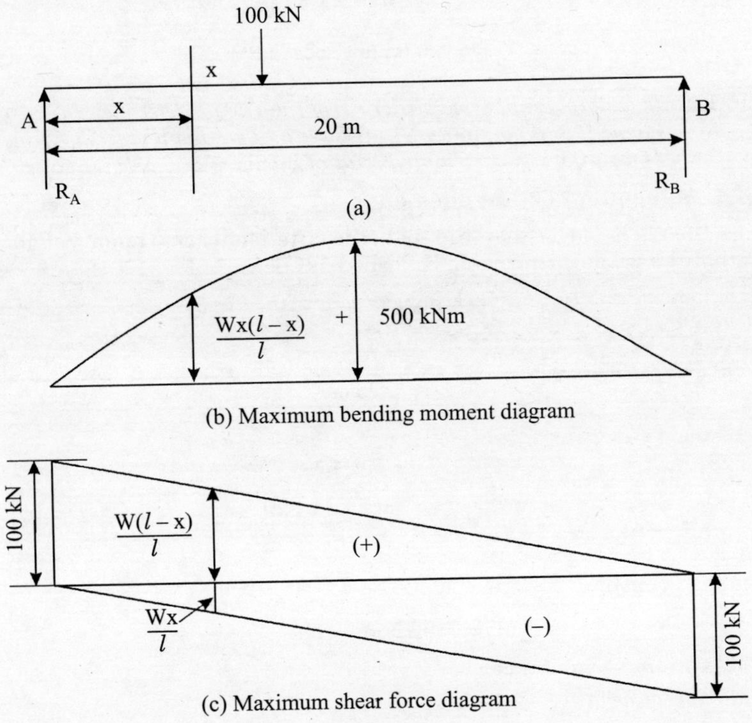

Fig. 1.6

(*i*) Absolute maximum bending moment :

$$M_{max\ max} = \frac{Wl}{4} \text{ and it occurs at mid span when the load is at mid span.}$$

$$M_{max\ max} = \frac{100\,(20)}{4} = 500 \text{ kNm}$$

(*ii*) Absolute maximum positive shear force = W = 100 kN. This occurs at A.

Note that Fig. 1.6(*b*) is not a BMD. It is an envelope of BMDs [Fig. 1.6(*d*)] for different values of '*a*'.

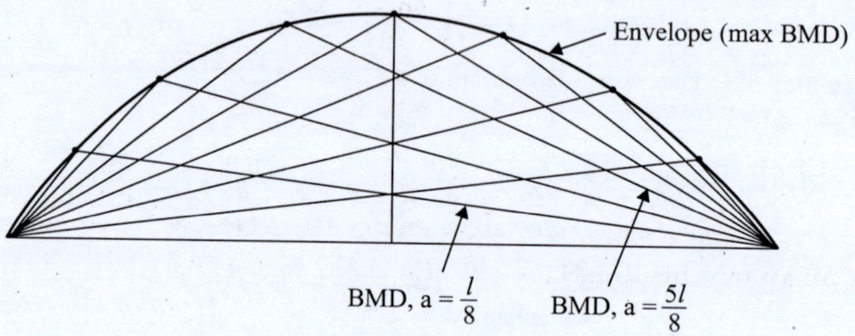

Fig. 1.6 (*d*) Envelope of BMDs.

Example 1.2 *In a simply supported girder AB of span 20 m, determine the maximum bending moment and maximum shear force at a section 5 m from A, due to the passage of a uniformly distributed load of intensity 20 kN/m, longer than the span.*

Solution 1. Maximum bending moment:

Since the udl is longer than the span, the criterion for maximum bending moment at a section is that the whole span should be loaded as shown in Fig. 1.7.

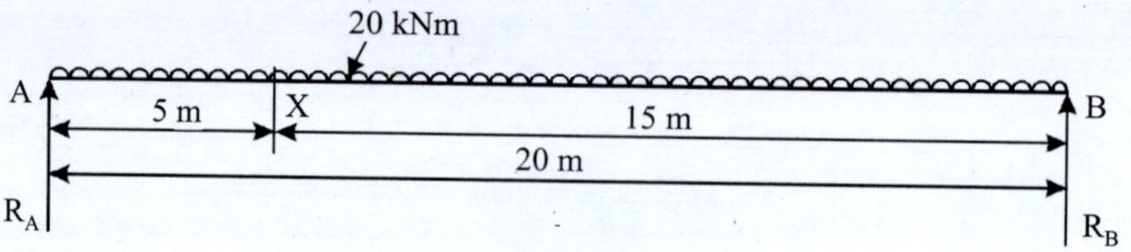

Fig. 1.7

$$R_A = R_B = \frac{wl}{2} = \frac{20 \times 20}{2} = 200 \text{ kN}$$

$$\text{Bending moment at } x = 5 \text{ m} = R_A\,(x) - \frac{20\,(x)^2}{2} = 200(5) - \frac{20\,(5)^2}{2}$$

$$M_x = 750 \text{ kNm}$$

2. Maximum shear force:

(*i*) *Maximum negative shear force :*

Maximum negative shear force at a section occurs when the head of the load reaches the section. (*i.e.,* when the left portion AX is loaded and the right portion XB is empty)

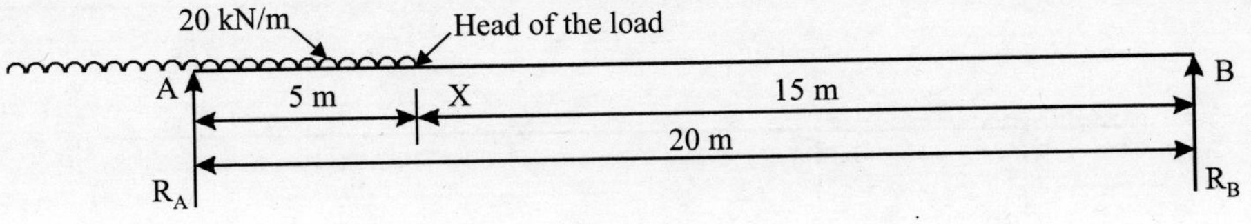

Fig. 1.8

Maximum negative shear force = – R_B

$$R_B = \frac{wx}{l} \cdot \frac{x}{2} = \frac{20 \times 5 \times 2.5}{20} = 12.5 \text{ kN}$$

Maximum negative shear force = – R_B = – 12.5 kN

(*ii*) *Maximum positive shear force :*

Maximum positive shear force occurs at X when the tail of the load is at X as it moves from left to right. (*i.e.,* AX is empty and the portion XB is loaded)

Maximum positive shear force = R_A

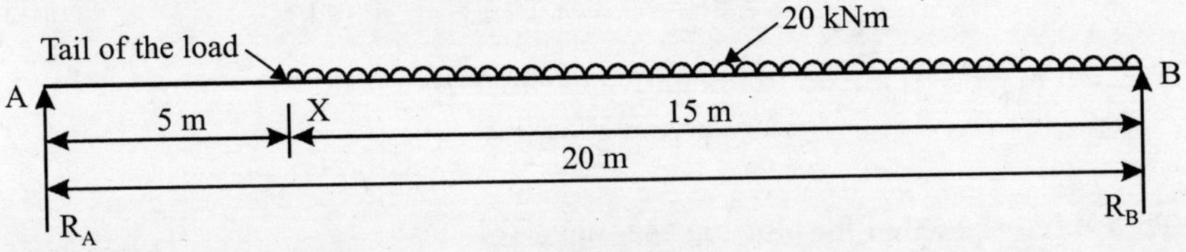

Fig. 1.9

$$R_A = w\,\frac{(l-x)}{l}\,\frac{(l-x)}{2} = \frac{20 \times 15 \times 15}{20 \times 2} = 112.5 \text{ kN}$$

Maximum positive shear force = R_A = 112.5 kN.

Example 1.3 *A uniform load of 40 kN/m run, 6 m long crosses a girder of 30 m span. Calculate the maximum shear force and bending moment at sections 5, 10 and 15 m from the left hand support. Construct the maximum shear force and bending diagrams.*

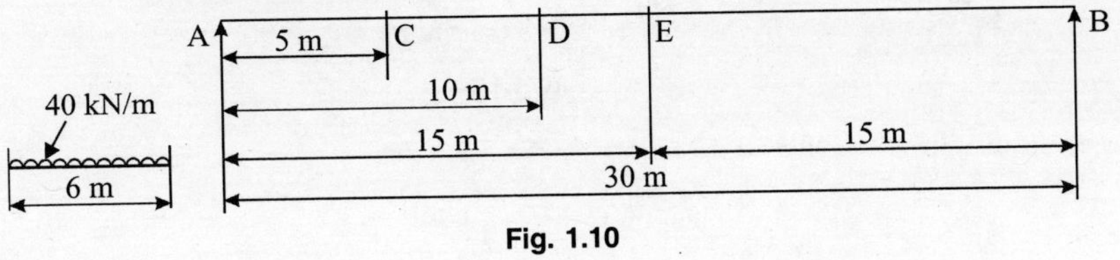

Fig. 1.10

c = 6 *m* ; *l* = 30 *m* ; *x* = 5 *m*, 10 *m* and 15 *m*.

Solution Let "*a*" be the distance of load head from A

Let "*x*" be the distance of the section X at which SF and BM are considered.

1. Maximum shear force:

(*a*) *Negative shear force:*

Maximum negative shear force at a section occurs when the head of the load is at the section as it traverses from left to right.

(i) when $a \leq 6$

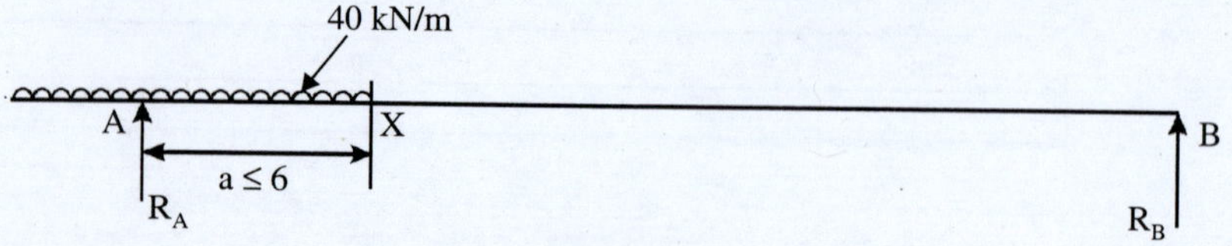

Fig. 1.11

Maximum negative shear force at X = – R_B

$$R_B = \frac{wa^2}{2l} = \frac{wx^2}{2l}, \text{ since } a = x \qquad \text{...(1.22)}$$

As equation (1.22) contains an x^2 term, R_B is a parabolic function of x. (*i.e.*, R_B varies parabolically when $x \leq 6$)

At x = 5 m, maximum negative shear force = – R_B

$$F_{c\,max\,-} = -\frac{wx^2}{2l} = -\frac{40 \times (5)^2}{2 \times 30} = -16.67 \text{ kN}$$

At x = 6 m, (x = a) maximum negative shear force = – R_B

$$= -\frac{wx^2}{2l} = -\frac{25 \times (6)^2}{2 \times 30} = -24 \text{ kN}$$

This value is needed for plotting the max. SFD

(ii) when $a \geq 6$ m

Maximum negative shear force at X = – R_B

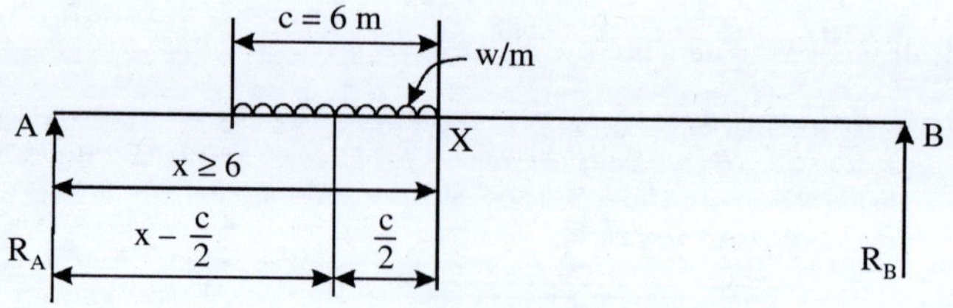

Fig. 1.12

Taking moments about A,

$$R_B \,.\, l - w \,.\, c \left(x - \frac{c}{2}\right) = 0$$

$$\text{Maximum negative shear force} = R_B = \frac{wc}{l}\left(x - \frac{c}{2}\right) \qquad \text{...(1.23)}$$

Equation (1.23) is a linear function of x for $x \geq 6$ m

At x = 6 m, maximum negative shear force

$$-R_B = -\frac{40 \times 6}{30}\left(6 - \frac{6}{2}\right) = -24 \text{ kN}$$

At x = 10 m, maximum negative shear force

$$-R_B = -\frac{40 \times 6}{30}\left(10 - \frac{6}{2}\right) = -56 \text{ kN}$$

At x = 15 m, maximum negative shear force

$$-R_B = -\frac{40 \times 6}{30}\left(15 - \frac{6}{2}\right) = -96 \text{ kN}$$

At x = 30 m, maximum negative shear force,

$$-R_B = -\frac{40 \times 6}{30}\left(30 - \frac{6}{2}\right) = -216 \text{ kN}$$

(*b*) *Positive shear force:*

Maximum positive shear force at a section occurs when the tail of the load is on the section as the load traverses from left to right.

(*i*) when $(l - x) \leq 6$ m (or) $x \geq 24$ m

Maximum positive shear force = R_A

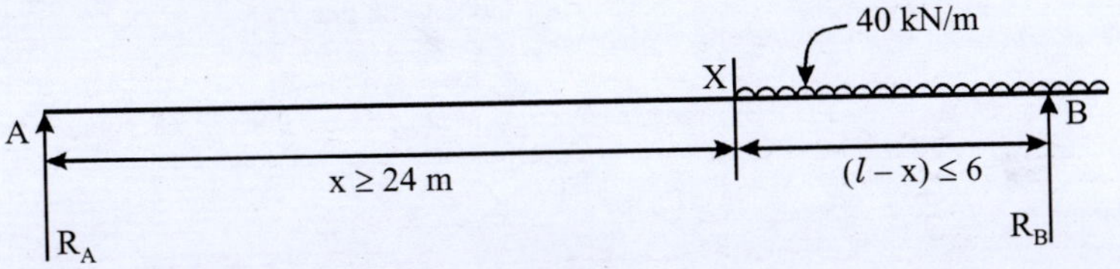

Fig. 1.13

$$R_A = \frac{w(l-x)^2}{2l}$$

As the above equation contains an x^2 term, R_A is a parabolic function of x for $(l - x) \leq 6$ m.

At x = 30 m, maximum positive shear force = R_A = 0

At x = 24 m, maximum positive shear force = $\dfrac{40\,(30-24)^2}{2 \times 30} = \dfrac{720}{30}$ = 24 kN

(*ii*) when $(l - x) \geq 6$ m *i.e.*, $x \leq 24$ m

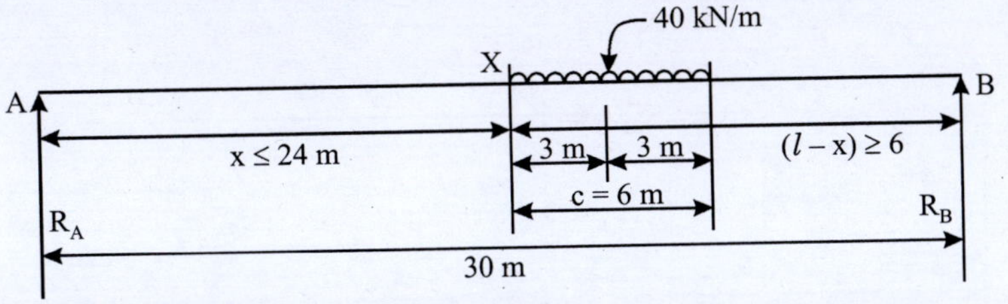

Fig. 1.14

Maximum positive shear force,

$$R_A = \frac{wc}{l}\left(l - x - \frac{c}{2}\right)$$

R_A varies linearly with x

At $x = 0$, $R_A = \dfrac{40 \times 6}{30}\left(30 - 0 - \dfrac{6}{2}\right) = 216$ kN

At $x = 5$ m, $R_A = \dfrac{40 \times 6}{30}\left(30 - 5 - \dfrac{6}{2}\right) = 176$ kN

At $x = 10$ m, $R_A = \dfrac{40 \times 6}{30}\left(30 - 10 - \dfrac{6}{2}\right) = 136$ kN

At $x = 15$ m, $R_A = \dfrac{40 \times 6}{30}\left(30 - 15 - \dfrac{6}{2}\right) = 96$ kN

At $x = 24$ m, $R_A = \dfrac{40 \times 6}{30}\left(30 - 24 - \dfrac{6}{2}\right) = 24$ kN

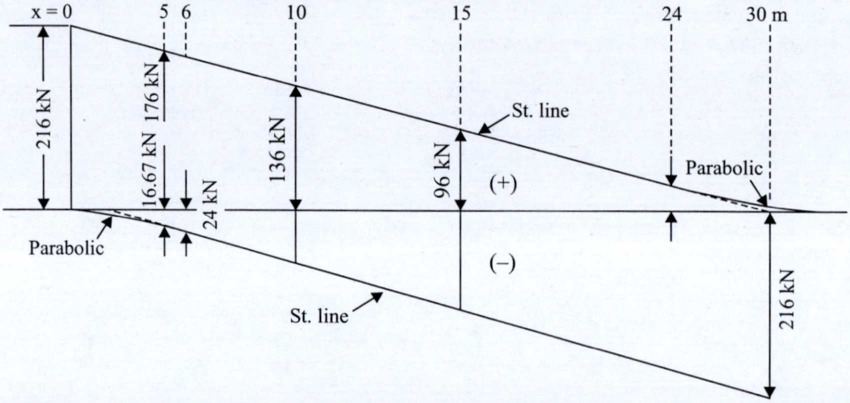

Fig. 1.15. Maximum shear force diagram.

2. Maximum bending moment diagram:

For getting the maximum bending moment at a section, the load should be placed in such a way that the section divides the load in the same ratio as it divides the span.

(*i*) At $x = 5$ m, *i.e.*, at C :

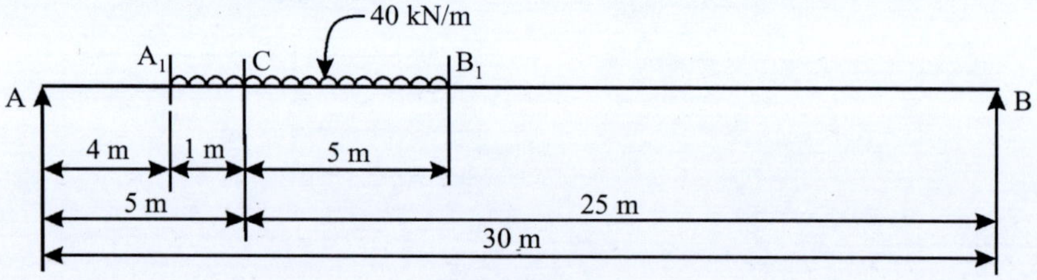

Fig. 1.16

$$\frac{A_1C}{B_1C} = \frac{AC}{BC} = \frac{5}{25} = \frac{1}{5}$$

$A_1C = 1$ m, $CB_1 = 5$ m

Hence, a = 10 m

Reactions at A and B :

$$R_B = \frac{wc}{l}\left(4+\frac{c}{2}\right) = \frac{40 \times 6}{30}\left(4+\frac{6}{2}\right) = 56 \text{ kN}$$

R_B = 56 kN

R_A = Total load – R_B = 40 × 6 – 56

R_A = 184 kN

Bending moment at $C = R_A\,(AC) - w\,\dfrac{(A_1C)^2}{2}$

$$(BM)_C = 184\,(5) - \frac{40\,(1)^2}{2} = 900 \text{ kNm}$$

(*ii*) At x = 10 m *i.e.*, at D

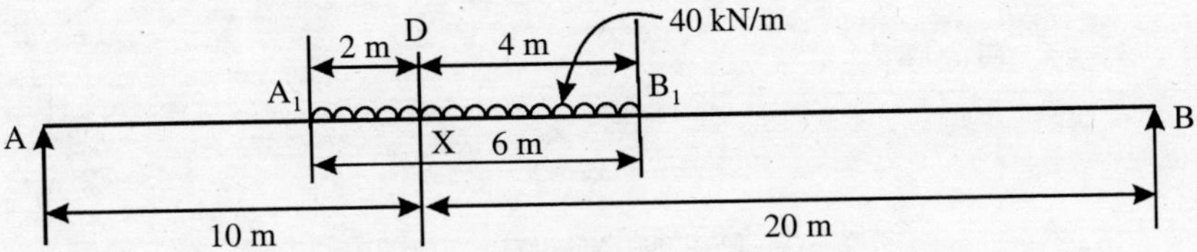

Fig. 1.17

To find the critical position of the load

$$\frac{AD}{BD} = \frac{A_1D}{B_1D} = \frac{10}{20}$$

A_1D = 2 m

B_1D = 4 m

Reactions at A and B:

$$R_B = \frac{wc}{l}\left((AD - A_1D) + \frac{c}{2}\right) = \frac{40\,(6)}{30}\left((10-2)+\frac{6}{2}\right) = 88 \text{ kN}$$

R_B = 88 kN

R_A = 152 kN

Bending moment at $D = R_A(AD) - w\,\dfrac{(A_1D)^2}{2} = 152\,(10) - \dfrac{40}{2}\,(2)^2$

$(BM)_D$ = 1440 kNm

(*iii*) At x = 15 m *i.e.*, at E :

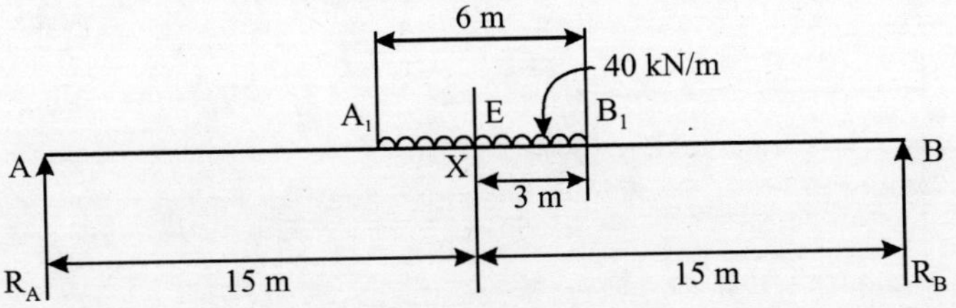

Fig. 1.18

$$\frac{A_1E}{B_1E} = \frac{AE}{BE} = \frac{15}{15}$$

So, $A_1E = B_1E$ = 3 m

Reactions at A and B :

$$R_A = R_B = 120 \text{ kN (by symmetry)}$$

$$\text{Bending moment at E, } (BM)_E = R_A\,(15) - \frac{w\,(A_1E)^2}{2}$$

$$= 120(15) - \frac{40(3)^2}{2} = 1012.5 \text{ kNm}$$

$$(BM)_E = 1620 \text{ kNm}$$

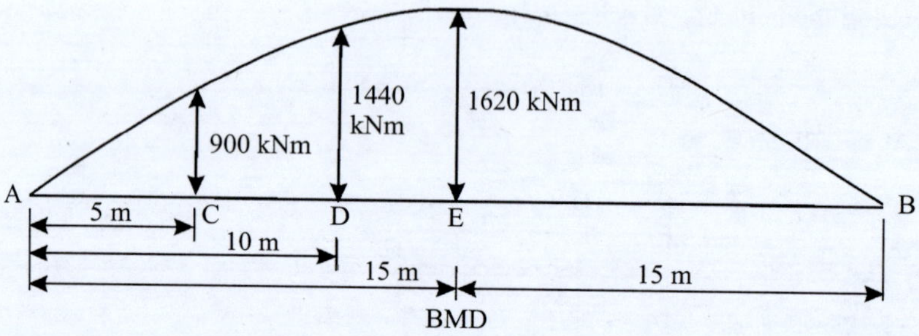

Fig. 1.19. Maximum bending moment diagram.

Example 1.4 *An udl of 10 kN/m, 4 m long rolls over a girder of 10 m span. Find the absolute maximum bending moment in the span.*

Solution

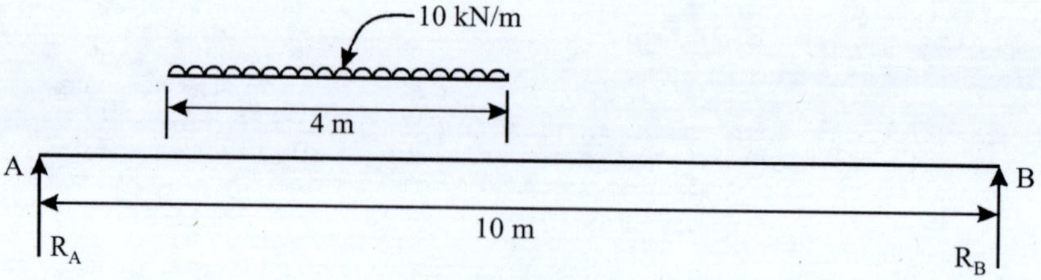

Fig. 1.20

Since the beam is subjected to udl shorter than the span the absolute maximum bending moment will occur at the centre when the udl is symmetrically placed at centre.

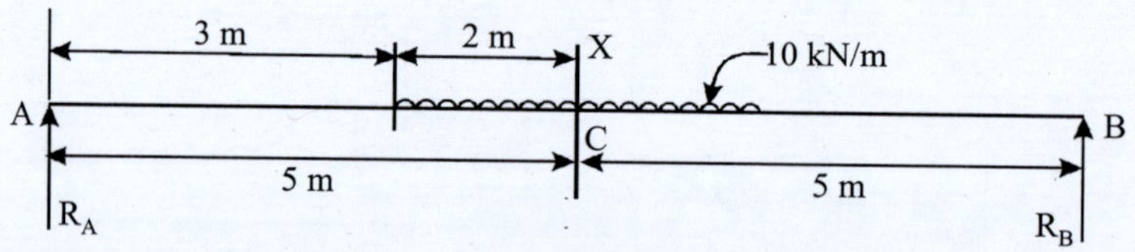

Fig. 1.21

VED
L-Struct\2str1-2.pmd IInd 21-6-13 3rd 25-9-14 4th 21-10-14
(Vol-II)

$$R_A = R_B = 20 \text{ kN}$$

From equation (1.20)

$$M_{x\,max} = \frac{wcx}{l}(l - x)\left(1 - \frac{c}{2l}\right)$$

$$c = 4,\ w = 10$$

This is a parabola with the central ordinate (at $x = l/2$) as $M_{max\,max}$.

$$M_{max\,max} \text{ (at C)} = R_A(5) - \frac{10(2)^2}{2}$$

$$M_{max\,max} = 100 - 20 = 80 \text{ kNm.}$$

Example 1.5 *A simple girder of 20 m span is traversed by a moving udl of 6 m length with an intensity of 20 kN/m, from left to right. Find the maximum bending moment and maximum positive and negative shear forces at sections 4 m from left support and 5 m from right support. Also find the absolute maximum bending moment that may occur anywhere in the girder.*

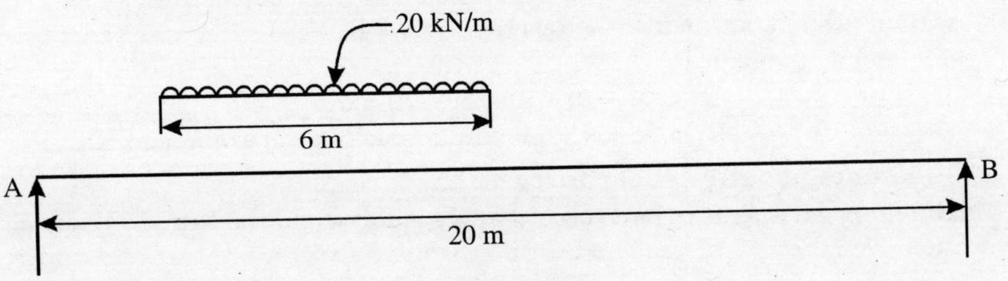

Fig. 1.22

Solution Case I. Section at 4 m from left support

(*a*) **Maximum bending moment :**

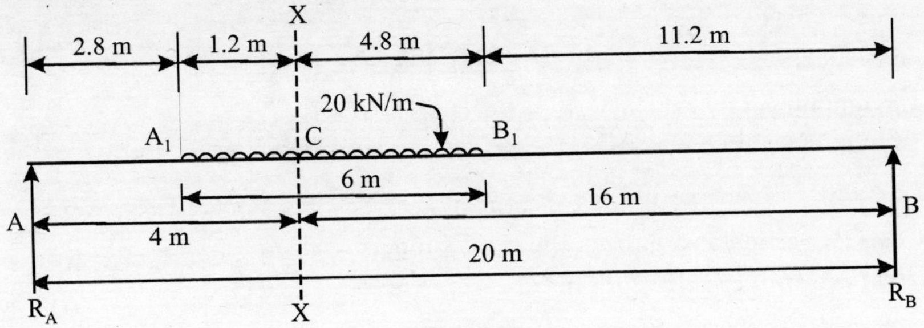

Fig. 1.23

In this case, the length of udl is shorter than the span. The maximum bending moment at a given section occurs when the section divides the load in the same ratio as it divides the span.

$$\frac{AC}{BC} = \frac{A_1C}{B_1C} = \frac{4}{16}$$

Hence, $\quad A_1C = 1.2 \text{ m},\ B_1C = 4.8 \text{ m}$

Taking moments about B,

$$R_A(20) - 20 \times 6 \times \left(11.2 + \frac{6}{2}\right) = 0$$

$$R_A = 85.2 \text{ kN}$$

Maximum bending moment at XX = R_A (4) – 20 × 1.2 × 0.6 = 85.2 (4) – 14.4

$$M_{max} = 326.4 \text{ kNm}.$$

(*b*) **Maximum positive shear force :**

Maximum positive shear force occurs at a section when the tail of the load is placed on the section.

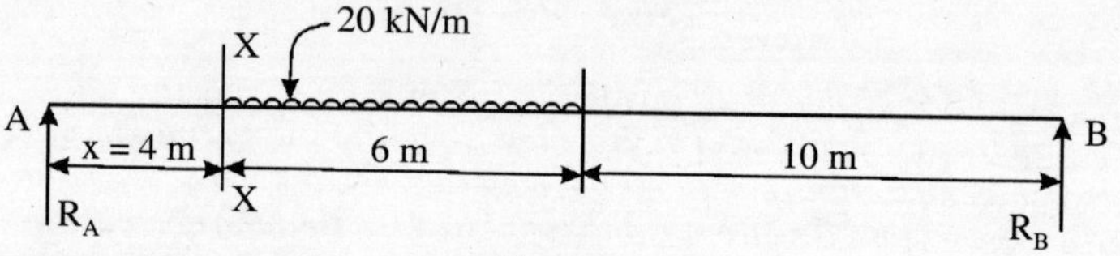

Fig. 1.24

Maximum positive shear force = Reaction at A = R_A

Taking moments about B,

$$R_A (20) - 20 \times 6 \times (10 + 3) = 0$$

R_A = 78 kN = Maximum positive shear force at XX.

(*c*) **Maximum negative shear force :**

Maximum negative shear force occurs at a section when the head of the load is placed on the section.

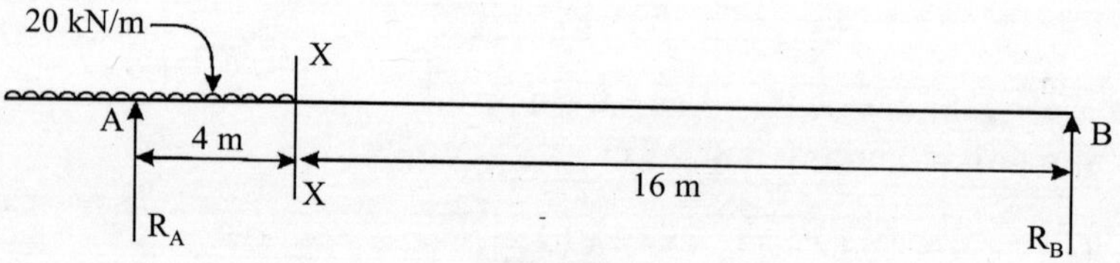

Fig. 1.25

Maximum negative shear force = Reaction at B = R_B

Taking moments about A,

$$R_B (20) - 20 \times 4 \times 2 = 0$$

R_B = 8 kN = Maximum negative shear force at XX.

Case II. Section at 5 m from the right support

(*a*) **Maximum bending moment :**

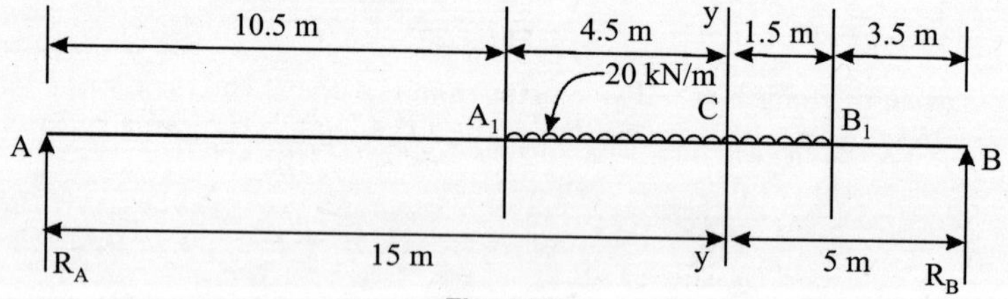

Fig. 1.26

$$\frac{A_1C}{B_1C} = \frac{AC}{BC} = \frac{15}{5}$$

$$A_1C = 4.5 \text{ m}, B_1C = 1.5 \text{ m}$$

Taking moments about A,

$$R_B(20) - 20 \times 6 \times \left(10.5 + \frac{6}{2}\right) = 0$$

$$R_B = 81 \text{ kN}$$

Maximum bending moment at $yy = R_B(5) - 20 \times 1.5 \times \frac{1.5}{2} = 81(5) - 22.5$

$$M_{max} \text{ at } yy = 382.5 \text{ kNm}$$

(*b*) **Maximum positive shear force :**

Maximum positive shear force occurs when the tail of the load is placed at the section.

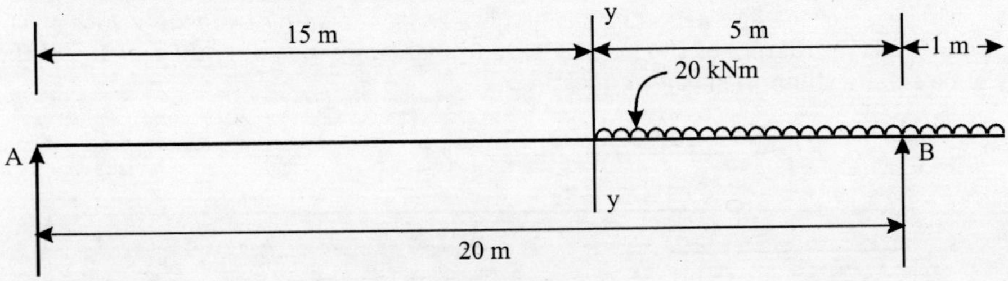

Fig. 1.27

Maximum positive shear force = Reaction at A = R_A

Taking moments about B,

$$+R_A(20) - \frac{20 \times (5)^2}{2} = 0$$

$$R_A = 12.5 \text{ kN}$$

$$F_{max} + = 12.5 \text{ kN}$$

(*c*) **Maximum negative shear force :**

Maximum negative shear force occurs when the head of the load is on the section.

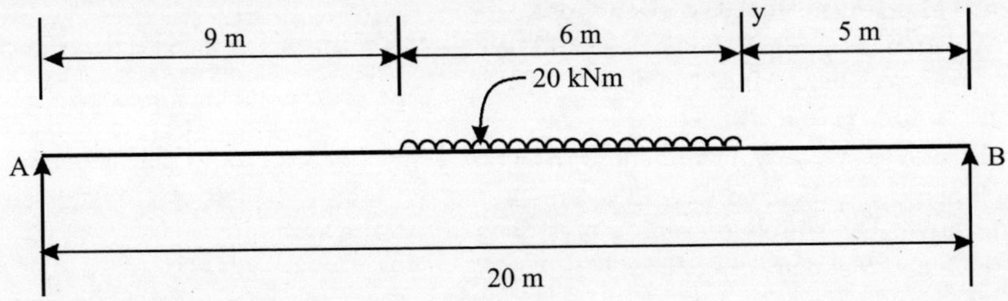

Fig. 1.28

Maximum negative shear force = Reaction at B = R_B

Taking moments about A,

$$R_B\,(20) - 20 \times 6 \times \left(9 + \frac{6}{2}\right) = 0$$

$$R_B = 72 \text{ kN}$$

$$F_{max} - = -\,72 \text{ kN}$$

Case III. Absolute maximum bending moment $M_{max\,max}$.

$M_{max\,max}$ would occur at **midspan** when the load is symmetrical.

$$R_A = R_B = \frac{120}{2} = 60 \text{ kN}$$

$$M_{max\,max} = 60 \times 10 - 20 \times 3 \times \frac{3}{2} = 510 \text{ kNm.}$$

1.2.5 S.S. Beams with Two Concentrated Loads

This case of loading arises in bridge girders traversed by two-axle vehicles. This also occurs in gantry girders. For the loading case to make any meaning, the spacing of the axles has to be smaller than the span in question.

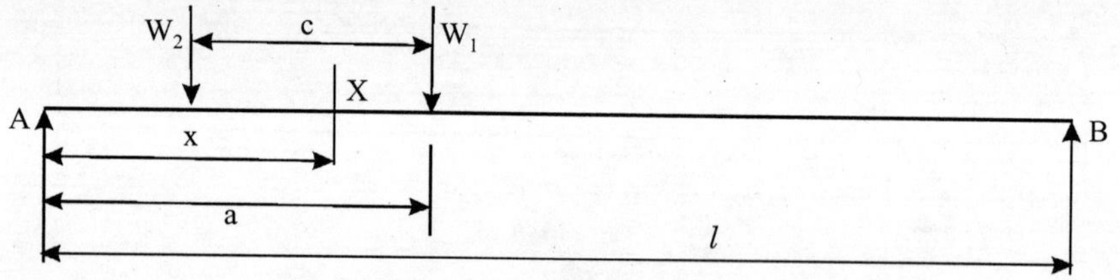

Fig. 1.29

Figure 1.29 shows a pair of rolling loads W_1 and W_2 traversing a span l. We will investigate the SF and BM at X, x from A.

We can readily identify 3 cases.

(*i*) $a \le x$, both loads to the left of X, the section under consideration.

(*ii*) $a \ge x$ and $(a - c) \le x$, when the loads are on either side of X, and

(*iii*) $(a - c) \ge x$, when both loads are to the right of X.

We can now discuss the maximum shears and maximum bending moment at X.

(*a*) **Maximum positive shear at X :**

(*i*) *When* $a \le x$. Both the loads would be on AX if a is $< x$. In this case, section X, will not get any positive shear, since the shear

$F_{x_1} = R_A - (W_1 + W_2)$

R_A due to $(W_1 + W_2)$ will be less than $(W_1 + W_2)$; hence F_x will not be positive.

When $a = x$, then, on W_2 will be on AX, and W_1 just to the left of X. Then also F_x will be only negative. The simple reasoning process is left to the student.

(*ii*) *When the loads are on either side of X :*

Now $(x + c) \ge a \ge x$

$$F_x = R_A - W_2 = \frac{W_1\,(l - a)}{l} + \frac{W_2\,(l - a + c)}{l} - W_2$$

$$F_x = \frac{W_1(l-a)}{l} + \frac{W_2(c-a)}{l}$$

When W_1 is just to the right of X, $a = x$, and

$$F_{x2} = \frac{W_1(l-x)}{l} + \frac{W_2(c-x)}{l} \quad ...(1.24)$$

The second part of the RHS is negative. So F_x will be positive when the W_2 related part is small. The values of F_{x2} are plotted in Fig. 1.30.

(*iii*) When both loads are on XB. $a > (x + c)$

$$F_{x3} = R_A = \frac{W_1(l-a)}{l} + \frac{W_2(l-a+c)}{l} \quad ...[1.24(a)]$$

Obviously F_x will be maximum when the negative quantity 'a' in equation (1.24) is least. *i.e.*, when $a = x + c$

$$F_{x\,max} = \frac{W_1(l-x+c)}{l} + \frac{W_2(l-x)}{l} \quad ...(1.25)$$

$F_{x\,max}$ is plotted in Fig. 1.31 (*a*). The plot is in 2 parts. The W_1 part ranging from $-c$ to $(l - c)$ and the W_2 part from 0 to l.

(*b*) **Maximum negative shear F_x:** This obviously would occur in either case (*i*) or case (*ii*) above

(*i*) When both W_1 and W_2 are on AX,

$$F_x - = -R_B = -\frac{W_1 a}{l} - \frac{W_2(a-c)}{l}$$

This gets its maximum negative value when a is maximum. *i.e.*, when $a = x$

$$F_{x\,max} - = -\frac{W_1 x}{l} - \frac{W_2(x-c)}{l} \quad ...(1.26)$$

This is plotted in Fig. 1.31 (*b*). The W_1 part ranges from 0 to l and the W_2 part from c to $(l + c)$.

(*ii*) When W_1 and W_2 are on either side of X.

$$F_x = -R_B + W_1 = -\frac{W_1 a}{l} - \frac{W_2(a-c)}{l} + W_1$$

$$F_x = -\frac{W_2(a-c)}{l} + \frac{W_1(l-a)}{l}$$

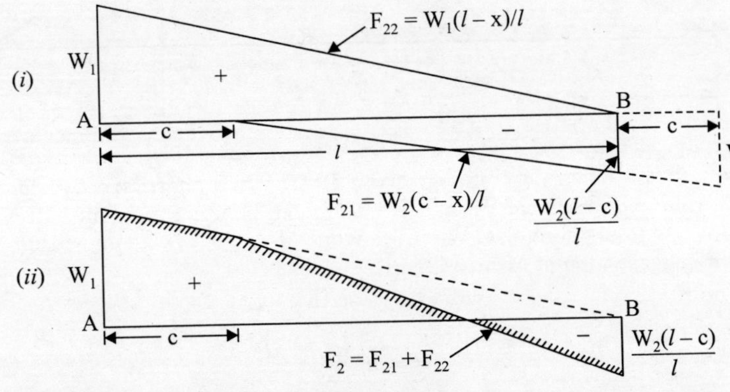

Fig. 1.30

F_x would be max negative only if W_2 is dominant and when a has the highest value ; *i.e.*, $a = x$. The plot will be similar to Fig. 1.30 and is not of much significance.

(*iii*) When both W_1 and W_2 are on XB. There will be no negative shear.

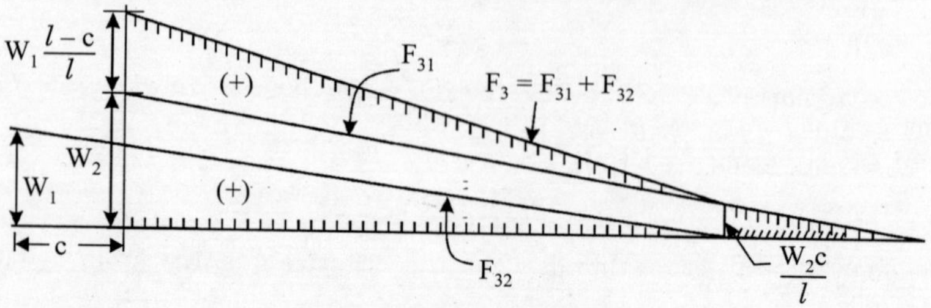

(a) Max positive SF diagram.

$F_{x\,max}$ is the bigger of F_2 and F_3

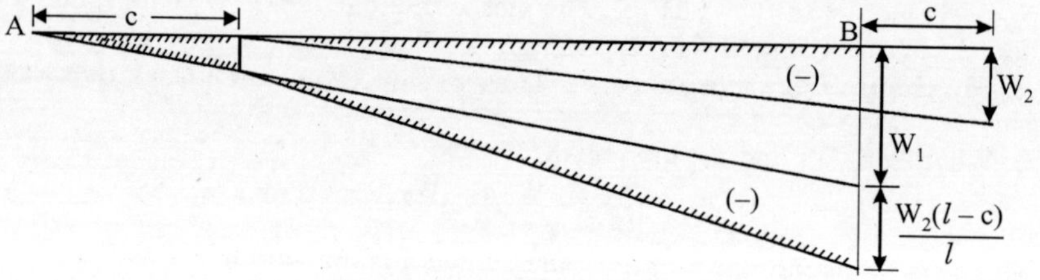

(b) Max negative SF diagram.

Fig. 1.31

(*c*) **Maximum bending moment at X :**

(*i*) $a - c = x$ and (*ii*) $a = x$

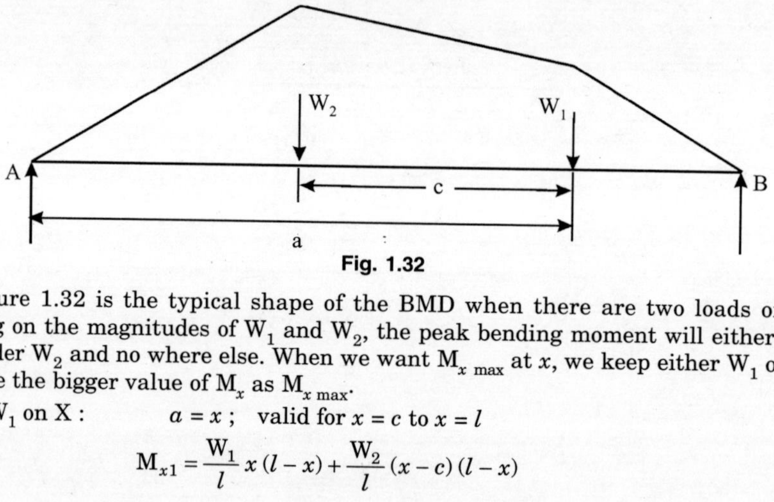

Fig. 1.32

Figure 1.32 is the typical shape of the BMD when there are two loads on a span. Depending on the magnitudes of W_1 and W_2, the peak bending moment will either be under W_1, or under W_2 and no where else. When we want $M_{x\,max}$ at x, we keep either W_1 or W_2 at X and choose the bigger value of M_x as $M_{x\,max}$.

(*i*) W_1 on X : $a = x$; valid for $x = c$ to $x = l$

$$M_{x1} = \frac{W_1}{l} x (l - x) + \frac{W_2}{l} (x - c)(l - x) \qquad ...(1.27)$$

(*ii*) W_2 on X : $(a - c) = x$; valid for $x = 0$ to $x = (l - c)$

$$M_{x2} = \frac{W_1(x)}{l}(l - x - c) + \frac{W_2(x)}{l}(l - x) \qquad ...(1.28)$$

It is again obvious that among M_{x_1} and M_{x_2} above, the bigger value will be the one occurring under the larger load between W_1 and W_2. However, at some regions in the beam, BM under the smaller load can be more. So when it comes to designing the beam, we may have to super pose M_{x_1} and M_{x_2} and draw the envelope of the two curves in Fig. 1.33 and Fig. 1.34.

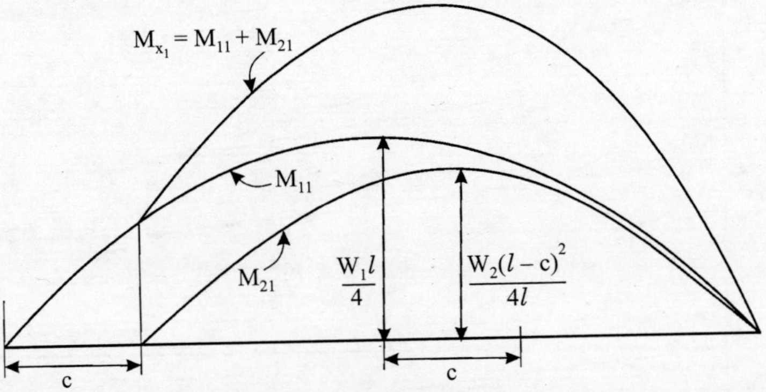

Fig. 1.33

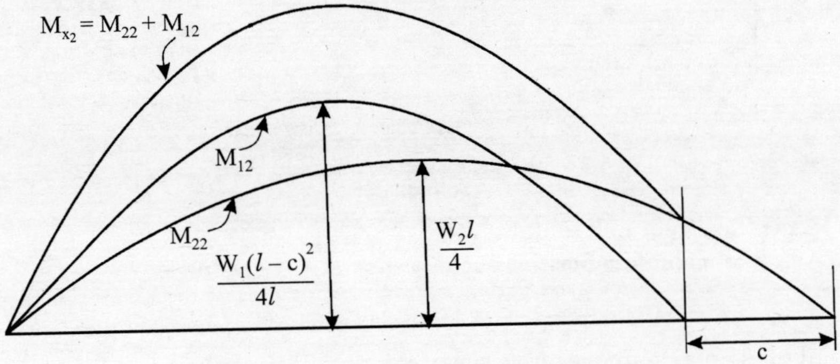

Fig. 1.34

Example 1.6 *Two wheel loads of 80 kN and 160 kN cross a span of 10 m from left to right. The 80 kN wheel is 2 m ahead of the 160 kN wheel. Find the following force values and their locations.*

(*i*) *The absolute maximum positive shear*

(*ii*) *The absolute maximum negative shear*

(*iii*) *The absolute maximum bending moment.*

Solution We can directly plot the maximum shear force diagrams based on equations (1.25) and (1.26).

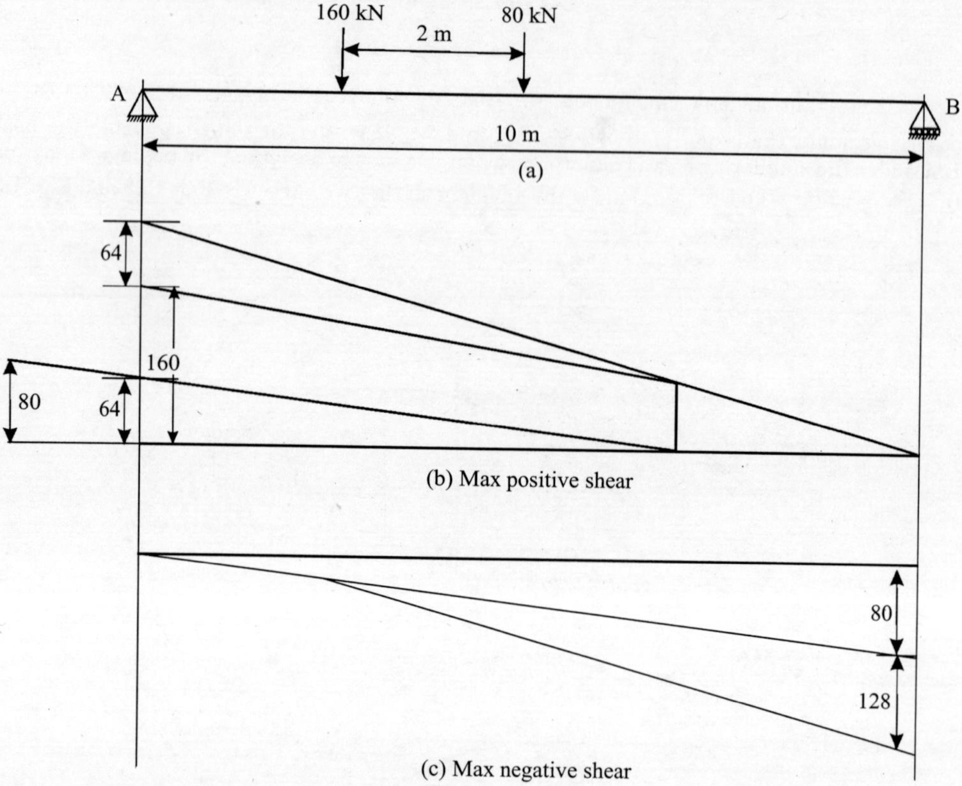

Fig. 1.35

$F_{x\,max}$ is the bigger of F_2 and F_3.

(*i*) Absolute maximum positive shear occurs at A. Its value is

$$W_2 + \frac{W_1(l-c)}{l} = 160 + 64 = 224 \text{ kN}$$

(*ii*) Absolute maximum negative shear occurs at B. Its value is

$$W_1 + \frac{W_2(l-c)}{l} = 80 + 128 = 208 \text{ kN}$$

(*iii*) Absolute maximum BM occurs either under W_1 (80 kN) or under W_2 (160 kN)

$$M_{max\,1} = W_1 \frac{x\,(l-x)}{l} + W_2 \frac{(x-c)\,(l-x)}{l}$$

$$c = 2 \text{ m} \quad l = 10 \text{ m}, \quad W_1 = 80 \text{ kN}, \quad W_2 = 160 \text{ kN}$$

$$M_{max\,1} = \frac{80\,.\,x(10-x)}{10} + \frac{160\,(x-2)\,(10-x)}{10}$$

$$M_{max\,1} = 80x - 8x^2 - 16x^2 + 192x - 320$$

This is maximum when $\dfrac{\partial M_{max}}{\partial x} = 0 ; -48x + 272 = 0 ; x = 5.67$ m.

$$M_{max\ max1} = 450.6 \text{ kNm}$$

$$M_{max\ 2} = W_1 \frac{x(l-x-c)}{l} + W_2 \frac{x(l-x)}{l} = 224\,x - 24x^2$$

$$\frac{\partial M_{max\ 2}}{\partial x} = 224 - 48x = 0$$

$$x = 4.67$$

$$M_{max\ max\ 2} = 522.67 \text{ kNm}$$

This occurs at $x = 4.67$ m, under the 160 kN load.

1.2.6 Equivalent Uniformly Distributed Load (EUDL)

As discussed earlier, moving live loads on a structure are of five types. They may also be a combination of distributed and concentrated loads. Such loadings involve tedious computations to determine the required stress components. A given system of loading crossing a girder/ structure can always be replaced by a uniformly distributed load, longer than the span, such that the bending moment or S.F. due to this static load, every where, is at least equal to that caused by the actual system of moving loads. Such a static load is known as equivalent uniformly distributed load. This practice is often adopted in the specifications for railway loads in order to simplify analytical computations.

The EUDL is derived by equating the maximum bending moment (or) shear force due to actual loading to the bending moment (or) shear force in the same span due to a fictitious udl covering the entire span. The EUDL may be different for bending moment and shear force for the same span, and depends upon the span and the disposition of actual loading.

The bending moment diagram for EUDL will be a parabola and this must completely envelope the maximum bending moment diagram due to the given moving load.

The EUDL for bending moment will be denoted by w_{em} and for shear force will be denoted by w_{ef}

Generally, the EUDL of bending moment will be obtained by equating the $M_{max\ max}$ due to the given load system and the max. Bending moment due to EUDL as,

$$\frac{w_{em}\, l^2}{8} = M_{max\ max} \quad \therefore \quad w_{em} = \frac{8M_{max\ max}}{l^2} \qquad ...(1.29)$$

The EUDL for few cases of loading are obtained as below :

(*i*) **Single concentrated load**

(*a*) *EUDL for bending moment:*

Due to a single concentrated load, W on a span l

Maximum bending moment, $M_{max\ max} = \dfrac{Wl}{4}$

Maximum bending moment due to EUDL = $\dfrac{w_{em}\, l^2}{8}$

Equating the two,

$$\frac{w_{em}\, l^2}{8} = \frac{Wl}{4}, \quad w_{em} = \frac{Wl}{4} \times \frac{8}{l^2}, \quad w_{em} = \frac{2W}{l} \qquad ...(1.30)$$

(*b*) *EUDL for shear force:*

Maximum shear force due to a single concentrated load, W = W

Maximum shear force due to EUDL = $\frac{w_{ef}\, l}{2}$

Equating the two, $\frac{w_{ef}\, l}{2} = W, \quad w_{ef} = \frac{2W}{l}$

Hence, in the case of a single concentrated load, the EDUL for bending moment and shear force are the same.

(*ii*) **UDL shorter than the span**

(*a*) *EUDL for bending moment:*

Maximum bending moment at the centre of the span, in a beam of span l and subjected to a udl shorter than the span and of length c [From eqn. (1.21)].

$$M_{\text{max max}} = \frac{wc}{4}\left(l - \frac{c}{2}\right)$$

Maximum bending moment at the centre of the span due to

$$\text{EUDL} = \frac{w_{em}\, l^2}{8}$$

Equating,

$$\frac{w_{em}\, l^2}{8} = \frac{wc}{4}\left(l - \frac{c}{2}\right)$$

$$w_{em} = \frac{8}{l^2} \cdot \frac{wc}{4}\left(l - \frac{c}{2}\right)$$

$$w_{em} = \frac{2wc}{l^2}\left(l - \frac{c}{2}\right) \qquad \ldots(1.31)$$

(*b*) *EUDL for shear force:*

Maximum shear force at the supports due to udl shorter than the span = $\frac{wc}{l}\left(l - \frac{c}{2}\right)$

From eqn. (1.17) with $x = l$

Maximum shear force due to EUDL = $\frac{w_{ef}\, l}{2}$

Equating,

$$\frac{w_{ef}\, l}{2} = \frac{wc}{l}\left(l - \frac{c}{2}\right)$$

$$w_{ef} = \frac{2}{l}\, \frac{wc}{l}\left(l - \frac{c}{2}\right)$$

$$w_{ef} = \frac{2wc}{l^2}\left(l - \frac{c}{2}\right)$$

In this case also, the EUDL for bending moment and shear force are the same.

VED
L-Struct\2str1-2.pmd IInd 21-6-13 3rd 25-9-14 4th 21-10-14
(Vol-II)

Example 1.7 *Find the equivalent uniformly distributed load corresponding to*
(i) a single rolling load of 100 kN crossing a girder of 20 m span.
(ii) a uniform load of 25 kN/m, 6 m long crossing a girder of 30 m span.

Solution **Case** (*i*) Single concentrated load

$$W = 100 \text{ kN}, l = 20 \text{ m}$$

$$w_{em} = w_{ef} = \frac{2W}{l}, \text{ from equation (1.30)} = \frac{2 \times 100}{20} = 10 \text{ kN/m}$$

Therefore, equivalent uniformly, distributed load

$$w_e = 10 \text{ kN/m}$$

Case (*ii*) UDL shorter than the span

$$w = 25 \text{ kN/m}$$

Length of the udl, $c = 6 \text{ m}, l = 30 \text{ m}$

$$w_{em} = w_{ef} = \frac{2wc}{l^2}\left(l - \frac{c}{2}\right), \text{ from equation (1.31)}$$

$$= \frac{2 \times 25 \times 6}{30^2}\left(30 - \frac{6}{2}\right) = 9 \text{ kN/m}$$

Therefore, equivalent uniformly distributed load = 9 kN/m.

Example 1.8 *Two point loads of 40 kN and 60 kN, spaced at 6 m apart cross a girder of 16 m span, from left to right, the 40 kN load leading.*
(a) Check the possibility of the max shear force occurring between the loads.
(b) Construct the maximum shear force and bending moment diagrams.

Solution (*a*) **Equation (1.24) says**

$$F_{x2} = \frac{W_1(l-x)}{l} + \frac{W_2(c-x)}{l}, \text{ when } W_1 \text{ and } W_2 \text{ are on either side of X.}$$

Equation [1.24 (*a*)] gives

$$F_{x3} = \frac{W_1(l-x-c)}{l} + \frac{W_2(l-x)}{l}, \text{ when } W_1 \text{ and } W_2 \text{ are to the right of X.}$$

If $F_{x2} > F_{x3}$

$$\frac{W_1(l-x)}{l} + \frac{W_2(c-x)}{l} > \frac{W_1(l-x-c)}{l} + \frac{W_2(l-x)}{l}$$

This gives the condition

$$c > \frac{W_2\, l}{W_1 + W_2}$$

In this example $c = 6$ m ; $\dfrac{W_2\, l}{W_1 + W_2} = \dfrac{60 \times 16}{100} = 9.6$

c is less than $\dfrac{W_2\, l}{W_1 + W_2}$

Hence, max positive shear will not occur when the loads are on either side of X.

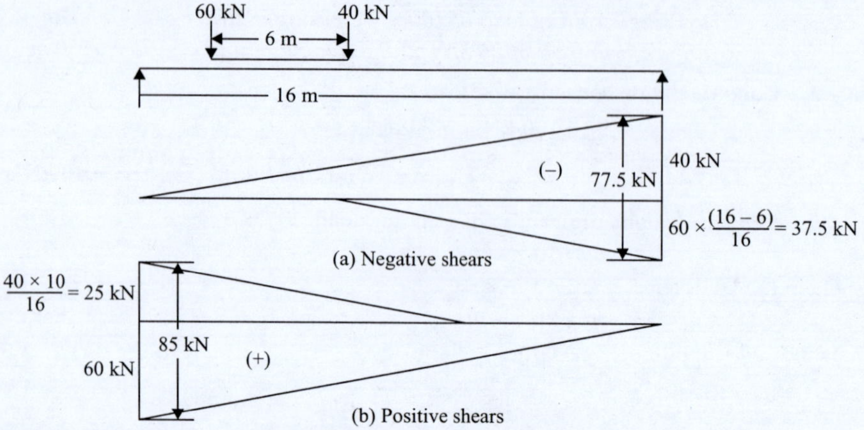

Fig. 1.36. Maximum SF diagrams.

(*b*) (*i*) Maximum negative shear (F_{max} – 1)

$F_{max\,-}$ will occur under the 40 kN, when it is just to the left of a section X. The diagram is as shown in Fig. 1.36 (*a*), corresponding to equation

$$F_{max\,-} = -\frac{W_1 x}{l} - \frac{W_2 (x - c)}{l}$$

$$W_1 = 40 \; ; W_2 = 60 \; ; l = 16 \; ; c = 6$$

Absolute max F – occurs at B and is 77.5 kN

(*ii*) The maximum positive shear ($F_{max\,+}$)

$F_{max\,+}$ will occur under the 60 kN load when the load is just to the right of X. The diagram, corresponding to equation (1.25) is given in Fig. 1.36 (*b*)

$$F_{max\,+} = \frac{W_1 (l - x - c)}{l} + \frac{W_2 (l - x)}{l} \qquad \text{[Use eqn. (1.25)]}$$

Absolute maximum F+ occurs at A and is 85 kN.

(*iii*) Maximum bending moment diagram

The maximum bending moment will occur under either W_1 or W_2. The diagrams, corresponding to equations (1.27) and (1.28) are shown in Figs. 1.37 (*a*) and (*b*).

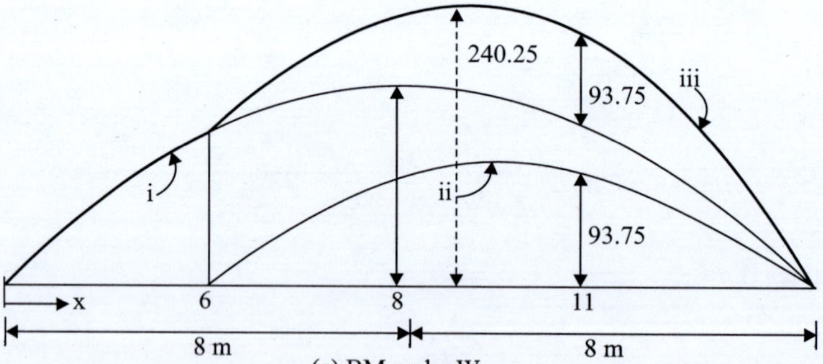

(a) BM under W_1

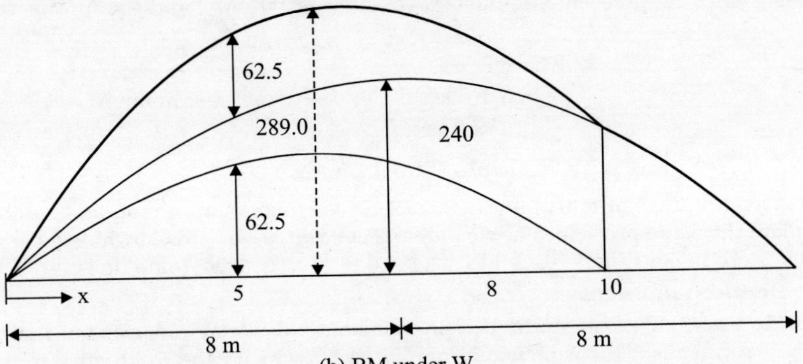

(b) BM under W_2

Fig. 1.37. Maximum bending moment diagrams.

$$M_{x1} = \frac{40}{16}x\,(16 - x) + \frac{60}{16}\,(x - 6)\,(16 - x)$$

$$M_{x2} = \frac{40}{16}x\,(10 - x) + \frac{60}{16}x\,(16 - x)$$

M_{x1} is a combination of 2 parabolas. The parabola due to W_1 (the 40 kN load) has a range of 16 m.

Its mid ordinate = $\frac{40}{16} \times 8 \times 8 = 160$ kNm

The second parabola due to W_2 ranges from $x = 6$ to $x = 16$. The max ordinate occurs at $x = 11$ m.

Its mid ordinate = $\frac{60}{16} \times 5 \times 5 = 93.75$ kNm

The ordinates of these 2 parabolas are added up and shown as curve (*iii*)

When we differentiate M_{x1} and equate to zero, we get the value of x that gives max M_{x1} (under the 40 kN load)

This gives $12.5x = 122.5$

$x = 9.8$. This is where M_{x1} is maximum

$M_{x1\,max} = 240.25$ kNm

M_{x2} is again a combination of 2 parabolas.

The curve due to the 60 kN load ranges from 0 to 16.

Its mid ordinate = $\frac{60}{16} \times 8 \times 8 = 240$ kNm

The second parabola due to the 40 kN load ranges between $x = 0$ and $x = 10$.

Its mid ordinate = $\frac{40}{16} \times 5 \times 5 = 62.5$ kNm

These 2 parabolas are superposed to give M_{x2}.

To get the location of maximum M_{x2}, we differentiate M_{x2} and equate this to zero. This gives,

$$12.5x = 85$$

$$x = 6.8. \text{ This is the position of maximum } M_{x2}$$

$$M_{x2\ max} = 289.0 \text{ kNm.}$$

1.2.7 S.S. Spans with Multiple Wheel Loads

This loading case of multiple wheel loads is very common in road and rail bridges. We can not adopt the same procedure as we did for 2 concentrated loads for the simple reason that it is arduous. To make things tractable, we have to evolve some guide lines.

(*a*) **Bending moments :**

Guide line 1. The maximum bending moment will occur under one of the wheel loads and not between loads. This is obvious from Fig. 1.32, which can be extended to more than 2 concentrated loads.

Guide line 2. When a train of wheel loads move on a S.S. span the maximum bending moment occurs under a load when the mid span section divides the distance between that load and the resultant of the load train equally. (vide Fig. 1.38)

The meaning of the statement is as under. If R is the resultant of the *n* loads on span AB, maximum bending moment will occur under W_3 when mid section C is right in the middle of resultant R and load W_3.

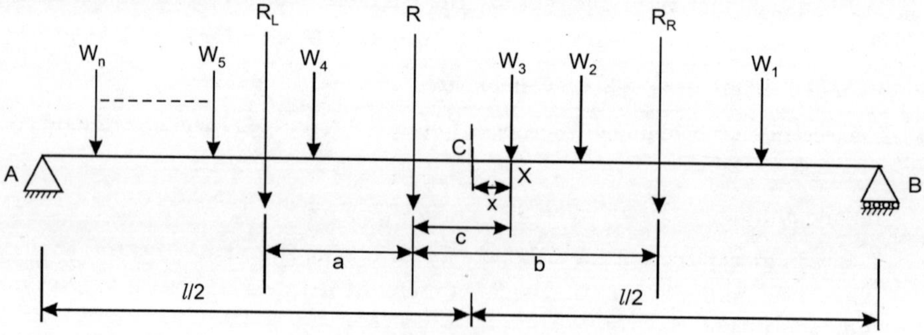

Fig. 1.38

Proof : Let R_R be the resultant of loads to the right of R

Let R_L be the resultant of loads to the left of R

a and *b* are distance of R_L and R_R from R. *c* is the distance of X the location of the critical load W_3 from R.

x is the distance of W_3 from mid span C. For a given loading train, *a*, *b* and *c* are constants. *x* decides the position of the load train from mid span.

Now, $$R_A = R \cdot \left(\frac{l}{2} - x + c\right)\Big/ l$$

M_x the bending moment at X is given by

$$M_x = R_A \left(\frac{l}{2} + x\right) - R_L(a + c)$$

$$M_x = \frac{R}{l}\left(\frac{l}{2} - x + c\right)\left(\frac{l}{2} + x\right) - R_L(a + c)$$

For M_x to be the maximum , $\frac{dM_x}{dx} = 0$

$$-2x + c = 0 ; \quad x = c/2$$

So C is mid way between the resultant and the critical load W_3. So guide line 2 has been proved valid.

But we still have a problem. How to decide which load is critical ? That takes us to guide line 3.

Guide line 3. The critical load, when moved from left of a beam section to the right, can change the average intensity on left side (compared to the average intensity on right side) from a higher value to lower value.

We need a numerical example to explain this point.

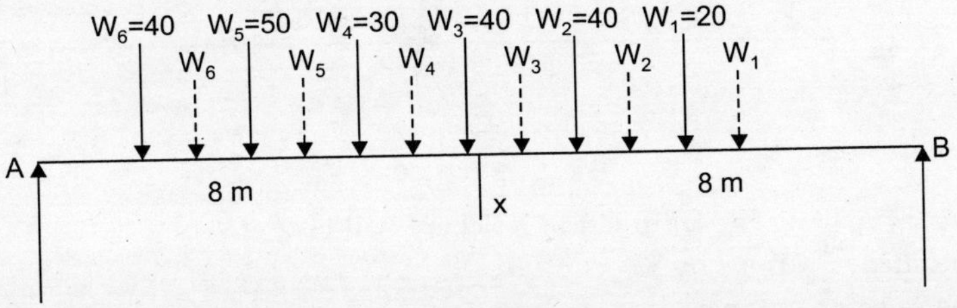

Fig. 1.39

When W_4 is to the left of X.

$$w_l = \text{average load on AX} = \frac{40 + 50 + 30}{8} = 15 \text{ kN/m}$$

$$w_r = \text{average load on XB} = \frac{40 + 40 + 20}{8} = 12.5 \text{ kN/m} ; w_l > w_r$$

Let us now move the train to the right so that W_4 moves to the right of X. (shown dotted on Fig. 1.39)

After shifting

$$w_l = \frac{40 + 50}{8} = 11.25 \text{ kN/m}$$

$$w_r = \frac{30 + 40 + 40 + 20}{8} = 16.25 \text{ kN/m}$$

Now w_l is smaller than w_r

Shifting W_4 to the right of X has done the trick. It has changed, w_l from the higher (than w_r) to lower (than w_r) value. So, W_4 is the critical load.

Now we can use guide line 2 to fix the position of X, where the bending moment is highest in the span.

(*b*) **Shear forces :**

For a load train like in Fig. 1.39 to cause the maximum positive shear at X, we would expect all the loads to be right of X, the last load just to the right of X. Then R_A is the shear at X [Fig. 1.40 (*a*)].

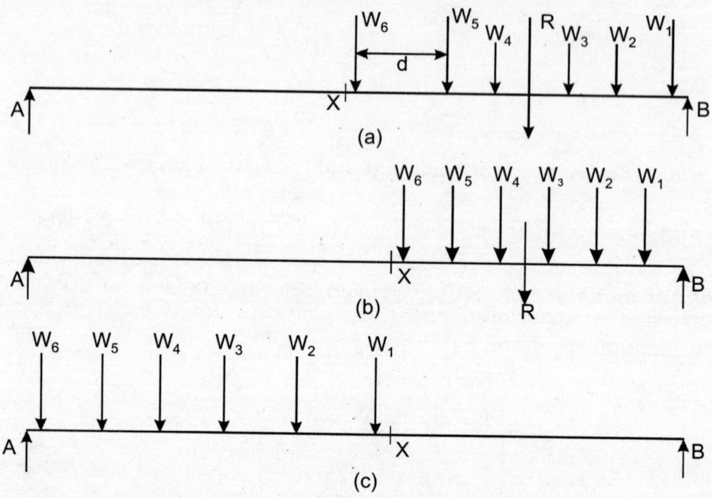

Fig. 1.40

Let us try moving the train left of X. There will be 2 effects.

(*i*) The shear F_x drops by W_6

(*ii*) The reaction R_A increases.

This continues till W_5 comes just to the right of X. By now the train has moved through a distance *d*. The change (increase) in F_x is

$$\Delta F_x = \Delta R_A - W_6 = R \cdot \frac{d}{l} - W_6$$

where R is the resultant. If ΔF_x works out to be positive, then our move (to the left) is right and W_5 is the critical load. We may have to continue this to check on W_4. But usually (*a*) is the candidate for maximum F_x and W_6 shall be at A, just to its right.

For maximum negative shear F_x at X, try Fig. 1.40 (*c*), with W_1 just to the left of *x*. The trial movements shall be to the right.

Example 1.9 *A girder of 15 m span is traversed by a moving load as shown in Fig. 1.41. Determine the maximum bending moment at D, 7 m from the left hand support.*

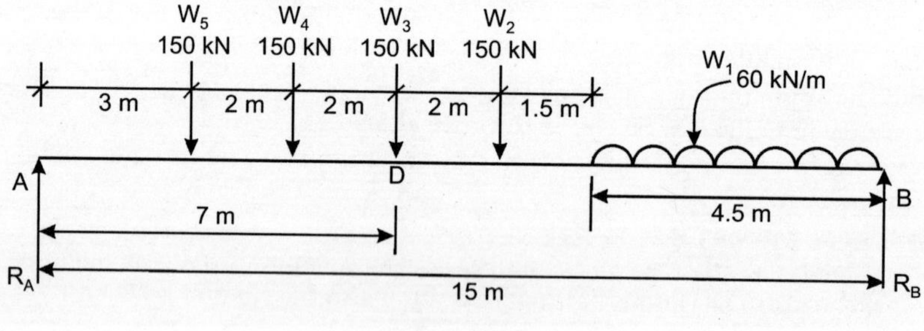

Fig. 1.41

Solution **Maximum bending moment :**

Maximum bending moment at D occurs under a critical load W_c which when moved from left to right of D changes the sign of E, where

$$E = \frac{W_{left}}{x} - \frac{W_{right}}{l - x}.$$

To find the critical load W_c, let us try with W_2.

Firstly keep this 150 kN load to the left of D,

$$W_{left} = 150 + 150 + 150 + 150 = 600 \text{ kN}$$

$$W_{right} = 60 \times 4.5 = 270 \text{ kN}$$

$$x = 7 \text{ m}$$

$$E = \frac{W_{left}}{x} - \frac{W_{right}}{l - x}$$

$$E = 51.96 \text{ (positive)}$$

Now, move W_2 to the right of D,

$$E = \frac{W_{left}}{x} - \frac{W_{right}}{l - x} = \frac{150 + 150 + 150}{7} - \frac{150 + 270}{8} = 11.78 \text{ (positive)}$$

$$E = +11.78$$

There is no change in the sign of E. Hence the maximum bending moment at D does not occur under W_2. Let us try with W_3.

Move W_3 to the right of D

$$\frac{W_{left}}{x} - \frac{W_{right}}{l - x} = \frac{150 + 150}{7} - \frac{150 + 150 + 270}{8} = 28.39 \text{ (negative)}$$

Since the sign of E changes, W_3 is the critical load.

Maximum bending moment at D occurs when W_3 is placed at D.

Taking moments about A,

$$-R_B(15) + 150(3) + 150(3 + 2) + 150(3 + 2 + 2) + 150(3 + 2 + 2 + 2) + 60 \times 4.5 \times \left(15 - \frac{4.5}{2}\right) = 0$$

$$R_B = 469.5 \text{ kN}$$

$$R_A = \text{Total load} - R_B = 150 + 150 + 150 + 150 + 270 - 469.5$$

$$R_A = 400.5 \text{ kN}$$

Maximum bending moment at D = $R_A(7) - 150(4) - 150\ (2) = 400.5\ (7) - 600 - 300$

$$(BM)_D = 1903.5 \text{ kNm.}$$

Example 1.10 *A system of moving loads crosses a girder of 36 m span simply supported at its ends. The loads and the distance are as follows :*

<table>
<tr><td>Wheel load (kN)</td><td colspan="2">100</td><td colspan="2">100</td><td colspan="2">200</td><td colspan="2">200</td><td colspan="2">160</td></tr>
<tr><td>Distance (m)</td><td></td><td colspan="2">3</td><td colspan="2">4.5</td><td colspan="2">4</td><td colspan="2">3.5</td><td></td></tr>
</table>

Determine the maximum bending moment in the girder.

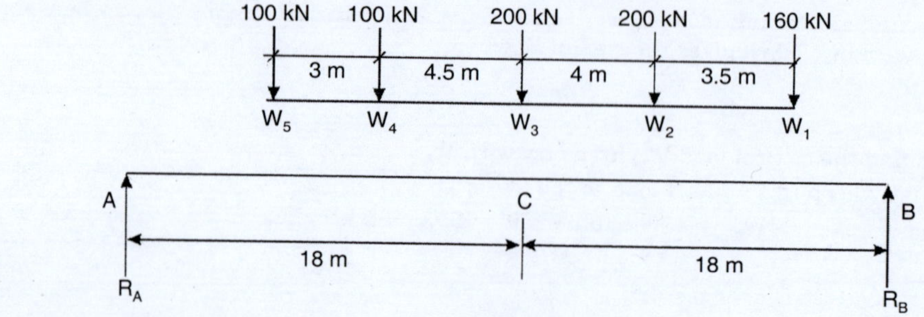

Fig. 1.42

Solution 1. To find the location of resultant of loads

Taking moments about W_5, $100 \times 3 + 200\,(7.5) + 200\,(11.5) + 160(15) = 760\,\bar{x}$

$\bar{x} = 8.55\,\text{m}$ from W_5

2. Maximum bending moment

Absolute maximum bending moment occurs under the load, W_3 (= W_C) nearest to the resultant R, when W_C and R are at equal distance from the centre of span (C).

In this case W_3 is nearest the resultant R. Hence bending moment under W_3 is the maximum bending moment

Distance between C and W_3 = Distance between C and R.

Distance between C and R = $\frac{1}{2}$ (Distance between W_3 and R)

$= \frac{1}{2}\,(1.05) = 0.525$ m.

Figure 1.43 gives the load position to produce max bending moment (under W_3)

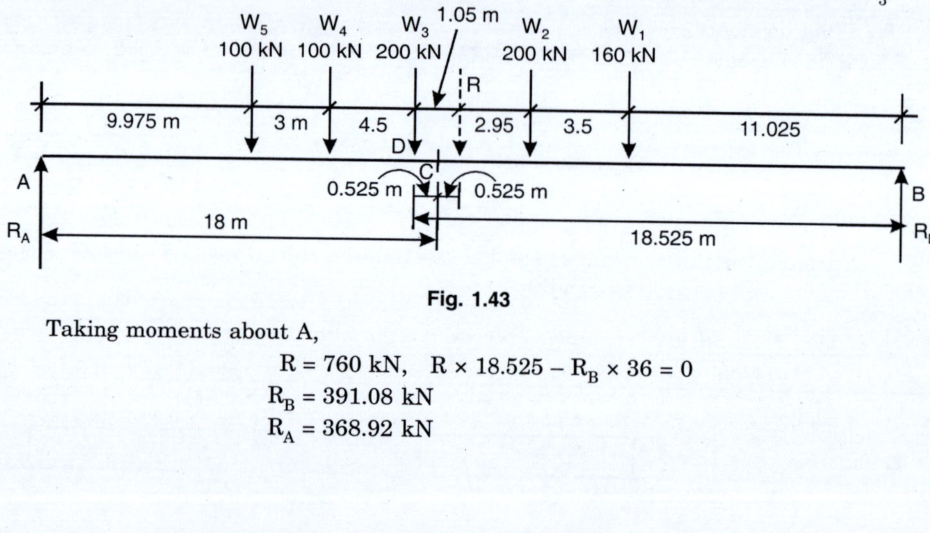

Fig. 1.43

Taking moments about A,

$$R = 760\text{ kN},\quad R \times 18.525 - R_B \times 36 = 0$$

$$R_B = 391.08\text{ kN}$$

$$R_A = 368.92\text{ kN}$$

Maximum bending moment,

$$(BM)_D = 368.92\ (17.475) - 100\ (7.5) - 100\ (4.5)$$

$$BM_{max\ max} = 5246.88\ \text{kNm.}$$

Example 1.11 *The following system of wheel loads move from left to right on a 15 m span, the 40 kN load leading.*

Loads (kN)	20	60	60	50	40
Distance (m)	1.5	1.5	2.0	1.0	

For a section C, 4 m from the left support, determine the
(i) Maximum bending moment *(ii) Maximum shear force.*

Solution

Figure 1.44 shows the span and the loads

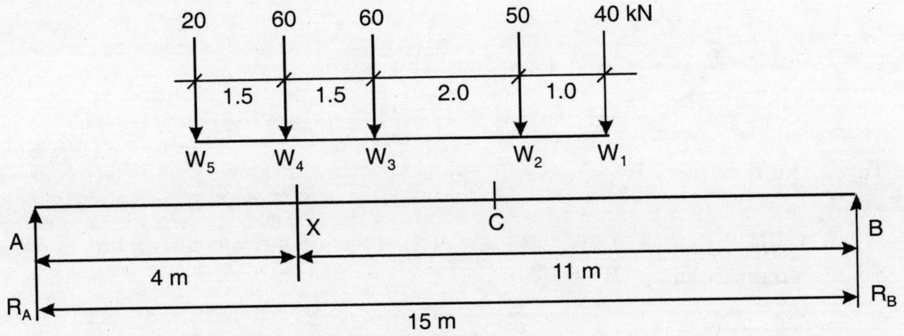

Fig. 1.44

(*i*) **Maximum bending moment at X**

We get maximum bending moment at X when the critical load is on X. We have to find what is the critical load. The critical W_c is that load which when moved from left to right of X, changes the sign of L_r, the differential loading rate.

$$L_r = \frac{W_{left}}{x} - \frac{W_{right}}{l - x}$$

Table 1.1, below gives the calculations for L_r when we try if W_4 can be critical.

Table 1.1

Trail No.	*Load position*	W_l	W_r	L_r	*Remarks*
1.	W_5 left of x	20	210	$\frac{20}{4} - \frac{210}{11} = -14.09$	Negative
2.	W_5, W_4 left of x	80	150	$\frac{80}{4} - \frac{150}{11} = +6.36$	Changes sign. W_4 is critical

Figure 1.45 gives the load position for maximum bending moment at X.

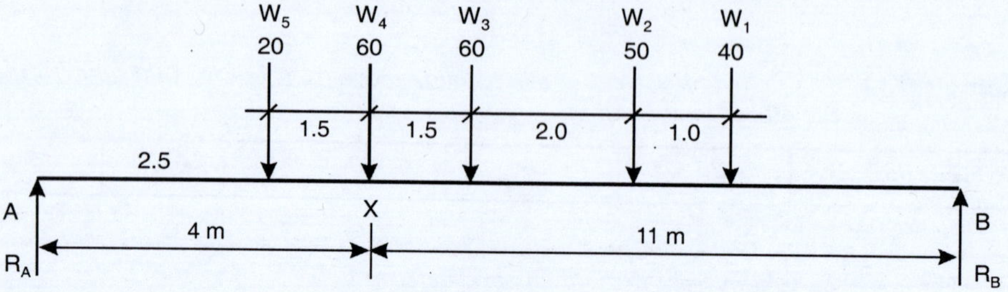

Fig. 1.45

$$R_B = \frac{20 \times 2.5 + 60 \times 4 + 60 \times 5.5 + 50 \times 7.5 + 40 \times 8.5}{15} = 89 \text{ kN}$$

$$R_A = 141 \text{ kN}$$

$$M_x = R_A \times 4 - 20 \times 1.5 = 534 \text{ kNm}$$

(*ii*) **Maximum shears**

(*a*) *Negative shear:*

To get the max negative shear at X, keep W_1 just left of X. 3 loads will be on the beam, W_1, W_2 and W_3.

When we let W_1 cross to the right of X, the shear at X drops by 40, but R_B will increase. So, increment in shear S_i is given by

$$S_i = \frac{Wd}{l} - W_1,$$

where W is the total load on span and d the distance moved. In this case d can be 1 m. At that moment W_2 will be on X.

$$S_i = \frac{(40 + 50 + 60) \times 1}{15} - 40 = -30$$

There is only decrement. So we better keep W_1 at X, just to the left of X.

$$S_{x-} = R_B = \frac{60 \times 1 + 50 \times 3 + 40 \times 4}{15} = 24.67 \text{ kN}$$

(*b*) *Positive shear at X:*

For max S_x we will try W_5 at X, just to the right. A trial movement of the train to the left will decrease the shear by 20 kN, but increase R_A. We will move the train by 1.5 m till W_4 is on X.

$$S_i = \frac{Wd}{l} - 20 = \frac{230 \times 1.5}{15} - 20 = 3$$

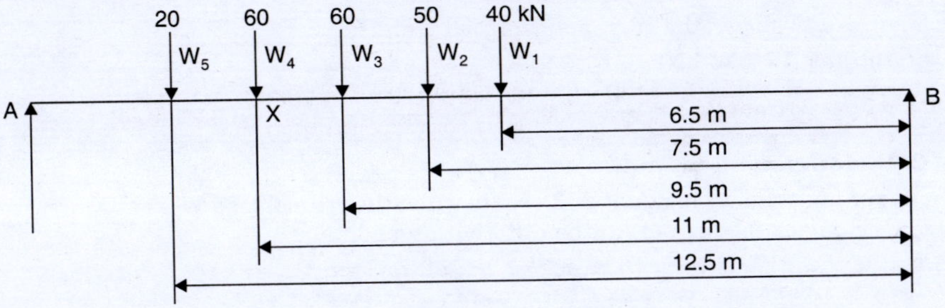

Fig. 1.46

There is an increment. We will again try moving the train to the left by 1.5 m, till W_3 is at X.

$$S_i = \frac{230 \times 1.5}{15} - 60 = -37. \text{ This is a decrease in } S_x$$

So, W_4 is the critical load for positive shear. And making use of Fig. 1.46,

$$S_{x+} = 141 - 20 = 121 \text{ kN}$$

Example 1.12 *Four point loads of 120 kN, 160 kN, 160 kN and 80 kN spaced 2 m between consecutive loads move on a girder of 25 m span, from left to right with the 120 kN load leading.*

Calculate the maximum bending moment at a point 10 m from left support. Also calculate the position and value of the absolute maximum bending moment.

Solution

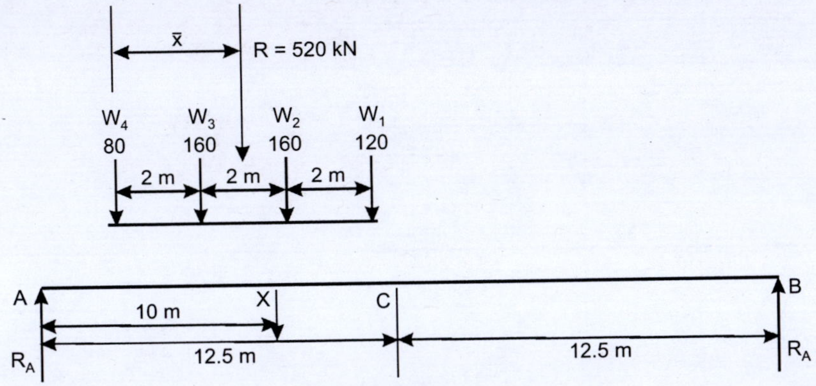

Fig. 1.47

1. Resultant

Figure 1.47 shows the span and the load train.

Resultant R = 520 kN

$$\bar{x} \text{ (from } W_4) = \frac{160 \times 2 + 160 \times 4 + 120 \times 6}{520} = \frac{1680}{520} = 3.23 \text{ m.}$$

2. Bending moment at X

Maximum bending moment at X occurs when the critical load W_c is at X. Critical load is that which changes the sign of L_r, the differential loading rate.

Let us try if W_2 is critical. Keep W_2 left of X.

$$W_{left} = 80 + 160 + 160 = 400 \text{ kN ;}$$

$$W_{right} = 120 \text{ kN}$$

$$L_r = \frac{400}{10} - \frac{120}{15} = 32 \text{ (\textit{positive})}$$

when W_2 is moved right of X,

$$W_{left} = 80 + 160 = 240$$

$$W_{right} = 160 + 120 = 280$$

$$L_r = \frac{240}{10} - \frac{280}{15} = 5.33 \text{ (\textit{positive})}$$

With W_2, L_r does not change sign. So, W_2 is not critical. Now try with W_3. Keep W_3 left of X.

$$L_r = \frac{80 + 160}{10} - \frac{160 + 120}{15} = 15 \text{ (\textit{positive})}$$

shift W_3 right of X

$$L_r = \frac{80}{10} - \frac{160 + 160 + 120}{15} = -21.33 \text{ (\textit{negative})}$$

L_r changes sign due to shift of W_3. W_3 is critical.

To get max. BM at X, keep W_3 at X (Fig. 1.48).

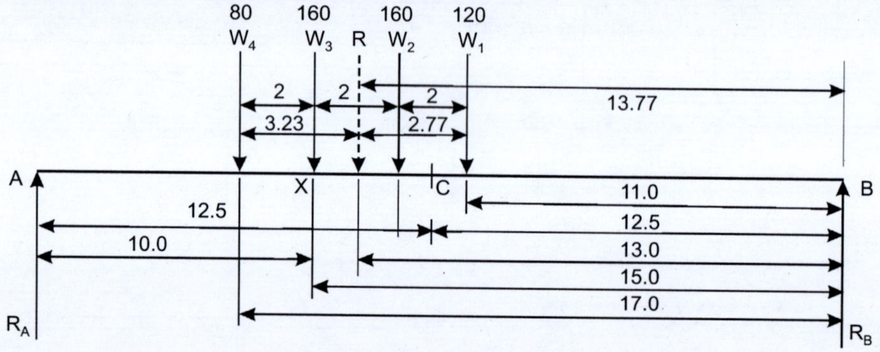

Fig. 1.48

$$R_A = \frac{520 \times 13.77}{25} = 286.4 \text{ ; } \quad R_B = 233.6 \text{ kN}$$

$$M_x = R_A \times 10 - 80 \times 2 = 2704 \text{ kNm.}$$

3. Absolute maximum bending moment

To get $M_{max\ max}$, we have to move load train such that R and W_2 are on either side of C the mid-point of the beam AB. The resulting load position is seen in Fig. 1.49.

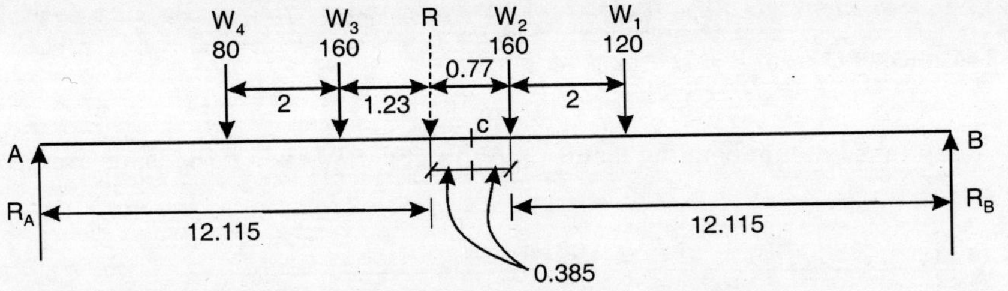

Fig. 1.49

Reactions :

$$\text{Resultant} = 520 \text{ kN}$$

$$R_B = \frac{R \times 12.115}{25} = 252 \text{ kN} ; \quad R_A = 268 \text{ kN}$$

$$M_x = M_{max\ max} = R_B \times 12.115 - 120 \times 2$$

$$M_{max\ max} = 2813 \text{ kNm}.$$

Example 1.13 *A girder having a span of 18 m is simply supported at its ends. It is traversed by a train of loads as shown in Fig. 1.50 with 100 kN load leading. Find the equivalent udl.*

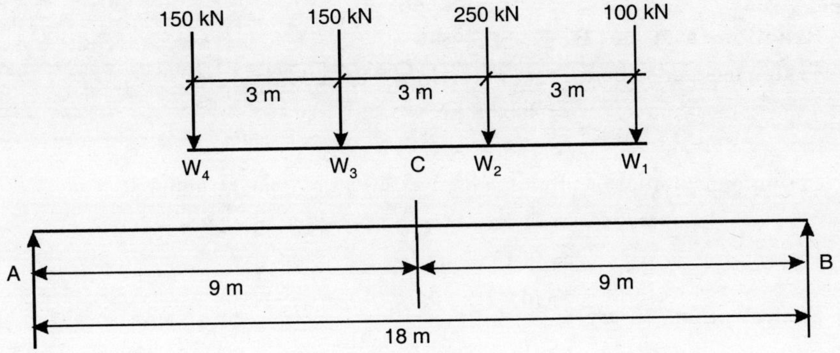

Fig. 1.50

Solution **Absolute maximum bending moment :**

$M_{max\ max}$ occurs under a load when the load, which is nearest to the resultant R, and the resultant R are at equal distances from the centre of span.

Position of resultant of all loads :

$$R = 650 \text{ kN}$$

Taking moments about W_4 and equating to zero.

$$150 \times 3 + 250(6) + 100(9) = 650\ \bar{x}$$

$$\bar{x} = 4.38 \text{ m from } W_4$$

Distance of C from R = Distance of C from W_3

$$= \tfrac{1}{2}\,(1.38) = 0.69 \text{ m.}$$

Absolute maximum bending moment occurs under W_3, when the loads are arranged as shown in Fig. 1.51.

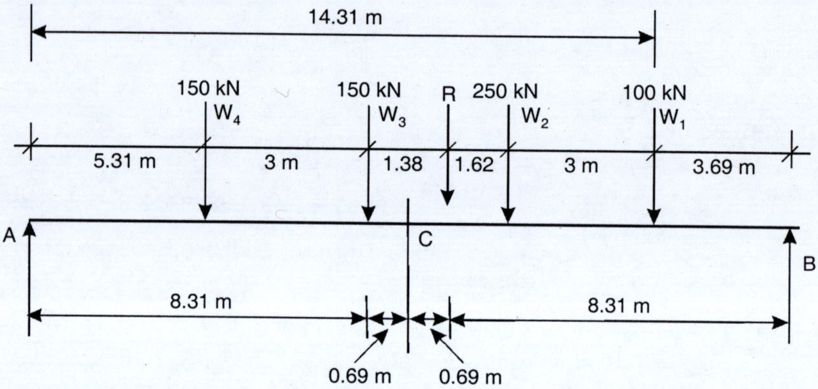

Fig. 1.51

Reactions at A and B

Taking moments about B, $R_A \times 18 - R \times 8.31 = 0$

$$R_A = 300.08 \text{ kN}$$

$$R_B = 349.92 \text{ kN}$$

Absolute maximum bending moment = Bending moment under W_3

$$M_{max\ max} = R_A\,(8.31) - 150\,(3) = 300.08\,(8.31) - 450$$

$$M_{max\ max} = 2043.67 \text{ kNm}$$

Equivalent udl, $w_e = \dfrac{8\,M_{max\ max}}{l^2} = \dfrac{8 \times 2043.67}{(18)^2} = 50.46 \text{ kN/m}$

$$w_e = 50.46 \text{ kN/m.}$$

Example 1.14 *A girder having a span of 18 m is simply two supported at the ends. It is traversed by a train of loads as shown in Fig. 1.53. The 50 kN load is leading. Find the maximum bending moment which can occur*

(i) Under the 200 kN load *(ii) Under 50 kN load.*

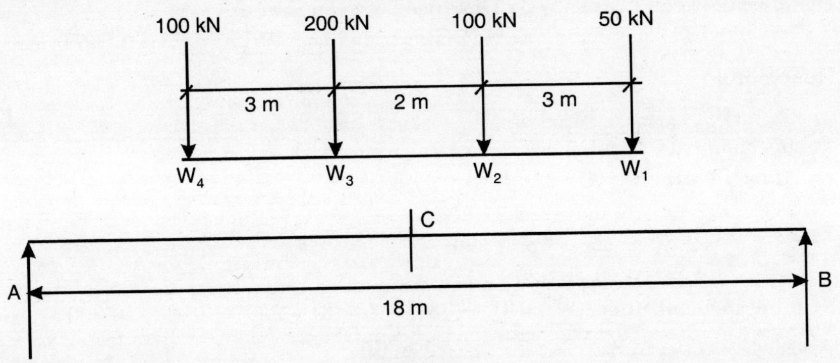

Fig. 1.52

Solution

(*i*) **Maximum bending moment (under W_3) :**

To get the maximum bending moment under W_3, the resultant (R) of all the loads and W_3 should be at equal distances from the centre of the span (C).

Location of resultant of loads :

Taking moments about W_4

$$100(0) + 200\,(3) + 100\,(3 + 2) + 50\,(3 + 2 + 3) = (100 + 200 + 100 + 50)\bar{x}$$

$$\bar{x} = 3.33 \text{ m from } W_4$$

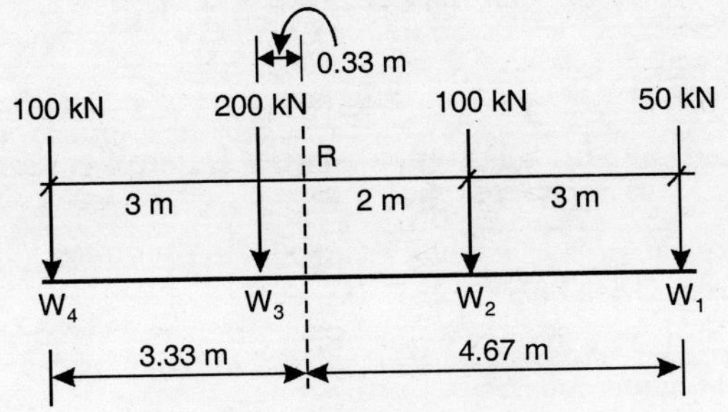

Fig. 1.53

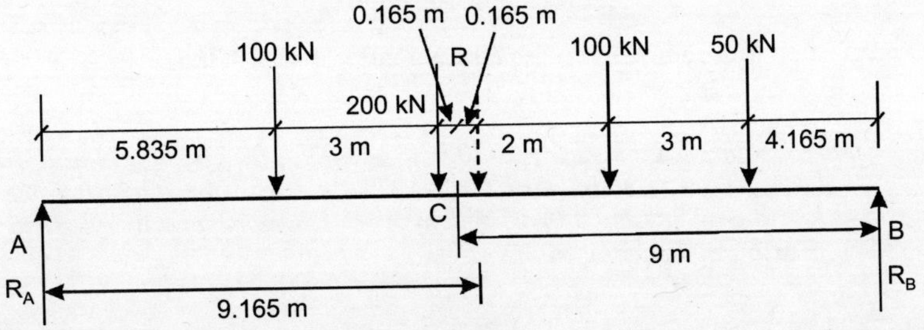

Fig. 1.54

Distance between C and W_3 = Distance between C and R

= ½ (Distance between W_3 and R) = 0.165 m

Reactions :

$$R = 450 \text{ kN}$$

Taking moments about A,

$$R_B \times 18 - R \times 9.165 = 0$$

$$R_B = 229.21 \text{ kN}$$

$$R_A = \text{Total load} - R_B$$

$$R_A = 220.79 \text{ kN}$$

Bending moment under $W_3 = R_A$ (8.835) – 100(3) = 220.79 (8.835) – 300

$$M_{max} = 1650.68 \text{ kNm}$$

(*ii*) **Bending moment under W_1 :**

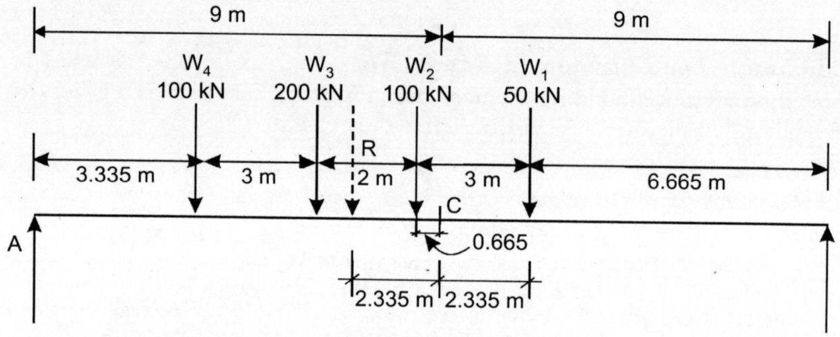

Fig. 1.55

Distance between C and R = Distance between C and W_1

$$R = 450 \text{ kN}$$

Figure 1.55 shows the load position corresponding to this.

Distance of R from C = 2.335 m

Reactions at A and B have to be determined.

Taking moments about A,

$$18\ R_B = R \times 6.665$$

$$R_B = 166.71 \text{ kN}$$

$$R_A = 283.29 \text{ kN}$$

Maximum bending moment under 50 kN load = + R_B (6.665)

$$M_{max} = +\ 1111.12 \text{ kNm.}$$

Example 1.15 *A system of 5 wheel loads 80 kN, 140 kN, 160 kN, 50 kN and 40 kN crosses a beam of 15 m span with the 80 kN load leading. The distances between the loads are 2.4 m, 3.0 m, 2.4 m and 1.6 m respectively. Find the absolute maximum bending moment.*

Solution **1. Resultant of the load system**

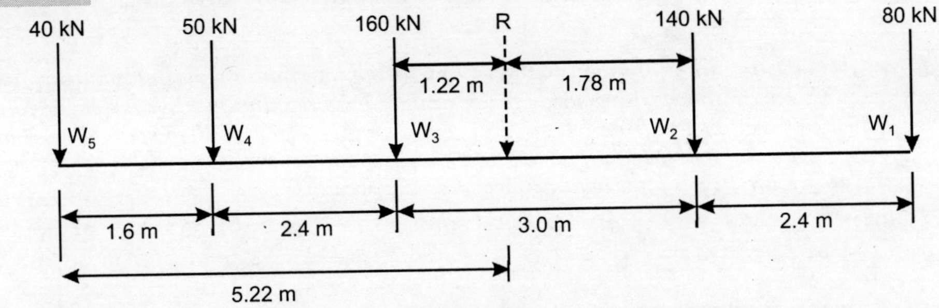

Fig. 1.56

Taking moments about W_5,

$$40(0) + 50\,(1.6) + 160\,(4.0) + 140\,(7.0) + 80\,(9.4) = (40 + 50 + 160 + 140 + 80)\,\bar{x}$$

$\bar{x} = 5.22$ m from W_5.

The load nearest to the resultant load R is 160 kN (W_3).

To get the absolute maximum bending moment, the load system should be placed in such a way that the resultant load, R and the load nearest to R are at equal distance from the centre of the span. Then the absolute maximum bending moment is equal to the bending moment under the load nearest to R (*i.e.,*) the bending moment under W_3.

2. Absolute maximum bending moment

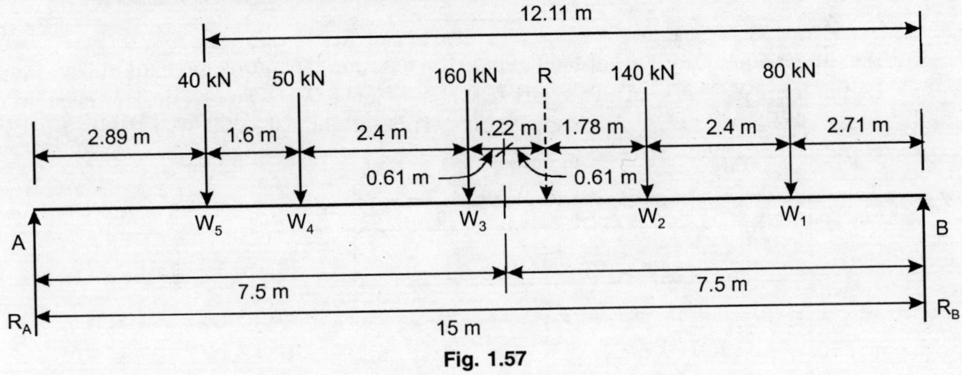

Fig. 1.57

Distance of 160 kN from R = 1.22 m

When R and W_3 are at equal distance from C ; (centre of span) distance of R from

$$C = \frac{1.22}{2} = 0.61 \text{ m.}$$

To find R_A :

Taking moments about B,

$$R_A(15) - 40\,(12.11) - 50(10.51) - 160(8.11) - 140\,(5.11) - 80\,(2.71) = 0$$

$$R_A = \frac{3239.7}{15} = 215.98 \text{ kN}$$

Absolute maximum bending moment = bending moment under 160 kN load.

Absolute maximum bending moment = R_A (7.5 – 0.61) – 40 (4.0) – 50 (2.4)

$$M_{max\ max} = 1208.10 \text{ kNm.}$$

Example 1.16 *Using an analytical method, find (i) the maximum bending moment, (ii) the maximum positive shear force, and (iii) the maximum negative shear force at a section 4 m from left support A of a simple girder of 10 m span when 4 wheel loads 10 kN, 15 kN, 30 kN and 30 kN spaced at 2 m, 3 m and 3 m respectively with 10 kN load leading, traverse the span.*

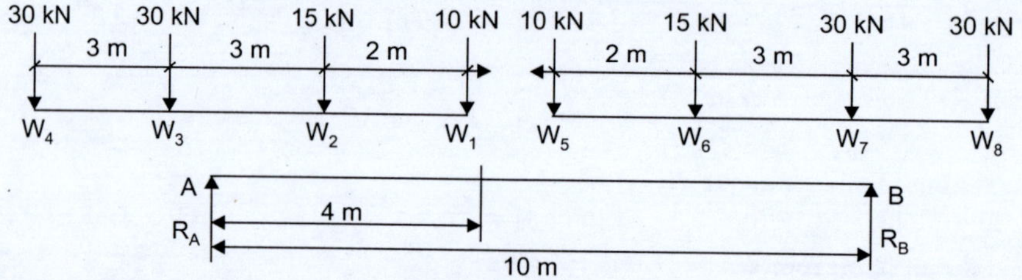

Fig. 1.58

Solution (*i*) **Maximum bending moment**

Maximum bending moment at a given section occurs when the critical load, which when moved either from left to right or from right to left changes the sign of the differential loading rate L_r, is at the section.

Here,
$$L_r = \frac{W_{left}}{x} - \frac{W_{right}}{l - x}$$

Bending moment under that load will be the maximum bending moment at that section.

Let us try if W_3 is critical. To verify that, let us move W_3 (= 30 kN) just to the right of C and then just to the left of C. If the sign of changes, it will indicate that W_3 is the critical load.

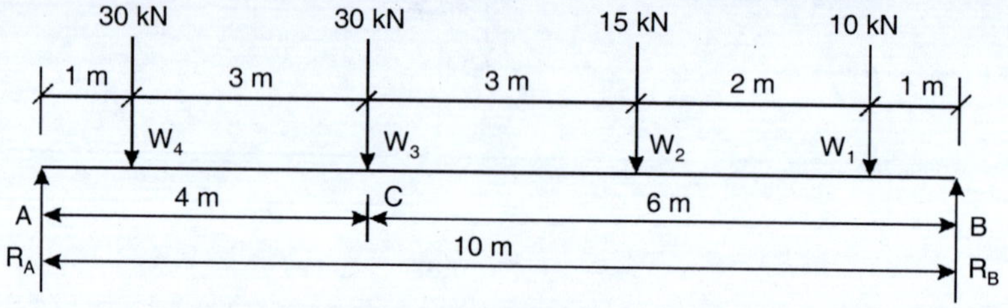

Fig. 1.59

The load system can traverse the span either from left to right or from right to left. But to get the maximum bending moment the load system should traverse from left to right with 10 kN load leading. Because when the load system moves from right to left and to have the 3rd 30 kN load at C, 10 kN load goes out of the span. Hence the bending moment is lesser than in the case which includes all the loads within the span.

Place W_3 just to the left of C ; then,

$$L_r = \frac{W_{left}}{x} - \frac{W_{right}}{l - x} = \frac{30 + 30}{4} - \frac{15 + 10}{10 - 4} = \frac{60}{4} - \frac{25}{6} = 10.83 \text{ (positive)}$$

Then place W_3 just to the right of C.

$$L_r = \frac{W_{left}}{x} - \frac{W_{right}}{l-x} = \frac{30}{4} - \frac{30+15+10}{6} = -1.67 \text{ (negative)}$$

As the sign of L_r changes, maximum bending moment occurs at C when W_3 is placed on the section C.

Maximum bending moment = bending moment under $W_3 = R_A (4) - 30 (3)$

To find R_A, take moments about B,

$$R_A (10) - 10(1) - 15 (3) - 30 (6) - 30 (9) = 0$$

$$R_A = 50.5 \text{ kN}$$

$$M_{max} = 50.5 (4) - 30(3)$$

$$M_{max} = 112 \text{ kNm}$$

(ii) **Maximum positive shear force**

Maximum positive shear force occurs when the loads are at the right of the section and the load system traverses from left to right. To get the maximum positive shear force, place W_4 as shown in Fig. 1.60.

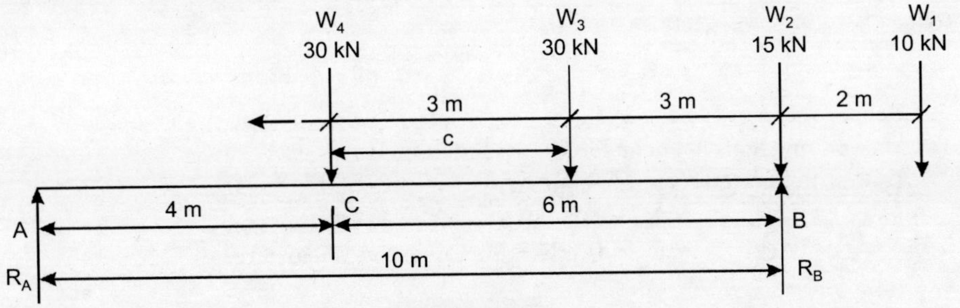

Fig. 1.60

Maximum positive shear force = R_A

To verify the validity of the above positive shear force, move the W_4 with the load train through d to get W_3 on C. If the sign of S_i (the shear increment) is negative, it will indicate that W_4 should not have been moved to the left but should be to the right of C.

Here, $S_i = \frac{Wd}{l} - W_4$, where W is the total load on the span and W_4 the load passing the section C.

In the above expression for S_i

W = total load on the span = 30 + 30 + 15 =75

d = distance through which the load is moved = 3 m

W_4 = the load passing the section (30 kN)

Hence, $$S_i = \frac{Wd}{l} - W_1 = \frac{75 \times 3}{10} - 30 = -7.5 \text{ (negative)}$$

Hence W_4 should be to the right side of C.

To find R_A, take moments about B,

$$R_A (10) - 15 (0) - 30(3) - 30 (3+3) = 0$$

$$R_A = 27 \text{ kN}$$

Hence the maximum positive shear force at C = 27 kN.

(*iii*) **Maximum negative shear force**

Maximum negative shear force occurs when the loads are to the left of C, while traversing from right to left. To get the maximum negative shear force, W_8 would be on the section C as shown.

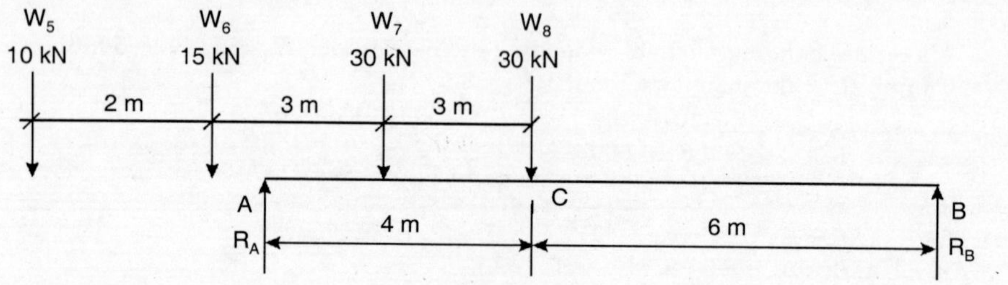

Fig. 1.61

To verify the occurrence of maximum negative shear force, move W_8 just to the right of C through a distance of 3 m. If the sign of S_i is negative it indicates that W_8 should not be moved to the right.

$$S_i = \frac{Wd}{l} - W_1 = \frac{60 \times 3}{10} - 30 = -12 \ (\textit{negative})$$

Hence W_8 should to the left of C.

Maximum negative shear force = Reaction at B

To find R_B, take moments about A,

$$R_B \times 10 - 30 \times 4 - 30 \times 1 = 0$$

$$R_B = 15 \text{ kN} = \text{Max negative shear force at C.}$$

PROBLEMS

1. A single rolling load of 20 kN rolls along a girder of 25 m span. Draw the diagrams of maximum B.M. and maximum S.F. positive and negative. What will be the absolute maximum (+) S.F. and B.M.?

   [**Ans.** S.F. = + 20 kN, B.M. = 125 kNm]

2. A simply supported beam AB of span 18 m is subjected to a uniformly distributed load of 25 kN/m, longer than the span. Determine:

   (*i*) maximum positive shear and maximum bending moment at a section 6 m from the left hand support.

   (*ii*) Absolute maximum bending moment.

   [**Ans.** $F_{6\ max}$ +ve = 100 kN; $M_{6\ max}$ = 900 kNm; $M_{max\ max}$ = 1012.5 kNm]

3. A simply supported beam of span 10 m is subjected to a uniform dead load of 60 kN/m and a uniform live load (longer than the span) of 100 kN/m. Determine the maximum and the minimum shear forces at the left hand quarter point ; and the range over which the shearing force may have only positive or negative values.

   [**Ans.** Maximum shear = 431.3 kN (–ve) Minimum shear = 118.7 kN (–ve) Range : Middle 2.4 m]

4. A live load of 20 kN/m, 6 m long moves on a simply supported girder of 10 m span. Find the maximum bending moment that can occur at a section 4 m from the left end.

   [**Ans.** 201.6 kNm]

5. Two wheel loads of 20 kN and 10 kN spaced 3 m apart cross a girder of 10 metres span, with the 10 kN load leading, from left to right. Draw the maximum shear force and bending moment diagrams.

[**Ans.** F_{max} (+ve) = 27 kN F_{max} (–ve) = 24 kN
$M_{max\ max}$ = 60.75 kNm at 4.5 m from the left end]

6. A system of moving loads cross a girder of 36 m span which is simply supported at its ends. The loads and their distances are as follows :

Wheel loads (kN)	10		10		20		20		16
Distance between Centres (metres)		3		4.5		4		3.5	

Determine the maximum bending moment at the left and right quarter span points.

For each case, make a sketch of the girder showing clearly the section where the maximum bending moment occurs and the corresponding position of the loads.

[**Ans.** M_{max} at left quarter span = 373.8 kNm
M_{max} at right quarter span = 396.8 kNm]

7. A train of concentrated loads shown in Fig. 1.62 move from left to right on a simply supported girder of span 16 m. Determine the absolute maximum shear force and bending moment developed in the beam.

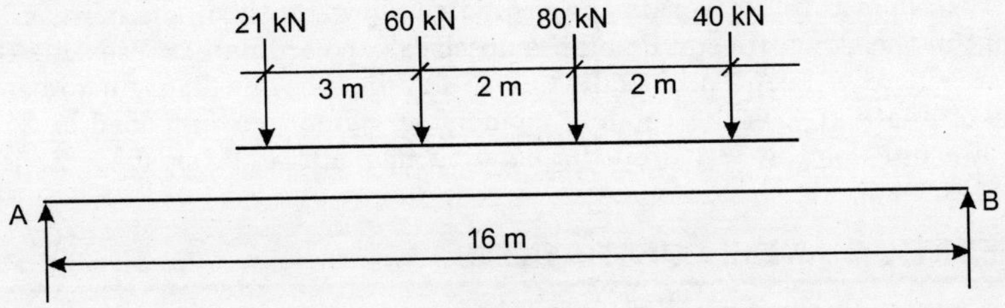

Fig. 1.62

[**Ans.** $F_{max\ max}$ = 166.25 kN (+ve) $M_{max\ max}$ = 651.524 kNm]

8. A train of concentrated loads shown in Fig. 1.63 move from left to right on a simply supported girder of span 25 m. Determine the maximum bending moment that will occur under the 80 kN load.

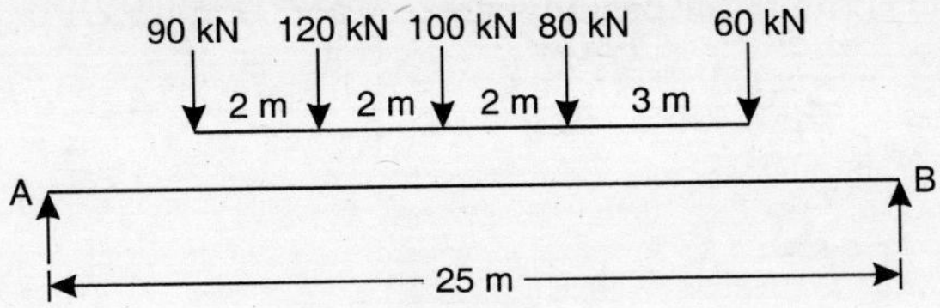

Fig. 1.63

[**Ans.** $M_{80\ max}$ = 2136.63 kNm]

9. A UDL of 10 kN/m, 4 m long rolls over a girder of 10 m span. Find the Equivalent Uniformly Distributed Load (EUDL). [**Ans.** 6.4 kN/m]

10. A girder having a span of 20 m is simply supported at the ends. It is traversed by a train of loads as shown in Fig. 1.64. The 100 kN load is leading. Find the EUDL.

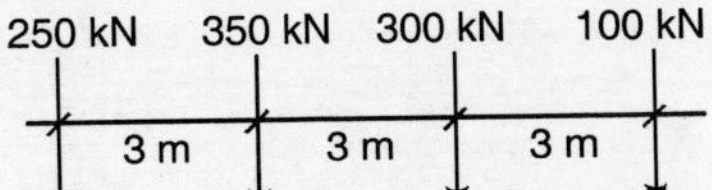

Fig. 1.64

[**Ans.** 77.75 kN/m]

CHAPTER 2
INFLUENCE LINES

2.1 INTRODUCTION

If you are working in a design office analyzing and designing bridge girders for rolling wheel loads, your life will be made much easier if you use the concept of influence lines. Influence lines are graphs drawn with the structure axis as X-axis and taking on the Y-axis any desired force quantity like reaction, thrust, internal shear, support moment, internal bending moment, etc. You can also think of influence lines for displacements (including slope) though these are rarely used.

In an influence line diagram the ordinates show the BM, shear, reaction, etc., at a chosen point in the structure say P and the abscissa (x co-ordinate) represents the point X on the structure where we keep a unit load. Thus, in an influence line diagram for bending moment at P, y the ordinate at X is the bending moment at P due to a unit load at X. It shows the influence of a unit load at any x on the bending moment at P (or shear at P or any force quantity at P).

2.2 DEFINITION AND ILLUSTRATIONS

An influence line is a graph showing, for any given beam, frame or truss, the variation of any force or displacement quantity (such as shear force, bending moment, tension, deflection) for all positions of a moving unit load as it crosses the structure from one end to the other.

For statically determinate structures the ILD for force quantities are made up of displaced configurations of the underformed parts of the structure. For example, in Fig. 2.1 (b) is the IL for R_A. Figure 2.1 (c) is the IL for bending moment at P. Figure 2.1 (d) is the IL for R_B.

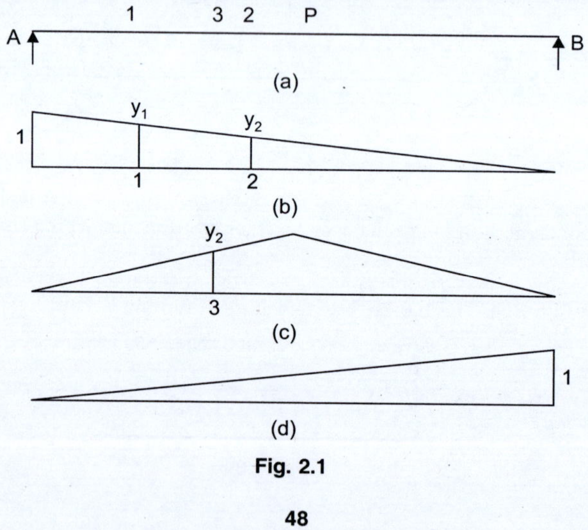

Fig. 2.1

Influence lines are very useful in quickly determining the force component at any given point due to set of moving loads. Thus in Fig. 2.1 (*b*), y_1 is the reaction at A due to a unit load at 1, y_2 is the reaction at A due to a unit load at 2 and $(W_1y_1 + W_2y_2)$ is the reaction at A due to loads W_1 at 1 and W_2 at 2. Extending this reasoning, y_3 is the bending moment at P due to a unit load at 3. $W_3 \times y_3$ is the bending moment at P due to a load W_3 at 3.

2.3 GETTING THE INFLUENCE LINES

2.3.1 Reactions in a SS Beam

External forces like reactions are the easiest force components for which influence lines can be sketched easily.

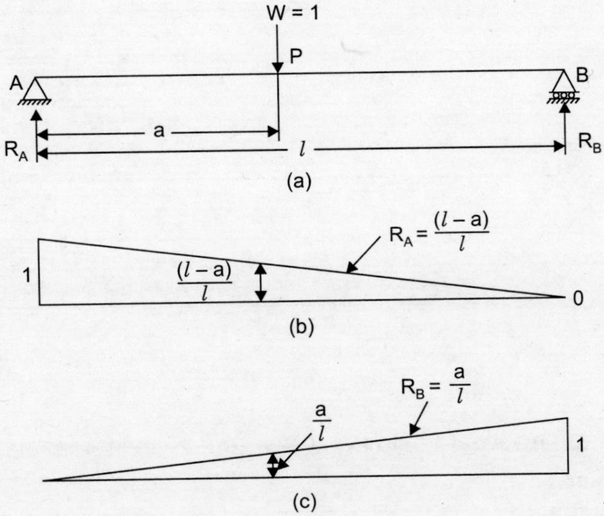

Fig. 2.2

Let us try to get the IL for R_A for the beam AB in Fig. 2.2 (*a*). Let a unit load act at P, distance *a* from A. Then

$$R_A = \frac{(l - a)}{l} \quad ...(2.1)$$

When we plot equation (2.1), that is the influence line for R_A Fig. 2.2 (*b*). As simple as that.

We will extend this reasoning to R_B and land at $R_B = \frac{a}{l}$ ...(2.2)

plotted in Fig. 2.2 (*c*)

2.3.2 Reactions in a Cantilever

Let us try to sketch the ILD for reactions R_A and M_A in a cantilever in Fig. 2.3.

R_A due a load 1 on the span is 1 irrespective of the load position a. Hence Fig. 2.3 (*b*) gives the ILD for R_A.

M_A due to a unit load at P, '*a*' from the free end is $M_A = (l - a)$ ...(2.3)

Figure 2.3 (*c*) plots this curve.

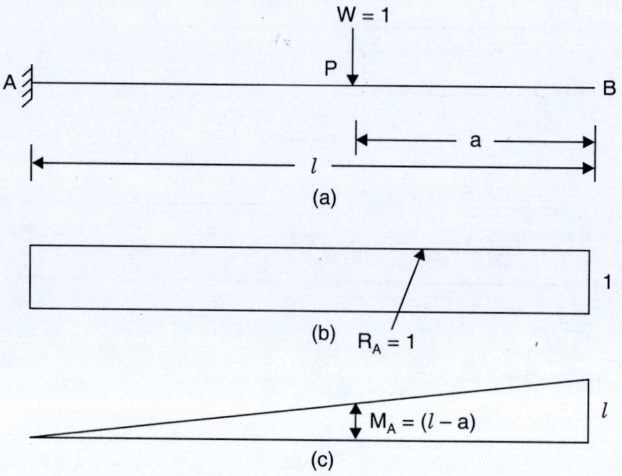

Fig. 2.3

2.3.3 ILD for Internal Shears and Bending Moments

(*a*) **SS beams**

(*i*) Let us investigate the SF and BM at X, at distance x from A. Let a be the coordinate position of a unit load.

For $a < x$,

$$F_x = R_A - 1 = \frac{(l-a)}{l} - 1 = \frac{-a}{l} \quad ...(2.4)$$

For $a > x$,

$$F_x = R_A = \frac{(l-a)}{l} \quad ...(2.5)$$

(*ii*) For $a < x$,

$$M_x = R_B \times (l - x) = \frac{a}{l}(l - x) \quad ...(2.6)$$

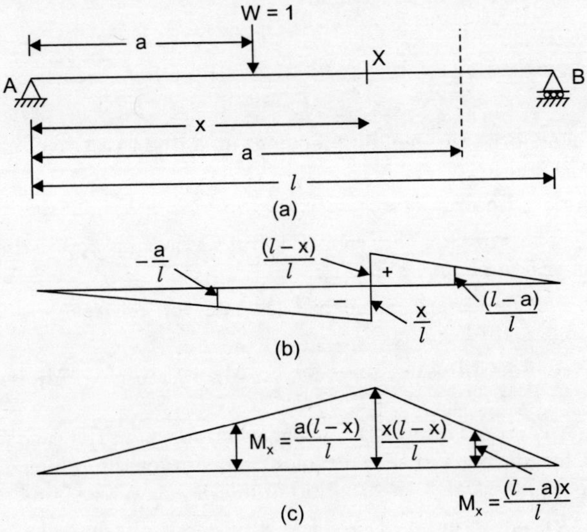

Fig. 2.4

For $a > x$,

$$M_x = R_A X, \quad M_x = \frac{(l-a)}{l} x \qquad ...(2.7)$$

In the plots in Fig. 2.4 (*b*) and (*c*), '*a*' varies from zero to x in AX and $(l - a)$ varies from 0 to $(l - x)$ in XB. These plots are the influence lines for F_x and M_x respectively.

(*b*) **Cantilevers**

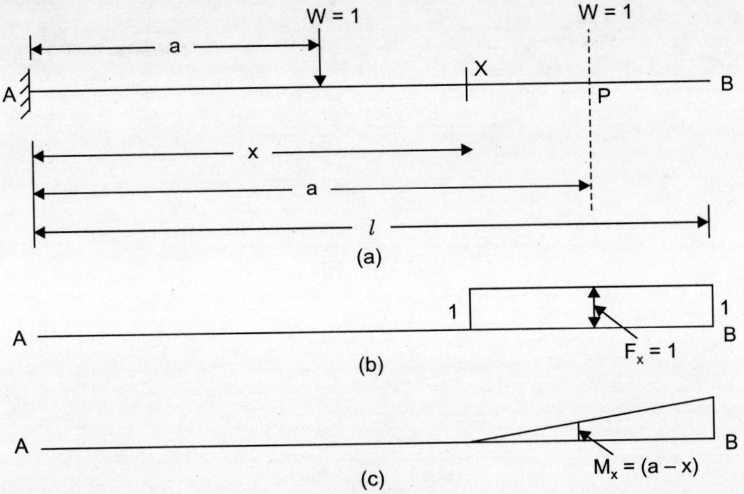

Fig. 2.5

(*i*) Shear force at X, x from A, when W = 1

$$a < x,\quad F_x = 0,\quad a > x,\quad F_x = 1 \qquad ...(2.8)$$

(*ii*) Bending moment at X, x from A when W = 1

$$a < x,\quad M_x = 0,\quad a > x,\quad M_x = (a - x) \qquad ...(2.9)$$

The above shear force and bending moment at X due to a travelling unit load at 'a' are plotted in Fig. 2.5 (*b*) and (*c*).

(*c*) **Overhanging beams**

Overhanging beams are not very common. But we do come across them in balcony beams, balanced cantilever bridge beams, etc.

In buildings, over hangs are considered as space you get free. Let us investigate a beam with double overhangs as in Fig. 2.6.

Let us try sketching influence lines for F_x, M_x, R_B, R_C, M_A, M_B for the 3 zones $a < 0$; $0 < a < l$; $a > l$

(*i*) $\mathbf{F_x}$ **:** When X is on AB or on CD, the shear will be as in the case of a cantilever. When X is on BC, the shear will be as in the case of a SS beam for the region BC ; but loads on the overhangs do produce shears at X (Fig. 2.6 (*b*) and (*c*)).

(*ii*) $\mathbf{M_x}$ **:** When X is on AB or on CD, the BM will be as in the case of a cantilever Fig. 2.6 (*d*). When X is on BC, the IL for M_x will be similar to the case of a SS beam for the region BC ; but even when the unit load is on the overhangs, BC will suffer bending moments. For example, when a unit load is on D,

$$R_B = -\frac{1 \times l_2}{l} \qquad ...(2.10)$$

and

$$M_x = -\frac{l_2}{l}x \qquad ...(2.11)$$

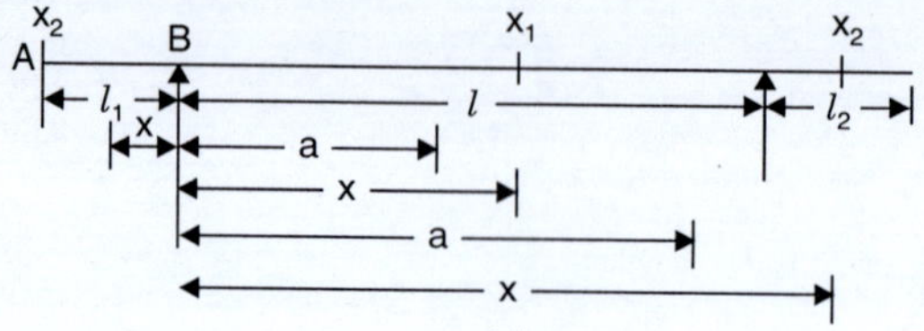

(a)

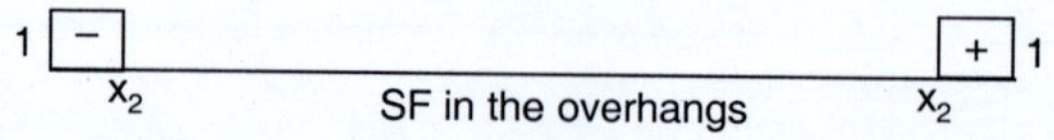

(b)

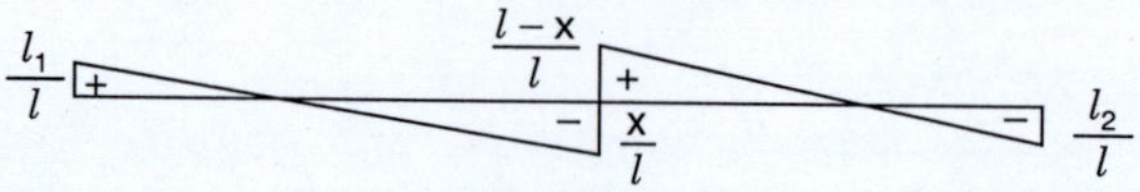

SF in the span

(c)

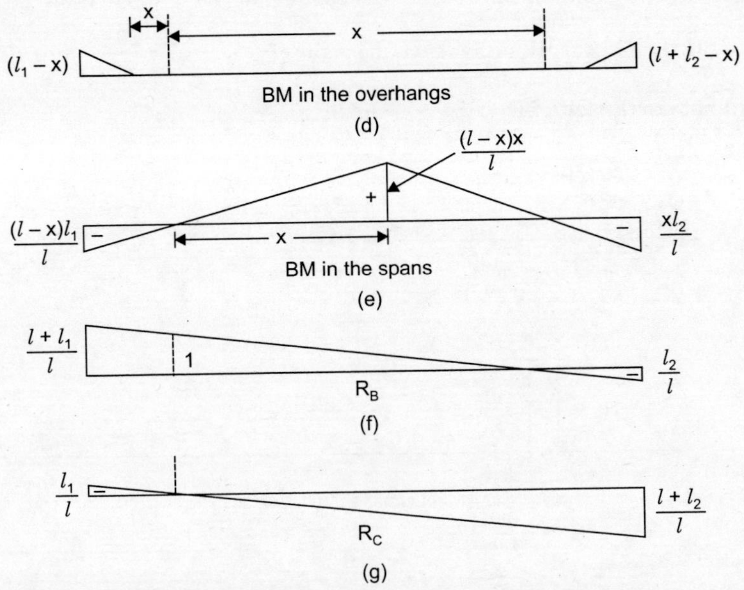

Fig. 2.6

Figure 2.6 (*c*) shows this influence.

(*iii*) **Reactions R_A and R_B:** We know that a unit load at B produces unit reaction R_B. A load at C produces zero reaction at B and a unit load at D produces $-\frac{l_2}{l}$ as in eqn. (2.10). This gives sufficient data to plot 2.6 (*f*) and 2.6 (*g*), the influence lines for R_B and R_C.

2.4 USING INFLUENCE LINES

In Chapter 1, we dealt with rolling loads without the benefit of influence lines. That was like going to a warfront without guns. We found the going difficult as we had to plod through several possibilities. Now, we will experience the pleasure of putting influence lines to work.

Influence lines duly plotted can help us get the total influence of all the loads on a span in inducing shear or bending moment at a point, or a support reaction or a support bending moment. They can also help us decide upon the strategic load positions to create maximum shear or bending moment *at a given section* for which the ILDs are drawn. The italicized phrase is a limitation. It may require some effort to locate the critical point for which we shall construct the influence line. Example 2.1 illustrates the use of influence lines.

Example 2.1 *A single rolling load of 100 kN moves on a girder of span 20 m. (a) Construct the influence lines for (i) shear force and (ii) bending moment for a section 5 m from the left support. (b) Construct the influence lines for points at which the absolute maximum shears and absolute maximum bending moment develop. Determine these absolute maximum values.*

Solution (*a*) To find maximum shear force and bending moment at 5 m from the left support :

For the ILD for shear, IL ordinate to the right of D = $\frac{l-x}{l} = \frac{20-5}{20} = 0.75$

IL ordinate to the left of D = $\frac{x}{l} = \frac{5}{20} = 0.25$

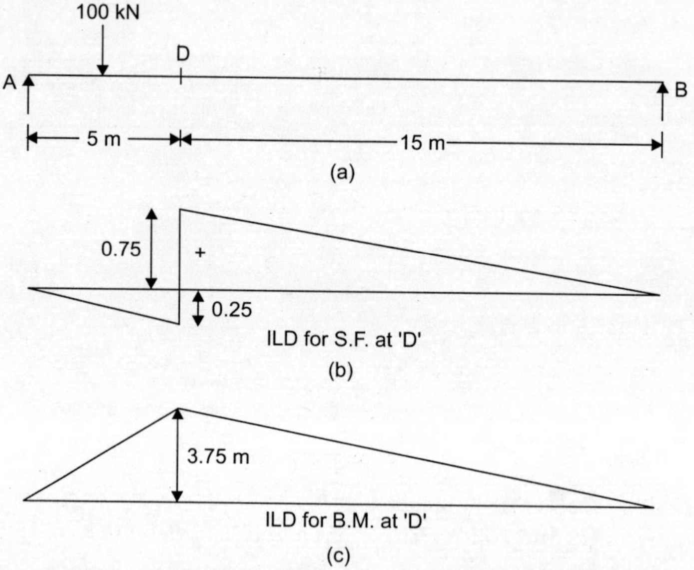

Fig. 2.7

For the IL for bending moment, IL ordinate at D = $\frac{x(l-x)}{l} = \frac{5 \times 15}{20} = 3.75$ m.

(*i*) **Maximum positive shear force**

By inspection of the ILD for shear force, it is evident that maximum positive shear force occurs when the load is placed just to the right of D.

Maximum positive shear force = load × ordinate = 100 × 0.75

At D, SF_{max} + = 75 kN.

(*ii*) **Maximum negative shear force**

Maximum negative shear force occurs when the load is placed just to the left D.

Maximum negative shear force = load × ordinate = 100 × 0.25

At D, SF_{max} – = 25 kN.

(*iii*) **Maximum bending moment**

Maximum bending moment occurs when the load is placed on the section D itself.

Maximum bending moment = load × ordinate = 100 × 3.75 = 375 kNm.

(*b*) Absolute maximum positive shear force will occur at A. Absolute maximum negative shear force will occur at B. Maximum bending moment will occur at midspan. The ILs are sketched in Fig. 2.8.

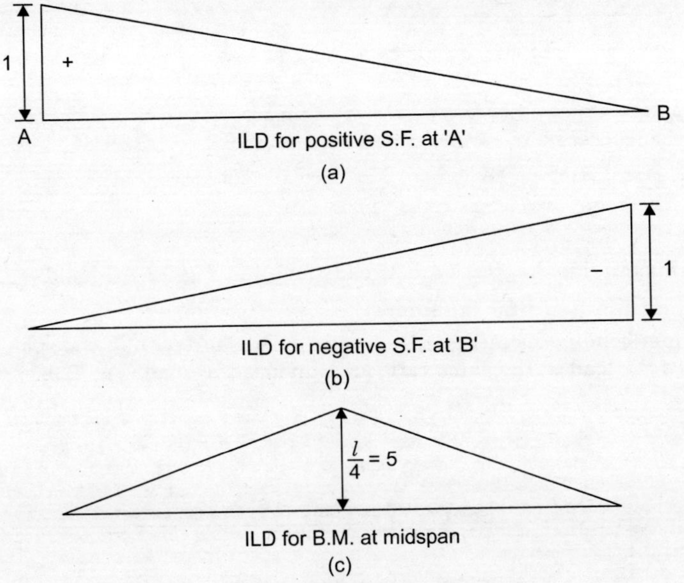

Fig. 2.8

(*i*) **Positive shear force**

Absolute maximum positive shear force occurs when the load is placed at A.

Absolute maximum positive shear force = load × ordinate = 100 × 1

$$SF_{max\ max} + = 100 \text{ kN}.$$

(*ii*) **Negative shear force**

Absolute maximum negative shear force occurs when the load is placed at B.

Absolute maximum negative shear force = load × ordinate = 100 × (– 1)

$$SF_{max\ max} = -100 \text{ kN}.$$

(*iii*) **Absolute maximum bending moment**

Absolute maximum bending moment occurs when the load is at midspan

Maximum bending moment = $M_{max\ max}$ = load × ordinate = 100 × 5 = 500 kNm.

Example 2.2 *Draw the ILD for shear force and bending moment for a section at 5 m from the left hand support of a simply supported beam, 20 m long. Hence, calculate the maximum bending moment and shear force at the section, due to an uniformly distributed rolling load of length 8 m and intensity 10 kN/m run.*

Solution

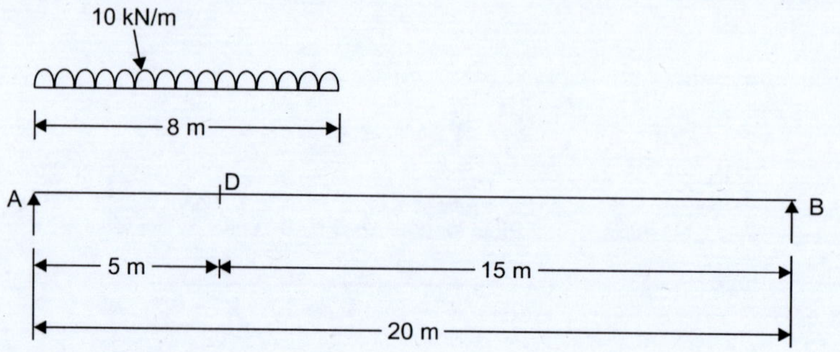

The influence line diagrams are already found in Figs. 2.7 (*b*) and (*c*).

(*a*) **Maximum bending moment**

Maximum bending moment at D due to a UDL shorter than the span occurs when the section divides the load in the same ratio as it divides the span (Fig. 2.9).

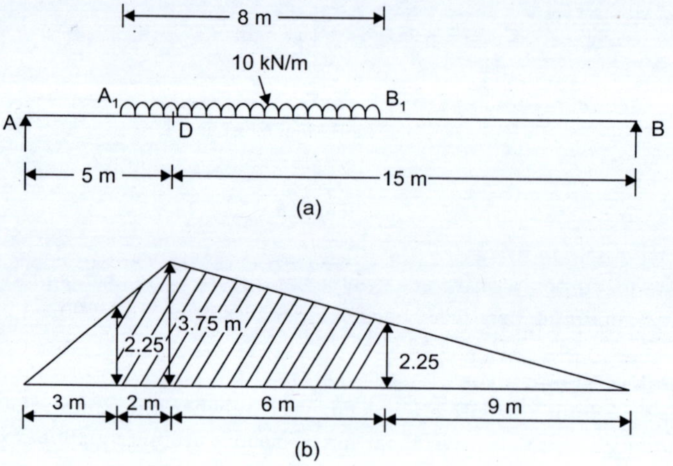

Fig. 2.9

In the above figure, $\frac{A_1D}{B_1D} = \frac{AD}{BD}$, $A_1D = 2$ m, $B_1D = 6$ m

Ordinates :

$$\text{Ordinate under } A_1 = \frac{3.75}{5} \times 3 = 2.25$$

$$\text{Ordinate under } B_1 = \frac{3.75}{15} \times 9 = 2.25$$

Maximum bending moment = Intensity of load × Area of ILD under the load

$$= 10 \times \frac{(3.75 + 2.25) \times 8}{2}$$

At D, M_{max} = 240 kNm.

(*b*) **Maximum positive shear force**

Maximum positive shear force occurs when the tail of the UDL is at D as it traverses from left to right (Fig. 2.10).

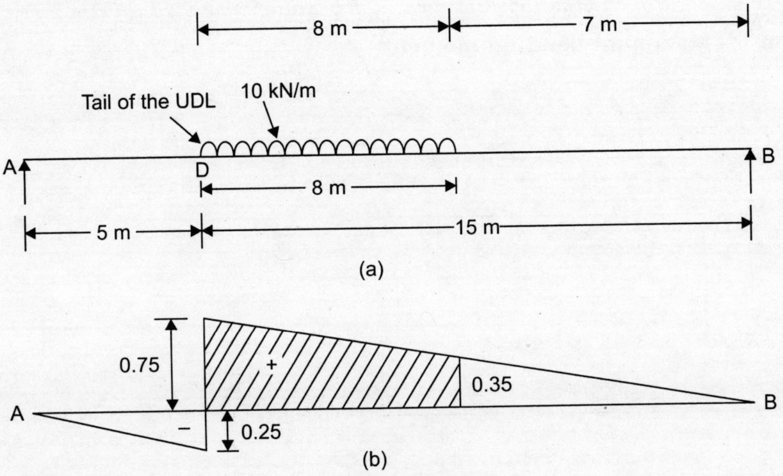

Fig. 2.10

Ordinate under $B_1 = \frac{0.75}{15} \times (15 - 8) = 0.35$

Maximum positive shear force = Intensity of load × Area of ILD under load

$$= 10 \times \frac{(0.75 + 035) \times 8}{2}$$

SF_{max} + = 44 kNm.

(*c*) **Maximum negative shear force**

Maximum negative shear force due to an UDL occurs when the head of the load is at D as it traverses from left to right (Fig. 2.11).

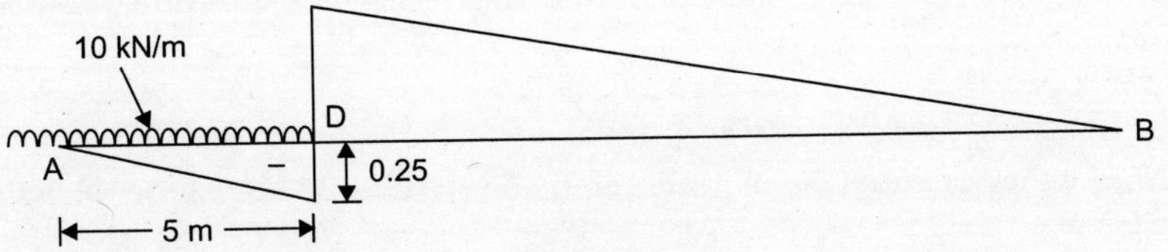

Fig. 2.11

Maximum negative shear force = Intensity of load × Area of ILD under the load

$$= 10\left(\tfrac{1}{2} \times 0.25 \times 5\right)$$

Negative $SF_{Max} = 6.25$ kN.

Example 2.3 *Two point loads of 100 kN and 200 kN spaced 3 m apart cross a girder of span 15 m from left to right with the 100 kN load loading. Draw the influence line for shear force and bending moment and find the value of maximum shear force and bending moment at a section D, 6 m from the left hand support. Also, find the absolute maximum bending moment due to the given load system.*

Solution (*a*) **Maximum bending moment**

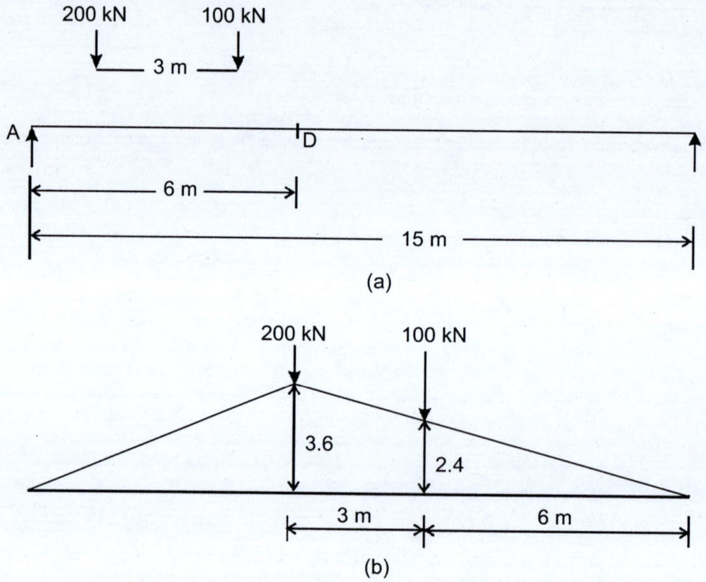

Fig. 2.12

$$\text{Max. ordinate of ILD} = \frac{x(l-x)}{l} = \frac{6 \times 9}{15} = 3.6 \text{ m.}$$

Maximum bending moment at D occurs under the critical load. This load, when it moves from left to right or from right to left of 'D' changes the sign of L_r, the differential loading rate, where,

$$L_r = \frac{W_{left}}{x} - \frac{W_{right}}{l-x}$$

Now, let us try with 200 kN load. Firstly, keep this 200 kN load to the left of 'D'.

$$L_r = \frac{W_{left}}{x} - \frac{W_{right}}{l-x} = \frac{200}{6} - \frac{100}{9} = 22.22 \text{ (+ve)}$$

Moving this load to the right of 'D',

$$L_r = \frac{W_{left}}{x} - \frac{W_{right}}{l - x} = \frac{0}{6} - \frac{300}{9} = -33.33 \text{ (-ve)}$$

Since, the sign of L_r changes, the 200 kN load is critical. The maximum bending moment occurs at D when this 200 kN load is placed at D.

$$\text{Ordinate under 100 kN load} = \frac{3.6}{9} \times 6 = 2.4$$

$$\text{Maximum bending moment} = \Sigma \text{ (Load} \times \text{Ordinate)} = 200 \times 3.6 + 100 \times 2.4$$

$$M_{max} = 960 \text{ kNm.}$$

(*b*) **Maximum shear force**

(*i*) *Positive shear force*

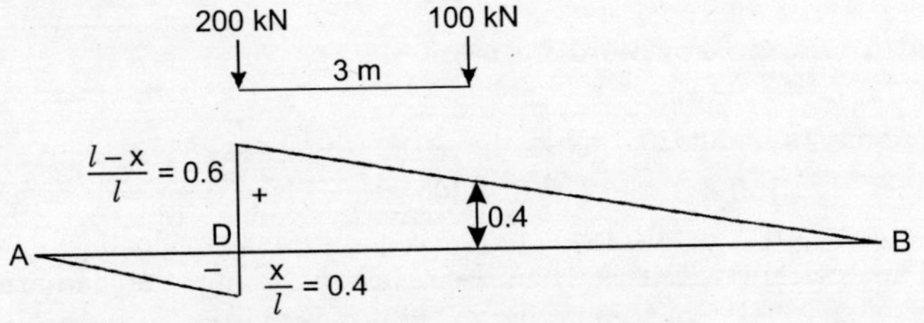

Fig. 2.13. ILD for shear force at 'D'.

Figure 2.13 shows the load position for the absolute maximum positive shear force. If the load train is moved to the left, the positive contribution due to the bigger (200 kN) load is lost and a negative contribution is obtained.

If the load train moves to the right, the ordinates under to the loads decrease. Hence, the indicated load position is the critical one.

Ordinate under 200 kN load = 0.6

$$\text{Ordinate under 100 kN load} = \frac{0.6}{9} = 0.4$$

Maximum positive shear force = 200 (0.6) + 100 (0.4)

$$SF_{max} + = 160 \text{ kN.}$$

(*ii*) *Negative shear force*

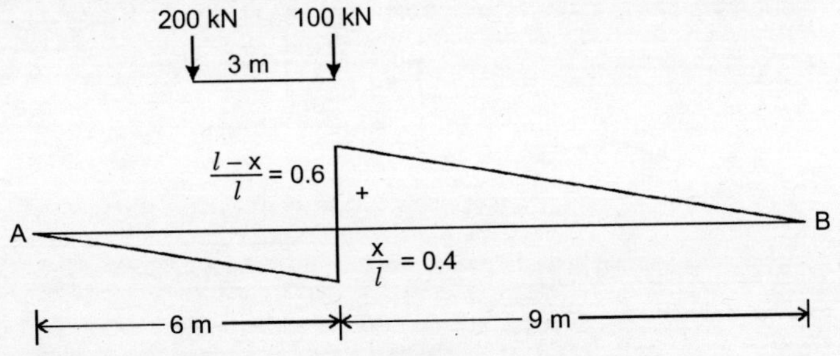

Fig. 2.14

Trying with 100 kN load, first keep this 100 kN load to the left of D. Then move this load to the right of 'D' by 3 m. If the value of the shear increment S_i is a negative value, it indicates a decrease in negative shear force.

$$S_i = \frac{Wc}{l} - W_1 = \frac{300 \times 3}{15} - 100 = -40 \text{ (–ve)}.$$

Hence, the negative shear force decreases when this load train is moved to the right of 'D'. Hence, to get maximum negative shear force, this 100 kN load should be kept just to the left of D.

Ordinate under 100 kN load $= 0.4$

Ordinate under 200 kN load $= \frac{0.4}{6} \times 3 = 0.2$

Maximum negative shear force $= 100 \times 0.4 + 200 \times 0.2 = 80$

$$SF_{max}- = 80 \text{ kN}.$$

(*c*) **Absolute maximum bending moment**

(*i*) *Resultant of the loads*

Taking moments about 200 kN load,

$$100 \times 3 = R.\bar{x}, \quad R = 300 \text{ kN}, \quad \bar{x} = 1.0 \text{ m}.$$

Absolute maximum bending moment occurs under the load which is nearer to the resultant 'R'. The critical position is when the resultant 'R' and the load are at equal distance from the centre of span (C). (In this problem, 200 kN is nearer to the resultant 'R').

Distance of this 200 kN from C = Distance of 'R' from C.

$= 1/2 \times 1.0 = 0.5$ m

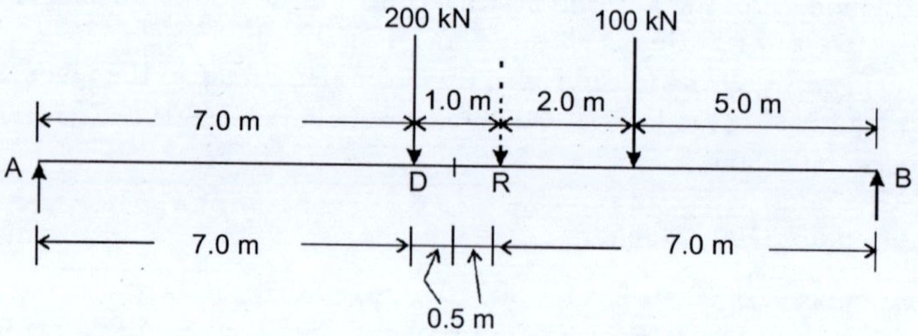

(a)

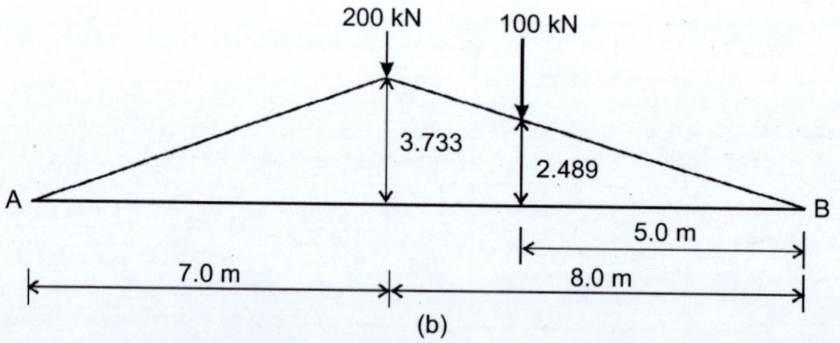

(b)

Fig. 2.15

Maximum ordinate of ILD *i.e.*, ordinate under 200 kN load = $\dfrac{x(l-x)}{l}$ = 3.733 m

Ordinate under 100 kN load = $\dfrac{3.733 \times 5}{8}$ = 2.333 m

Absolute maximum bending moment, $M_{max\ max}$ = Σ (Load × Ordinate)

$= 200 \times 3.733 + 100 \times 2.333 = 979.9$ kNm

Example 2.4 *A train of 5 wheel loads crosses a simply supported beam of span 22.5 m. Using influence lines, calculate the maximum positive and negative shear forces at midspan and absolute maximum bending moment anywhere in the span.*

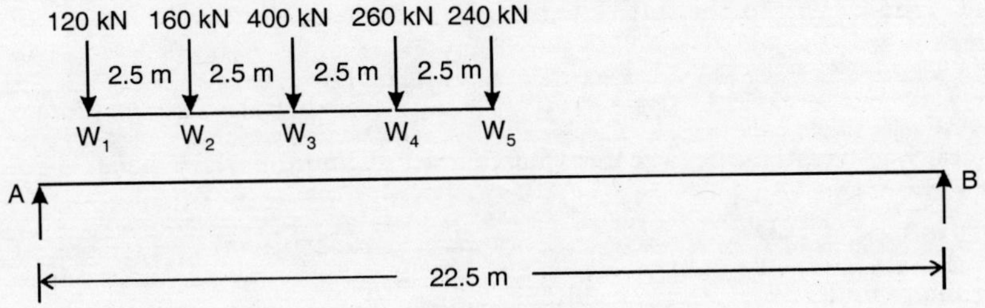

Fig. 2.16

Solution (*a*) **Maximum shear force**

(*i*) *Positive shear force*

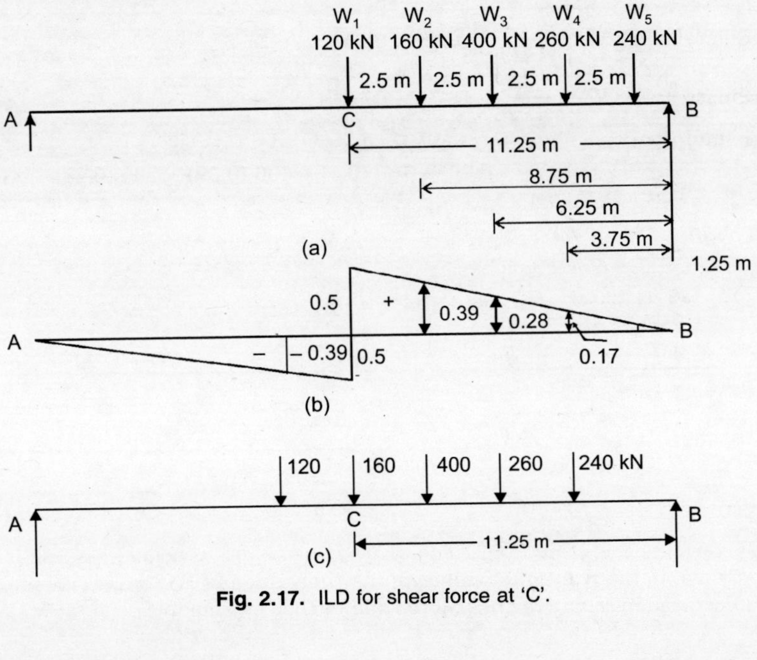

Fig. 2.17. ILD for shear force at 'C'.

To determine the load position to get the maximum positive shear force, let us keep all the loads to the right of C. Then move W_1 load to the left of 'C' by 2.5 m. If the sign of shear increment S_i is negative, it will indicate that W_1 shall be retained at C.

$$S_i = \frac{Wc}{l} - W_1$$

W = Total load on the span = 120 + 160 + 400 + 260 + 240 = 1180 kN

c = Distance through which the load train is moved = 2.5 m

$$S_i = \frac{1180 \times 2.5}{22.5} - 120 = 11.11 \text{ (+ve)}$$

Since S_i is positive, the shear force increases due to the shifting of W_1 to the left of C. Again, let us move W_2 to the left of C by 2.5 m to check whether the shear force further increases or not.

$$S_i = \frac{Wc}{l} - W_2 = \frac{1180 \times 2.5}{22.5} - 160 = -28.89 \text{ (–ve)}$$

Since S_i is negative, it indicates that to get maximum positive shear force, W_2 should stay just right of C.

Ordinates of ILD

Ordinate under $W_1 = \frac{-0.5}{11.25} \times (11.25 - 2.5) = -0.39$

Ordinate under $W_2 = \frac{22.5 - 11.25}{22.5} = 0.5$

Ordinate under $W_3 = \frac{0.5}{11.25} \times 8.75 = 0.39$

Ordinate under $W_4 = \frac{0.5}{11.25} \times 6.25 = 0.28$

Ordinate under $W_5 = \frac{0.5}{11.25} \times 3.75 = 0.17$

Maximum positive shear force = Σ (Load × Ordinate)

= 120 (– 0.39) + 160 (0.5) + 400 (0.39) + 260 (0.28) + 240 (0.17)

At C, SF_{max} + = 302.8 kN.

(ii) Negative shear force

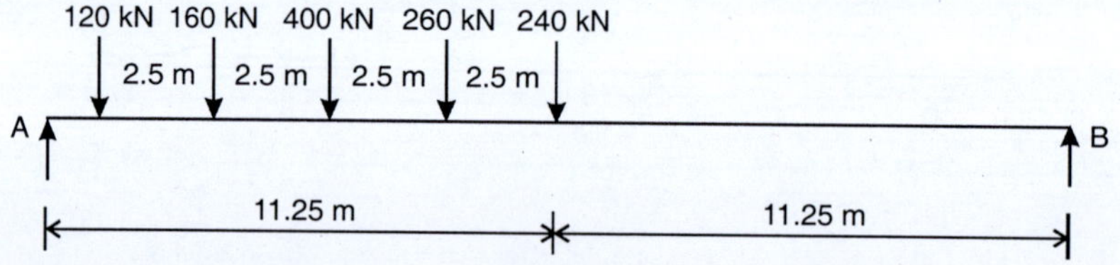

Fig. 2.18

To determine the position of loads to get the maximum negative shear force, move the loads one by one to the right of C and compute the value of S_i. If S_i becomes negative, it will indicate a decrease in negative shear force due to that movement.

First let us move the leading W_5 to the right of 'C' by 2.5 m and calculate S_i.

$$S_i = \frac{Wd}{l} - W_5$$

$$W_5 = 240 \text{ kN} ; d = 2.5 \text{ m}$$

$$S_i = \frac{1180 \times 2.5}{22.5} - 240 = -108.89 \text{ (–ve)}$$

Since S_i is –ve, it indicates that W_5 should stay just to the left of C.

Ordinates of ILD

$$\text{Ordinate under } W_5 = \frac{x}{l} = \frac{11.25}{22.5} = 0.5$$

$$\text{Ordinate under } W_4 = \frac{0.5}{11.25} \times (11.25 - 2.5) = 0.39$$

$$\text{Ordinate under } W_3 = \frac{0.5}{11.25} \times (11.25 - 5.0) = 0.278$$

$$\text{Ordinate under } W_2 = \frac{0.5}{11.25} \times (11.25 - 7.5) = 0.167$$

$$\text{Ordinate under } W_1 = \frac{0.5}{11.25} \times (11.25 - 10) = 0.056$$

Maximum negative shear force at C = 240 (– 0.5) + 260 (– 0.39) + (400 (– 0.278)) + 160 (– 0.167) + 120 (– 0.056)

$$F_{max} = 366.04 \text{ kN}$$

(*b*) **Absolute maximum bending moment**

(*i*) *Position of resultant of all loads*

Taking moments about W_1, 120 (0) + 160 (2.5) + 400 (5.0) + 260 (7.5) + 240 (10.0) = $R.\bar{x}$

$$R = 1180 \text{ kN}$$

$$\bar{x} = 5.72 \text{ m from } W_1$$

(*ii*) *Location of absolute maximum bending moment*

Absolute maximum bending moment occurs under the load, which is nearest to the resultant 'R'. (In this problem, W_3 is nearest to the resultant 'R'). The distance between C and R and the distance between C and W_3 shall be equal.

Distance between R and centre of span (C) = 1/2 (0.72) = 0.36 m.

Figure 2.19 shows the IL for bending moment at the critical spot D, 10.89 m from A.

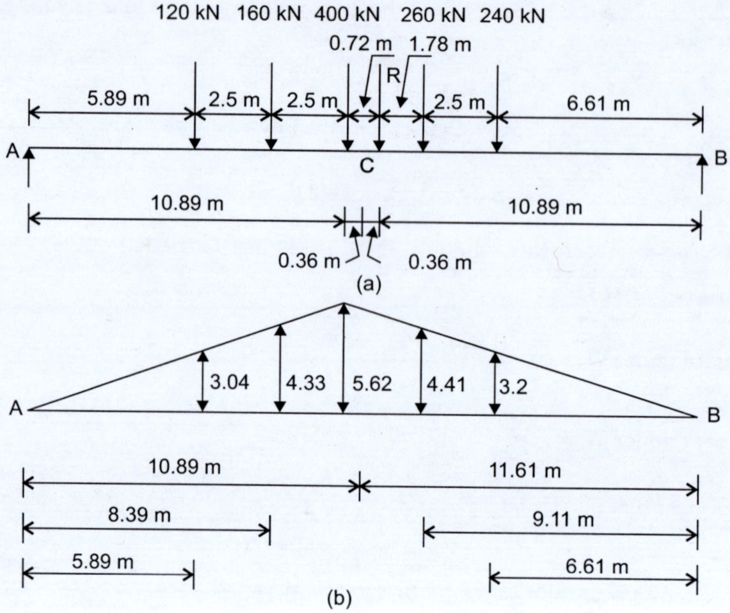

Fig. 2.19. ILD for bending moment at 'D'.

Ordinates of ILD

Maximum ordinate of ILD

i.e., ordinate under $W_3 = \dfrac{x(l-x)}{l} = \dfrac{10.89(22.5 - 10.39)}{22.5} = 5.62$

Ordinate under $W_2 = \dfrac{5.62}{10.89} \times 8.39 = 4.33$

Ordinate under $W_1 = \dfrac{5.62}{10.89} \times 5.89 = 3.04$

Ordinate under $W_4 = \dfrac{5.62}{11.61} \times 9.11 = 4.41$

Ordinate under $W_5 = \dfrac{5.62}{11.61} \times 6.61 = 3.20$

Absolute maximum bending moment = 120 (3.04) + 160 (4.33) + 400 (5.62)
+ 260 (4.41) + 240 (3.2)

$M_{max\ max} = 5220.2$ kNm.

Example 2.5 *A girder having a span of 18 m is simply supported at the ends. It is traversed by a train of loads as shown in Fig. 2.20, the 50 kN load leading. Find the maximum bending moment which can occur (i) under the 200 kN load, and (ii) under 50 kN load, using influence line diagrams (ILD).*

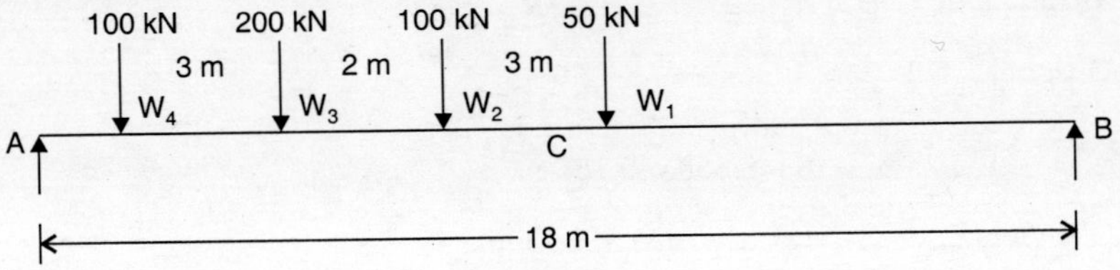

Fig. 2.20

Solution **Maximum bending moment**

(i) Under 200 kN load

To get the maximum bending moment under W_3, the resultant (R) and W_3 should be at equal distances from the centre of the span (C). For that, the point of action of resultant (R) should be determined first.

Resultant of loads :

$$R = 450 \text{ kN}$$

Taking moments about W_4

$$200\ (3) + 100\ (3 + 2) + 50\ (3 + 2 + 3) = 450\ \bar{x}$$

$$\bar{x} = 3.33 \text{ m from } W_4.$$

Bending moment under 200 kN load

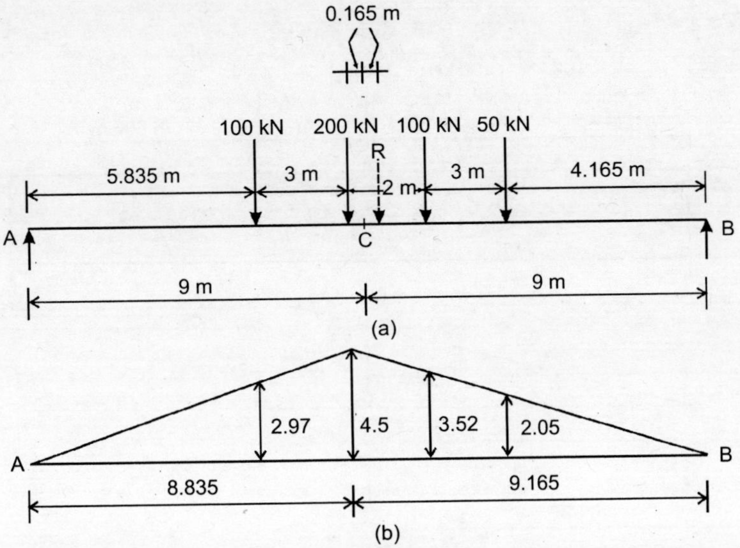

Fig. 2.21

Distance between C and 200 kN load = Distance between C and R = 0.33 / 2 = 0.165 m.

Ordinates of ILD

ILO under $W_3 = \dfrac{x(l-x)}{l}$

$x = 8.335$ m

$l - x = 18 - 8.835 = 9.165$ m

ILO under $W_3 = \dfrac{8.835 \times 9.165}{18} = 4.50$ m

ILO under $W_4 = \dfrac{4.5}{8.835} \times 5.835 = 2.97$ m

ILO under $W_2 = \dfrac{4.5}{9.165} \times (3 + 4.165) = 3.52$ m

ILO under $W_1 = \dfrac{4.5}{9.165} \times 4.165 = 2.05$ m

Bending moment under the 200 kN load

$= 200\ (4.5) + 100\ (2.97) + 100\ (3.52) + 50\ (2.05) = 1651.5$ kNm.

(*ii*) *Bending moment under 50 kN load*

To get the maximum bending moment under W_1, W_1 and 'R' must be at equal distances from the centre of span (C). Distance between C and R = Distance between C and W_1

$= 1/2\ (2 - 0.33 + 3) = 1/2\ (4.67) = 2.335$ m.

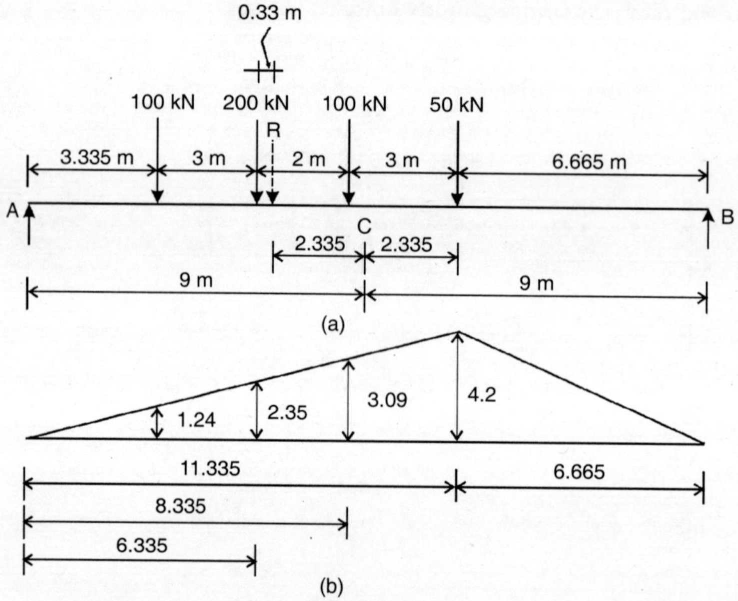

Fig. 2.22

Ordinates of ILD

ILO under $W_1 = \dfrac{x(l-x)}{l} = \dfrac{11.335 \times 6.665}{18} = 4.20$

$$\text{ILO under } W_2 = \frac{4.2}{11.335} \times 8.335 = 3.09 \text{ m}$$

$$\text{ILO under } W_3 = \frac{4.2}{11.335} \times 6.335 = 2.35$$

$$\text{ILO under } W_4 = \frac{4.2}{11.335} \times 3.335 = 1.24$$

Bending moment under the 50 kN load

$$= 4.2\,(50) + 3.09\,(100) + 2.35\,(200) + 1.24\,(100) = 1113 \text{ kNm.}$$

Example 2.6 *Using the relevant influence line diagrams, find (i) the maximum bending moment, (ii) the maximum positive and negative shears at 4 m from the left support A of a SS girder of span 10 m, when a train of 4 wheel loads of 10 kN, 15 kN, 30 kN and 30 kN spaced at 2 m, 3 m, and 3 m respectively cross the span with the 10 kN load leading.*

Solution The influence line diagrams for bending moment and shear are sketched below (Fig. 2.23).

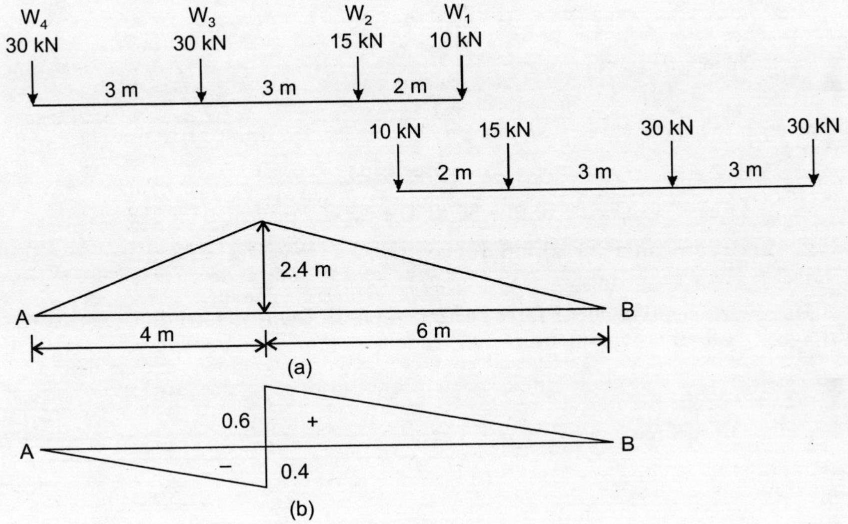

Fig. 2.23

$$\text{ILO at C for bending moment is } \frac{ab}{l} = \frac{4 \times 6}{10} = 2.4 \text{ m}$$

$$\text{ILO at C for shear right of C} = \frac{x}{l} = \frac{6}{10} = 0.6$$

$$\text{ILO at C for shear left of C} = \frac{4}{10} = 0.4$$

1. Maximum bending moment occurs at C when the critical load is placed at C. The critical load is that which when moved from left of C to the right changes the sign of the loading rate L_r given by $\left(\frac{W_{left}}{x} - \frac{W_{right}}{l-x}\right)$.

The load system can traverse the span either from left to right or from right to left. But to get the maximum bending moment, the load system should traverse from left to right, with W_1 leading.

(Otherwise, with W_3 at C, W_1 will be off the span).

When W_3 is at C (just to the left of C),

$$L_r = \left(\frac{W_{left}}{x} - \frac{W_{right}}{l-x}\right) = \frac{60}{4} - \frac{25}{6} = 10.83 \text{ (+ve)}$$

When W_3 has crossed to the right of C,

$$L_r = \frac{30}{4} - \frac{55}{6} = -1.67 \text{ (–ve)}$$

L_r changes sign when W_3 crosses C. Hence, W_3 is critical and should be placed at C for maximum bending moment at C. Figure 2.24 shows the load positions and the corresponding influence line ordinates (ILO).

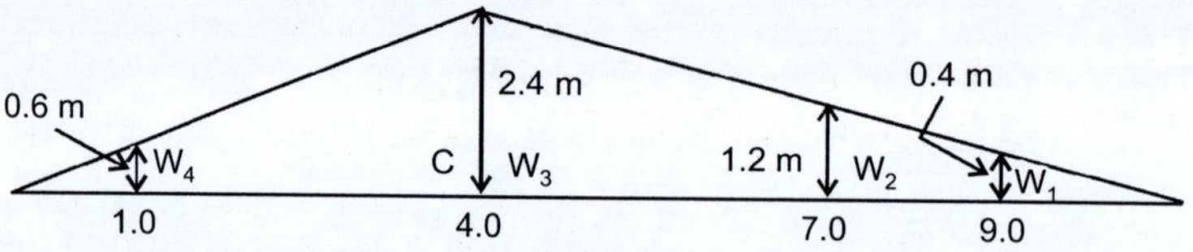

Fig. 2.24

$$M_{max} \text{ at C} = 30\,(0.6) + 30\,(2.4) + 15\,(1.2) + 10(0.4) = 112 \text{ kNm.}$$

2. Maximum positive shear force

The ILD for shear is shown in Fig. 2.25.

Maximum positive shear force occurs when all the loads are to the right of the section and the load system traverses from right to left.

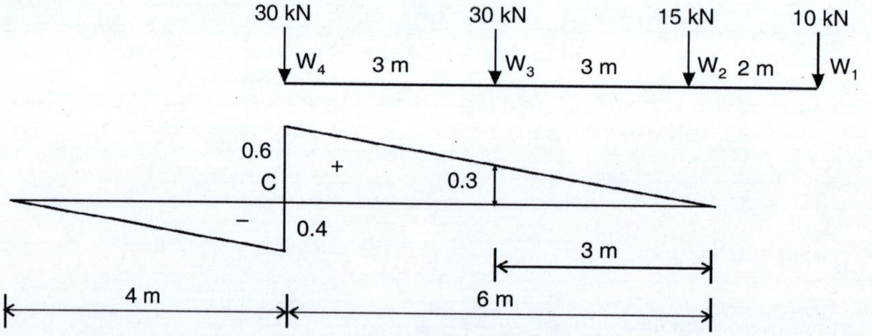

Fig. 2.25

To get maximum positive shear force, place W_4 as shown in Fig. 2.25. To ascertain that this is the load position for maximum positive shear force, move W_4 to the left through a distance of 3 m so that W_3 is at C. If the sign of shear increment S_i, given by $\frac{Wd}{l}$ – W m is negative, it indicates that W_4 should be at the right of 'C'. (W_m is the load moved to the other side)

In the above expression,

$$S_i = \frac{Wd}{l} - W_m = \frac{75 \times 3}{10} - 30 = -7.5 \text{ (–ve)}$$

The shear increment happens to be negative.

Hence, W_4 should not be moved to the left.

Ordinate under W_4 = 0.6

Ordinate under $W_3 = \frac{0.6}{6} \times 3 = 0.3$

Ordinate under W_2 = 0

Maximum positive shear force = 0.6 (30) + 0.3 (30) = 27 kN.

3. Maximum negative shear force

Maximum negative shear force occurs when the loads are to the left of the section while traversing from left to right.

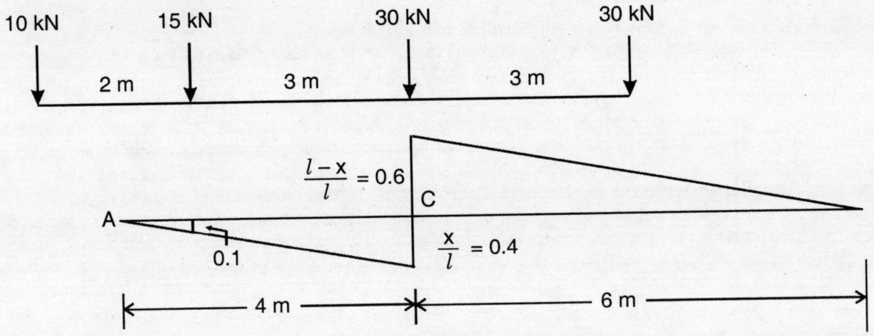

Fig. 2.26

To ascertain if this is the maximum negative shear force position, move W_4 to the right of C through a distance of 3 m. If the sign of S_i is negative, it indicates that the W_4 should not be moved.

$$S_i = \frac{Wd}{l} - W_1 = \frac{60 \times 3}{10} - 30 = -12 \text{ (–ve).}$$

Since the sign is negative, W_4 should stay at the left of 'C'.

Ordinate under $W_2 = \frac{0.4}{4} \times 1 = 0.1$

Ordinate under W_1 = 0.4

Maximum negative shear force = 0.1 × 30 + 0.4 × 30 = 15 kN.

Example 2.7 *Four equal loads of 150 kN each equally spaced at 2 m apart followed by a u.d.l. of 60 kN/m at a distance of 1.5 m from the last 150 kN load cross a girder of 20 m span from right to left. Using influence lines, calculate the shear force and bending moment at a section 8 m from the left hand support when the leading 150 kN load is at 5 m from the left hand support.*

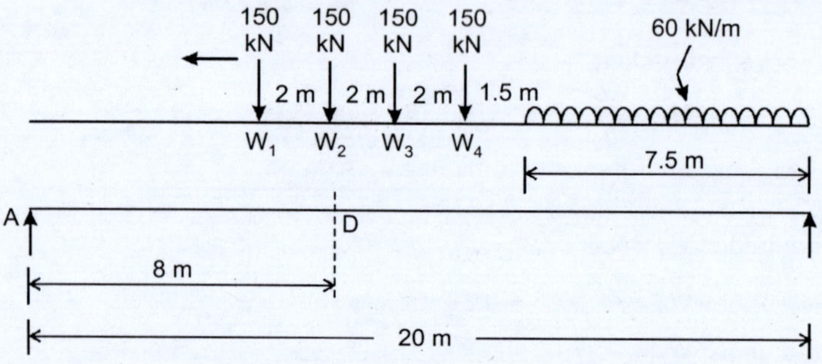

Fig. 2.27

Solution (*a*) **Bending moment for the given load position**

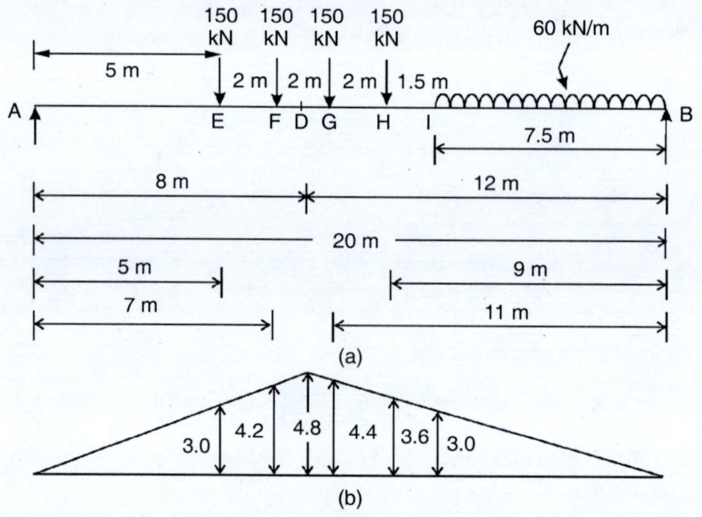

Fig. 2.28

Ordinates of ILD

Maximum ordinate at $\text{D} = \dfrac{x(l-x)}{l} = \dfrac{8x(20-8)}{20} = 4.8 \text{ m}$

Ordinate under $\text{E} = \dfrac{4.8}{8} \times 5 = 3.0 \text{ m}$

Ordinate under $\text{F} = \dfrac{4.8}{8} \times 7 = 4.2 \text{ m}$

Ordinate under $G = \frac{4.8}{12} \times 11 = 4.4$ m

Ordinate under $H = \frac{4.8}{12} \times 9 = 3.6$ m

Ordinate under $I = \frac{4.8}{12} \times 7.5 = 3.0$ m

Bending moment at

D = Σ (Load × ordinate) for point loads + (Load intensity × area of ILD below UDL)

$= 150\ (3) + 150\ (4.2) + 150\ (4.4) + 150\ (3.6) + 60\ (1/2 \times 7.5 \times 3) = 2955$ kNm.

(*b*) **Shear force at this section**

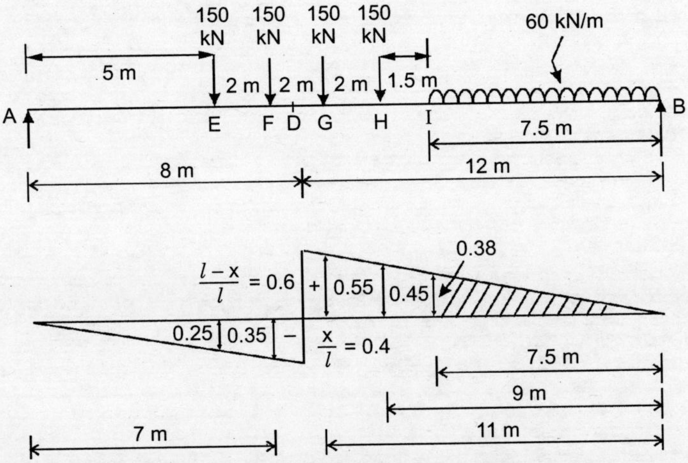

Fig. 2.29

Ordinates of ILD

Maximum positive ordinate of ILD $= \frac{(l - x)}{l} = \frac{(20 - 8)}{20} = \frac{12}{20} = 0.6$

Maximum negative ordinate of ILD $= \frac{x}{l} = \frac{8}{20} = 0.4$

Ordinate under $E = \frac{-0.4}{8} \times 5 = -0.25$

Ordinate under $F = \frac{-0.4}{8} \times 7 = -0.35$

Ordinate under $G = \frac{0.6}{12} \times 11 = 0.55$

Ordinate under $H = \frac{0.6}{12} \times 9 = 0.45$

Ordinate under $I = \frac{0.6}{12} \times 7.5 = 0.38$

Shear force at D for the given load arrangement

$= \Sigma$ (load × ordinate) + (load intensity × area of ILD below UDL)

$= 150\,(-0.25) + 150\,(-0.35) + 150\,(0.55) + 150\,(0.45) + 60\,(1/2 \times 7.5 \times 0.38)$

$= 145.5$ kN SF_D

2.5 INFLUENCE LINES FOR OVERHANGING BEAMS

Drawing influence lines for beams with overhangs does not need any new principles. We need extend the concept we are already familiar with. Loads on overhangs can produce downward reactions at supports.

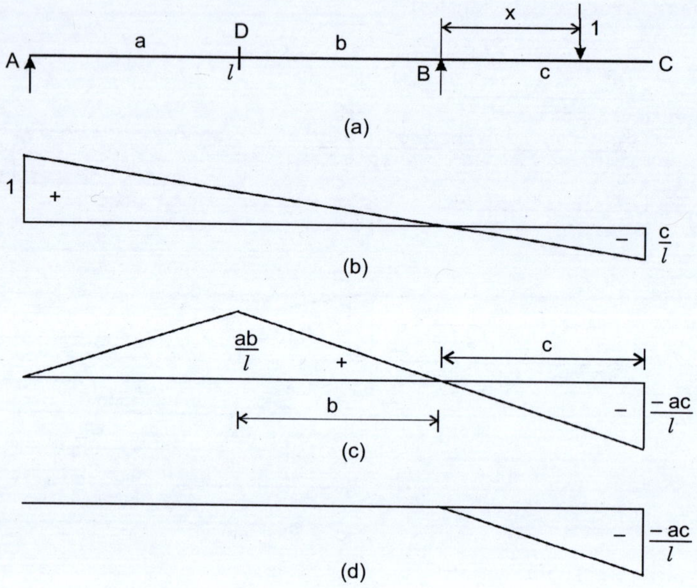

Fig. 2.30

These in turn can produce hogging bending moments. We need be, therefore, careful about the sign of the influence line ordinates.

For the beam ABC in Fig. 2.30 (*a*) the influence line for R_A is shown in Fig. 2.30 (*b*). There is a negative part in the overhang BC. This of course means, that a load on BC produces a negative (downward) reaction at A.

Figure 2.30 (*c*) is the influence line for bending moment at D on segment AB is just like the one for a SS beam. For the overhang, when the unit load is at x from B,

$$R_A = -1\,\frac{x}{l}\ ; \qquad M_D = R_A \,.\, a = -\frac{ax}{l}$$

Hence at C, where $x = c$ the ordinate is $-\dfrac{ac}{l}$

Regarding bending moment at B, we can use the same formula as for a cantilever.

Knowing that load on AB does not produce any moment at B, bending moment ordinates for segment AB = 0.

Bending moment ordinate at C = $-\dfrac{ac}{l}$.

Figure 2.30 (*d*) is the ILD for bending moment at B.

Example 2.8 *Draw the influence lines for*

(*a*) R_A (*b*) *Shear at D*

(*c*) *Bending moment at D* (*d*) *Shear to the right of B*

(*e*) *Bending moment at B for the overhanging beam shown in following figure.*

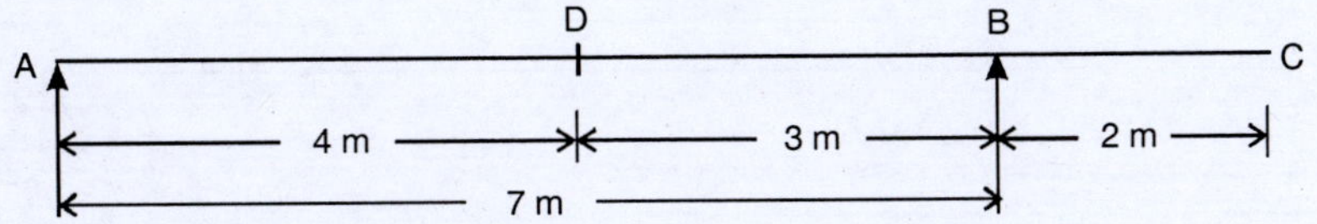

Solution (*a*) **IL for R_A**

Let us consider a section X at a distance x from C. When a unit load is at X,

$R_A = \dfrac{x-2}{7}$. This is valid for all x from C to A. The influence line ordinates are as under.

x	0	2	5	9
ILO	$\dfrac{-2}{7}$	0	$\dfrac{3}{7}$	1

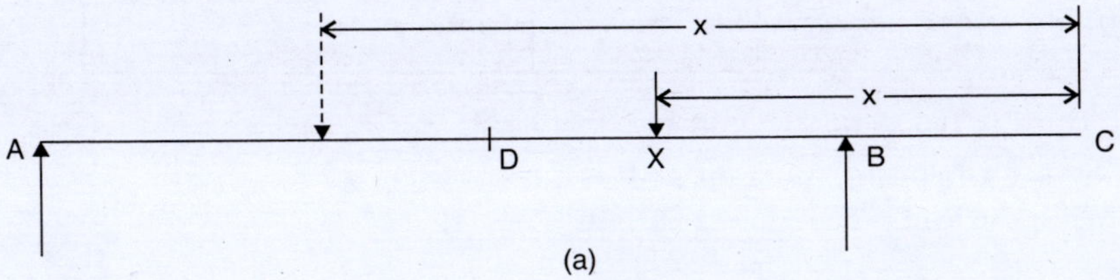

(a)

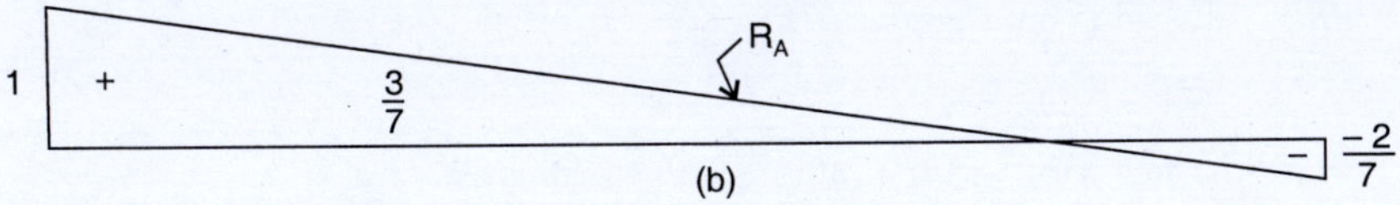

(b)

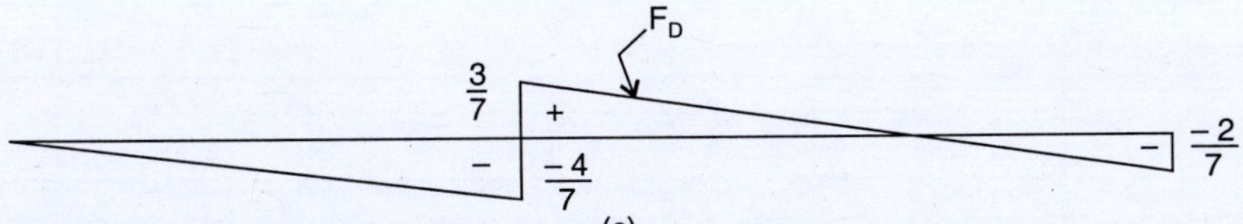

(c)

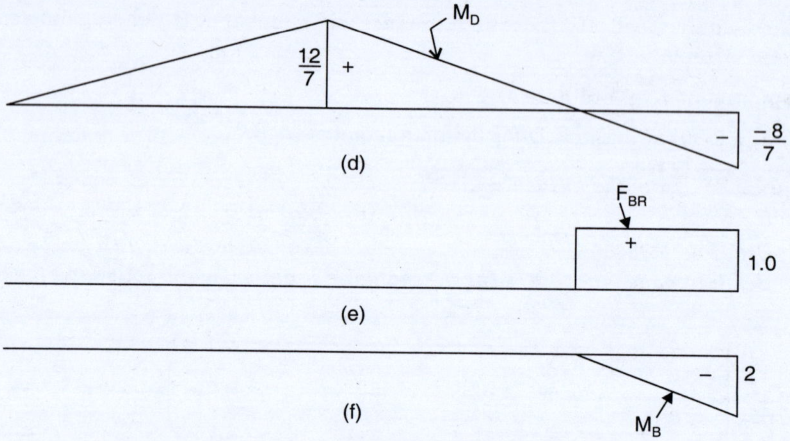

Fig. 2.31

Figure 2.31 (*b*) is the influence line diagram for R_A.

(*b*) **IL for shear at D (F_D)**

when the unit load is on DC, $F_D = R_A = \dfrac{x-2}{7}$

when the unit load is on AD, $F_D = -R_B = -(1 - R_A) = -1 + \dfrac{x-2}{7} = \dfrac{x-9}{7}$

The ordinates are tabulated under:

x	0	2	5 (Right)	5 (Left)	9
ILO	$\frac{-2}{7}$	0	$\frac{3}{7}$	$\frac{-4}{7}$	0

Figure 2.31(*c*) shows the influence line for S_D.

(*c*) **IL for bending moment at D (M_D)**

When the load is on segment CD,

$$M_D = R_A \times 4 = \frac{x-2}{7} \times 4 \qquad \text{...(A)}$$

When the load is on segment AD,

$$M_D = R_B \times 3 = \frac{9-x}{7} \times 3 \qquad \text{...(B)}$$

Hence the following are the influence line ordinates:

x	0	2	5	5	9
ILO	$\frac{-8}{7}$	0	$\frac{12}{7}$	$\frac{12}{7}$	0

Using (A) (columns x = 0, 2, 5) — Using (B) (columns x = 5, 9)

Figure 2.31 (*d*) shows the influence line for M_D.

(*d*) **Shear to the right of B (F_{BR})**

$F_{BR} = 1$, since the unit load will be the shear at B. When the load is on AB, there will be no shear S_{BR}. Figure 2.31 (*e*) is the IL diagram for F_{BR}.

(*e*) **Bending moment at B (M_B)**

Again only when there is a load on the overhang, there will be M_B and $M_B = (2 - x)$

Figure 2.31 (*f*) is the IL for M_B.

Example 2.9 *Draw influence lines for*

(*a*) *Reaction at A(R_A)*

(*b*) *Reaction at B(R_B)*

(*c*) *Shear force at D(F_D)*

(*d*) *Bending moment at D(M_D)*

(*e*) *Shear force at E (F_E) and*

(*f*) *Bending moment at E(M_E).*

The beam is overhanging at both ends as shown in following figure.

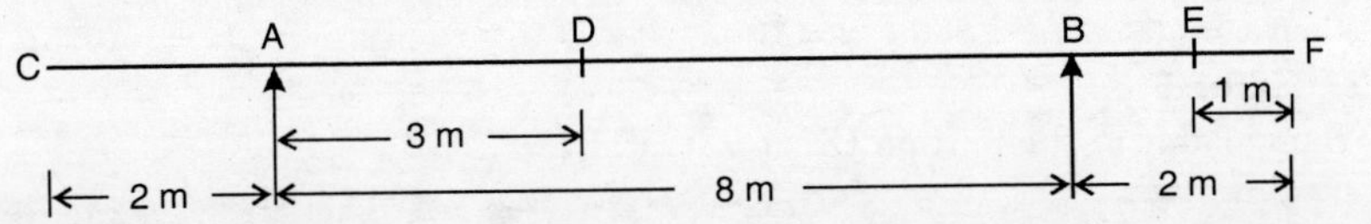

Solution (*a*) **IL for R_A**

Consider a section X at a distance x from C and apply a unit load at X.

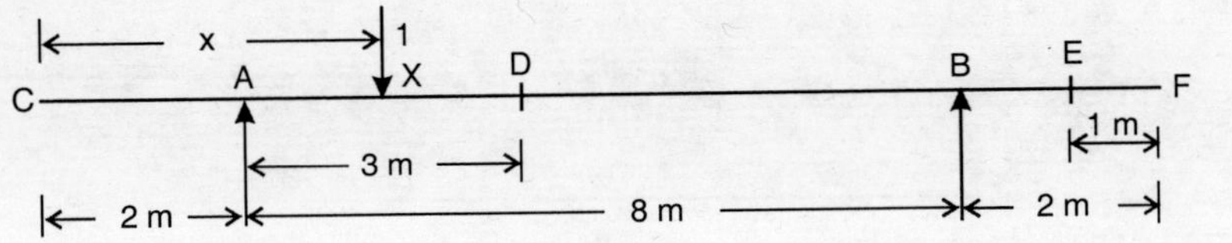

Fig. 2.32

Taking moments about B,

$$R_A \times 8 = 1 \times (10 - x)$$

$$R_A = \frac{10 - x}{8}$$

This equation is valid for all positions of unit load from C to F.

When $x = 0$, $R_A = 1.25$, $x = 2$ m, $R_A = 1$, $x = 10$ m, $R_A = 0$, $x = 12$ m, $R_A = -0.25$

(*b*) **IL for R_B**

Taking moments about A,

$$R_B \times 8 = 1 \times (x - 2)$$

$$R_B = \frac{x - 2}{8}$$

When $x = 0$, $R_B = -0.25$, $x = 2$ m, $R_B = 0$, $x = 10$ m, $R_B = 1$, $x = 12$ m, $R_B = 1.25$

(*c*) **IL for F_D**

When the unit load is in portion CD, (x = 0 to 5 m)

$$F_D = -R_B = \frac{2-x}{8}$$

When x = 0, F_D = 0.25, x = 2 m, F_D = 0, x = 5 m, F_D = – 0.375

When the unit load is in portion DF, (x = 5 to 12 m)

$$F_D = R_A = \frac{10-x}{8}$$

When x = 5, F_B = 0.625, x = 10 m, F_B = 0, x = 12 m, F_B = – 0.25

(*d*) **IL for M_D**

When the unit load is in portion CD, (x = 0 to 5 m)

$$M_D = R_B \times 5 = \frac{5(x-2)}{8}$$

When x = 0, M_D = – 1.25, x = 2 m, M_D = 0, x = 5 m, M_D = 1.875

When the unit load is in portion DF, (x = 5 to 12 m)

$$M_D = R_A \times 3 = \frac{3(10-x)}{8}$$

When x = 5, M_D = 1.875, x = 10 m, M_D = 0, x = 12 m, M_D = – 0.75

(*e*) **IL for F_E**

When load is in the portion CE (x = 0 to 11 m)

$$F_E = 0$$

When load is in the portion EF (x = 11 to 12 m)

$$F_E = 1$$

(*f*) **IL for M_E**

When load is in the portion CE (x = 0 to 11 m)

$$M_E = 0$$

When load is at F, ME = – 1 × 1 = – 1 unit.

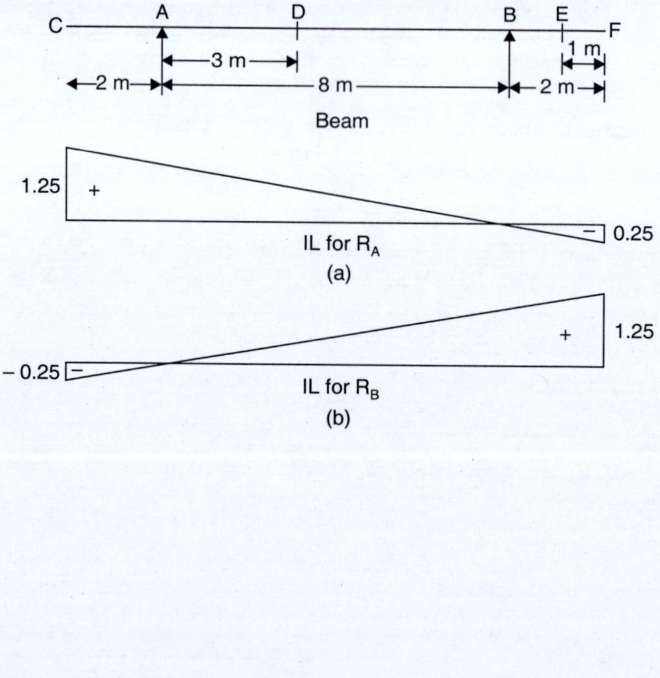

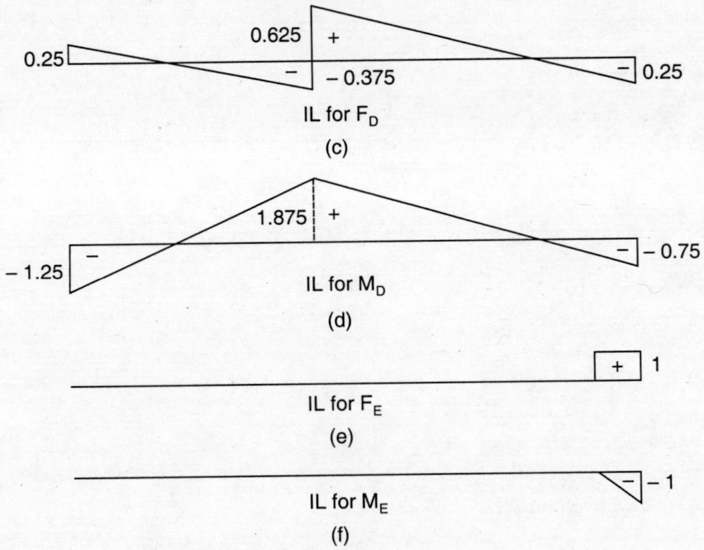

Fig. 2.33

Example 2.10 *A beam ABC is supported at A, B and C as shown in Fig. 2.34.*

Draw influence lines for

(a) Reactions at A, B and C

(b) Shear to the right of B

(c) Bending moment at E.

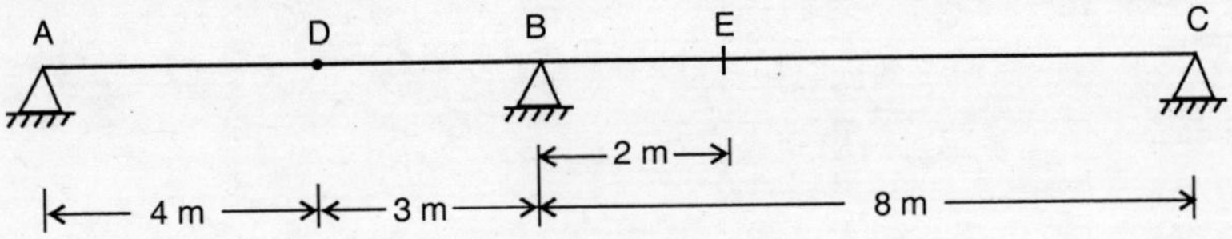

Fig. 2.34

Solution (*a*) **Influence lines for R_A, R_B and R_C**

(*i*) *IL for R_A*

Due to the hinge at D, AD will behave as a simply supported beam. The reaction at D, when it acts upward at D on AD, will produce a downward action at D of DBC (Fig. 2.35).

Any load on DBC, will have no effect on AD.

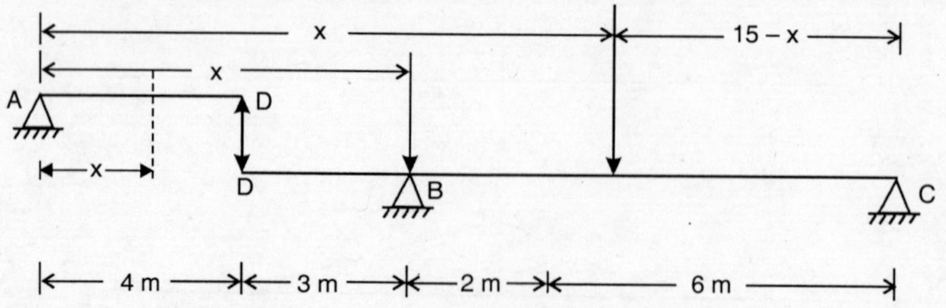

Fig. 2.35

When a unit load is on AD,

$$R_A = \frac{4-x}{4}$$

At $x = 0$, $R_A = 1$; at $x = 4$, $R_A = 0$

When x is > 4, R_A will be zero.

(ii) *IL for R_B*

When a unit load is on AD,

$$R_A = 1 - \frac{x}{4} \quad \text{and} \quad R_D = 1 - R_A = \frac{x}{4}$$

This would produce a 'load' at D for the portion DBC, equal to R_D.

Taking moment about C.

$$R_B \times 8 - R_D \times 11 = 0$$

$$R_B = \frac{R_D \times 11}{8} = \frac{x}{4} \times \frac{11}{8} = \frac{11x}{32}$$

When $x = 0$, $R_B = 0$

When $x = 4$, $R_B = 1.375$

When the load is on DBC, the reaction at D is zero.

Taking moments about C,

$$R_B \times 8 - 1\,(15 - x) = 0; \; R_B = \frac{15 - x}{8}$$

When $x = 4$ $\quad R_B = 1.375$

When $x = 7$ $\quad R_B = 1.0$

When $x = 15$ $\quad R_B = 0$

(iii) *IL for R_C*

When the unit load is on AD, 'load' at D = $\frac{x}{4}(\downarrow)$

Taking moments about B,

$$-\frac{x}{4} \times 3 - R_C \times 8 = 0 \; ; \; R_C = -\frac{3x}{32}$$

When $x = 0$, $R_B = 0$

When $x = 4$, $R_B = -0.375$ (Negative sign indicates downward reaction)

When the load is over DBC, $R_D = 0$

Taking moments about B,

$$-1(7-x) - R_C \times 8 = 0 \; ; \; R_C = \frac{x-7}{8}$$

When $x = 4$ $\quad R_C = -0.375$

When $x = 7$ $\quad R_C = 0$

When $x = 15$ $\quad R_C = 1$

The 3 influence lines are plotted shown in Fig. 2.36.

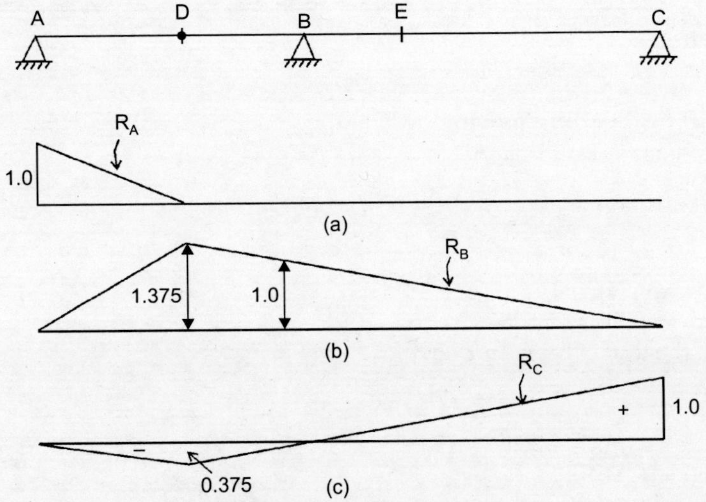

Fig. 2.36. ILD for R_A, R_B, R_C

(*b*) **ILD for shear to the right of B (F_B)**

When the load is on AD, $F_B = -R_C$

$$R_C = -\frac{3x}{32} \; ; \; F_B = \frac{3x}{32}$$

When $x = 0$, $F_B = 0$

When $x = 4$, $F_B = 0.375$

When the load is over DB, $\quad F_B = -R_C$

$$R_C = \frac{x-7}{8} \; ; \; R_B = \frac{7-x}{8}$$

When $x = 4$, $F_B = 0.375$

When $x = 7$, $F_B = 0$

When the load is over BC, $F_B = R_B$

$$R_B \times 8 - 1(15 - x) = 0$$

$$F_B = R_B = \frac{15 - x}{8}$$

When $x = 7$, $F_B = 1$

When $x = 15$, $F_B = 0$

The ILD for F_B is plotted shown in Fig. 2.37.

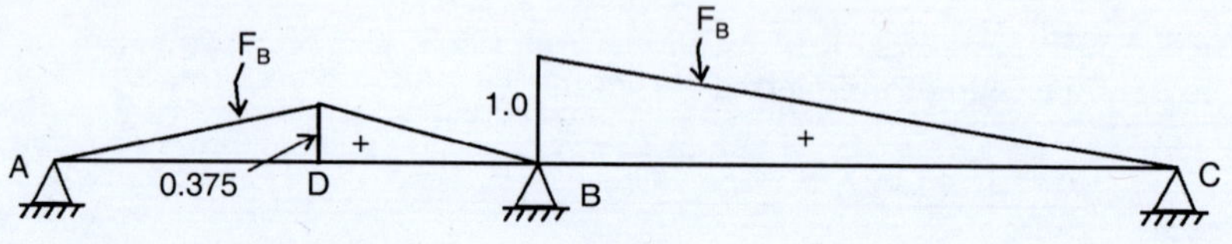

Fig. 2.37

(c) **ILD for bending moment at E (M_E)**

When the unit load is on AD,

$$M_E = R_C \times 6 = -\frac{3x}{32} \times 6$$

$$M_E = -\frac{18x}{32}$$

When $x = 0$, $M_E = 0$

When $x = 4$, $M_E = -2.25$

When the load is between D and E,

$$M_E = R_C \times 6 = \frac{(x - 7)}{8} \times 6 = \frac{3}{4}(x - 7)$$

When $x = 4$, $M_E = -2.25$

When $x = 7$, $M_E = 0$

When $x = 9$, $M_E = 1.50$

When the load is between E and C,

$$M_E = R_C \times 6 - 1(x - 9)$$

$$R_C = \frac{x - 7}{8}$$

Hence, $$M_E = (x - 7)\frac{6}{8} - (x - 9)$$

When $x = 9$ m, $M_E = 1.5$ m

When $x = 15$, $M_E = 0$

The ILD is shown in Fig. 2.38.

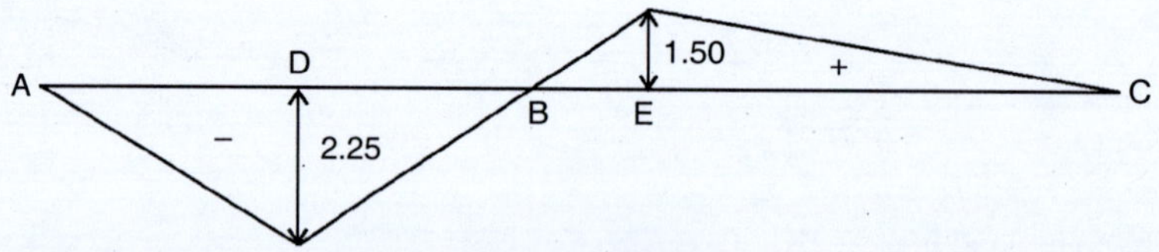

Fig. 2.38. ILD for M_E.

2.6 INFLUENCE LINES FOR FORCES IN TRUSS MEMBERS

2.6.1 Trusses

Pin jointed determinate trusses are frequently employed as bridge girders. As vehicles move over bridges, the various truss members suffer fluctuating forces, some times changing from compression to tension and vice versa.

Every truss configuration requires some degree of innovation to identify member forces in them as loads roll along the deck supported by the truss.

Truss bridges always come in pairs. Their nodes are interconnected by cross girders.

Cross girders are interconnected by longitudinal stringers parallel to the traffic direction.

A wheel load is transmitted from deck slab to stringers.

Stringers transfer the load to 2 cross girders one in front and another behind. The cross girders apportion the load to the nodes of the bridge truss one to the left and another to the right.

Thus, the wheel load never rests on a truss directly. The effect of the wheel load gets transferred to one or more nodes of each truss.

2.6.2 A Truss is a Beam Too

A truss is also a beam. In a beam the top fibre takes maximum compression and bottom fibre maximum tension. We use I sections for beams so that material (flange) is concentrated where the stress is maximum for full utilization of material. In an I section the flanges take tension and compression and the web takes shear.

A truss is an improved version of I beam. The top and bottom chord members function as compression and tension members respectively (like the flanges in an I section). The vertical and inclined members account for the beam shear.

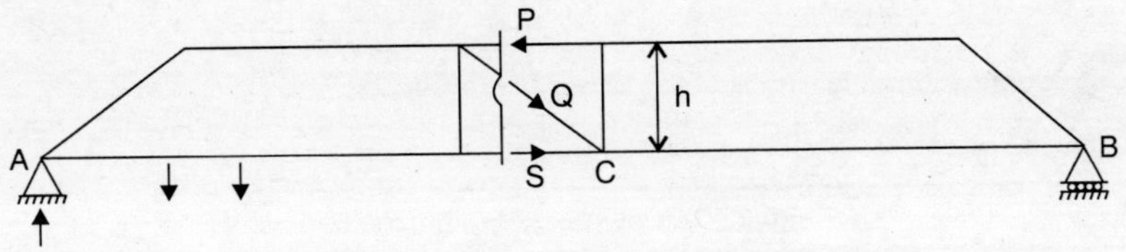

Fig. 2.39. Free body of part of a truss.

Figure 2.39 shows a long truss of span l, with a large number of panels. Let us take a section cutting 3 members of the truss as shown and *remove* the right part of the truss. Each of the 3 members cut by us must have had an internal force. Let the forces be P, Q and S. Having removed the right part of the girder, let us at least put back the 3 forces exerted by the right part on the left part. These forces P, Q and S as shown in the figure should keep the left part of the truss and what ever external forces were on it in equilibrium.

Let us take moments of all forces on the free body about C (where 2 unknown forces meet). This must be zero. The moment due to the forces other than P is actually the *bending moment* at C of the beam AB (M_C).

Then, $M_C - Ph$ must be zero. (– Ph is used since it is anti-clockwise about C and sagging or positive bending moment M_C would be clockwise about C)

So, $$P = \frac{M_C}{h}$$

The force P in the top chord member resists the beam bending moment at C with a lever arm of h. h is infact the structural depth of the truss (the distance between the CG of the top chord and the CG of the bottom chord).

The aim of providing a truss instead of a beam is to increase the value of h without covering the entire depth. 'h' might vary from 1.5 m to 12 m, depending upon the span of the bridge.

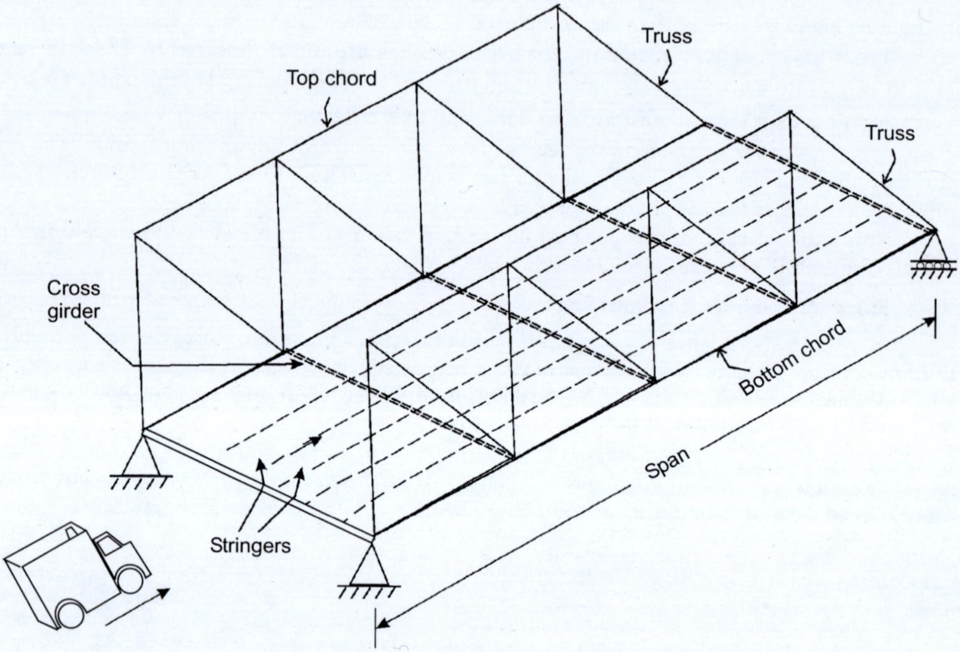

Fig. 2.40. A warren girder through bridge.

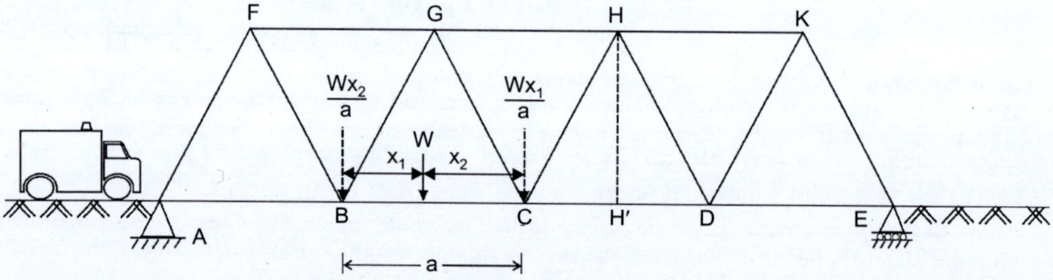

(a) Through bridge

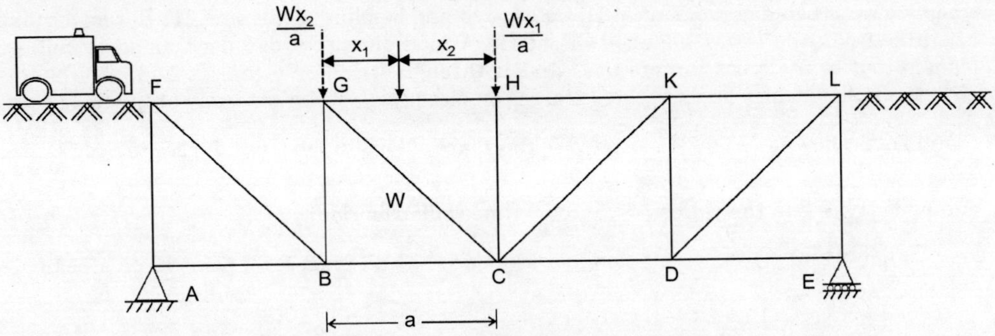

(b) Deck bridge

Fig. 2.41. Transfer of loads to truss nodes.

2.6.3 Through and Deck Bridges

Figure 2.40 shows a 'through bridge'. A through bridge is one in which the road is at the bottom chord level. When you go through the bridge, you would see the web members of the bridge on your left and right. When the span is large we may have to have a large height, as much as 6 to 8 metres. It may be inconvenient to have the road way *over* the truss.

For smaller spans (and lower truss heights of 1.5 to 2.5 m), we might use a *deck bridge*. Here, the road is at the top chord level. We would not see the trusses when we ride the bridge. We might see only the hand rails.

Some times we also have a half through bridge in which the bridge deck is at mid height of the trusses. In these bridges the cross girder loads would be transferred to the bottom nodes, like in through bridges.

It is necessary to know if a bridge receives loads at bottom chord nodes or top chord nodes if we want to draw influence lines for bending moment at any node. Bending moments at nodes are important because forces in the chord members are equal to the bending moment divided by h.

2.6.4 Influence Lines for Bending Moments and Shears in Trusses

(*a*) Through bridges

(*i*) *Bending moment*

For nodes B, C and D in the through bridge in Fig. 2.41 (*a*), the influence lines for bending moments are sketched below. These cases do not need explaining.

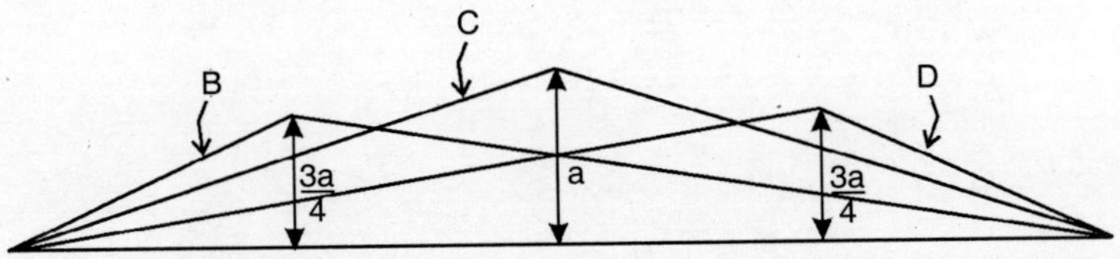

Fig. 2.42. ILD for bending moment at B, C and D.

But nodes F, G, H and K would present a problem. Loads can not act on these nodes. So when we want bending moment at H, we have to find bending moment at H′. But load cannot act directly on the chord member CD at H′. A load on the bridge deck along H′ will get transmitted to the cross girders at C and D through stringers (see Fig. 2.40). The 2 cross girders would transfer loads to the 2 trusses on either side of the road way at the node points C and D. As shown in Fig. 2.41 (a), if W is the share of load to the truss, it goes to node B $\frac{Wx_2}{a}$ and node C $\frac{Wx_1}{a}$ in the ratios $\frac{x_2}{a}$ and $\frac{x_1}{a}$ (linear distribution).

Figure 2.43 (a) shows the nominal influence line diagram for a point H′ on a beam.

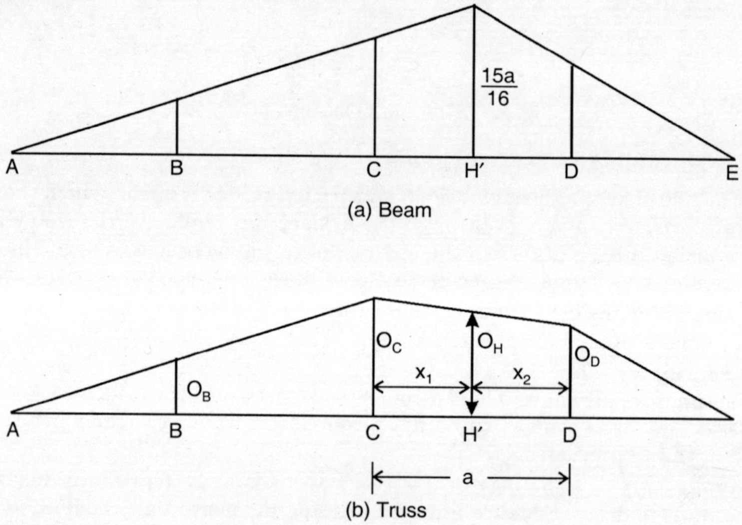

Fig. 2.43. Influence line for M_H.

The ordinate at H′ is $\frac{2.5a \times 1.5a}{4a} = \frac{15a}{16}$. In this diagram, the ordinates at nodes A, B, C, D and E are really significant since loads can act there. The ordinate at H′ can not have a meaning since no load acts directly at H′. The load at H′, if any, gets divided linearly between C and D. So the nodal ordinates zero, O_B, O_C, O_D, and zero can all be joined by straight lines to form a polygon as shown in Fig. 2.43 (b).

This is the influence line for M_H.

For a load W at H′, we get

$$M_H = \frac{O_C\, Wx_2}{a} + \frac{O_D \,.\, Wx_1}{a}$$

This is the same as saying

$$M_H = W\left(\frac{O_C x_2}{a} + \frac{O_D x_1}{a}\right) = W\ O_H.$$

So, by this method of constructing a polygon, we get the ILO we need at points between nodes.

We can also see that the only difference between Fig. 2.43(*a*) and (*b*) is in the panel CD.

(*ii*) *Shear forces*

Let us go back to the free body of the truss in Fig. 2.39. The vertical shear in the panel to the left of C has to be balanced by the vertical component of Q. Hence, we want to investigate the shear in the *panel*.

Let us construct for the beam AE, the influence line for shear in BC.

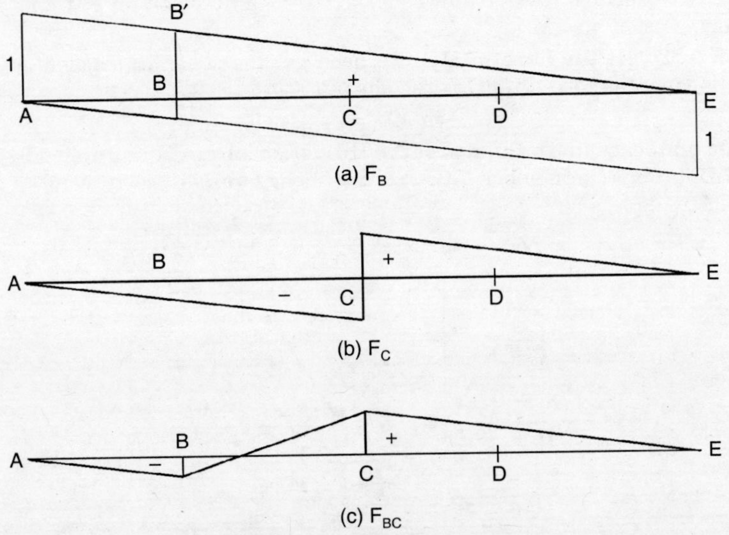

Fig. 2.44. IL for shear at B, C and BC.

Figure 2.44 (*a*) and (*b*) show the ILD for shear in a beam at points B and C. But in a truss, for the panel BC, the ILD is got by joining by straight lines the significant IL ordinates at nodes A, B, C, D and E. Figure 2.44 (*c*) is the ILD for the panel BC in the truss.

This construction takes into account the fact that a load in between nodes B and C distributes itself between B and C depending on its distance from B.

(*b*) **Deck bridges**

Figure 2.41 (*b*) is a deck bridge. There would be no problem in drawing influence line diagrams for bending moment for any of the nodes, G, H, K, B, C or D. If we want shear in panel GH, the ILD would look exactly like Fig. 2.44 (*c*).

2.6.5 Influence Lines for Member Forces

Member forces are expressible in terms of bending moments and shears. Hence, their influence lines would look like the ILDs for moment/shear. Only, the ordinates have to be scaled up or down by a suitable factor (such as h, sec θ, etc.).

We will examine a series of cases of trusses and decide which would be the most advantageous free body diagram to get the desired member forces.

In the N girder deck bridge in Fig. 2.45 (*a*) let us say we want the ILD for forces in CD, FM and MG.

(*a*) For CD, let us take the free body left of section (1). It cuts CD, CK and HK. The forces in CK and HK meet at K. So, we take moments about K of forces to the left of (1). This means, bending moment at K or at D. This ILD is easy to sketch. If we divide the ordinates by the truss height h, we get the ILD for force in CD.

(*b*) For FM, we have to take the free body to the left of section (2). The force in FM is equal to the sum of all the forces to the left of (2). This means, shear force on panel FG. This can be drawn as in Fig. 2.44(*c*). The same would be the ILD for force in FM. Positive shear force (left upwards) would mean tension on FM and negative shear, compression on FM.)

(*c*) **MG** : To get the force in MG, we need get the shear in segment FG. The vertical component of force in MG is the shear. (F_{FG})

$$\text{Force in MG} = F_{FG} \sec \theta.$$

The influence line ordinates of the ILD for shear in FG multiplied by sec θ gives the ILD for force in member MG. Positive shear force will correspond to compression in MG.

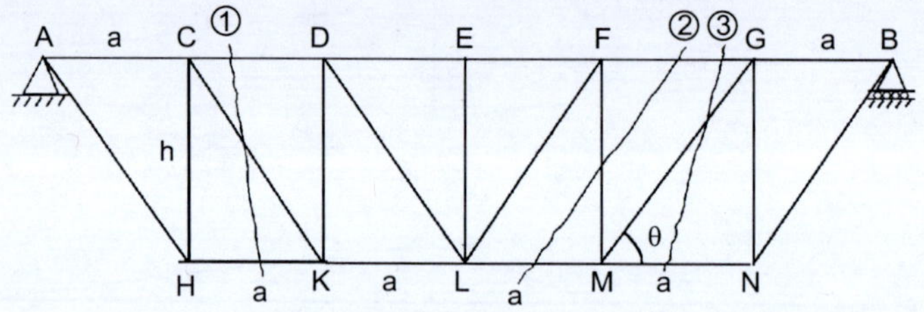

(a) N Girder deck bridge

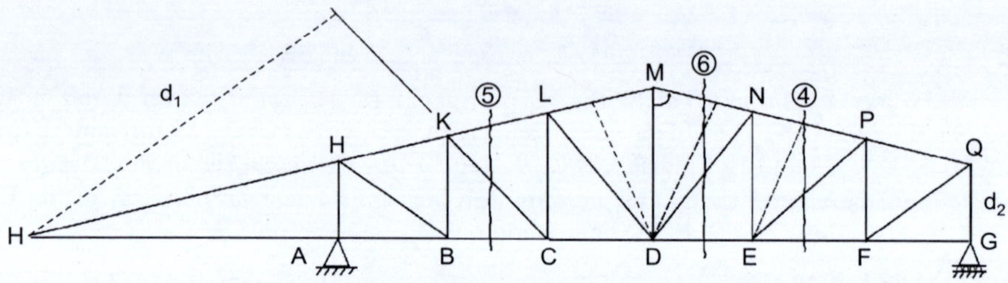

(b) N Girder through bridge with sloping chord

Fig. 2.45. How to use method of sections.

(*d*) Force in member NP in Fig. 2.45 (*b*). Section 4 cuts 3 members. If we take the free body of the part left of (4), we get 3 forces, one each on NP, EP and EF. Since we are investigating NP we can take moments of forces in the free body about E. That will give us compression on NP = sagging moment at E ÷ d, where d is the normal distance of E from NP. So we only have to sketch the ILD for M_E and divide the ordinates by d. That would give us the ILD for compression in NP.

(*e*) **Force in KC** : Section (5) cuts KL, KC and BC. If we take the free body left of (5), we may have to take moments about H, where CB and LK meet. This moment ÷ *d*, is the tension in KC. But how do we designate the moment about H and how do we draw the influence line for the same ? We have to change our strategy.

To get the ILO for F_{CK} when a unit load is at B, consider the free body left of (5). Take moments about H. That will give us ILO at B, for force in CK. ILO at A and ILO at G are zero.

When we keep a unit load at C, consider the free body left of (5). Take moments about H. This moment ÷ *d*, is the ILO at C. The ILD will look like an ILD for shear on panel BC. (vide Fig. 2.46 below).

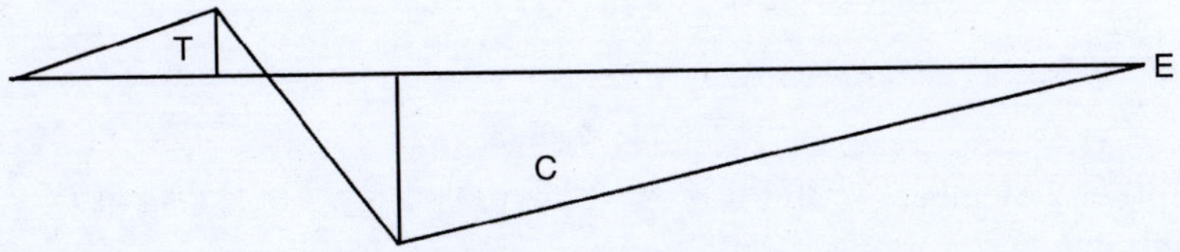

Fig. 2.46. IL for force in CK.

(*f*) **Force in MD** : The force in MD is, by method of joints, given by

$$F_{MD} = 2\,F_{MN}\,.\,\cos\alpha$$

So, we need draw the ILD for F_{MN} (and multiply the ordinates by 2 cos α). A section along (6) will take care of this. It is easy to conclude that

$$F_{MN} = M_D \div d_2$$

Hence, $$F_{MD} = M_D \times 2\cos\alpha \div d_2.$$

Example 2.11 *Draw the ILD for the forces in members P, Q, R, S of the truss shown in Fig. 2.47.*

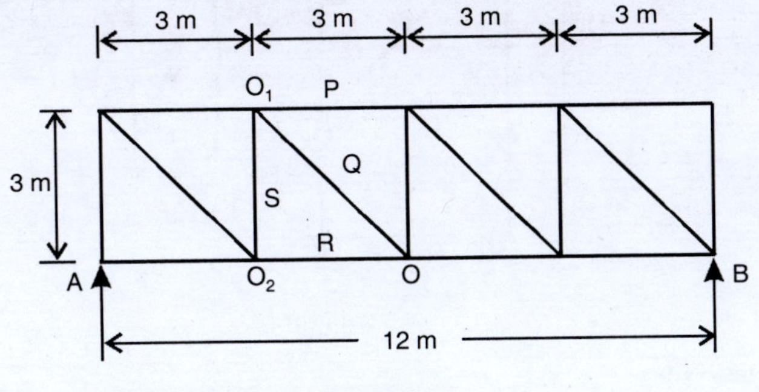

Fig. 2.47

Solution (*a*) **ILD for the member 'P'**

Pass a section (1) cutting the member 'P' and passing through not more than 3 members.

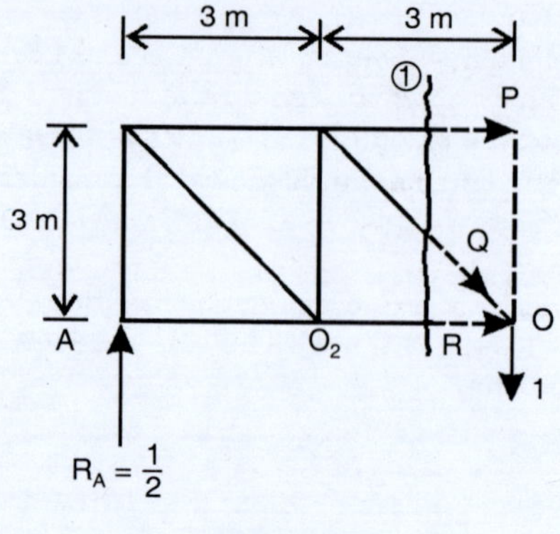

Fig. 2.48 (*a*)

The other 2 members Q and R meet at joint O. Apply a unit load at O. Since O is the point at midspan,

$$R_A = R_B = \frac{1}{2}$$

For the free body left of (1), taking moments about O,

$$P(3) + R_A \times 6 = 0, \quad P = -1, \quad P = 1 \text{ (Compression)}$$

Hence, the ILD is a triangle having a maximum ordinate of 1 at O.

(*b*) **ILD for the force in member 'Q'**

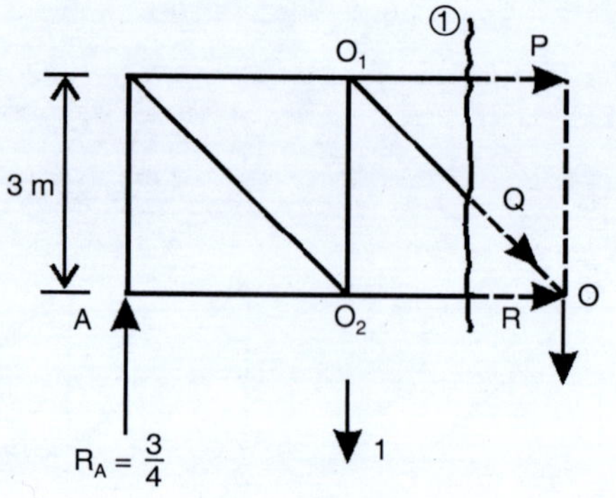

Fig. 2.48 (*b*)

(*i*) *Unit load at* O_2 :

Taking moments about B,

$$R_A \times 12 - 1 \times 9 = 0$$

$$R_A = \frac{3}{4}$$

Since $\Sigma V = 0$, for the free body left of (1),

$$R_A - 1 - Q \cos 45 = 0$$

$$Q = \frac{-0.25}{0.707}$$

At O_2, ILO for Q = 0.353 (Compression)

(*ii*) *Unit load at O :*

$$R_A = \frac{1}{2}$$

Since $\Sigma V = 0$, for the free body,

$$R_A + (-Q \cos 45) = 0$$

$$\frac{1}{2} = Q \times \frac{1}{\sqrt{2}}$$

At O, ILO for $Q = \frac{1}{\sqrt{2}} = 0.707$ (Tension)

(*c*) **ILD for the member 'R'**

To find the ILO at O_1 (or O_2), keep a unit load at O_2.

Then $R_A = \frac{3}{4}$

Refer to the free body in Fig. 2.48 (*b*). If we want R, we must take moments of the forces in the free body about O_1.

$$R = \frac{3}{4} \text{ (tension) at } O_1.$$

Hence, ILD for 'R' is a triangle having a maximum ordinate of $\frac{3}{4}$ at O_2.

(*d*) **ILD for the member 'S'**

Force in the member 'S' = shear in the panel O_2O. The nature of shear force in the panel O_2O changes as the load moves from O_2 to O. When the unit load is at O_2, shear force is negative and the force in 'S' is tension. When it is at O, the force in 'S' is compression. Hence, we have to find the ILO at O_2 and at O and join them to get the IL for shear in the panel O_2O.

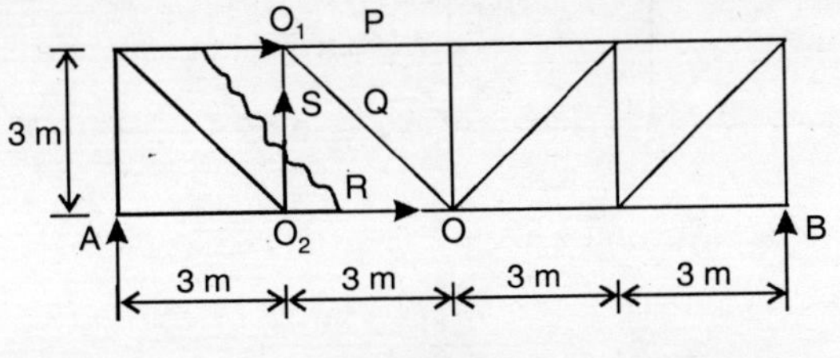

Fig. 2.48 (*c*)

When the unit load is at O_2, shear in the panel $O_2O = -R_B = -0.25$.

When the load is at O, shear in the panel $= R_A = 0.5$

ILOs at A and B are zero. Hence, Fig. 2.49 (*d*) is the ILD for S. The other influence lines are also plotted in Fig. 2.49.

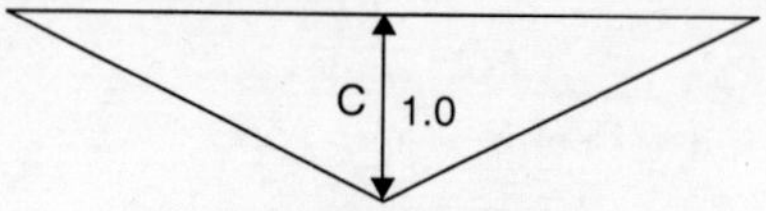

ILD for the member 'P'
(a)

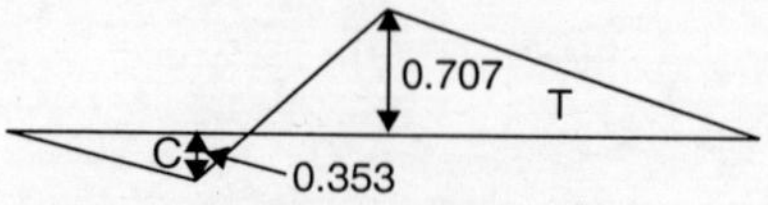

ILD for the member 'Q'
(b)

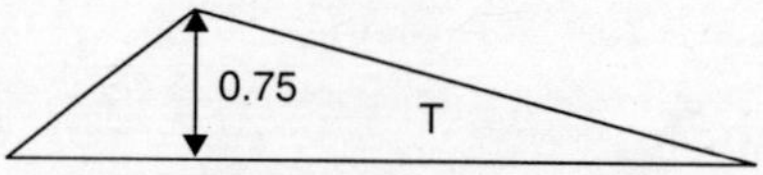

ILD for the member 'R'
(c)

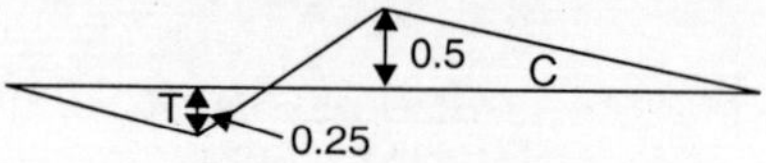

ILD for the member 'S'
(d)

Fig. 2.49. ILD for the members.

Example 2.12 *For the sloping chord through type truss in Fig. 2.50, draw the influence lines for forces in members LM and LE.*

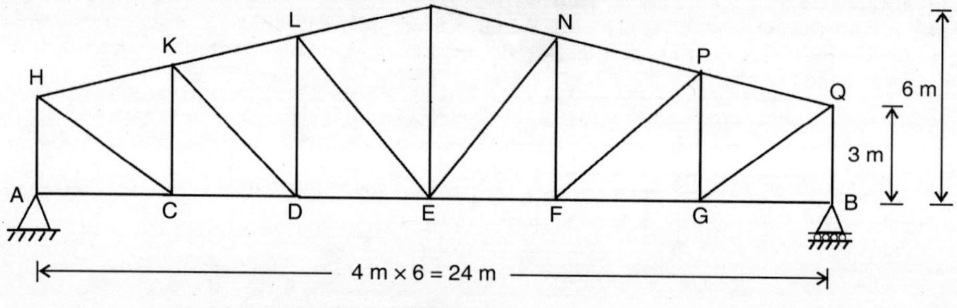

Fig. 2.50

Solution To get the force F_1 on LM, let us cut the truss along (1) and substitute the member forces F_1, F_2, F_3 on the 3 cut members. We will consider the free body (Fig. 2.51) and its equilibrium to evaluate forces F_1 and F_2.

(a) **Member LM (force F_1)**

Consider the moment equilibrium of the free body about E (where F_2 and F_3 intersect).

$$M_E + F_1 \times D = 0$$

$F_1 = -\dfrac{M_E}{D}$, where M_E is the moment of all the forces in the free body about E. M_E is actually the bending moment at E. Hence, ILD for force in LM is got by dividing the ordinates of the ILD for M_E by D.

Now, $\tan \theta = \dfrac{1}{4}$; $\theta = 14° \, 02' \, 11''$

$$D = M_E \cos \theta = 6 \times \cos 14° \, 02' \, 11'' = 5.82 \text{ m.}$$

$$\text{ILO for } F_1 \text{ at E} = \frac{ab}{l} \times \frac{1}{5.82} = -\frac{12 \times 12}{24} \times \frac{1}{5.82} = -1.031.$$

Negative sign indicates compression, since F_1 as marked in figure is tension. The ILD for F_1 is drawn in Fig. 2.52 (a).

(b) **Member LE (force F_2)**

Observing Fig. 2.51, it is obvious that to evaluate F_2 we need take moment equilibrium condition about O where the lines of F_1 and F_3 meet.

O is obviously 12 m to the left of A. The normal distance of O from the line of action of F_2 is D_1 and is given by D_1 = OE sin α

$$\tan \alpha = \frac{5}{4} \; ; \; \alpha = 51° \, 20' \, 25''$$

$$OE = 24 \text{ m}$$

$$D_1 = 24 \sin \alpha = 18.74 \text{ m}$$

To get the ILO for F_2 ; at points D and E, let us apply unit loads at D and E one by one.

When a unit load acts at D, $R_A = \dfrac{2}{3}$.

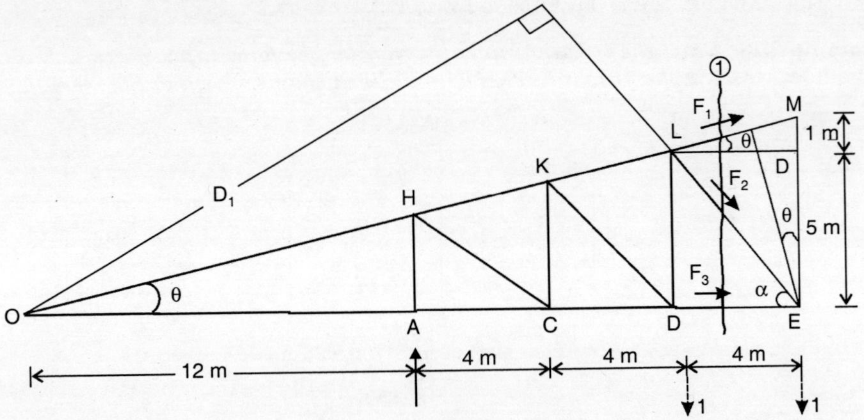

Fig. 2.51. Free body of truss.

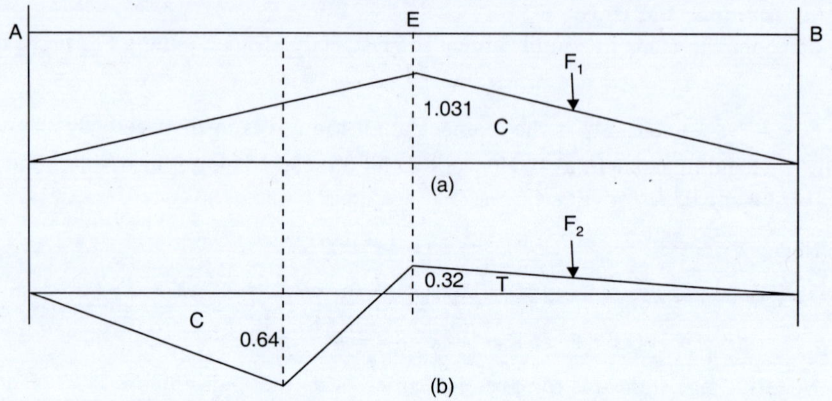

Fig. 2.52. ILD for forces F_1 and F_2.

Equating moments about O to zero, $-R_A \times 12 + 1 \times 20 + F_2 \times D_1 = 0$

$$F_2 = \frac{8 - 20}{18.74} = -0.64$$

This is the ILO for F_2 at D.

When a unit load acts at E, $R_A = \dfrac{1}{2}$

Moment about O is $-R_A \times 12 + F_2 \times D = 0$

$$F_2 = \frac{6}{18.74} = 0.32 \quad \text{(Tension)}$$

This is the ILO at E for F_2.

The ILOs at A and B are zero.

Thus, we have the IL for force in LE in Fig. 2.52 (*b*).

Example 2.13 *Draw the influence line diagram for the force in members U_1U_2, U_1L_2 and L_1L_2 for the pratt truss with inclined chords shown in Fig. 2.53.*

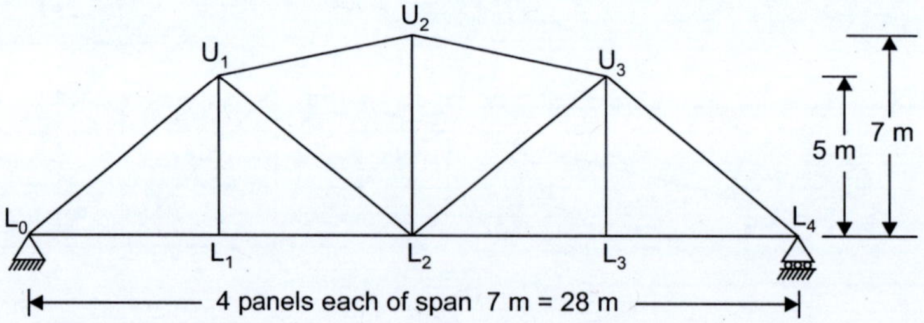

Fig. 2.53

Solution Let us name the members U_1U_2, U_1L_2 and L_1L_2 as P, Q and R respectively.

ILD for P

To get the influence line for P, pass a section through P and cutting not more than 3 members.

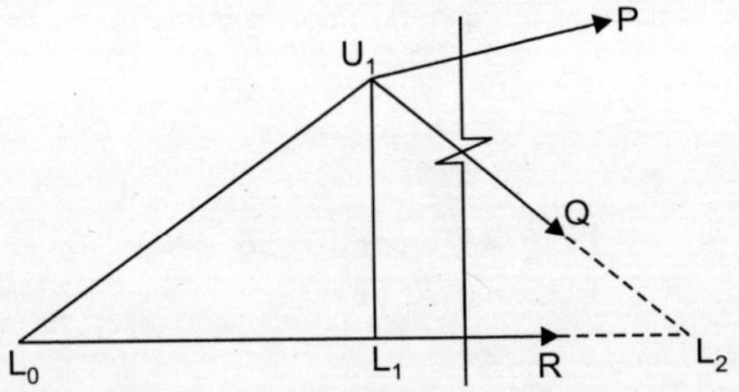

Here the other two cut members are Q and R. They meet at L_2. Apply unit load at L_2.

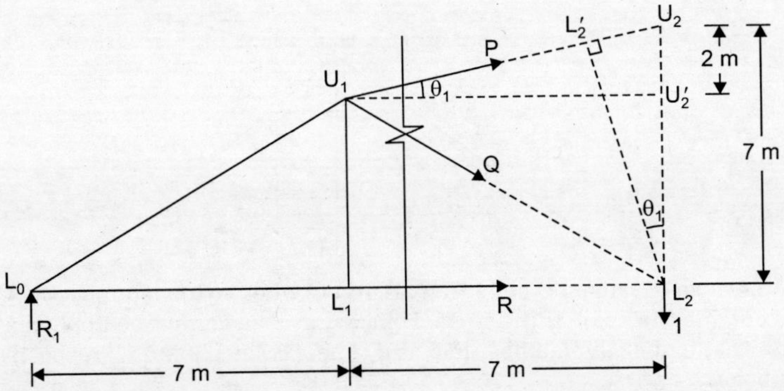

Reaction at the support "L_0".

$$R_1 \times 28 - 1 \times 14 = 0$$

$$R_1 = \frac{1}{2}$$

From figure
$$\tan\theta_1 = \frac{2}{7}$$

$$\theta_1 = 15°\ 56'\ 44''$$

Consider the equilibrium of free body diagram to the left of the cut-section. Taking moment about L_2.

$$\frac{1}{2} \times 14 + P \times L_2L_2' = 0$$

$$\frac{1}{2} \times 14 + P \times 6.73 = 0$$

$$P = -1.04$$

i.e.,
$$P = 1.04 \quad \text{(Compression)}$$

$$\frac{U_2U_2'}{U_1U_2'} = \frac{2}{7} = \tan\theta_1$$

$$\theta_1 = \tan^{-1}\left(\frac{2}{7}\right) = 15°\ 56'\ 44''$$

$$\cos\theta_1 = \frac{L_2L_2'}{L_2U_2} = \frac{L_2L_2'}{7}$$

$$L_2L_2' = 7 \times \cos 15°\ 56'\ 44''$$

$$L_2L_2' = 6.73 \text{ m}$$

ILD for Q

To get the influence line for Q, pass a section through member Q, such that the section cuts not more than 3 members.

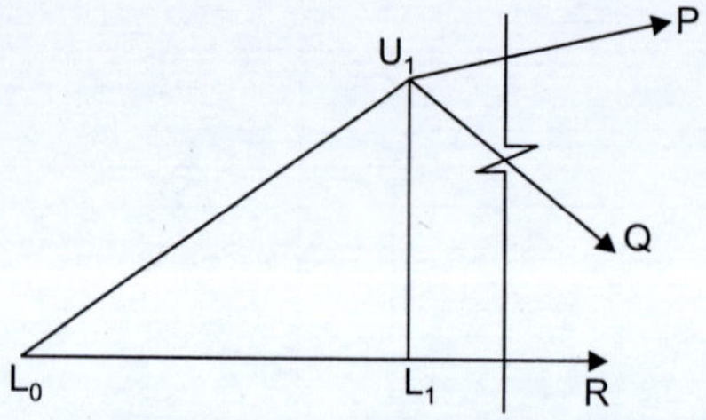

The other two cut members P and Q. P and Q when extended meet at "O", to the left of L_0. This lies outside the span of the truss. Hence unit load cannot be applied at "O". Applying unit load at either end of the member one at a time, *i.e.*, at U_1 and L_2 respectively the force is obtained as below.

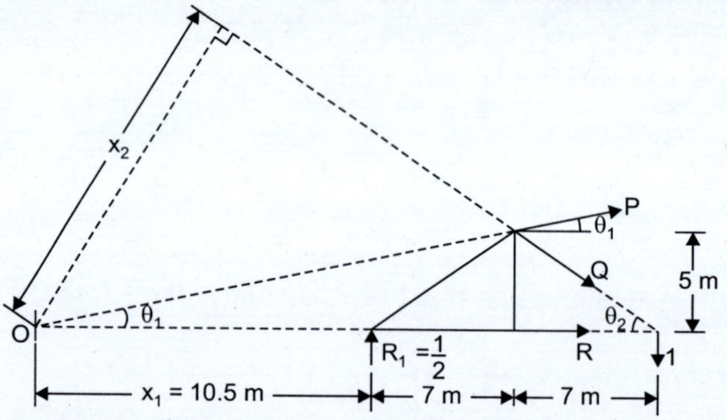

$$\tan \theta_2 = \frac{5}{7}$$

$$\theta_2 = 35° \, 32' \, 15''$$

$$\tan \theta_1 = \frac{5}{x_1 + 7}$$

$$\tan 15° \, 56' \, 44'' = \frac{5}{x_1 + 7}$$

$$x_1 = 10.5 \text{ m}$$

$$\sin \theta_2 = \frac{x_2}{(10.5 + 14)}$$

$$x_2 = 14.24 \text{ m}$$

Apply unit force at the right end of the member *i.e.,* at L_2.

$$R_1 \times 28 - 1 \times 14 = 0$$

$$R_1 = \frac{1}{2}$$

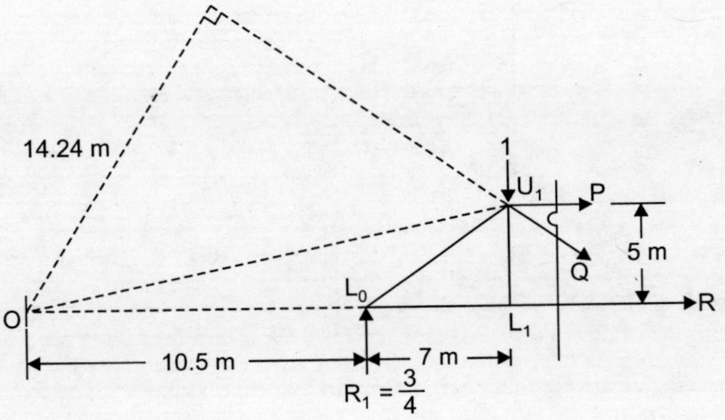

Taking moment about "O".

$$-\frac{1}{2} \times 0.5 + Q \times 14.24 = 0$$

$$Q = \frac{5.25}{14.24}$$

$$Q = 0.369 \quad \text{(Tension)}$$

Applying unit force at left end of the member *i.e.,* at U_1.

then, $$R_1 \times 28 - 1 \times 21 = 0$$

$$R_1 = \frac{3}{4}$$

Taking moment about "O".

$$-\frac{3}{4} \times 10.5 + Q \times 14.24 + 1 \times 17.5 = 0$$

$$Q = -0.676$$

i.e., $$Q = 0.676 \quad \text{(Compression)}$$

ILD for R

To get the influence line diagram for R, pass a section through "R" such that the section cuts not more than 3 members.

The other two cut members are P and Q. They meet at U_1. So apply unit load at U_1.

Take moment about U_1:

$$\frac{3}{4} \times 7 - R \times 5 = 0$$

$$R \times 5 = 5.25$$

$$R = 1.05 \quad \text{(Tension)}$$

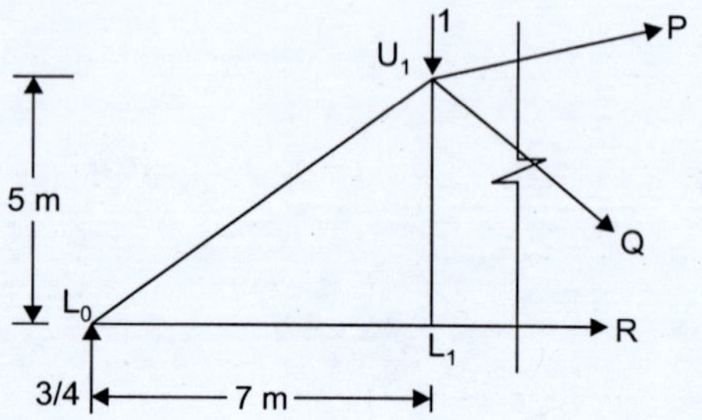

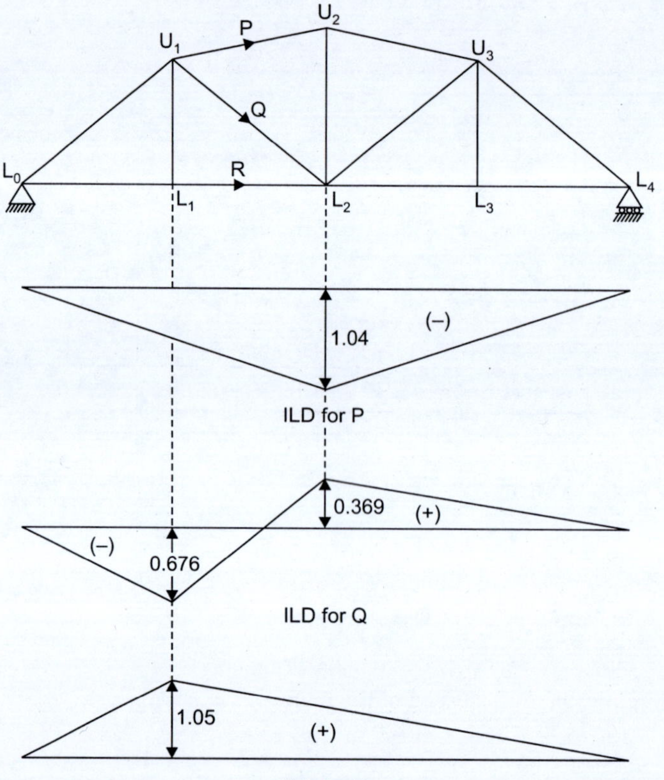

Example 2.14 *A warren girder consisting of 4 panels of 5 m each is shown in Fig. 2.54. Obtain the influence line for the force in the members P, Q and R in the second panel from the left.*

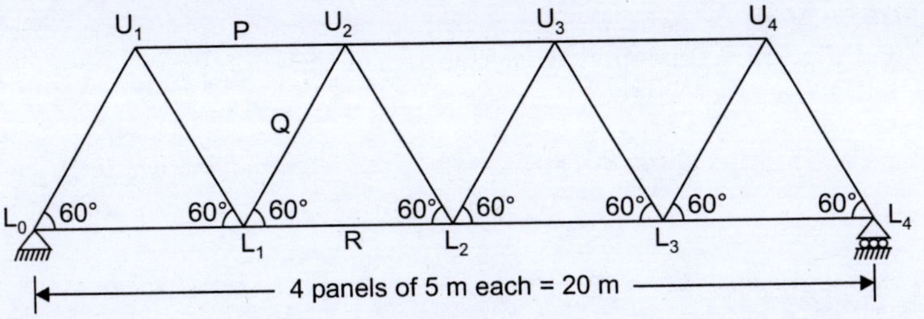

Fig. 2.54

Solution ILD for P

To get the influence line for P, pass a section through 'P' and cutting not more than 3 members.

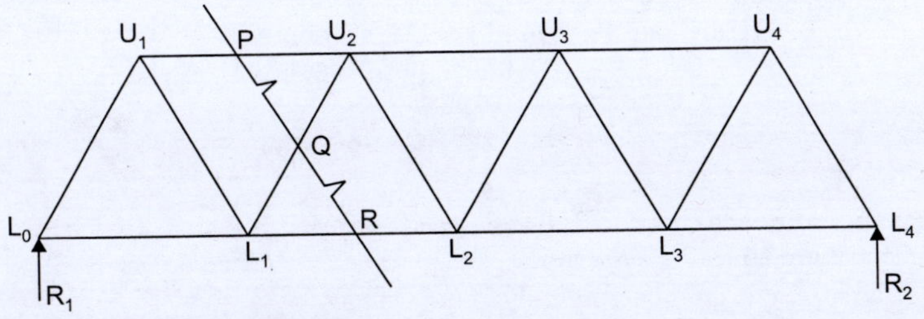

The other two cut members namely Q and R meet at node L_1. Apply unit load at L_1. Consider the free body to the left of the cut section.

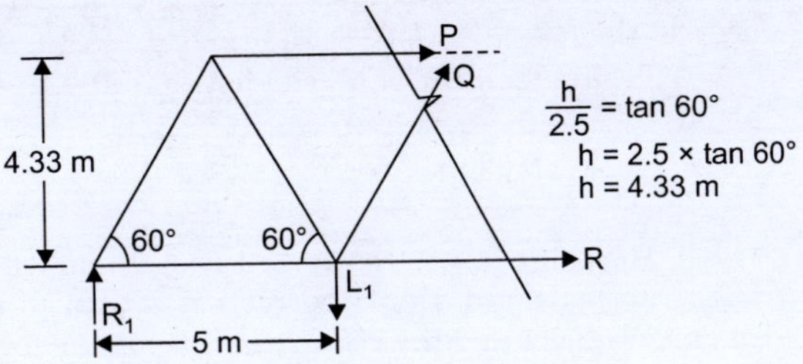

Taking moment about L_4.

$$R_1 \times 20 - 1 \times 15 = 0$$

$$R_1 = \frac{3}{4}$$

Taking moment of all the forces in the free body about L_1.

$$\frac{3}{4} \times 5 + P \times 4.33 = 0$$

$$P = -\,0.866$$

$$P = 0.866 \text{ (Compression)}$$

ILD for Q

To get the influence line for Q, pass a section through Q and cutting not more than 3 members.

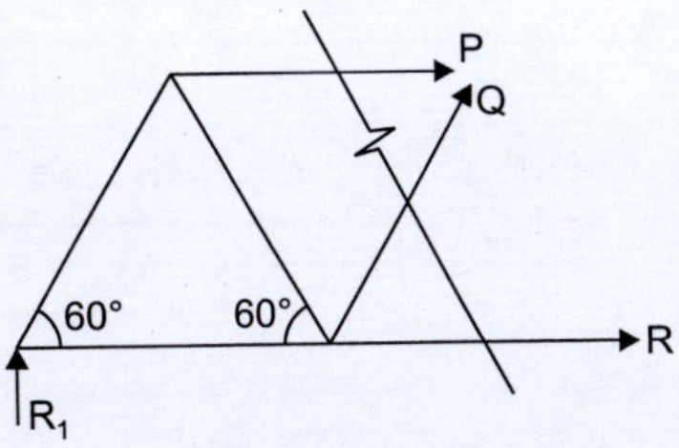

The other two cut members P and R are parallel members. They will not meet. Therefore applying unit force at left end of the member *i.e.,* at L_1, and applying unit force at the right end of the same member *i.e.,* at U_2, one by one and using the condition $\Sigma V = 0$, the forces will be obtained.

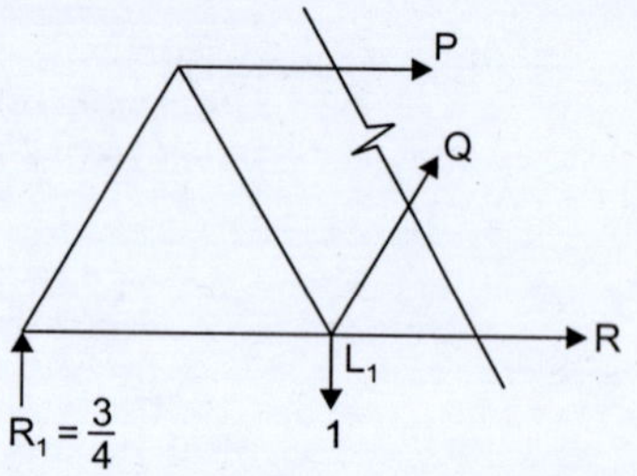

Unit force at L_1

The condition $\Sigma V = 0$ is applied for the part of the truss to the left of the cut section.

$$\frac{3}{4} - 1 + Q \cos 30° = 0$$

$$Q = 0.288$$
$$= 0.288 \text{ (Tension)}$$

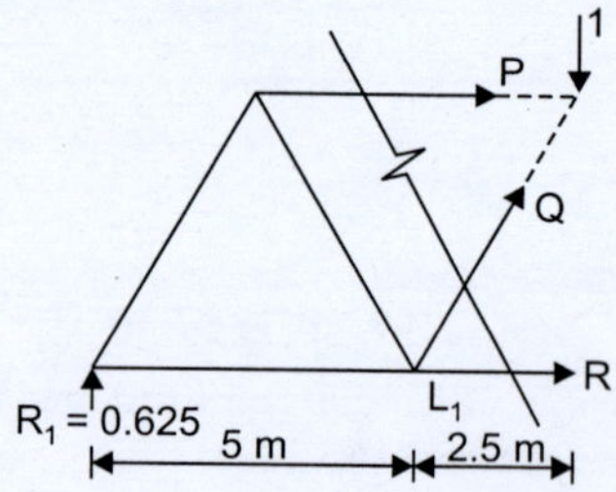

Unit force at U_2

Take moment about L_4

$$R_1 \times 20 - 1 \times 12.5 = 0$$
$$R_1 = 0.625$$

The condition $\Sigma V = 0$ is applied for the part of the truss to the left of the cut section.

$$0.625 + Q \cos 30° = 0$$

(Unit load is beyond the cut section and hence not considered)

$$Q = -0.722$$
$$= 0.722 \quad \text{(Compression)}$$

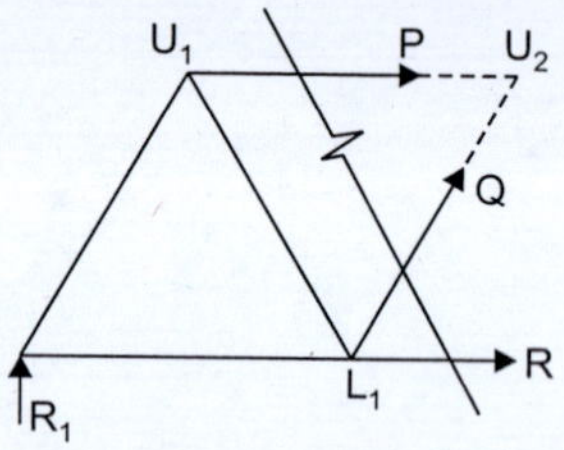

ILD for R

To get the influence line for R, pass a section through R and cutting not more than 3 members.

The others two cut members namely P and Q meet at node U_2. Apply unit load at U_2. Consider the free body to the left of the cut section.

Taking moment of all the forces in the free body about U_2.

$$0.625 \times 7.5 - R \times 4.33 = 0$$
$$R = 1.083 \text{ (Tension)}$$

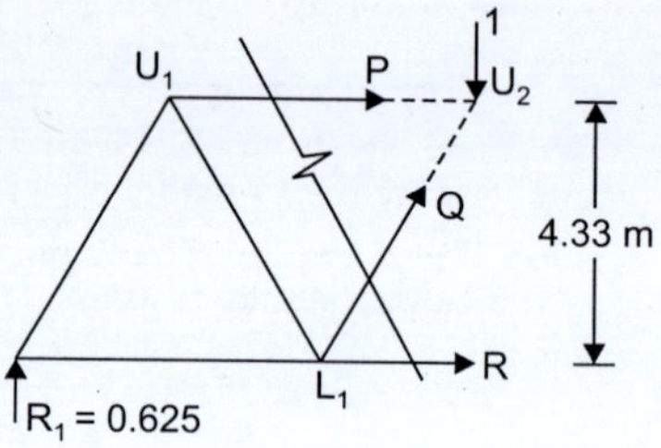

The point U_2 where the unit load is applied is not in the loaded chord. Hence correction is applied as per the article 2.6.4 and the corrected ILD for the force to member R is given below.

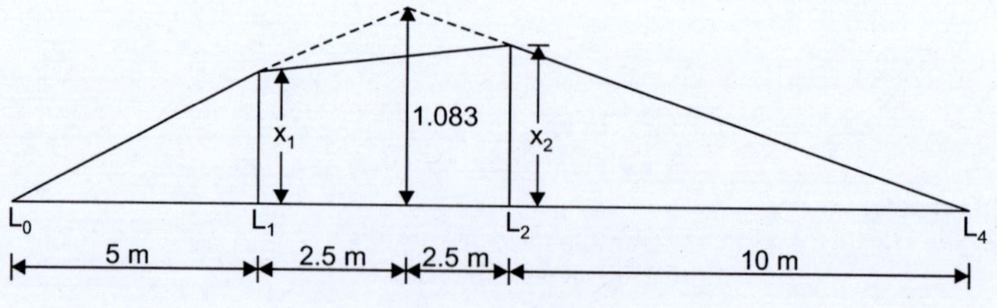

ILD for force is member R

Left ordinate $$x_1 = \frac{1.083}{7.5} \times 5 = 0.722$$

Right ordinate $$x_2 = \frac{1.083}{12.5} \times 10 = 0.866$$

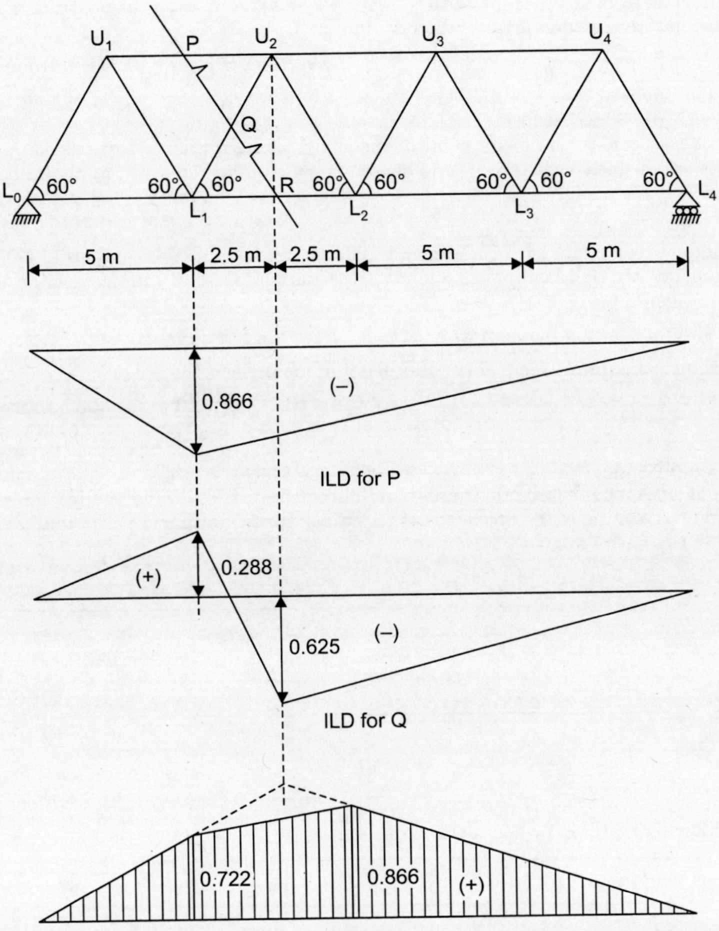

PROBLEMS

1. A simply supported beam has a span of 15 m and is subjected to a uniformly distributed load of 40 kN/m 5 m long, traversing from left to right. Draw the influence line diagram for shear force and bending moment at a section 6 m from left end. Use these diagrams to calculate the maximum shear force and bending moment at this section.

[**Ans.** F_{max} (+ve) = 86.67 kN, F_{max} (–ve) = 46.667 kN, M_{max} = 600 kNm]

2. Five wheel loads 10 kN, 20 kN, 15 kN, 16 kN and 24 kN spaced at 1 m intervals roll on a girder of span 20 m, from left to right, with the 10 kN load loading. Find the absolute maximum bending moment on the girder. Use influence lines.

[**Ans.** 373.1 kN m at 10.141 m from the left support]

3. Two point loads of 90 kN and 120 kN spaced 3.5 m apart cross a girder of span 20 m from left to right with the 90 kN load leading. Determine the maximum values of shear force and bending moment that can occur at any point of the girder using influence lines. Also plot the maximum positive and negative shear force and bending moment diagrams stating their absolute maximum values.

[**Ans.** F_{max} = 194.25 kN (left support), M_{max} = 898.4 kNm (9.25 from left support)]

4. A system of four loads 80, 160, 160 and 120 kN crosses a simply supported beam of span 25 m with the 120 kN load leading. The loads are equally spaced at 1 m. Determine the values of the following using influence lines.

(*i*) Maximum bending moment at a section 10 m from left support, and

(*ii*) Absolute maximum shear force and bending moment in the beam.

[**Ans.** (*i*) M_{max} at 10 m from left support = 2912 kNm, (*ii*) F_{max} = 491.2 kN ; $M_{max\ max}$ = 3030.7 kNm]

5. For a simply supported beam with an overhang as shown in Fig. 2.55, plot the influence lines for reactions at A, shear force at D and bending moment at D. Using the influence line diagrams, determine the value of the above stress functions due to a uniformly distributed load of 20 kN/m covering the entire span.

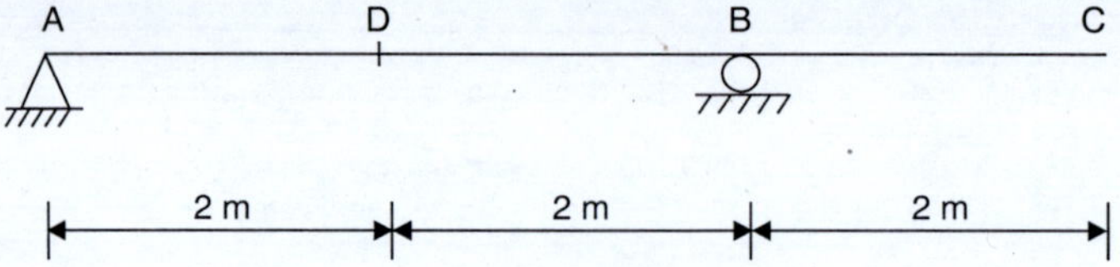

Fig. 2.55

[**Ans.** R_A = 30 kN ; R_B = 90 kN ; F_D = 10 kN ; M_D = 20 kNm]

6. The following system of wheel loads cross a plate girder of 30 m span :

Wheel load (kN)	8	18	18	15
Distance between centres (m)	4.5	3.5	4.0	

Using influence lines, determine the maximum value of shear force and bending moment at mid-span. Also, find the equivalent uniformly distributed load which could produce the same maximum bending moment at midspan.

[**Ans.** F_{max} = 19.4 kN ; M_{max} = 349.875 kNm, Equivalent UDL = 3.11 kNm]

7. A beam ABC is supported at A, B and C and has a hinge at D at a distance of 3 m from A. AB = 7 m and BC = 10 m. Draw the influence lines for

(*i*) Reactions at A, B and C.

(*ii*) Shear force at a point just to the right of B.

(*iii*) Bending moment at a section 1 m to the right of B.

Hence if a UDL of intensity 2 kN/m and length 3 m, travels from left to right, calculate above quantities for which IL are drawn. [**Ans.** R_A = 3 kN (↑) ; R_B = 7.65 kN (↑)

R_C = 5.1 kN (↑) ; F_B = 5.1 kN (↓)

B.M. (at 1 m to the right of B) = 16.98 kNm]

8. A warren girder of 25 metres span is made up of five panels of 5 m each. The diagonals are inclined at 60° to the horizontal. Draw the influence line for force in the lower chord member in the second panel from left. Hence, evaluate the force in it when there is a load of 100 kN at each lower joint.

$$\left[\textbf{Ans. } \frac{500}{\sqrt{3}} \text{ kN, tension}\right]$$

CHAPTER 3

INFLUENCE LINES FOR INDETERMINATE STRUCTURES

3.1 INTRODUCTION

We have already been introduced to influence lines in Chapter 2. But all those diagrams were for statically determinate structures. Herein we will extend the concept of influence lines to indeterminate structures also.

For determination of influence lines for, say reactions, shears and bending moments in determinate structures, we merely applied logic and we were able to get the diagrams. If we examine all those diagrams again we observe one thing in common between them. All the influence line diagrams (ILD) are made up of straight lines. We would find this is not the case with indeterminate structures. To understand why we must get familiar with Muller Breslau Principle.

3.2 MULLER BRESLAU PRINCIPLE

Muller Breslau principle states that, if we want to sketch the influence line for any force quantity (like thrust, shear, reaction, support moment or bending moment) in a structure.

(*i*) We remove from the structure the restraint to that force quantity, and

(*ii*) We apply on the remaining structure a unit displacement corresponding to that force quantity.

The resulting displacements in the structure are the influence line ordinates sought. Stated this way Muller Breslau principle is not very evocative ; not even meaningful. We will apply the principle to determinate structures first, for cases for which the ILD is already known to us. Then the principle would make a lot more sense and actually prove to be immensely valuable.

3.3 APPLICATION OF MULLER BRESLAU PRINCIPLE TO DETERMINATE STRUCTURES

3.3.1 S.S. Beams

Let us take the example of a S.S. beam of span (you guessed it) l. (Fig. 3.1)

(*a*) Let us try to get the IL for R_B. Muller Breslau principle says "remove the restraint corresponding to R_B". So take away support B. "The remaining structure" is a mechanism (Fig. 3.1(*b*)). On this, at B, we shall apply a unit displacement corresponding to R_B. In other words, we must push B up by one unit. This is easy, and requires no effort (meaning no force). Since we have removed the restraint at B. The beam AB is determinate, having just the required number of restraints to make it stable. If we remove one of them (actually we did, by removing support B) the structure becomes a mechanism having no resistance to some displacements. When we apply these displacements (as we did at B) the 'structure' moves as a rigid body. It does not get strained or bent in the process of displacement. This is very convenient for us. We have a resulting displacement diagram (Fig. 3.1 (*c*)) which, according to Muller Breslau principle, is the ILD for R_B. The graph is unbent and straight since the beam did not have to get bent in order to move through 1 unit at B.

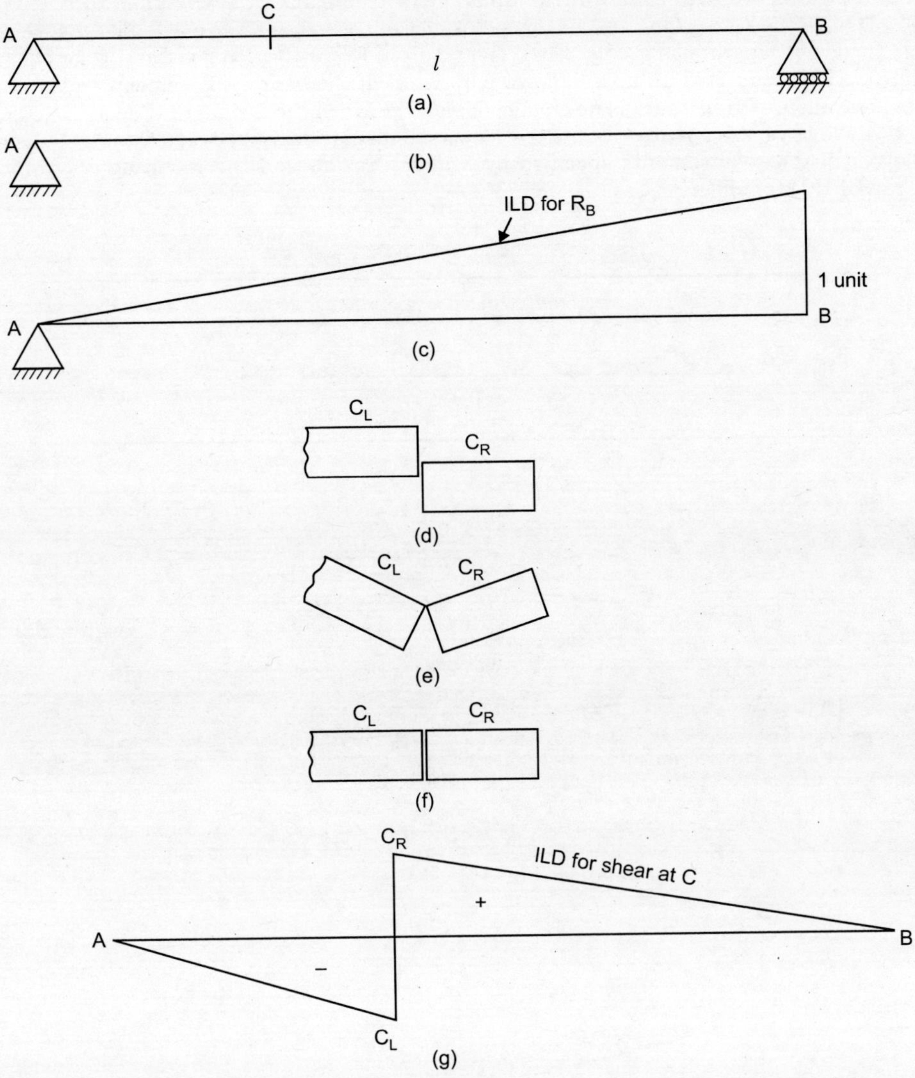

Fig. 3.1

(*b*) Now let us take an internal force, say positive shear at C. Who or What offers resistance to shear at C ? The beam section of course ! So cut that resistance. In other words cut the beam at C. Now, have we been hasty in cutting the beam ? The beam was offering resistance to bending, shear and tension. We must now remove the resistance to shear only and keep the resistance to rotation at C and tensile separation at C. (Does it sound like the court scene in Merchant of Venice ?) But we can make it. We will allow the 2 parts of the beam to slide up and down (Fig. 3.1 (*d*)). We will restrain any relative rotation such as in Fig. 3.1 (*e*). In other words even after allowing an up and down slide at C, we will make sure A – C_L and C_R – B have the same slope. (C_L means point just to the left of C). We will also ensure that C_L and C_R do not separate as in Fig. 3.1(*f*). The result in Fig. 3.1(*g*). This is the ILD for shear at C. Here we have imported an up and down movement at C (shear displacement) equal to 1 unit. We have ensured that there is no change of slope between the 2 sides of C (slope at C_L = slope at C_R). We have also ensured that there is no axial displacement between C_L and C_R. Hence we obeyed all the requirements specified by Muller Breslau and the resulting diagram is the required ILD for shear at C.

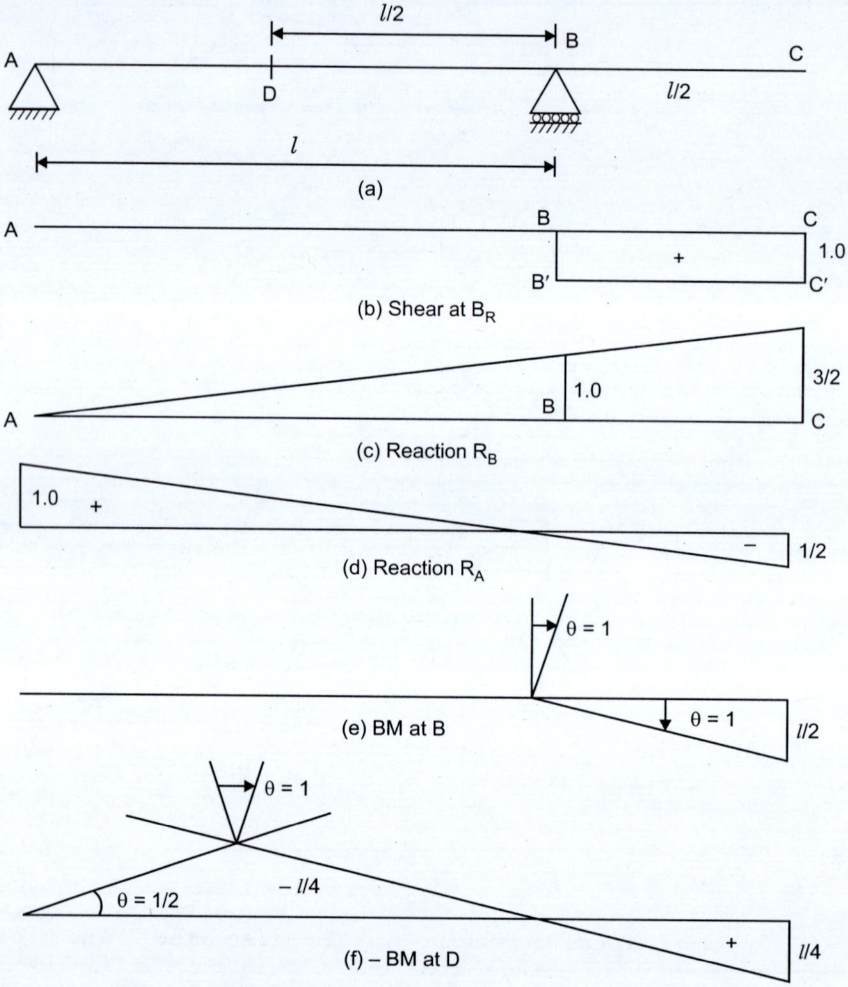

Fig. 3.2

In Figs. 3.2 (*a*) to (*f*) we deftly apply Muller Breslau procedure to sketch a few more influence lines for a determinate structure–an overhanging beam.

Figure 3.2 (*b*) is the IL for shear to the right of B. To get this we cut the beam at B_R and remove the restraint to shear displacement. We must ensure that restraints to flexural displacement and tensile displacement are retained. This means, in the displaced diagram, BA and BC cannot have different slopes at B. Now, can B_{left} and B_{right} be axially separated? Figure 3.2 (*b*) has honoured these constraints. Since we applied a shear displacement due to positive shears (left up, right down), B′C′ is below BC. The loadings on BC in this diagram correspond to downward loads. In other words, a downward unit load any where on BC, produces a unit positive shear at B_R.

Figure 3.2 (*c*) and (*d*) are self explanatory. So, IL for external reactions are relatively easy to obtain, since there is no question of cutting the member and retaining some restraints.

Figure 3.2 (*e*) is interesting. We want the hogging bending moment at B (negative bending moment). So we remove the resistance to rotation at B (and keep the resistances to shear displacement and axial separation). In other words, we introduce a hinge at B. Then we apply a unit displacement corresponding to M_B at B. This means we cause a 1 radian deviation between the directions of BA and BC (or one radian deviation between the normals at B_{left} and B_{right}). The resulting displaced diagram in Fig. 3.2 (*e*).

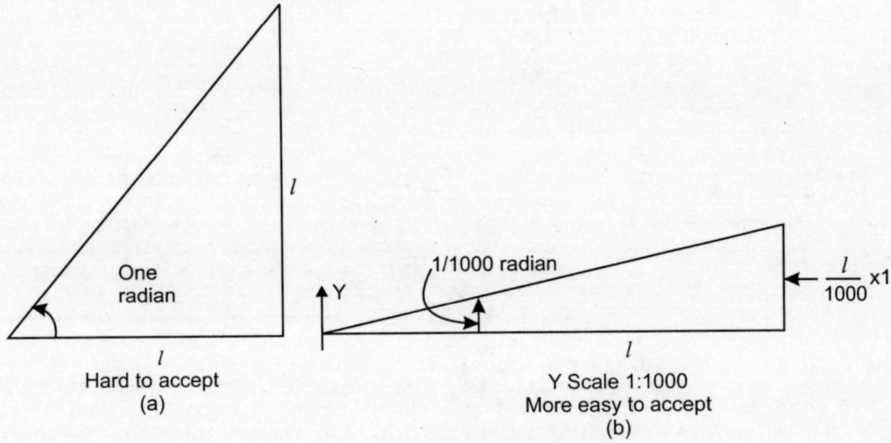

Fig. 3.3

Now a 1 radian rotation is an enormous rotation. There may be a mental resistance to think of applying such large displacements to any structure. This is overcome by choosing a scale of 1/1000 for IL ordinates (in relation to horizontal scale). So a 1 radian rotation can be reduced to 1/1000 radian.

The resulting displacement ordinates can now be scaled up by 1000 to get the ILOs.

Figure 3.2 (*f*) shows the ILD for negative bending moment at D, the mid-point of AB. We introduce a hinge at D and cause a unit angular deviation between DA and DB. This has resulted in Fig. 3.2 (*f*). If we want the IL for sagging bending moment at D, we can just interchange the signs of the 2 parts of the ILD.

3.4 MULLER BRESLAU PRINCIPLE EXTENDED TO INDETERMINATE STRUCTURES

With determinate structures, Muller Breslau Principle went like a cake walk. The main reason is when we removed one restraint the 'structure' become a 'mechanism' and the applied displacement did not create any stressing or deformation of members. Hence there was no elastic curve to deal with.

With structures with one degree of indeterminacy we are likely to get a determinate structure when we take away one restraint. Now any applied displacement is on this stable, determinate structure. There will be deformations to compute as we would see in example 3.1.

Example 3.1 *Draw the IL for reaction at B and for the support moment M_A at A for the propped cantilever in Fig. 3.4 (a). Compute the IL ordinates at 1.5 m intervals.*

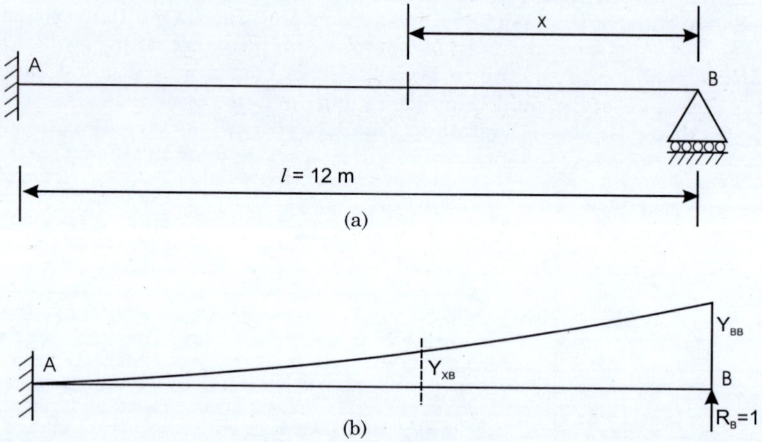

Fig. 3.4

Solution To generate the ILD for R_B

(*i*) Remove the restraint due to R_B (remove support B)

(*ii*) Apply a unit displacement (upward) at B. Instead, we can apply a unit load corresponding to R_B and compute the resulting displacement. Thus we can work out the force needed along R_B direction to cause a unit displacement at B by dividing 1 by y_{BB}. This is the multiplier for all other displacements that would yield the IL ordinates. The resulting elastic curve is the IL for R_B.

When $R_B = 1$, then Y_{XB} is the displacement at section x due to unit load applied at B.

$$M_x = -EI\frac{d^2y}{dx^2} = R_B \,.\, x = 1x\;;\quad EI\frac{d^2y}{dx^2} = -x$$

$$EI\frac{dy}{dx} = -\frac{x^2}{2} + C_1 \qquad ...(3.1)$$

$$EIy = -\frac{x^3}{6} + C_1x + C_2 \qquad \text{...(3.2)}$$

at $x = 12, y = 0,$ $\frac{dy}{dx} = 0$

Hence, $C_1 = 72, \quad C_2 = -576$

$$Y_{XB} = \frac{1}{EI}\left[-\frac{x^3}{6} + 72x - 576\right] \qquad \text{...(3.3)}$$

$$Y_{BB} \text{ (at } x = 0) = -\frac{576}{EI} \text{ ; ILO for } R_B \text{ at X} = \frac{Y_{XB}}{Y_{BB}} = \frac{\frac{1}{EI}\left[-\frac{x^3}{6} + 72x - 576\right]}{\frac{1}{EI}(-576)}$$

$$= \frac{-\frac{x^3}{6} + 72x - 576}{-576} \qquad \text{...(3.4)}$$

When we plot this against x, we get the IL for R_B.

Table 3.1. Ordinates of ILD for R_B at 1.5 m intervals.

x m	0	1.5	3	4.5	6	7.5	9	10.5	12
R_B	1	0.814	0.632	0.463	0.312	0.184	0.085	0.022	0.00

Figure 3.5 is the influence line diagram for R_B.

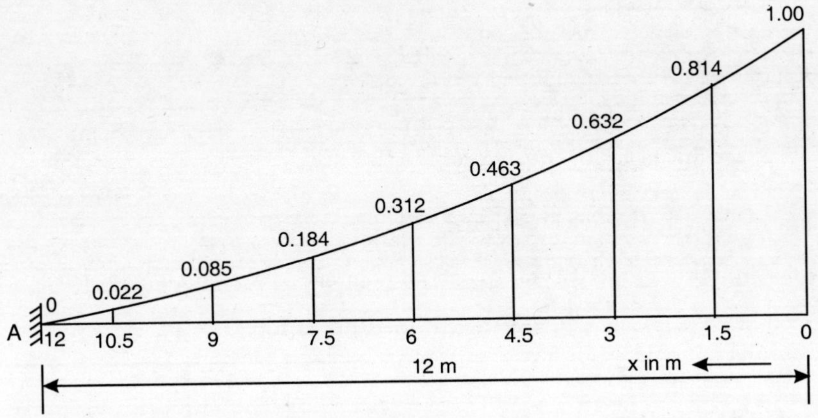

Fig. 3.5

To get the IL for M_A we have to

(*i*) Introduce a hinge at A, and

(*ii*) We have to apply a unit rotation at A. Instead we will apply a unit moment at A find the general displacement at any x from B. We will then divide the displacements by the actual rotation at A.

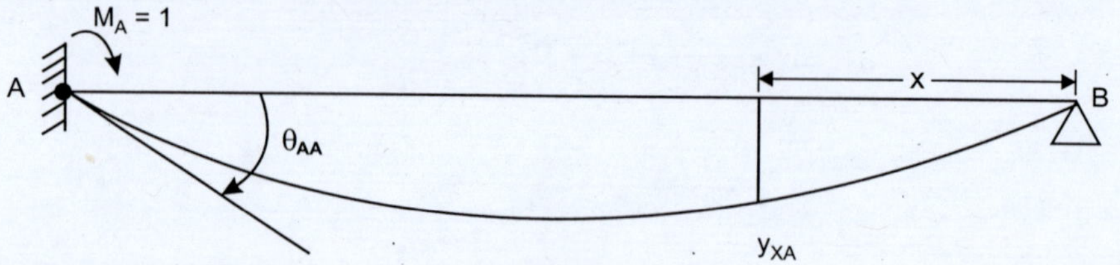

Fig. 3.6

Due to $M_A = 1$

$$R_B = -R_A = \frac{1}{12}$$

$$M_x = -EI\frac{d^2y}{dx^2} = \frac{x}{12}$$

$$EI\frac{dy}{dx} = -\frac{x^2}{24} + C_1$$

$$EI\,y = -\frac{x^3}{72} + C_1x + C_2$$

At $x = 0, y = 0$; at $x = 12, y = 0$

Hence, $C_2 = 0$; $C_1 = 2$.

$$y_{XA} = \frac{1}{EI}\left(-\frac{x^3}{72} + 2x\right) \quad \text{...(3.5)}$$

$$\frac{dy}{dx} = \theta_{XA} = \frac{1}{EI}\left(-\frac{x^2}{24} + 2\right)$$

$$\theta_{AA}\ (\text{at } x = 12) = \frac{-4}{EI}$$

When we divide y_{XA} by θ_{AA}, we get the ILO at x

$$\text{ILO for } M_A = \left(+\frac{x^3}{288} - \frac{x}{2}\right) \quad \text{...(3.6)}$$

The ordinates of the ILD for M_A at 1.5 m intervals are tabulated below.

Table 3.2. IL ordinates for M_A.

x	0	1.5	3.0	4.5	6.0	7.5	9.0	10.5	12.0
ILO	0.0	– 0.738	– 1.406	– 1.934	– 2.250	– 2.285	– 1.969	– 1.230	0.0

Figure 3.7 shows the plotted ILD for M_A.

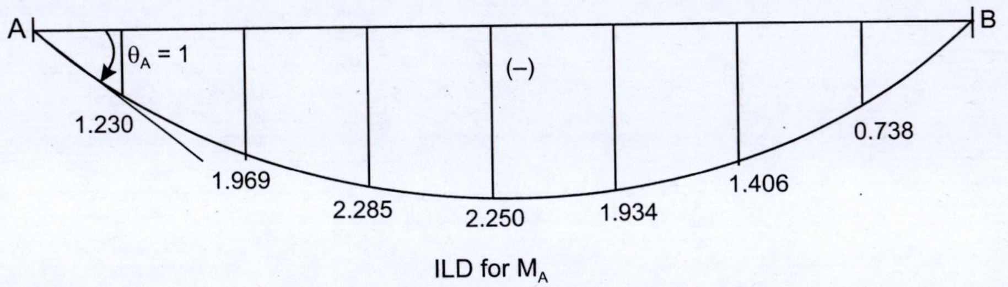

Fig. 3.7

Example 3.2 *Draw the influence line for the moment at C of the propped cantilever shown in Fig. 3.8 compute the ordinates at 1 m intervals.*

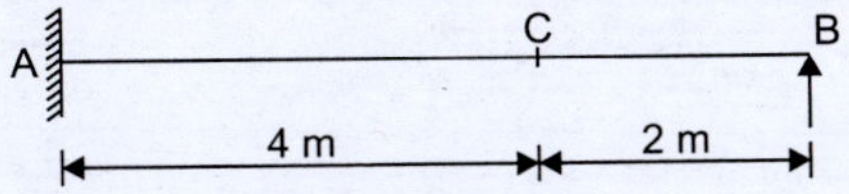

Fig. 3.8

Solution To get the influence line for M_C,

(*i*) Introduce a hinge at C.

(*ii*) Apply a unit bending moment at C.

(*iii*) Determine the deflection y_{XC} at any x and the slope θ_{CC} at C

(*iv*) $\dfrac{y_{XC}}{\theta_{CC}}$ is the influence line ordinate at any x.

Portion AC

$\Sigma V = 0$ gives

$$R_{CA} = \frac{1}{2}\ (\uparrow)$$

$$R_A = \frac{1}{2}\ (\downarrow)$$

$$M_A = \frac{1}{2} \times 4 + 1$$

$$M_A = 3\ (\curvearrowright)$$

Portion CD

$\Sigma M_B = 0$

$$R_{CB} \times 2 + 1 = 0$$

$$R_{CB} = \frac{-1}{2}$$

$$R_{CB} = \frac{1}{2}\ (\downarrow)$$

$$R_B = \frac{1}{2}\ (\uparrow)$$

Portion AC

Considering a section XX at a distance x from C,

$$M_{XX} = \frac{1}{2}x + 1$$

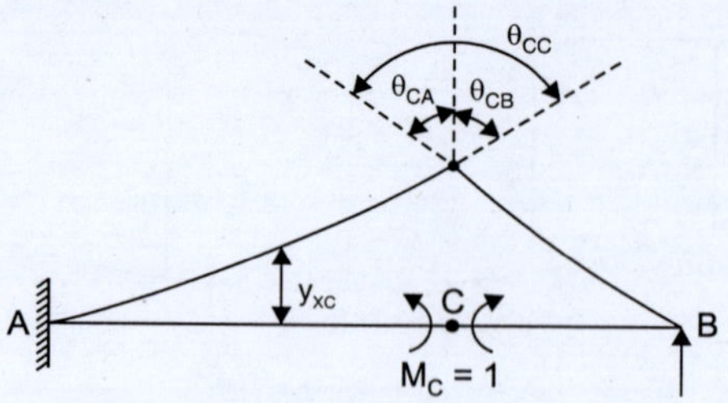

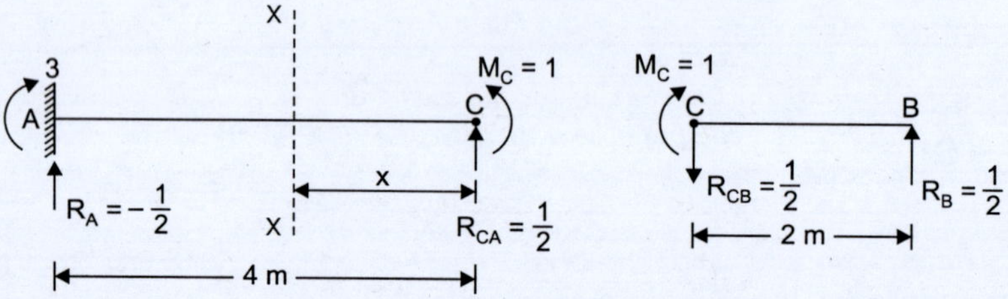

$$EI \cdot \frac{d^2y}{dx^2} = -M_{XX} = -\left(\frac{1}{2}x + 1\right)$$

$$EI \cdot \frac{d^2y}{dx^2} = -\frac{x}{2} - 1$$

$$EI \cdot \frac{dy}{dx} = -\frac{x^2}{2 \times 2} - x + C_1$$

At $x = 4$ m, $\dfrac{dy}{dx} = 0$

$$EI \times 0 = -\frac{1}{2} \times \frac{4^2}{2} - 4 + C_1$$

$$C_1 = 8$$

$\therefore$ $$EI \cdot \frac{dy}{dx} = -\frac{x^2}{4} - x + 8 \qquad \ldots(1)$$

$$EI.\, y = -\frac{1}{4} \times \frac{x^3}{3} - \frac{x^2}{2} + 8x + C_2$$

At $x = 4$ m, $y = 0$

$$EI \times 0 = -\frac{1}{4} \times \frac{4^3}{3} - \frac{4^2}{2} + 8 \times 4 + C_2$$

$$0 = -\frac{16}{3} - 8 + 32 + C_2$$

$$C_2 = -18.87$$

$\therefore \quad EI \, . \, y = -\dfrac{x^3}{12} - \dfrac{x^2}{2} + 8x - 18.67$ ...(2)

At $x = 0$

$$EI \, . \, y_C = 0 - 0 + 0 - 18.87$$

$$y_C = -\frac{18.87}{EI} \quad \text{(at C in CA)}$$

i.e., $y_{CA} = -\dfrac{18.87}{EI}$ ...(3)

At $x = 0$

$$EI \, . \, \frac{dy}{dx} = 0 - 0 + 8$$

$$\frac{dy}{dx} = \frac{8}{EI} \quad \text{(at C in CA)}$$

i.e., $\theta_{CA} = \dfrac{8}{EI}$

Portion CB

Considering a section XX at a distance 'x' from the end 'B'.

$$M_{XX} = \frac{1}{2} \, . \, x$$

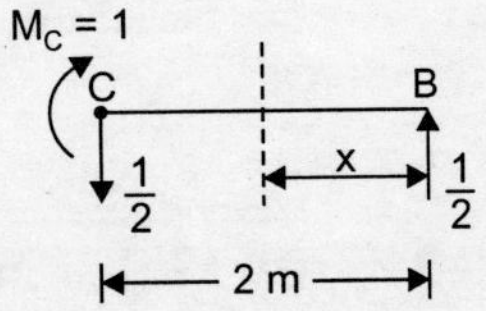

$$EI \, . \, \frac{d^2y}{dx^2} = -M_{XX} = -\frac{x}{2}$$

$$EI \, . \, \frac{dy}{dx} = -\frac{x^2}{2 \times 2} + C_1$$

$$EI \, . \, y = -\frac{x^3}{4 \times 3} + C_1 x + C_2$$

At $x = 0, y = 0$ (at B)

$$EI \, . \, 0 = 0 + 0 + C_2$$

$$C_2 = 0$$

At $x = 2 \text{ m}, y = \dfrac{-18.67}{EI}$ [from (3)]

$$EI\left(-\frac{18.67}{EI}\right) = -\frac{2^3}{12} + 2\,C_1 + 0$$

$$-18.67 = -0.67 + 2\,C_1$$

$$C_1 = -9$$

$\therefore \quad EI \dfrac{dy}{dx} = -\dfrac{x^2}{4} - 9$ ...(4)

$$\text{EI}.\, y = -\frac{x^3}{12} - 9x \qquad \ldots(5)$$

At $\quad x = 2$ m

$$\Rightarrow \quad \frac{dy}{dx} = \frac{1}{\text{EI}}\left[-\frac{2^2}{4} - 9\right]$$

$$\frac{dy}{dx} = -\frac{10}{\text{EI}}$$

i.e.,

$$\theta_{\text{CB}} = -\frac{10}{\text{EI}}$$

$$\theta_{\text{CC}} = \theta_{\text{CA}} - \theta_{\text{CB}}$$

$$= \frac{8}{\text{EI}} - \left[-\frac{10}{\text{EI}}\right]$$

$$\theta_{\text{CC}} = \frac{18}{\text{EI}} \qquad \ldots(6)$$

For region CA

ILO for

$$\text{M}_\text{C} = \frac{y_{\text{XC}}}{\theta_{\text{CC}}} = \frac{\frac{1}{\text{EI}}\left[-\frac{x^3}{12} - \frac{x^2}{2} + 8x - 18.67\right]}{\frac{1}{\text{EI}} \times 18}$$

$$\text{M}_\text{C} = \frac{-\frac{x^3}{12} - \frac{x^2}{2} + 8x - 18.67}{18} \qquad \ldots(7)$$

For region BC

ILO for

$$\text{M}_\text{C} = \frac{y_{\text{XC}}}{\theta_{\text{CC}}} = \frac{\frac{1}{\text{EI}}\left[-\frac{x^2}{12} - 9x\right]}{\frac{1}{\text{EI}} \times 18}$$

$$\text{M}_\text{C} = \frac{-\frac{x^2}{12} - 9x}{18} \qquad \ldots(8)$$

The influence line ordinates are tabulated below:

x from	*B*			*C*				
	0	1	2	0	1	2	3	4
x from C	0	1	2	2	3	4	5	6
ILO	0	– 0.505	– 1.019 – 1.037 ≈ – 1.00		– 0.625	– 0.296	– 0.079	0

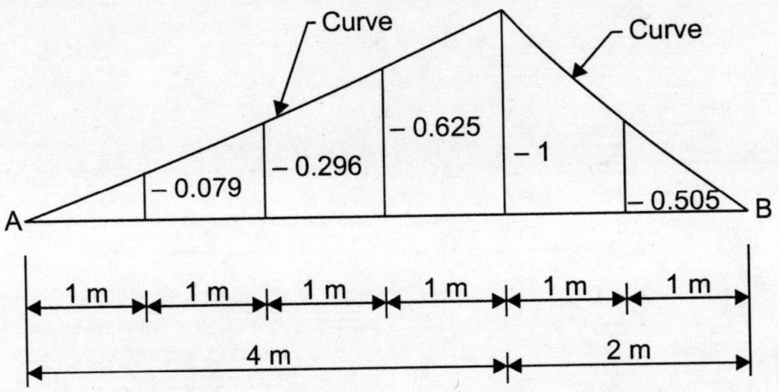

Influence line diagram for the moment at "C".

Example 3.3 *Determine the influence line for R_A for the continuous beam shown in Fig. 3.9. Compute the IL ordinates at 1 m intervals.*

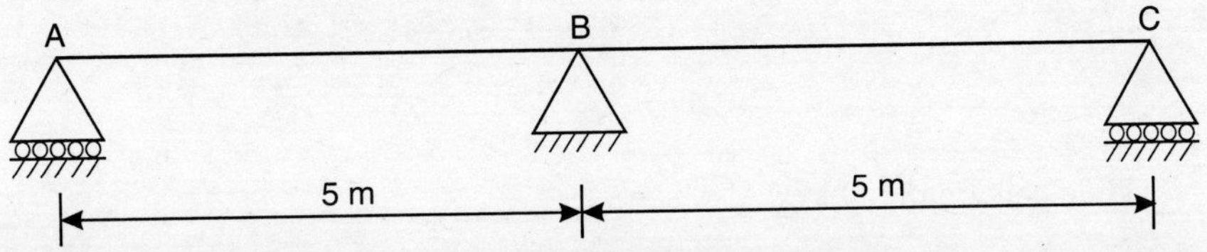

Fig. 3.9

Solution To get IL for R_A

(*i*) Remove support A.

(*ii*) Apply a unit force at A (along R_A) and compute the deflection at any x on CB and BA

(*iii*) Divide these deflections by the displacement at A.

Figure 3.10 shows the elastic curve due to $R_A = 1$.

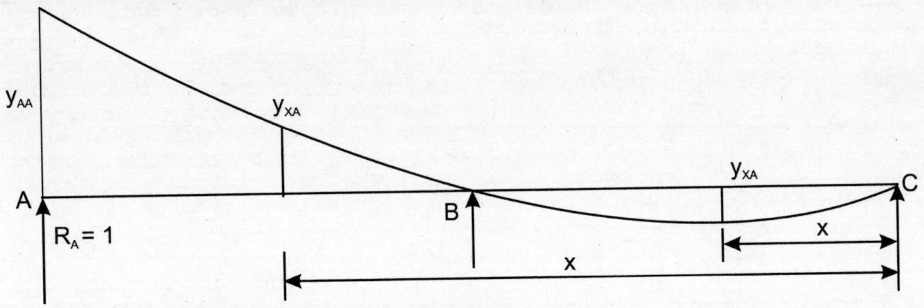

Fig. 3.10

By taking moments about C, R_B can be obtained and $\Sigma V = 0$ will give R_C.

Since $R_A = 1$, $R_B = -2$, and $R_C = 1$.

$$M_x = -\frac{d^2y}{dx^2} \cdot EI = 1 \cdot x \;\vdots\; -2(x-5)$$

$$EI\frac{d^2y}{dx^2} = -x \;\vdots\; + 2(x-5)$$

$$EI\frac{dy}{dx} = -\frac{x^2}{2} + C_1 \;\vdots\; + (x-5)^2$$

$$EI\,y = -\frac{x^3}{6} + C_1x + C_2 \;\vdots\; + \frac{(x-5)^3}{3}$$

At $x = 0, y = 0$

At $x = 5, y = 0$

Hence, $C_2 = 0$; $C_1 = 4.167$

$$y_{XA} = \frac{1}{EI}\left[-\frac{x^3}{6} + 4.167x \;\vdots\; + \frac{(x-5)^3}{3}\right] \quad ...(3.7)$$

at A(x = 10),

$$y_{AA} = \frac{1}{EI}[-166.67 + 41.67 + 41.67]$$

$$y_{AA} = \frac{-83.33}{EI} \quad ...(3.8)$$

Dividing equation (3.7) by y_{AA}, we get

$$\text{ILO at } x = -\frac{1}{83.33}\left[-\frac{x^3}{6} + 4.167x \;\vdots\; + \frac{(x-5)^3}{3}\right] \quad ...(3.9)$$

Table 3.3. Ordinates of ILD for R_A.

x m	Support C 0	1	2	3	4	Support B 5	6	7	8	9	Support A 10
R_A	0	– 0.048	– 0.084	– 0.096	– 0.072	0	0.128	0.304	0.516	0.752	1

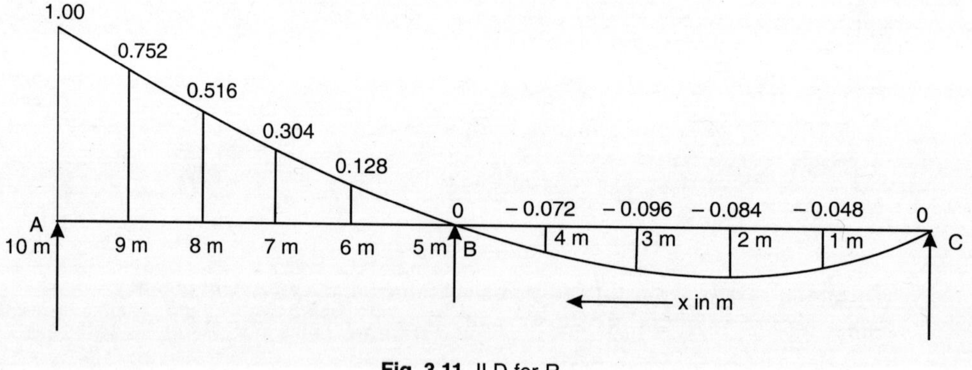

Fig. 3.11. ILD for R_A.

Example 3.4 *Determine the influence line for the reaction at the middle support B of continuous beam shown in Fig. 3.12. Compute the ordinates at every 1 m interval.*

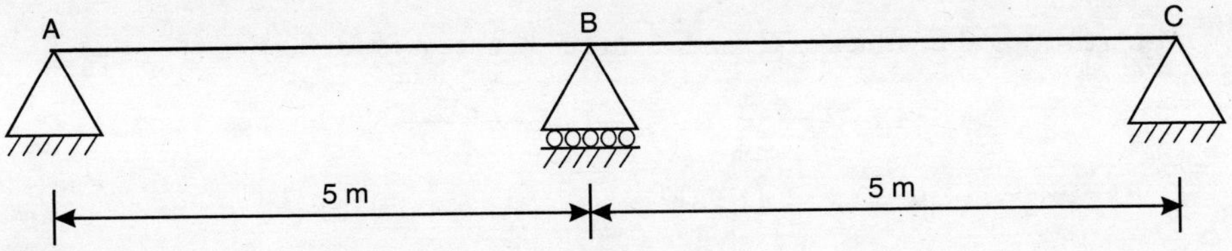

Fig. 3.12

Solution To draw ILD for R_B

(*i*) Remove the support B.

(*ii*) Apply a unit load at B ; determine the general deflection y_{XB} at x and the deflection y_{BB} at B.

(*iii*) $\frac{y_{XB}}{y_{BB}}$ is the ILO at any x.

In Fig. 3.13,

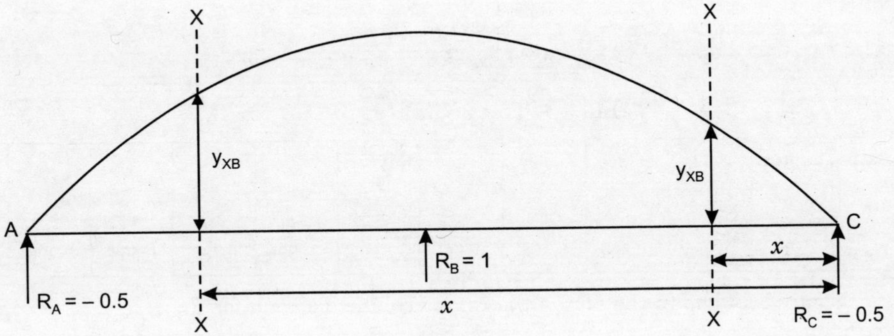

Fig. 3.13

When $R_B = 1$, $R_A = R_C = -0.5$ (Using $\Sigma V = 0$ and symmetry)

Considering a section at XX

$$M_x = -0.5x \;\vdots\; + 1(x-5)$$

$$EI\frac{d^2y}{dx^2} = 0.5x \;\vdots\; -(x-5)$$

Integrating

$$EI\frac{dy}{dx} = \frac{0.5x^2}{2} + C_1 \;\vdots\; -\frac{(x-5)^2}{2}$$

Again integrating

$$EIy = \frac{0.5x^3}{6} + C_1x + C_2 \;\vdots\; -\frac{(x-5)^3}{3\times 2}$$

The boundary conditions are,

When $x = 0$; $y = 0$

When $x = 10$; $y = 0$

Hence, $C_2 = 0$; $C_1 = -6.25$

Substituting the values of C_1 and C_2 in the equation for deflection

$$EIy = \frac{0.25x^3}{3} + C_1x + C_2 \; \vdots \; - \frac{0.5(x-3)^3}{3}$$

$$y_{XB} = \frac{1}{EI}\left[\frac{0.25x^3}{3} - 6.25x \; \vdots \; - \frac{0.5(x-5)^3}{3}\right]$$

$$y_{BB} = (\text{at } x = 5 \text{ m}) = \frac{1}{EI}\left[\frac{0.25 \times 5^3}{3} - 6.25 \times 5\right] = -\frac{20.83}{EI}$$

$$\text{ILO for } R_B = \frac{y_{XB}}{y_{BB}} = \frac{\frac{1}{EI}\left[\frac{0.25x^3}{3} - 6.25x \; \vdots \; - \frac{0.5(x-5)^3}{3}\right]}{\frac{1}{EI} \times (-20.83)}$$

$$\text{ILO for } R_B = -\frac{1}{20.83}\left[\frac{0.25x^3}{3} - 6.25x \; \vdots \; - \frac{0.5(x-5)^3}{3}\right] \quad ...(3.10)$$

Table 3.4. The ordinates of ILD for R_B.

x m	Support C 0	1	2	3	4	Support B 5	6	7	8	9	Support A 10
R_B	0	0.296	0.568	0.782	0.994	1.00	0.994	0.782	0.568	0.296	0

Figure 3.14 shows the plot of the influence line diagram for R_B.

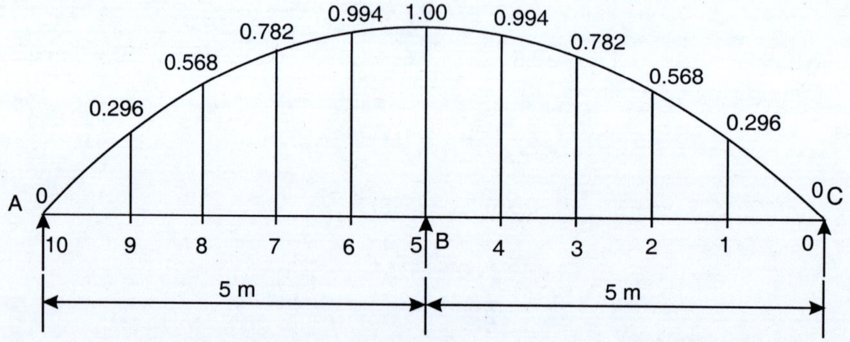

Fig. 3.14

Example 3.5 *Using Muller Breslau principle, draw the influence line for bending moment at the mid-point D of span AB of the continuous beam ABC shown in Fig. 3.15. Determine the influence line ordinates at suitable intervals and plot them.*

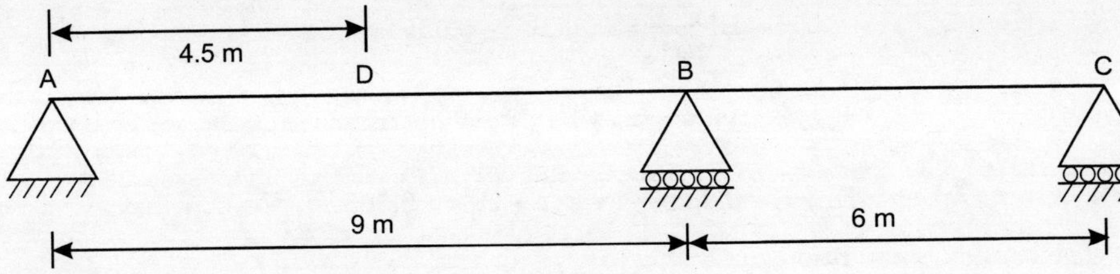

Fig. 3.15

Solution To get the influence line for M_D

(*i*) introduce a hinge at D.

(*ii*) apply a unit bending moment at D

(*iii*) determine the deflection y_{XD} any x and the slope θ_{DD} at D (Fig. 3.16)

$\dfrac{y_{xD}}{\theta_{DD}}$ is the influence line ordinate at any x.

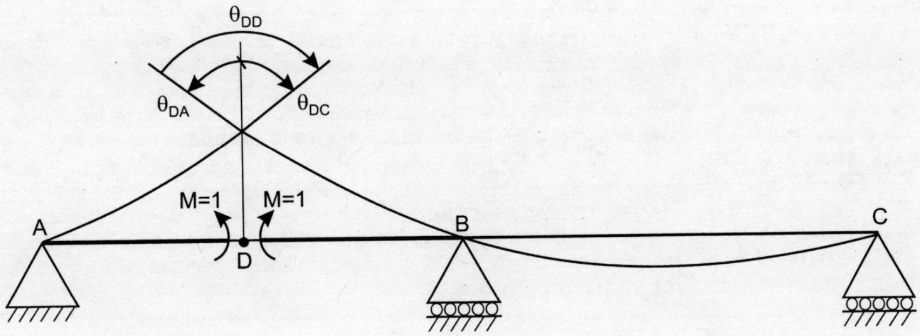

Fig. 3.16

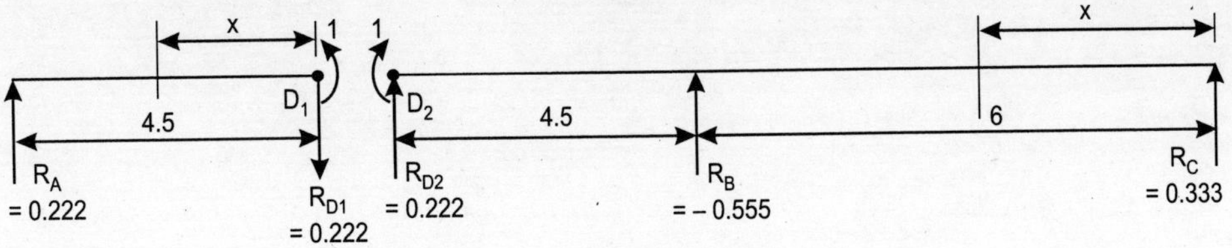

Fig. 3.17

Due to $M = 1$ at D, $R_A \times 4.5 = 1$, $R_A = 1/4.5 = 0.222$, $R_{D1} = 0.222\downarrow$

Hence, $R_{D2} = 0.222\uparrow$

Taking moments about C,

$$0.222 \times 10.5 + 1 + R_B \times 6 = 0$$

$$R_B = -0.555,\ R_C = 0.333$$

The two regions AD and DBC will be considered separately (because of the discontinuity at D)

From Fig. 3.17, bending moment at any x is

$$+M_x = -EI\frac{d^2y}{dx^2} = 0.333x \;\vdots\; -0.555\,(x-6)$$

$$EI\frac{dy}{dx} = -0.333\,\frac{x^2}{2} + C_1 \;\vdots\; +0.555\,\frac{(x-6)^2}{2}$$

$$EI\,y = -0.333\,\frac{x^3}{6} + C_1x + C_2 \;\vdots\; +0.555\,\frac{(x-6)^3}{6}$$

The boundary conditions are,

$$y = 0 \text{ at } x = 0 \text{ and } x = 6$$

Hence, $C_2 = 0\,;\, C_1 = 2$

Substituting these

$$EI\frac{dy}{dx} = -0.333\,\frac{x^2}{2} + 2 \;\vdots\; +0.555\,\frac{(x-6)^2}{2} \qquad \text{...(3.11)}$$

at D, ($x = 10.5$) in Fig. 3.16

$$\theta_{DC} = \left(\frac{dy}{dx}\right)_{DC} = \frac{1}{EI}\left[-0.333\times\frac{10.5^2}{2} + 2 + 0.555\times\frac{4.5^2}{2}\right]$$

$$\theta_{DC} = -\frac{10.738}{EI}$$

$$y_D = y_{10.5} = \frac{1}{EI}\left[-0.333\times\frac{10.5^3}{6} + 2 + 0.555\times\frac{4.5^3}{6}\right] \qquad \text{...(3.12)}$$

$$y_D = -\frac{34.82}{EI} \qquad \text{...(3.13)}$$

For the zone AD

$$M_x = 1 - 0.222x$$

$$EI\frac{d^2y}{dx^2} = 0.222x - 1$$

$$EI\frac{dy}{dx} = 0.222\,\frac{x^2}{2} - x + C_3$$

$$EI\,y = 0.222\,\frac{x^3}{6} - \frac{x^2}{2} + C_3x + C_4$$

At $x = 0\,;\, y = -\dfrac{34.82}{EI}$

At $x = 4.5$, $y = 0$

Substituting these values

$$C_4 = -34.82$$

$$C_3 = 9.24$$

$$EI\frac{dy}{dx} = 0.222\,\frac{x^2}{2} - x + 9.24 \qquad \text{...(3.14)}$$

$$\theta_{DA} = \left(\frac{dy}{dx}\right)_{x=0} = \frac{9.24}{EI}$$

$$EI \cdot y = 0.222\frac{x^3}{3} - \frac{x^2}{2} + 9.24x - 34.82$$

$$y_{xD} = \frac{1}{EI}\left[0.222\frac{x^3}{3} - \frac{x^2}{2} + 9.24x - 34.82\right] \qquad ...(3.15)$$

$$\theta_{DD} = \theta_{DA} - \theta_{DC} = \frac{9.24}{EI} + \frac{10.738}{EI} = \frac{19.978}{EI}$$

For the region CD,

$$\text{ILO for } M_D = \frac{y_{xD}}{\theta_{DD}}\left[\frac{-0.333\frac{x^3}{6} + 2x \;\vdots + 0.555\frac{(x-6)^3}{6}}{19.978}\right] \qquad ...(3.16)$$

For the region DA

$$\text{ILO for } M_D = \left[\frac{0.222\frac{x^3}{6} - \frac{x^2}{2} + 9.24x - 34.82}{19.978}\right] \qquad ...(3.17)$$

The influence line ordinates are tabulated under:

x from C	0.0	3.0	6.0	9.0	10.5	12.0	15.0
ILO	0	0.225	0.0	– 0.999	– 1.743	– 1.099	0

The plot is shown in Fig. 3.18.

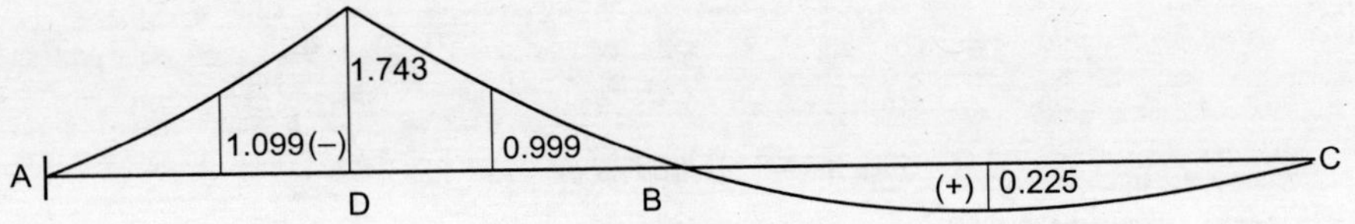

Fig. 3.18 ILD for M_D.

Example 3.6 *Determine the influence line for shear force at D, the mid-point of span BC of a continuous beam shown in Fig. 3.19. Compute the IL ordinates at 1.5 m intervals.*

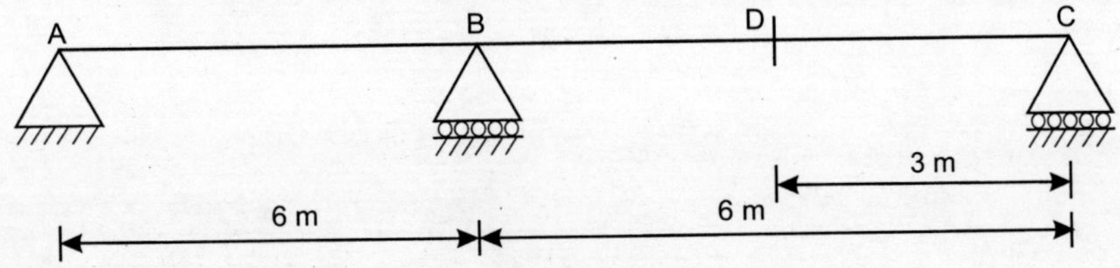

Fig. 3.19

Solution To plot the influence line for F_D, the shearing resistance of the beam at D is removed by installing a sliding device which permits shearing displacement at D but does not impair moment resistance at D.

When a unit shear (S = 1) is applied at D, DB and DC maintain the same slope at D. At any x, the deflection is y_{XD}. At D, the shear displacement is y_{DD}.

Then F_D, the shear force at D due to a unit load at x is given by $F_D = \dfrac{y_{xD}}{y_{DD}}$

Now we have to get the bending moment at any section x in order to form the differential equation in x and y.

From the free body Fig. 3.20 (*b*) $M_D = 3$

From the free body Fig. 3.20 (*c*) $R_C = 1$, $M_D = 3$

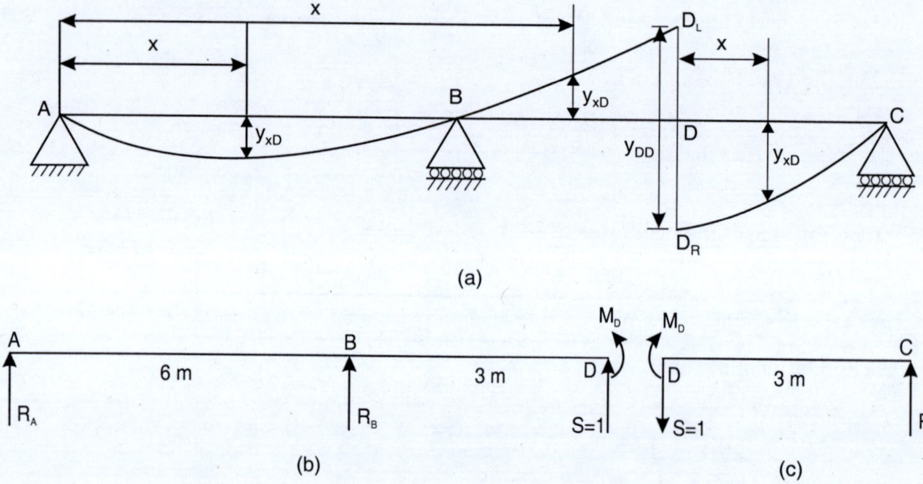

Fig. 3.20. Free bodies of AD and DC.

Moments about A = 0

$$R_B \times 6 + 1 \times 9 + 3 = 0\ ;\ R_B = -2$$

$$R_A + R_B + 1 = 0\ ;\ R_A = 1$$

For region AD

$$M_x = x \ \vdots\ - 2(x - 6)$$

$$EI\frac{d^2y}{dx^2} = - M_x = - x \ \vdots\ + 2(x - 6)$$

$$EI\frac{dy}{dx} = -\frac{x^2}{2} + C_1 \ \vdots\ + (x - 6)^2$$

$$EI\,y = -\frac{x^3}{6} + C_1x + C_2 \ \vdots\ + \frac{(x-6)^3}{3}$$

at $x = 0$, $y = 0$; at $x = 6$, $y = 0$

Hence $C_2 = 0$; $C_1 = 6$

$$\text{EI}\,\frac{dy}{dx} = -\frac{x^2}{2} + 6 \;\vdots\; + (x-6)^2 \qquad \text{...(3.18)}$$

at D($x = 9$)

$$\text{EI}\left(\frac{dy}{dx}\right)_{\text{DB}} = -\frac{81}{2} + 6 + 3^2$$

$$\theta_{\text{DA}} = \left(\frac{dy}{dx}\right)_{\text{DB}} = -\frac{25.5}{\text{EI}} \qquad \text{...(3.19)}$$

Again,
$$\text{EI}\,y = -\frac{x^3}{6} + 6x \;\vdots\; + \frac{(x-6)^3}{3} \qquad \text{...[3.19(a)]}$$

at D, $x = 9$ m

$$y_{\text{DL}} = \left(\frac{-\dfrac{9^3}{6} + 6\times 9 + \dfrac{(9-6)^3}{3}}{\text{EI}}\right) = \frac{-129.5 + 54 + 9}{\text{EI}} = -\frac{58.5}{\text{EI}}$$

For region DC, $M_x = 3 - x$

$$\text{EI}\frac{d^2y}{dx^2} = -3 + x$$

$$\text{EI}\,\frac{dy}{dx} = -3x + \frac{x^2}{2} + C_3$$

$$\text{EI}\,y = -\frac{3x^2}{2} + \frac{x^3}{6} + C_3x + C_4$$

From eqn. (3.19)

At $x = 0$, $\left(\text{EI}\dfrac{dy}{dx}\right)_{\text{DC}} = \left(\text{EI}\dfrac{dy}{dx}\right)_{\text{DB}} = -25.5$

So, $C_3 = -25.5$

At $x = 3$, $y = 0$, $C_4 = 85.5$

Hence,
$$(y)_{x\text{D}} = \frac{1}{\text{EI}}\left(-\frac{3x^2}{2} + \frac{x^3}{6} - 25.5x + 85.5\right) \qquad \text{...(3.20)}$$

at D, ($x = 0$)
$$y_{\text{DR}} = \frac{85.5}{\text{EI}}$$

$$y_{\text{DD}} = y_{\text{DL}} - y_{\text{DR}} = \frac{-58.5}{\text{EI}} - \frac{85.5}{\text{EI}} = \frac{-144}{\text{EI}}$$

$$\text{ILO for } F_{\text{D}} = \frac{y_{\text{DX}}}{y_{\text{DD}}}$$

For region AD, using eqn. (3.19)

$$\text{ILO} = \frac{1}{144}\left[-\frac{x^3}{6} + 6x \;\vdots\; + \frac{(x-6)^3}{3}\right] \qquad \text{...(3.21)}$$

For region DC, using eqn. (3.20)

$$\text{ILO} = \frac{1}{144}\left[-\frac{3x^2}{2} + \frac{x^3}{6} - 25.5x + 85.5\right] \quad ...(3.22)$$

Using the above two relations, the IL ordinates are tabulated below:

X	0 (A)	1.5	3.0	4.5	6.0 (B)	7.5	9.0 DL	0 DR	1.5	3.0 (C)
ILO for F_D	0.0	– 0.059	– 0.094	– 0.082	0	0.17	0.41	– 0.59	– 0.31	0.0

Figure 3.21 show the plot.

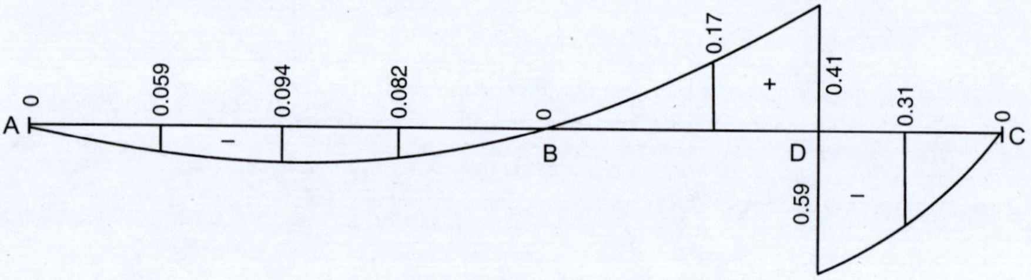

Fig. 3.21. IL for shear at D.

3.5 INDIRECT MODEL ANALYSIS

3.5.1 Introduction

Muller Breslau's principle is invaluable when applied to experimental methods. The constraints stated by M-B can be actually implemented in a table model and known displacements applied at predecided points. We only have to measure the resulting displacements and plot them to get the influence line we want.

3.5.2 Propped Cantilever Model

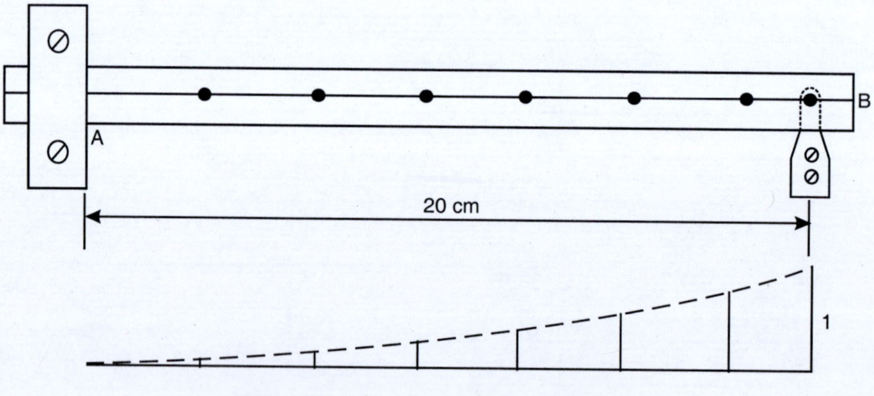

Fig. 3.22. Model of a propped cantilever.

If we want to analyze a propped cantilever of uniform cross-section, 8 m in span, first we make a model of it, say in Perspex or Plexiglas or similar thin, transparent easily workable material. Let us mount this on a drawing table preferably over a white paper.

We have to make a suitable arrangement to clamp the model at A.

Let us actually hinge it at B, by drilling a fine hole at B and inserting it in a pin mounted at B. The pin holder shall be clamped to the board.

If we want to get the ILD for reaction at B, what does M-B say we must do ?

(*i*) Remove the pin at B (restraint due to the reaction off).

(*ii*) Apply a vertical displacement (unit value) along R_B direction.

(*iii*) Plot the displaced configuration.

This plot is the ILD for R_B.

If our model is true and our measurements good, the ILD would be OK.

3.5.3 Practical Considerations

"Applying a unit displacement" and "Measuring the resulting displacement" are easier said than done. There are several problems.

1. How much is a unit ? If we make 20 cm long model (of the 8 m beam) we can not possibly apply a 1 cm displacement without breaking the strip. Possibly 1 mm might be more acceptable or 0.2 or 0.3 mm. The displacement shall be small enough to create stresses well within the elastic limit. It must be big enough to be measured.

2. How do we plot the displaced configuration ? We may have to mark some targets on the center line of the model. The displacements of the targets can be measured using micrometer microscopes kept trained over the targets.

The targets must be sharp circles filled in with blacking for easy and accurate measurement with microscopes.

3.5.4 Designing the Models

The model would normally be made of a transparent plastic sheet about 0.8 mm to 1.5 mm thick. The material can be Perspex , Plexiglas, Acrylic or other easily workable material. It is better to avoid materials that have internal stresses (since highly stressed regions may go into yielding during the application of displacements).

This is only an indirect model analysis. So a perfect 1 : 1 correspondence is not necessary. We can choose a convenient scale for the centre line dimensions. For the cross-section we need not use the same scale. Actually we must choose another convenient scale for moment of inertia. When we simulate an RC section we can scale down the equivalent EI value.

It is very difficult to vary the thickness of the model in this method. So we vary only the depth of the section to give us a scaled down M.I.

Example 3.7 shows how this is done in a specific case.

Example 3.7 *Figure 3.23 shows the profile of a R.C. rigid frame bridge. It has been designed by limit state method and checked by working stress method. Since this is going to be a type design for many bridges it is proposed to check its performance by an indirect model analysis.*

Design a suitable model to be tested on a drawing board of size 600 mm × 420 mm. The model is to be made of clear annealed Perspex sheet 1.25 mm thick. The width of the member in the model is limited to 18 mm. Taking the modular ratio as 12, the equivalent moment of inertia of 1 m width of the RC sections are tabulated below. The centre line locations are also indicated in the Table 3.5.

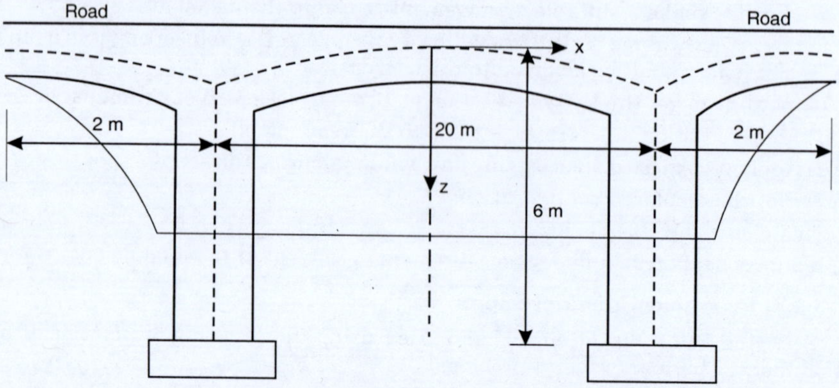

Fig. 3.23. Bridge dimensions.

Table 3.5. Bridge dimensions.

Co-ordinates of centre line		***Section dimensions***	
x (m)	*z (m)*	*Depth cm*	*I × 10^{-6} cm^4*
0	0	45	1.06
2	0.009	46.8	1.10
4	0.036	52.2	1.50
6	0.081	61.2	2.50
8	0.144	73.8	4.80
10	0.225	90.0	8.20
10	0.90	60.0 (Hor)	2.40
10	2.40	60.0	2.40
10	4.80	60.0	2.40
10	6.00	60.0	2.40
11	0.182	81.4	6.00
12	0.144	73.8	4.80

Solution The length of the bridge is 24 m

Maximum height = 6 m

Let us make the model 48 cm long. That makes the linear scale as $\frac{2400 \text{ cm}}{48 \text{ cm}} = 50$

Vertical dimension of the model = $\frac{600}{50} = 12$ cm

The x-z table has to be scaled down by 50 and retabulated (Table 3.6). The model will have a uniform thickness of 1.25 mm. For a depth of 18 mm, the M.I. in cm^4 is

$$I_m = \frac{1}{12}\, 0.125 \times 1.8^3 = 0.06075 \text{ cm}^4$$

The rigid frame bridge is a slab of appropriate width. Taking a 1 m width of the slab, its maximum depth is 90 cm. Its Moment of Inertia $I_0 = 8.2 \times 10^6$.

This makes the scale for the deepest section as

$$\text{Scale (MI)} = \frac{8.2 \times 10^6}{0.06075} = 135 \times 10^6$$

Our scale is thus already chosen when we take 18 mm width (w) to represent a MI of 8.2×10^6. For other sections we will choose a with w such that

$$\left(\frac{w}{18}\right)^3 = \frac{I}{8.2 \times 10^6}$$

$$w = \left(\frac{I}{8.2 \times 10^6}\right)^{1/3} \times 18$$

The chosen widths rounded to 0.01 mm are tabulated below. Figure 3.24 shows the model dimensions.

Table 3.6. Model dimensions.

Co-ordinates			
x ***(mm)***	z ***(mm)***	$\frac{I}{8.2 \times 10^6}$	$w = 18 \times \left[\frac{I}{8.2 \times 10^6}\right]^{1/3}$ ***mm***
0	0	0.129	9.10
40	0.2	0.142	9.40
80	0.7	0.213	10.76
120	1.6	0.305	12.12
160	2.9	0.585	15.06
200	4.5	1.000	18.00
200	18	0.293	11.96
200	48	0.293	11.96
200	96	0.293	11.96
200	120	0.293	11.96
220	3.6	0.732	16.22
240	2.9	0.585	15.06

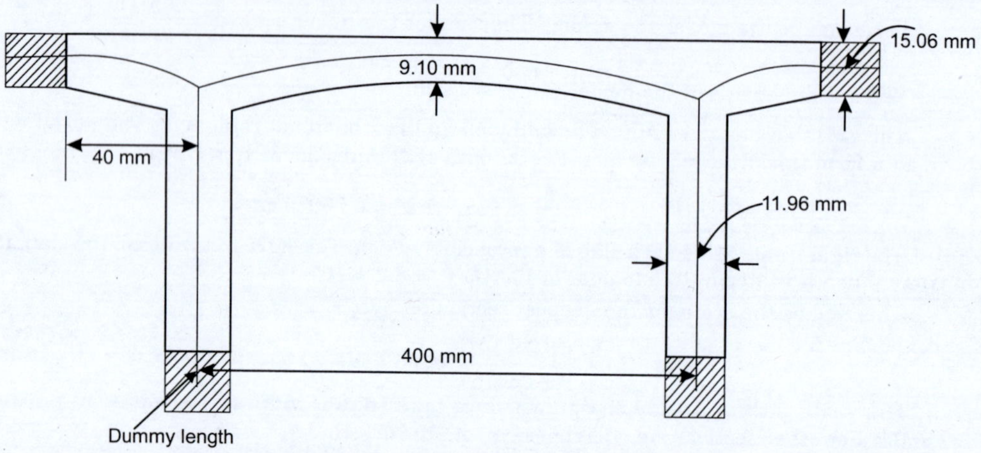

Fig. 3.24. Model dimensions.

At the extremities of the model an additional length of about 10 to 12 mm shall be left to enable any desired connection to be made with the gauges. (described in the next section).

3.5.5 The Beggs' Deformeter

Introduced by professor G.E. Beggs of Princeton University in 1922, Beggs' Deformeter addresses all the minute experimental considerations in applying M-B Principle for model analysis.

Figure 3.25 (*b*) shows an experimental set-up using Begg's Deformeter. Figure 3.25 (*c*) shows a single Beggs' Deformeter gauge.

The gauge is made up of 2 metal bars held together by a pair of spring loaded screws. The bars can be separated by a precise distance with the aid of several pairs of plugs (Fig. 3.26). Of the 2 bars, one is called the fixed bar. This has to be fixed to the drawing board with a pair of wood screws. To the other bar, the model, suitably designed and shaped to simulate any given structure, is attached. Three types of connections with the model are possible. (Fig. 3.27).

1. Hinged connection, in which the model is pivoted to the moving bar. A hole in the model engages into a pin on the moving bar.

2. Fixed connection in which the model is clamped to the moving bar using a fixing plate and 4 screws.

3. Floating connection in which neither of the beams is fixed to the drawing board but is kept afloat on a plate of glass, supported on steel balls which rest on another plate of glass. Here the model is attached to both the beams of the gauge using serrated metal strips about 3 mm wide and 25 mm long. To complete the floating gauge connection, the model has to be cut in the region, between the 2 bars of the gauge while normal plugs are in position.

(*a*) **Calibration of plugs**

For determining how much displacement is effected by each pair of plugs, normally a cantilever arrangement is employed. The arrangement is shown in Fig. 3.28 (*a*) to (*e*). The cantilever can be 100 mm long from the face of fixity to the target point.

The normal position is when both the slots in the gauge are fitted with normal plugs. A micrometer microscope is positioned over the target. This instrument is capable of measuring movements in the x and y co-ordinate directions correct to 0.004 mm. The initial readings are noted with normal plugs in position.

For x displacements we use the 2 pairs of thrust plugs. First remove the normal plugs, and introduce two large thrust plugs. The target points would move in x negative direction. The FM microscope would measure the x movement in the microscope units. (It is not even necessary to convert this into microns or millimetres since the units would cancel off.) Next we introduce two small thrust plugs. This would move the target in x-positive direction. The FMM readings would indicate the actual movement due to small thrust plugs. The net movement between the x positive and x negative extremes is the displacement effected by thrust plugs.

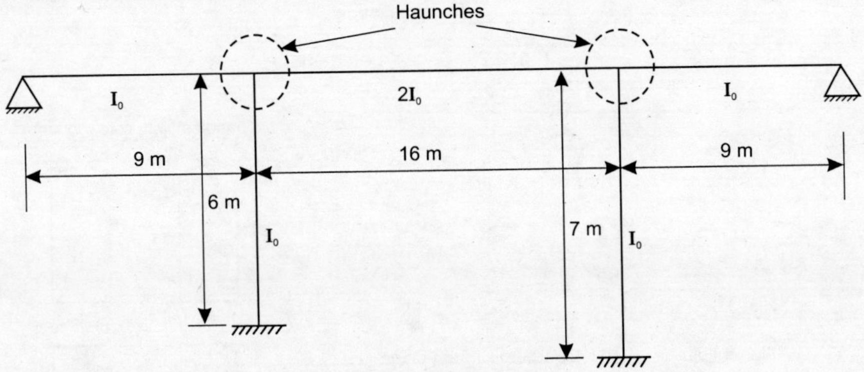

(a) Line diagram of structure

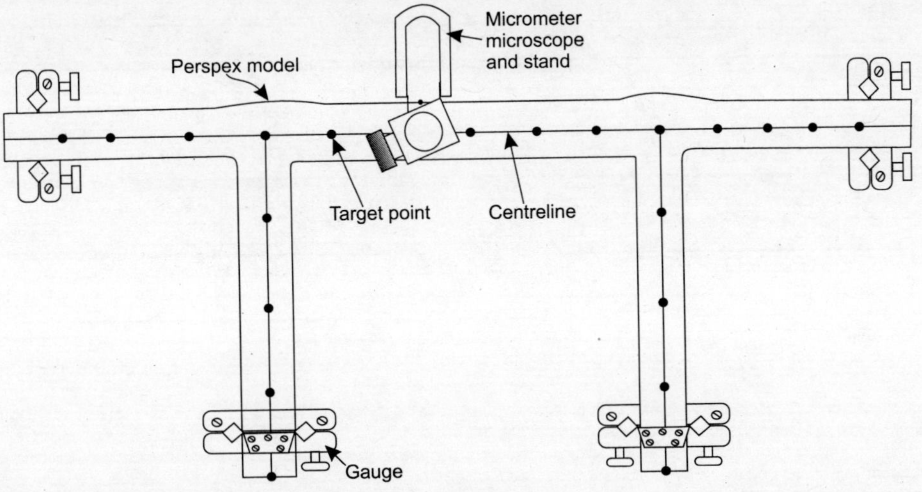

(b) Experimental set-up

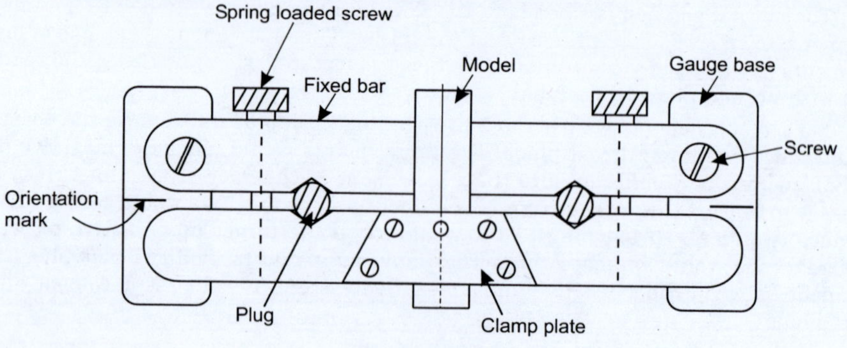

(c) Gauge

Fig. 3.25. Begg's deformeter.

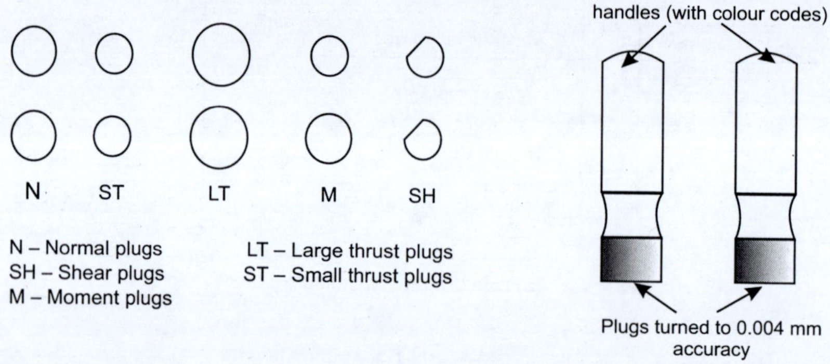

Fig. 3.26. Begg's deformeter plugs.

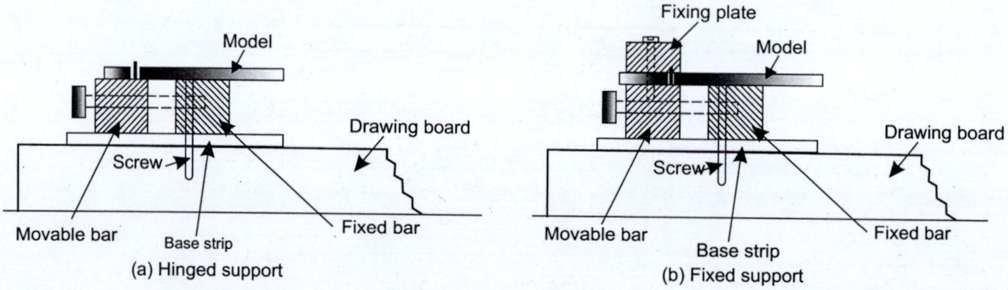

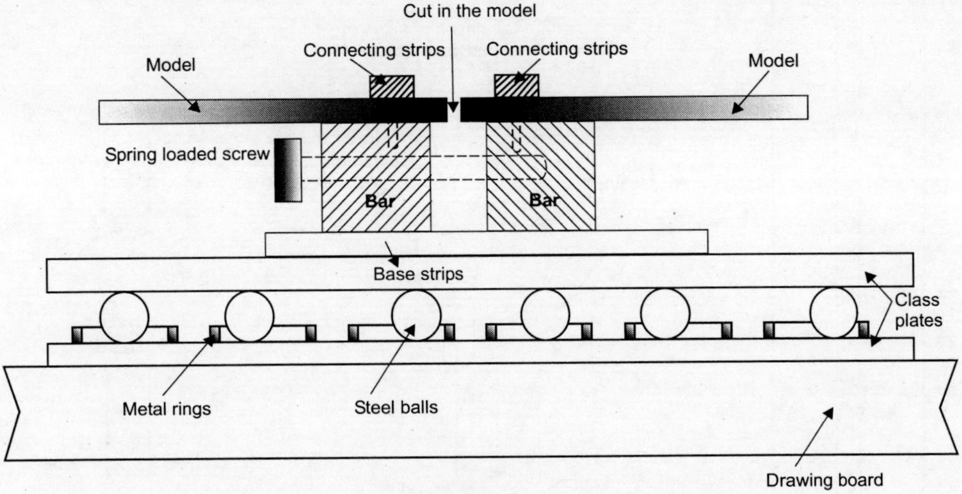

(c) Floating gauge

Fig. 3.27. Three types of connections with Begg's deformeter.

Now we have effectively calibrated the thrust plugs.

Likewise we keep the shear plugs to effect downward movement first (see Fig. 3.28 (*d*)) followed by another position of the pair of shear plugs to effect upward movement. These would be measured by the consequent *y* movement of the target point as observed by the F.M. Microscope. This would calibrate the shear plugs.

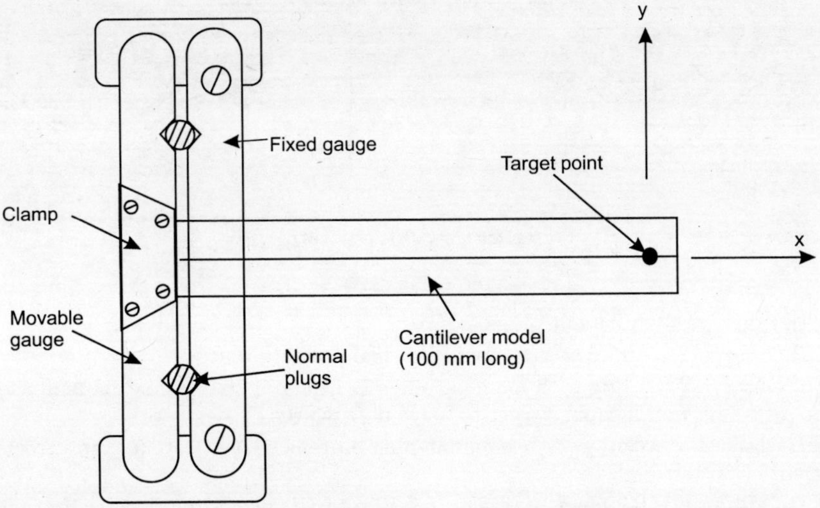

(a) Calibration set-up

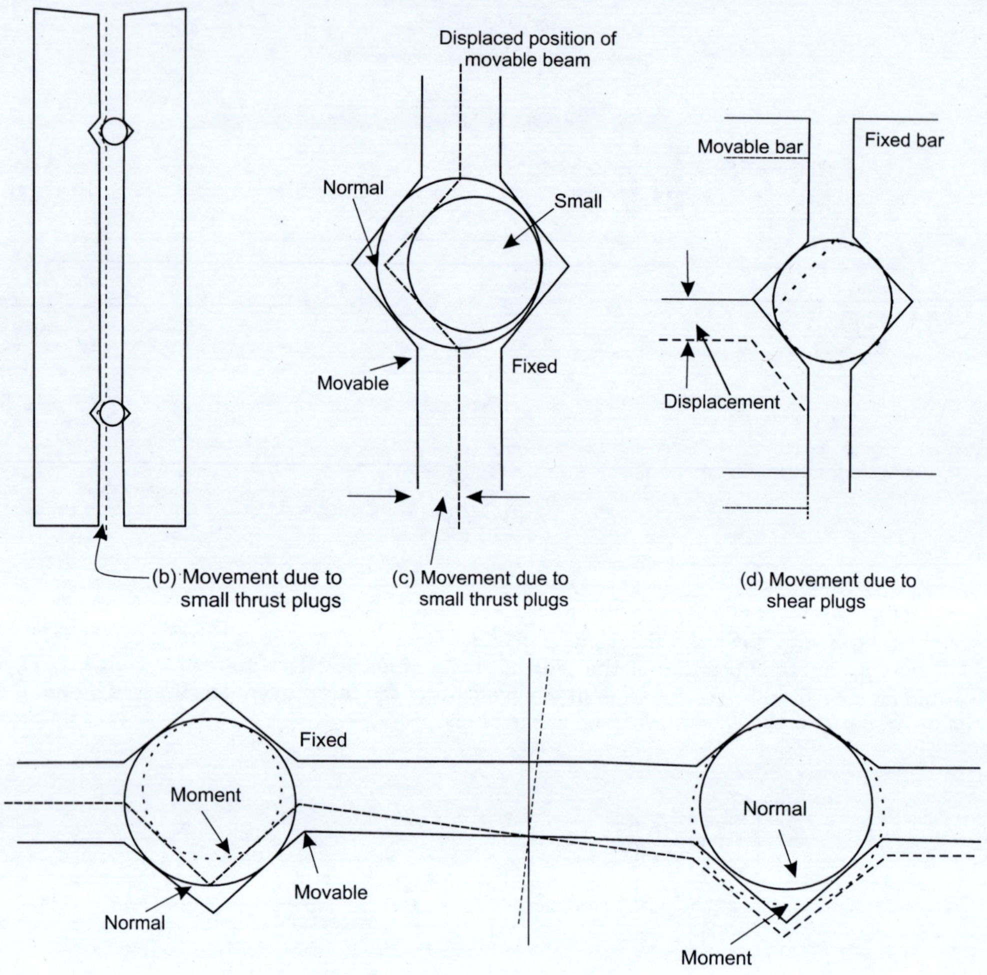

(e) Rotation due to moment plugs

Fig. 3.28

To calibrate the moment plugs (o, O),

(*i*) Observe Y reading of target with normal plugs in place

(*ii*) Observe Y reading with moment plugs as in Fig. 3.28 (*e*) (small plug above and large plug below). This would cause target to move downwards (Y-negative).

(*iii*) Observe Y reading with moment plugs inter charged causing upward (Y-positive) movement of target.

The difference between readings in (*iii*) and (*ii*) above divided by the length of the cantilever would give the calibration value of moment plugs.

(*b*) **Filar micrometer microscope**

This comes with a heavy metallic stand and can be set up above one target point on the model at a time.

Because of the large magnification what we think is a circular target looks like a figure with jagged edges. So special care must be taken in making targets in the form of black filled in circles.

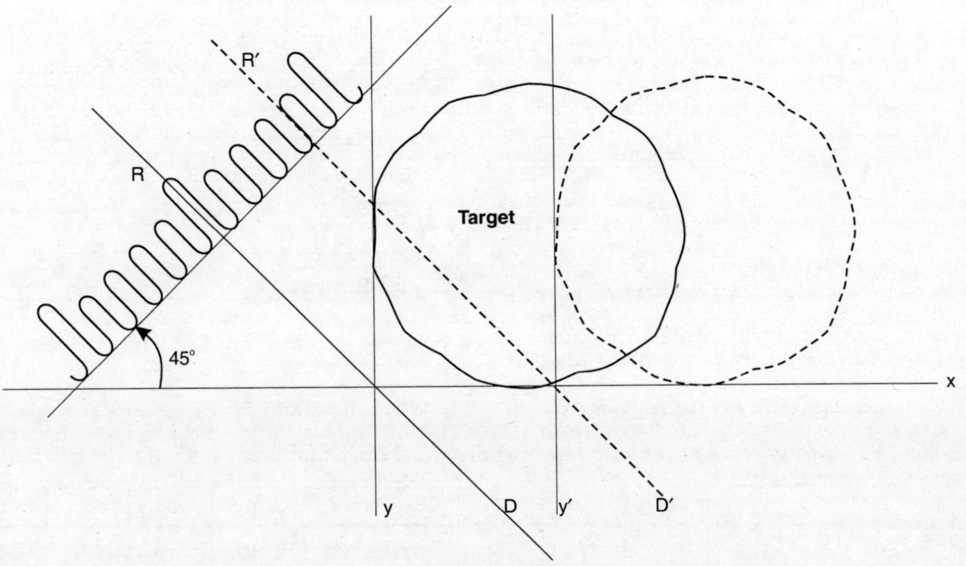

Fig. 3.29

In the field of a f.m. microscope a single diagonal scale serves for both x and y displacements. The diagonal scale is the main scale which is served by an outside drum scale.

The 3 intersecting lines x, y and D can be bodily moved in the field of view of the microscope.

If we want the x movement of the target we first make y line tangential to the target and take the main scale reading on D-line. After the target has shifted, we again bring y line tangential to the new target position shown dotted. This line is marked as y'. Now the D line has moved along the diagonal scale to D′. R′ is the new main scale reading. We can use the drum to bring the D′ line to the nearest whole main scale reading to get the fraction of the distance from the whole main scale reading.

Recent trends in f.m. microscope is to adopt digital indicators in which the displacement of the target can be read off a monitor attached to the f.m.m.

Example 3.8 *An indirect model analysis is made on a properly designed 0.9 mm thick Perspex model of a fixed arch. A floating gauge is attached at the crown and readings are taken with (i) small thrust plugs, (ii) normal plugs, and (iii) large thrust plugs at the crown. Readings on targets fixed on the left half alone are given in the table below. The targets are at equal x distances apart.*

From the readings, sketch the influence line for thrust at the crown.

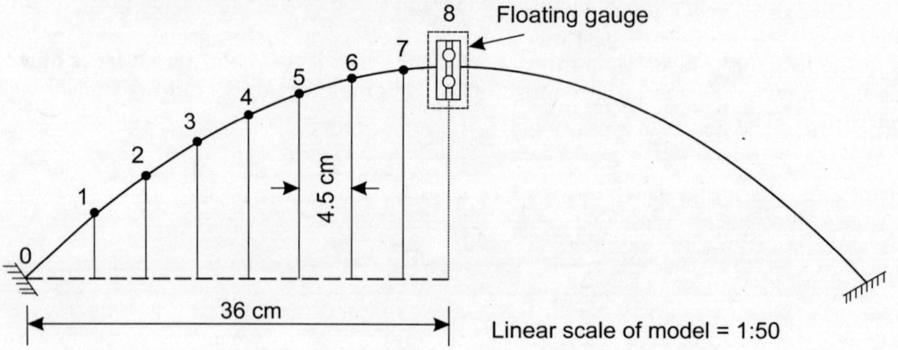

Fig. 3.30

Calibration :

X reading with small thrust plug (a)	*– 840*
X reading with normal plug (b)	*– 2*
X reading with large thrust plug (c)	*+ 841*
displacement for thrust plugs (c – a)	*1681 divisions*

Y. readings on the arch (all readings negative) are as under:

	0	1	2	3	4	5	6	7	8
Small thrust plug at crown	—	172	241	632	711	801	216	177	—
Normal plugs at crown	—	182	265	672	771	893	333	320	—
Large thrust plug at crown	—	192	291	711	829	986	451	463	—

Solution The crown has been given a tension displacement (small thrust plugs first and large thrust plugs last. This means an increase in gap at the crown). Hence the influence line ordinates must be considered as positive if the target movement is upwards and negative if downwards.

The normal thrust readings shall be for verification only. They must be at the median between the 2 thrust plug readings.

The applied displacement at the crown is = 1681 divisions.

Actually we need apply a unit displacement at the crown. Hence we treat 1681 as unity. In other words we divide all displacements (in divisions) by 1681 to get the ILOs. The table below gives the measured displacements at target points 1 to 7. The next line gives the ILO (values in the first line divided by 1681).

Point	1	2	3	4	5	6	7	8
Displacement (div)	– 20	– 50	– 79	– 118	– 185	– 235	– 286	– 302
ILO	– 0.012	– 0.030	– 0.047	– 0.070	– 0.110	– 0.140	– 0.170	– 0.180

The influence line for thrust at the crown is plotted in Fig. 3.31 for the left half of the arch. (The ordinate at 8 is written by extrapolation).

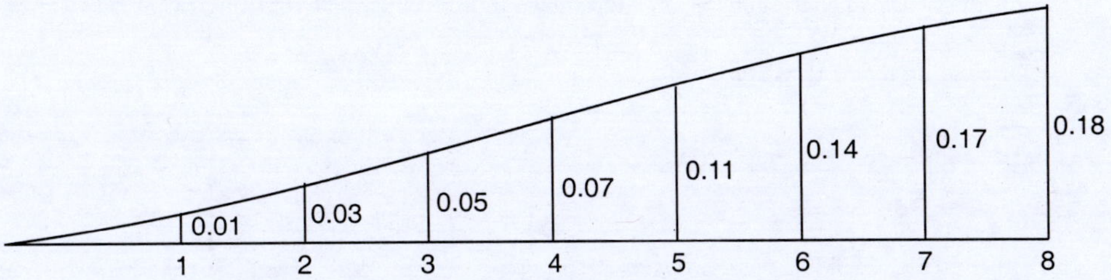

Fig. 3.31. Influence line for H using indirect model test.

PROBLEMS

1. Using Muller Breslau Principle, compute the influence line ordinates for (*i*) Reaction at B and (*ii*) Moment at A for the propped cantilever shown in Fig. 3.32 for a point 6.25 m from A.

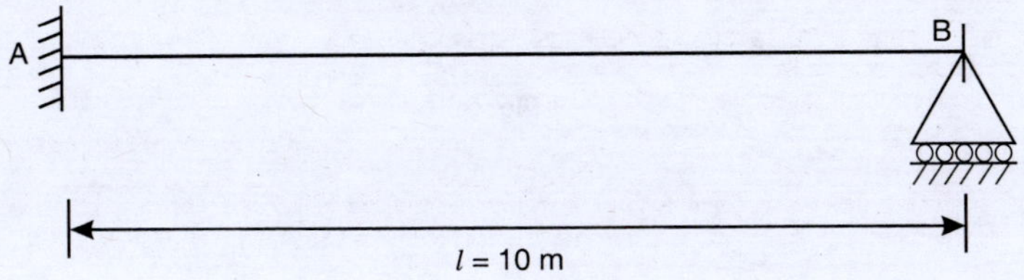

Fig. 3.32

[**Ans.** ILO for R_B = 0.0464 ; ILO for M_A = – 1.61]

2. Using Muller Breslau principle, compute the ILO for reaction at A of the continuous beam in Fig. 3.33 at point D and E shown.

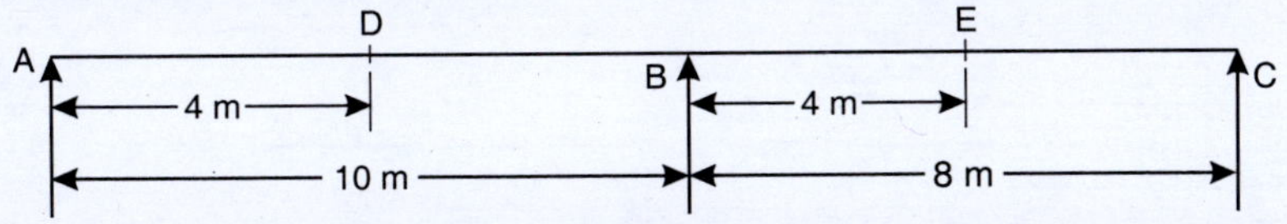

Fig. 3.33

[**Ans.** O_D = 0.51 ; O_E = – 0.07]

3. For the beam in Fig. 3.33, find the ILO for b.m. at E at D and E. [**Ans.** O_D = – 0.47 ; O_E = 1.67]

4. For the beam in Fig. 3.34, determine the ILO at E for reaction at C.

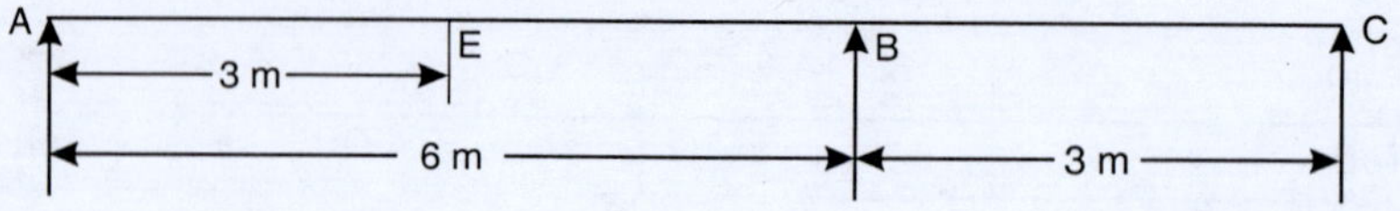

Fig. 3.34 [**Ans.** $O_E = -0.248$]

5. For the two span continuous beam ABC shown in Fig. 3.35, prove that the ILO at D $\left(\dfrac{l}{\sqrt{3}} \text{ from C}\right)$ for bending moment at B is $-\dfrac{l}{6\sqrt{3}}$.

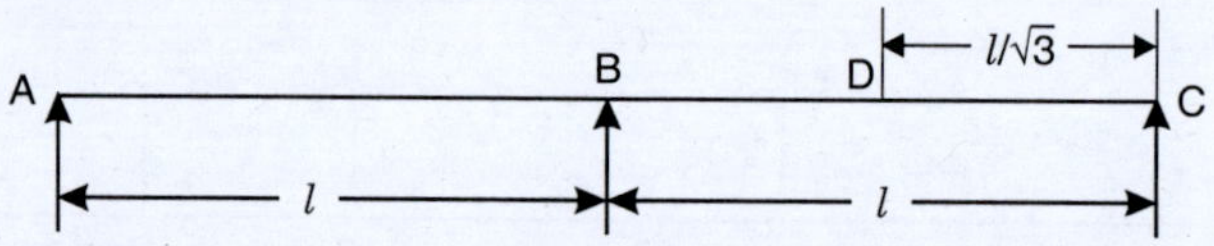

Fig. 3.35

6. Using Muller Breslau Principle develop the IL for B.M. at section D of the continuous beam shown in Fig. 3.36. Compute and plot the ordinate at 1 m intervals.

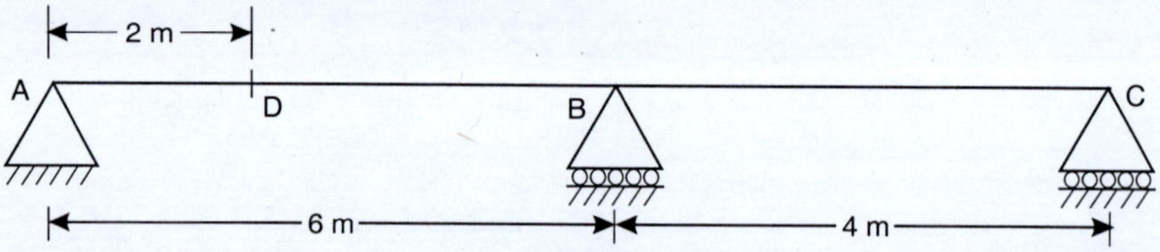

Fig. 3.36

Ans.

Distance in m (from A)	0	1	2	3	4	5	6	7	8	9	10
IL ordinate for M_D	0	0.569	1.156	0.742	0.444	0.180	0	– 0.088	– 0.100	– 0.063	0

7. A continuous beam ABC is shown in Fig. 3.37. Using Muller Breslau principle develop the ILD for the reaction at B. Compute and sketch the ordinates at every quarter of each span.

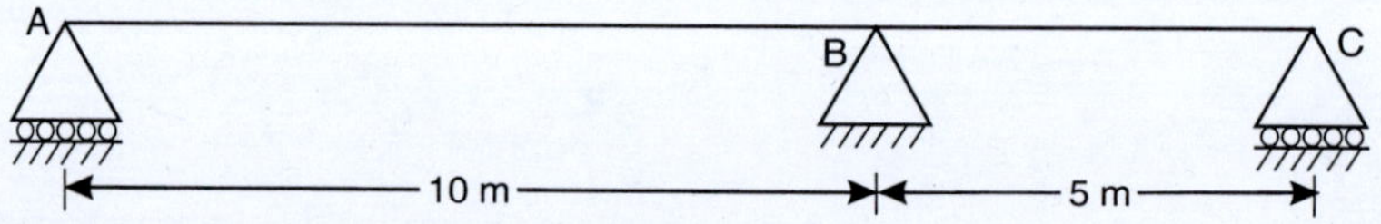

Fig. 3.37

Ans.

Ordinate	0_0	0_1	0_2	0_3	0_4	0_5	0_6	0_7	0_8
R_B	0.00	0.4844	0.8750	1.0781	1.00	0.832	0.5938	0.3086	0

CHAPTER 4
ARCHES

4.1 INTRODUCTION

Before the advent of steel and concrete in construction, architects had to make do with brick and stone arches to span distances. The ancient architects did it so well that they could span distances from 1 m to nearly 30 m with stone arches. In the bargain, the structures were also made aesthetically pleasing. Some of them have out lasted their eras and are preserved today as monuments.

Arches are shaped to take the load above them and develop only compression. We will see in this chapter that this is not strictly true and arches do develop bending moments and shear too. But they are predominantly in compression and this is a major advantage since bending is an inefficient mode of material use.

Arches can be :

1. Polygonal
2. Circular
3. Parabolic
4. Elliptical and
5. Of any other curved shape.

Arches can be built of masonry, reinforced concrete or steel. The era of masonry arches is gone for ever. R.C. and steel arches are still common on account of their architectural appearance and efficient use of material.

4.2 HINGES

To make analysis simple, it has been the practice to introduce structural hinges in arches. Thus there are 3 hinged arches with hinges introduced at springings and at the crown, and 2 hinged arches with hinges at the springings alone. This is applicable to both steel and RC arches.

Hinges, while they save the computational effort, make the arches structurally less efficient. The hinges are also difficult to fabricate and maintain. The justification for hinges at the springings is that the supports (piers or abutments) are kept free of any moments transmitted from the arches. Hinge at the crown is simply a case of "save the calculations and spoil the structure". Hinges are also known as articulations, since they allow the arches to swing a little to the left or right.

4.3 TRANSFER OF LOAD TO ARCHES

The curved shape of the arch makes it both beautiful and inconvenient. When an arch spans over a length AB (Fig. 4.1) of a bridge where do we have the road deck ? Not along the curve ACB! We have to make special efforts to support the deck by suitably filling the spandrel (spandrel filled arch) or bracing the spandrel (spandrel braced arch). The deck may also be supported at the level of the springings A and B, by suspending a stringer beam along AB from a series of suspenders from the arch to the stringer. Such an arch would be called (quite appropriately), a bow-string girder any case the deck load and the vehicular load on the deck get transferred to the arch. In the succeeding sections we will analyse the internal force components (thrust, b.m. and s.f.) developed in arches due to such loading. For analysis purposes we will pretend as if the loads come directly to the arch.

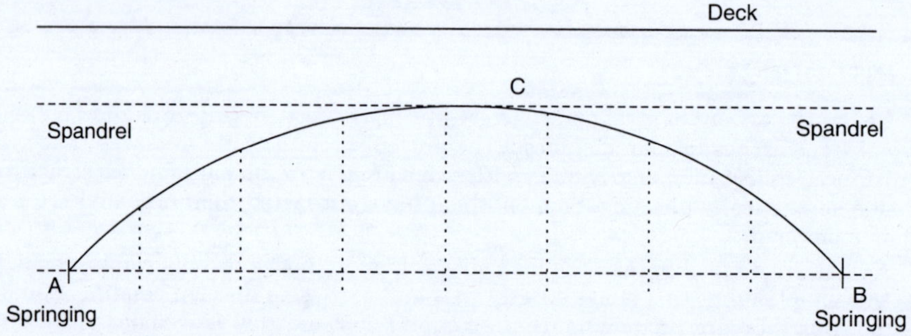

Fig. 4.1

4.4 LINEAR ARCH

Linear arch is a concept.

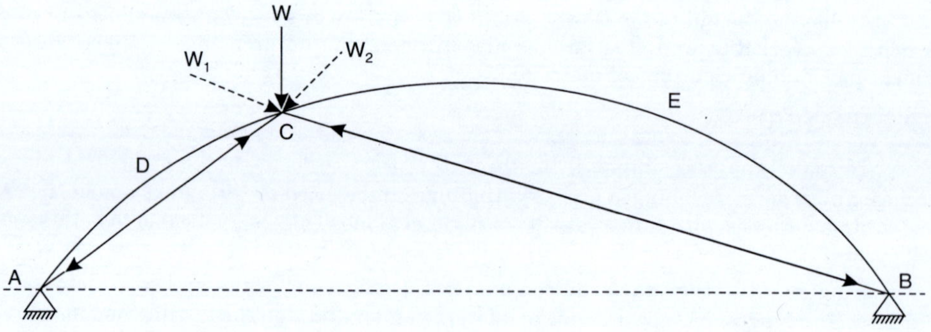

Fig. 4.2

Suppose we have a load W acting at a point C on an arch. If there were straight members CA and CB, the load W would be directly transmitted to the supports A and B in the form of thrusts along the notional members CA and CB. The arch made up of straight lines AC and CB is called the linear arch. If we had to deal only with a load at C, we could have constructed this linear arch itself and not bothered about the curve ACB. If W were a moving load, (and sure, it would be) we may have to have as many linear arches as there are load positions! Not a practical idea. Analytically, a linear arch is, in fact, a line of thrust.

In Fig. 4.2, let us resolve load W into W_1 (along CB) and W_2 (along CA). The thrust along BC takes on W_1. And the thrust along AC takes W_2.

The supports A and B will of course develop reactions along AC and BC respectively.

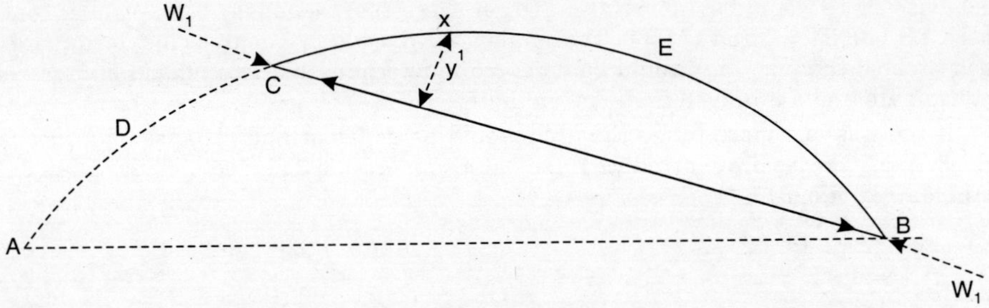

Fig. 4.3

Figure 4.3 is then the free body of the part CEB of the arch. Remember, there is no member along CB. The forces $W_1 - W_1$ will produce bending moments along the arch rib CEB. At any section X on the arch the bending moment is $W_1 \cdot y^1$. Thus the arch centre line itself becomes the bending moment diagram for the arch.

The same inference should be valid for the left portion A DC of the arch also. We will employ this idea later to evolve a theorem known as Eddy's theorem.

4.5 HINGES IN AN ARCH

Arches can have hinges introduced in them mainly to make its analysis manageable. Figures 4.4 (*a*) to (*e*) show different hinge positions in arches.

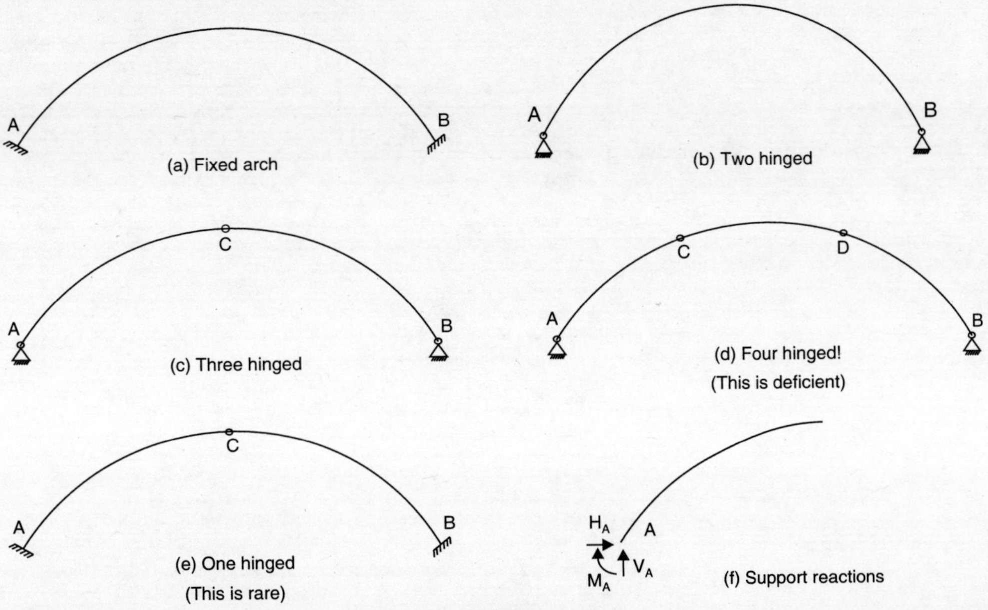

Fig. 4.4

The fixed arch is common in masonry arches. Its construction is hassle free. Its analysis is more complex since it is 3 degree indeterminate. Figures 4.4 (*f*) shows 3 support reactions at A. Likewise there would be 3 more at B. To get these six reactions, we get 3 static conditions namely, $\Sigma H = 0$, $\Sigma V = 0$ and $\Sigma M = 0$. That makes the fixed arch 3 degrees indeterminate. A one hinged arch (if there is one) would be 2 degree indeterminate a two hinged arch one degree indeterminate and a 3 hinged arch determinate.

If there is one more hinge the arch would be deficient and collapse as a mechanism. (Fig. 4.5 (*a*)). Likewise if we introduce rollers at a support in a 3 hinged arch, it will be deficient (unstable) (Fig. 4.5 (*b*)).

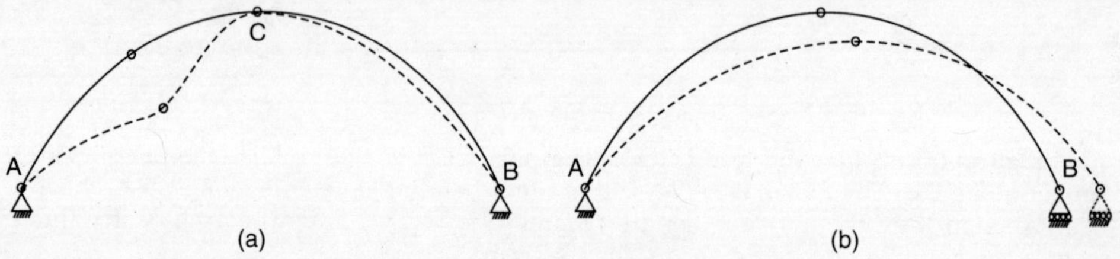

Fig. 4.5. Deficient arches.

4.6 ARCH ACTION

Arch action is a commonly encountered phenomenon even when there are no visible arches. In a human foot the weight from the leg is transmitted by arch action, partly to the heel

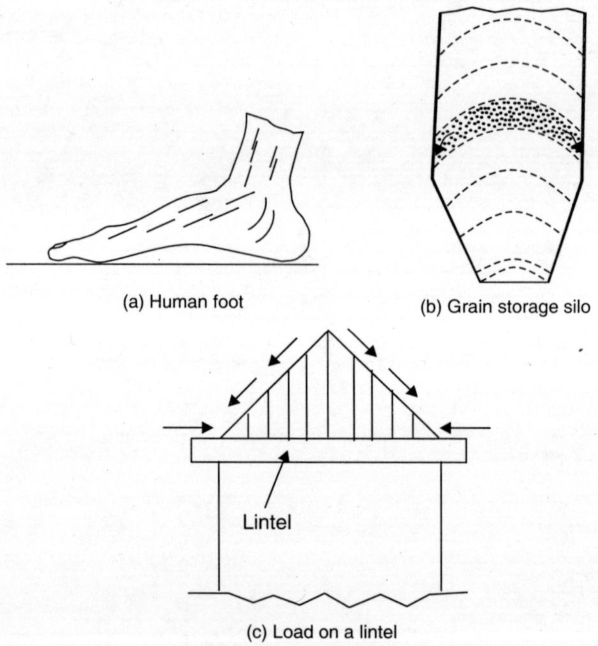

Fig. 4.6. Arch action.

and partly to the ball of the big toe (Fig. 4.6 (*a*)). In a grain storage silo, when the bottom shutter is opened the grain may not flow by gravity. By arch action (Fig. 4.6 (*b*)) the gain mass supports itself against the wall of the silo. The walls are seldom designed for fluid pressure due to the grains, but for the arch pressure from the grains, as they push against the wall of the silo. A typical arch layer is shown in Fig. 4.6 (*b*).

Likewise, the load coming on a lintel is not the load of masonry all the way up above the lintel but only the load from the triangular portion of masonry shown in Fig. 4.6 (*c*). The load from above this triangle gets transmitted to the edges of the lintel by arch action. This would imply there is a horizontal inward force on the notional arch, limited to the friction at the edges of the lintel.

4.7 HORIZONTAL FORCE

It appears that the horizontal inward force is important for arch action to set in.

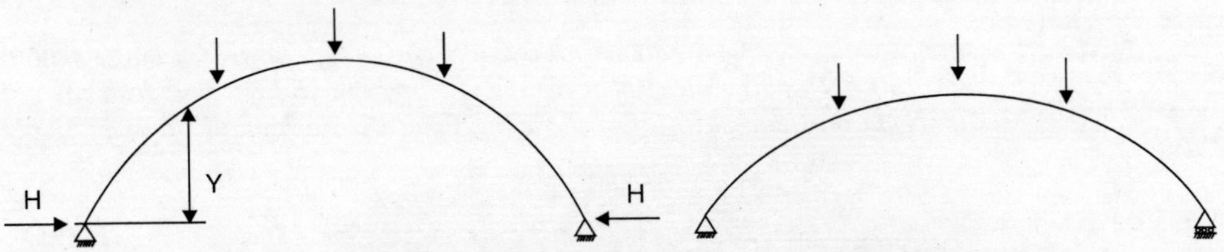

(a) H provides arch action (b) No H. No arch action

Fig. 4.7

In Fig. 4.7 (*a*) is an arch; (*b*) is not; although both look alike. The external load system in each will produce a bending moment (sagging) pattern just as in a beam of the same span. However in Fig. 4.7 (*a*), at every point on the arch there is a hogging bending moment of the value H.y. This will greatly reduce the net bending moment at every section in the arch. In addition there will be an axial thrust component due to H.

This, qualitatively describes arch action. This benefit is not available in the "arch" in Fig. 4.7 (*b*), since H is absent. For arches founded on yielding foundation or over weak masonry columns or walls where the springings are not firm enough to offer a horizontal force, arch action cannot be depended upon, and unless the arch rib is strong enough to take the entire beam moment, it will fail in flexure and collapse.

4.8 THREE HINGED ARCH

A three hinged arch usually has hinges at the two springings and at the crown. It is statically determinate. Figure 4.8 shows a 3 hinged, parabolic arch. Let us investigate the moments, shears and thrusts in the same.

4.8.1 Arch Geometry

Arches can be linear (rare), parabolic, circular, elliptical or rarely catenary in shape. Parabolic arch is the most common.

Let the span of a symmetric parabolic arch be l and its central rise y_C. The equation to the arch (centre line) can be written as

$$y = \frac{4\, y_C}{l^2}\, x(l - x) \qquad \text{...(4.1)}$$

This gives the rise of the arch at any x.

This, of course, has the left springing as the origin. It can be verified that $y = 0$ at $x = 0$ and at $x = l$ and $y = y_C$ at $x = l/2$. Differentiating, we get

$$\frac{dy}{dx} = \frac{4y_C}{l^2}(l - 2x) \quad ...(4.2)$$

This gives the slope of the arch at any x. Obviously slope $\frac{dy}{dx}$ is zero at $x = l/2$; at $x = 0$ slope is $\frac{4y_C}{l}$ and at $x = l$, slope is $-\frac{4y_C}{l}$.

4.8.2 Horizontal Thrust H

Example 4.1 shows how, in a 3 hinged arch, the force H is calculated by equating the bending moment at the central hinge to zero. Example 4.2 shows how the horizontal thrust H reduces the beam bending moment (called μ_x).

Actually in an arch $$M_x = \mu_x - Hy \quad ...(4.3)$$

Example 4.1 *A 3 hinged arch of span 40 m and rise 8 m carries concentrated loads of 200 kN and 150 kN at distances of 8 m and 16 m from the left end and an udl of 50 kN/m on the right half of the span. Find the horizontal thrust.*

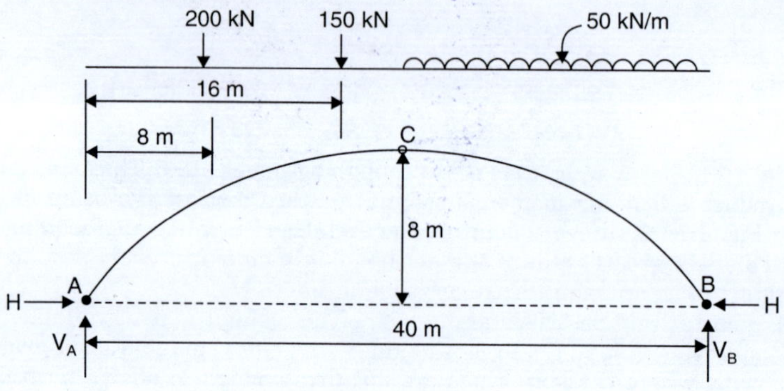

Fig. 4.8

Solution (*a*) **Vertical reactions V_A and V_B**

Taking moments about A,

$$200(8) + 150(16) + 50 \times 20 \times (20 + 20/2) - V_B(40) = 0$$

$$1600 + 2400 + 30000 - 40\,V_B = 0$$

$$V_B = 850 \text{ kN}$$

$$V_A = \text{Total load} - V_B = 200 + 150 + 50 \times 20 - 850$$

$$V_A = 500 \text{ kN}$$

(*b*) **Horizontal thrust (H)**

Taking moments about C,

$$-H \times 8 + V_A(20) - 200(20 - 8) - 150(20 - 16) = 0$$

$$-8H + 500 \times 20 - 200(12) - 150(4) = 0$$

$$H = 875 \text{ kN.}$$

Example 4.2 *A parabolic 3 hinged arch of span 'l' m is subjected to an u.d.l. of w/m run over the entire span. Find the horizontal thrust and bending moment at any section XX.*

Solution (*a*) **Reactions (Vertical) at the supports**

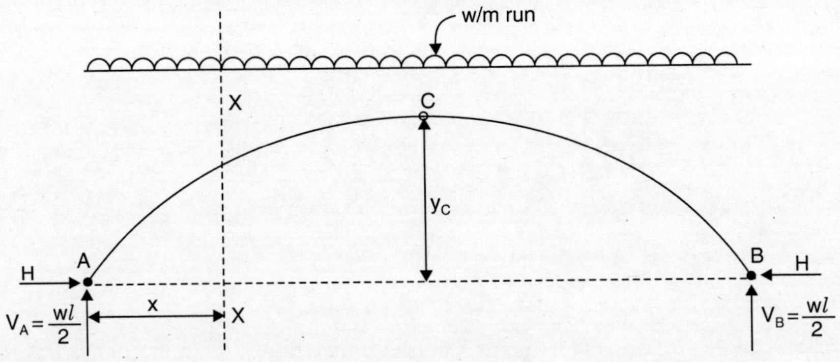

As the loading is symmetrical, vertical reactions at A and B are equal

$$V_A = V_B = \text{Total load}/2 = wl/2$$

(*b*) **Horizontal thrust**

Taking moments about the crown point C,

$$V_A\left(\frac{l}{2}\right) - \frac{w\left(\frac{l}{2}\right)^2}{2} - H(y_C) = 0$$

$$\frac{wl}{2} \times \frac{l}{2} - \frac{wl^2}{4 \times 2} - H(y_C) = 0$$

$$\frac{wl^2}{4} - \frac{wl^2}{8} - H(y_C) = 0$$

$$H(y_C) = \frac{wl^2}{8}$$

$$H = \frac{wl^2}{8y_C}$$

(*c*) **Bending moment at XX**

$$M_x = V_A(x) - \frac{wx^2}{2} - Hy_x$$

Since the arch is parabolic, $\quad y = \frac{4y_C}{l^2} x(l - x)$

$$M_x = \frac{wl}{2} x - \frac{w}{2} x^2 - \frac{wl^2}{8y_C} . \frac{4y_C}{l^2} . x(l - x) = \frac{wlx}{2} - \frac{wx^2}{2} - \frac{wxl}{2} + \frac{wx^2}{2}$$

$M_x = 0$ *i.e.*, the bending moment at any section is zero.

4.8.3 Normal Thrust and Radial Shear

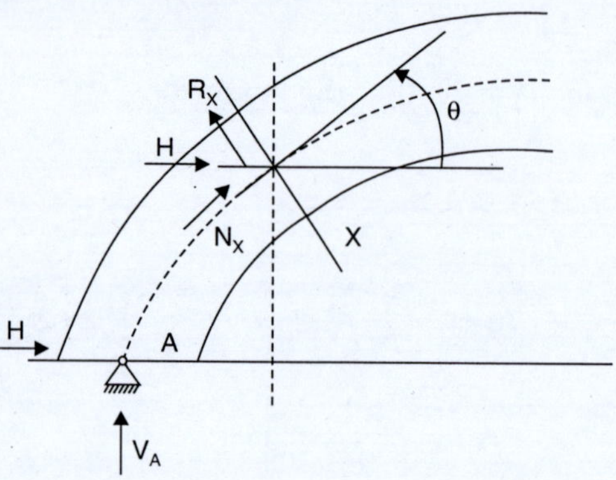

Fig. 4.9

If θ be the inclination of any point X on the centre line of an arch, then the vertical shear V_X and horizontal thrust H would combine to produce a radial shear R_X and a normal thrust N_X at the section.

Then $\quad F_x$ or $R_X = V_X \cos\theta - H \sin\theta$

and $\quad P_x$ or $N_X = V_X \sin\theta + H \cos\theta$ ...(4.4)

N_X and R_X are important for the design of the cross-section of the arch. Examples 4.3 to 4.5 deal with N_X and R_X in a parabolic arch.

Example 4.3 *A symmetrical three hinged parabolic arch of span l and central rise y_C Carries a u.d.l. of w / unit run over the left half of the span. The hinges are provided at the supports and at the centre of the arch. Calculate the location and magnitude of the maximum bending moments (positive and negative).*

Solution (*i*) **Vertical reactions at the supports**

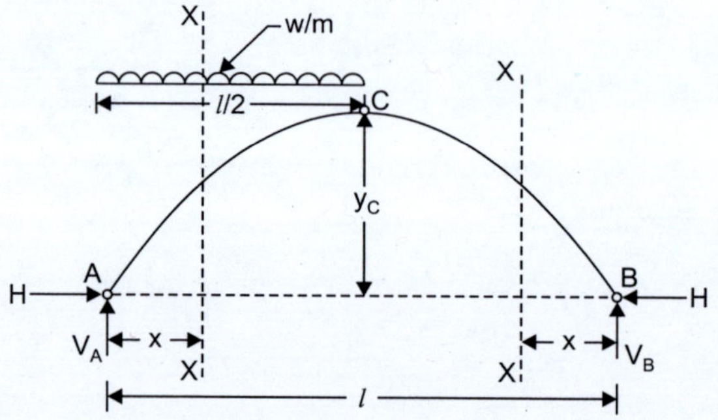

Fig. 4.10

Taking moments about A,

$$V_B \times l - w\,\frac{l}{2}\,\frac{l}{4} = 0$$

$$V_B \times l = \frac{wl^2}{8}$$

$$V_B = \frac{wl}{8}$$

$$V_A = \text{total load} - V_B$$

$$= w\,.\,\frac{l}{2} - \frac{wl}{8} = \frac{4\,wl - wl}{8}$$

$$V_A = \frac{3\,wl}{8}$$

(*ii*) **Horizontal thrust, H**

Taking moments of the forces to the right of crown C, about C,

$$V_B \times \frac{l}{2} - H\,.\,y_C = 0$$

$$-H\,.\,y_C = -V_B\,\frac{l}{2}$$

$$H = \frac{wl}{8} \times \frac{l}{2\,.\,y_C}$$

$$H = \frac{wl^2}{16\,y_C}$$

(*iii*) **Positive bending moment**

Considering a section XX at a distance x from the left hand support

Bending moment, $$M_X = V_A\,.\,x - w\,.\,x\,.\,\frac{x}{2} - H\,.\,y$$

$$= \frac{3}{8}\,wl\,.\,x - \frac{wx^2}{2} - \frac{wl^2}{16\,y_C} \times \frac{4\,y_C}{l^2}\,x(l-x)$$

$$M_X = \frac{3}{8}\,wlx - \frac{wx^2}{2} - \frac{w}{4}\,(lx - x^2) \qquad \text{...(1)}$$

For this moment to be maximum,

$$\frac{dM_x}{dx} = 0$$

$$\frac{dM_x}{dx} = \frac{3}{8}wl - \frac{2\,wx}{2} - \frac{w}{4}(l - 2x) = 0$$

$$\frac{3}{8}wl - wx - \frac{wl}{4} + \frac{w}{4}\,.\,2x = 0$$

$$-wx + \frac{wx}{2} = \frac{wl}{4} - \frac{3}{8}wl = \frac{2\,wl - 3\,wl}{8}$$

$$-\frac{wx}{2} = -\frac{wl}{8}$$

$$x = \frac{l}{4}$$

$\therefore$ Maximum positive B.M. will occur at $x = \frac{l}{4}$ from the left hand support,

Substituting the value of x in eqn. (1)

$$M_{max} = \frac{3}{8}wl \times \frac{l}{4} - \frac{w}{2}\left(\frac{l}{4}\right)^2 - \frac{w}{4}\left(l \cdot \frac{l}{4} - \frac{l^2}{16}\right)$$

$$= \frac{3}{32}wl^2 - \frac{w}{2} \cdot \frac{l^2}{16} - \frac{w}{4} \cdot \frac{l^2}{4} + \frac{w}{4} \cdot \frac{l^2}{16}$$

$$= \frac{wl^2}{64}(6 - 2 - 4 + 1) = \frac{wl^2}{6}$$

Maximum positive bending moment, $M_{max} = \frac{wl^2}{64}$

(*iv*) **Negative bending moment**

Considering a section XX at a distance x from the right support B,

$$M_x = V_B \cdot x - H \cdot y$$

$$= \frac{wl}{8}x - \frac{wl^2}{16\,y_C} \times \frac{4\,y_C}{l^2} \cdot x\,(l - x)$$

$$M_x = \frac{wl}{8} \cdot x - \frac{w}{4}(lx - x^2) \qquad \text{...(2)}$$

For M_x to be maximum,

$$\frac{dM_x}{dx} = 0$$

$$\frac{dM_x}{dx} = \frac{wl}{8} - \frac{w}{4}(l - 2x) = 0$$

$$\frac{wl}{8} - \frac{w}{4} + \frac{2\,wx}{4} = 0$$

$$\frac{wx}{2} = \frac{wl}{8}$$

$$x = \frac{l}{4}$$

Negative B.M. will be maximum at $x = \frac{l}{4}$ from right hand support.

Substituting $x = \frac{l}{4}$ in eqn. (2)

$$M_{max} = \frac{wl}{8} \cdot \frac{l}{4} - \frac{w}{4} \cdot l\frac{l}{4} + \frac{w}{4}\left(\frac{l}{4}\right)^2$$

$$= \frac{wl^2}{32} - \frac{wl^2}{16} + \frac{w}{4} \cdot \frac{l^2}{16} = \frac{wl^2}{64}(2 - 4 + 1) = -\frac{wl^2}{64}$$

Maximum –ve B.M., $M_{max} = -\frac{wl^2}{64}$.

Example 4.4 *A parabolic 3 hinged arch carries loads as shown in Fig. 4.11. Determine the resultant reactions at supports. Find the bending moment, normal thrust and radial shear at D, 5 m from A. What is the maximum bending moment ?*

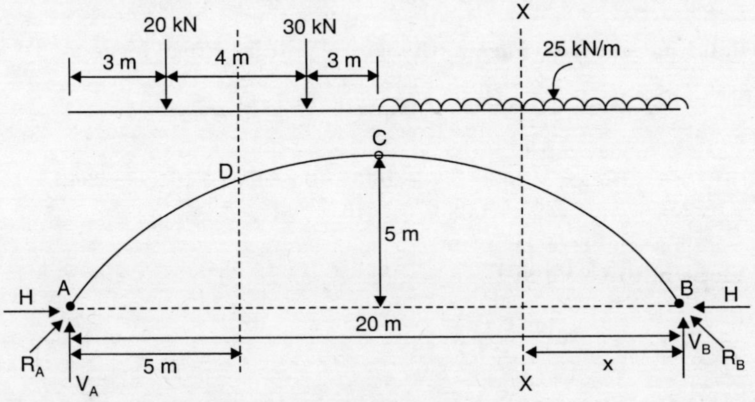

Fig. 4.11

Solution **(*a*) Reactions at supports (R_A and R_B)**

(*i*) **Vertical components of R_A and R_B (V_A and V_B):**

Taking moments about A,

$$20 \times 3 + 30(7) + 25 \times 10 \times (10 + 10/2) - V_B \times (20) = 0$$

$$V_B = 201 \text{ kN}$$

$$V_A = \text{Total load} - V_B = 20 + 30 + 25 \times 10 - 201$$

$$V_A = 99 \text{ kN}$$

(*ii*) **Horizontal thrust (H):**

Taking moments about the crown point C, considering the right side of 'C',

$$-V_B(20/2) + H(5) + 25 \times 10 \times 5 = 0$$

$$-201 \times (20/2) + 5H + 1250 = 0$$

$$H = 152 \text{ kN}$$

(*iii*) **Resultant reactions R_A and R_B:**

$$\theta_A = \tan^{-1}\frac{V_A}{H} = \tan^{-1}\frac{99}{152} = 33° \; 4' \; 36''.6$$

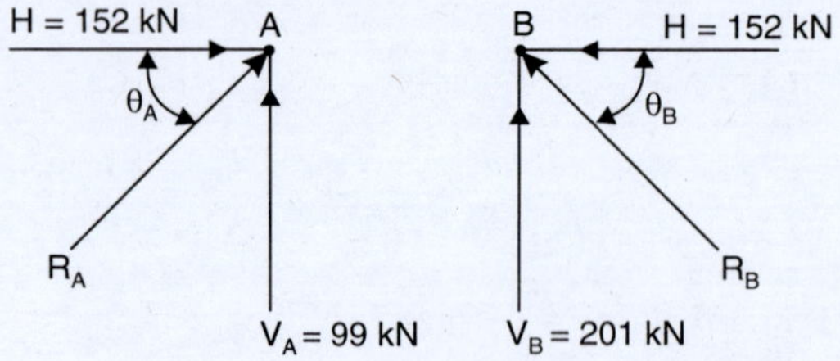

$$R_A = \sqrt{H^2 + V_A{}^2} = \sqrt{(152)^2 + (99)^2} = 181.39 \text{ kN}$$

$$R_B = \sqrt{H^2 + V_B{}^2} = \sqrt{(152)^2 + (201)^2} = 252 \text{ kN}$$

$$\theta_B = \tan^{-1}\frac{V_B}{H} = \tan^{-1}\frac{201}{152} = 52° \; 54' \; 9''.86$$

(*b*) **Bending moment, normal thrust and radial shear force (at D)**

(*i*) **BM at D:** $y_D = \frac{4r}{l^2}x\,(l - x) = \frac{4 \times 5}{(20)^2} \times 5(20 - 5)$

$$y_D = 3.75 \text{ m}$$

$$BM_D = V_A(5) - Hy_D - 20(5 - 3) = 99(5) - 152\, y_D - 20(2)$$

$$= 495 - 152(3.75) - 40$$

$$BM_D = -\,115 \text{ kNm}$$

(*ii*) **Slope of the arch at D:**

$$\tan\theta = \left(\frac{dy}{dx}\right)_D$$

$$\theta = \tan^{-1}\left(\frac{4r}{l^2}(l - 2x)\right)$$

$$\theta = \tan^{-1}\left(\frac{4 \times 5}{(20)^2}(20 - 2 \times 5)\right)$$

$$\theta = 26° \; 33' \; 55''.18$$

(*iii*) **Normal thrust:** $P = V_X \sin\theta + H\cos\theta$

V_X = Net beam shear force at D = $V_A - 20$

$V_X = 99 - 20 = 79$ kN

Substitute in (*iii*) $P = 79\sin\theta + 152\cos\theta = 171.28$ kN

(*iv*) **Radial shear force:** $F = V_X\cos\theta - H\sin\theta$

$F = 79\cos\theta - 152\sin\theta = 2.683$ kN

(*c*) **Maximum bending moment in CB**

Considering a section XX at a distance of '*x*' m from 'B'

$$BM_{XX} = V_B\,(x) - \frac{wx^2}{2} - Hy_x = 201x - \frac{25x^2}{2} - Hy_x$$

$$= 201x - 12.5x^2 - 152\left(\frac{4 \times 5}{20^2} \times x(20 - x)\right)$$

$$= 201x - 12.5\,x^2 - 7.6\,x(20 - x)$$

$$= 201x - 12.5\,x^2 - 152x + 7.6\,x^2$$

$$M_x = 49x - 4.9x^2$$

$$\frac{dM}{dx} = 49 - 9.8\,x$$

For bending moment to be maximum,

$$9.8x - 49 = 0$$

$$x = 5 \text{ m}$$

At $x = 5$ m, $M_x = 49(5) - 4.9\,x^2$

$$M_x = 122.5 \text{ kN.}$$

Example 4.5 *A symmetrical three hinged parabolic arch of span 40 m and rise 8 m carries an u.d.l. of 30 kN/m over the left half of the span. The hinges are provided at the supports and at the centre of the arch. Calculate the reactions at the supports. Also calculate the bending moment, radial shear and normal thrust at a distance of 10 m from the left support.*

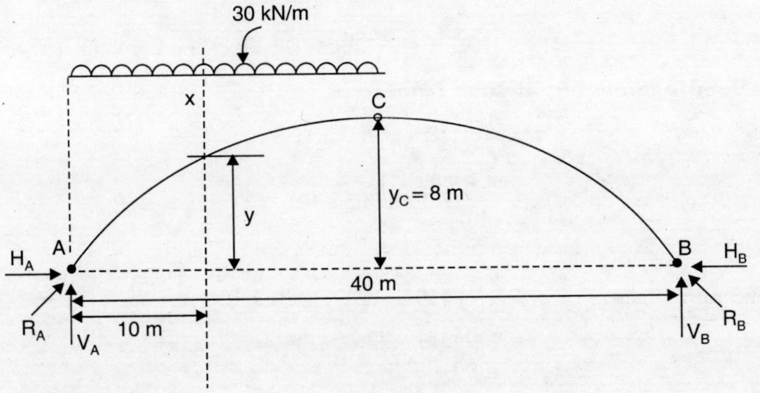

Fig. 4.12

Solution (*a*) **Reactions at the supports**

(*i*) **Vertical components :**

Taking moments about A,

$$30 \times \frac{(20)^2}{2} - V_B \times (40) = 0$$

Vertical component of R_B, $V_B = 150$ kN

$$V_A = \text{Total load} - V_B = 30 \times 20 - 150$$

Vertical component of reaction at A, $V_A = 450$ kN

(*ii*) **Horizontal components :**

Taking moments about the crown 'C',

$$V_A\left(\frac{l}{2}\right) - H_A(y_C) - 30 \times 20 \times \frac{20}{2} = 0$$

$$450(20) - H_A(8) - 30 \times \frac{(20)^2}{2} = 0$$

$$H_A = 375 \text{ kN}$$

Since $\Sigma H = 0\ ;\ H_A - H_B = 0$

$$H_A = H_B = 375 \text{ kN}$$

(*iii*) **Resultants R_A and R_B :**

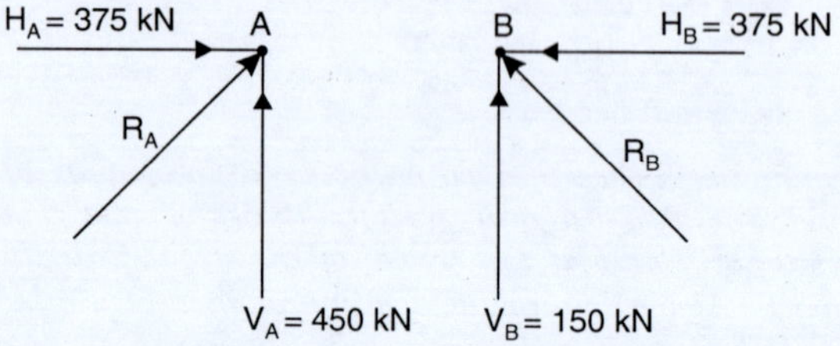

$$R_A = \sqrt{V_A{}^2 + H_A{}^2} = \sqrt{(450)^2 + (375)^2} = 585.77 \text{ kN}$$

$$R_B = \sqrt{V_B{}^2 + H_B{}^2} = \sqrt{(150)^2 + (375)^2} = 403.89 \text{ kN}$$

(*b*) **Bending moment at 10 m from A**

$$y = \frac{4r}{l^2} x(l - x)$$

$$y = \frac{4 \times 8}{(40)^2} \times 10\ (40 - 10)$$

$$y = 6 \text{ m at 10 m from A.}$$

Bending moment $= + V_A\ (10) - H_A(y) - 30 \times 10 \times \frac{10}{2}$

$$= 450\ (10) - 375y - 30(50)$$

$$BM_{XX} = 3000 - 375y$$

$$BM_{XX} = 3000 - 375\ (6)$$

$$\text{BM at 10 m} = 750 \text{ kNm}$$

(*c*) **Radial shear force at x = 10 m**

$$R = \text{Radial shear force} = V_X \cos\theta - H \sin\theta$$

where, V_X = Net vertical shear force at x = 10 m from A

H = Horizontal thrust.

$$\theta = \tan^{-1}\left(\frac{4r}{l^2}(l - 2x)\right)$$

$$\tan\theta = \frac{dy}{dx} = \frac{d}{dx}\left(\frac{4r}{l^2} x\ (l - x)\right)$$

$$\tan\theta = \left(\frac{4r}{l^2}(l - 2x)\right)$$

$$\theta = \tan^{-1}\left(\frac{4 \times 8}{(40)^2}(40 - 2(10))\right)$$

$\theta = 21° 48'$

$V_X = V_A - w \times 10 = 450 - 30 \times 10$

$V_X = 150$ kN

Radial shear force, $R = V_X \cos\theta - H \sin\theta$

$R = 150 \cos 21° 48' - 375 \sin 21° 48'$

$R = 0$

(*d*) **Normal thrust at x = 10 m from 'A'**

Normal thrust, $N = V_X \sin\theta + H\cos\theta = 150 \sin 21° 48' + 375 \cos 21° 48'$

$N = 403.89$ kN.

Example 4.6 *A three hinged parabolic arch of horizontal span 48 m has a central rise of 10 m. It carries a u.d.l. of 20 kN per horizontal metre run over the middle 16 m length of the span. Calculate the radial S.F., normal thrust and bending moment at 20 m from the left support.*

Solution

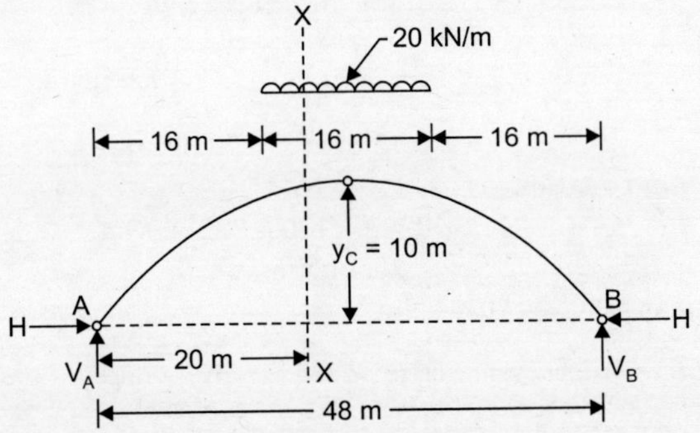

Fig. 4.13

Vertical reactions :

Taking moments about B

$$V_A \times 48 - 20 \times 16 \times \left(16 + \frac{16}{2}\right) = 0$$

$$48\, V_A = 7680$$

$$V_A = 160 \text{ kN}$$

$$V_B = 160 \text{ kN} \qquad \text{(by symmetry)}$$

Horizontal thrust:

Taking moments of forces to the left of C about the crown C,

$$V_A \times 24 - H \times 10 - 20 \times 8 \times 4 = 0$$

$$160 \times 24 - 640 = 10\, H$$

$$H = 320 \text{ kN}$$

Radial shear at $x = 20$ m from A

Radial shear force, $R_X = V_X \cos\theta - H \sin\theta$

where, V_X = Net vertical shear force at x = 20 m from A

$V_X = V_A - 20 \times 4 = 160 - 80 = 80$ kN

H = Horizontal thrust = 320 kN

$$\theta = \tan^{-1}\left[\frac{4r}{l^2}(l - 2x)\right] = \tan^{-1}\left[\frac{4\times10}{48^2}(48 - 2\times20)\right]$$

$$\theta = 7° \; 54' \; 26''$$

$$R_X = 80 \cos 7° \; 54' \; 26'' - 320 \sin 7° \; 54' \; 26''$$

$$R_X = 35.217 \text{ kN}$$

Normal thrust at x = 20 m from A :

Normal thrust, $P_N = V_X \,.\, \sin\theta + H \cos\theta$

$$= 80 \,.\, \sin 7° \; 54' \; 26'' + H \cos 7° \; 54' \; 26''$$

$$P_N = 327.963 \text{ kN}$$

Bending moment at x = 20 m from A :

$$M_X = V_A \times 20 - 20 \times 4 \times 2 - H \,.\, y$$

$$M_X = 160 \times 20 - 160 - 320\, y \qquad ...(1)$$

$$y = \frac{4r}{l^2}\, x\,(l - x) = \frac{4\times10}{48^2} \times 20\,(48 - 20)$$

$$y = 9.72 \text{ m at } x = 20 \text{ m from A}$$

Substituting y in equation (1)

$$M_X = 3200 - 160 - 320 \times 9.72$$

$$= -\,70.4 \text{ kNm}$$

Bending moment at X = – 70.4 kNm.

Example 4.7 *A three hinged symmetric parabolic arch of span l and central rise y_C has its hinges at either springings and at the crown. It is subjected to a uniformly varying load varying from zero at the crown to w / m run over either springings. Obtain the maximum bending moment and its location.*

Solution

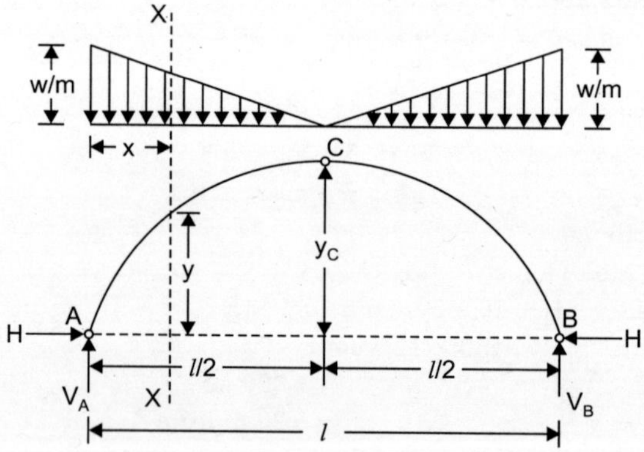

Fig. 4.14

Vertical reactions :

Since the arch and loading are symmetrical

$$V_A = V_B = \frac{\text{Total load}}{2} = \frac{1}{2}\left[2 \times \frac{1}{2} \times \frac{l}{2} \times w\right] = \frac{wl}{4}$$

To find the horizontal thrust, H :

Taking moment of forces to the left of C about the crown C,

$$V_A \times \frac{l}{2} - \left(\frac{1}{2} \times \frac{l}{2} \times w\right) \times \frac{2}{3} \times \frac{l}{2} - H \,.\, y_C = 0$$

$$\frac{wl}{4} \times \frac{l}{2} - \frac{wl^2}{12} = H \,.\, y_C$$

$$H = \frac{1}{y_C}\left(\frac{wl^2}{8} - \frac{wl^2}{12}\right) = \frac{1}{y_C}\left(\frac{3wl^2 - 2wl^2}{24}\right)$$

$$H = \frac{wl^2}{24\, y_C}.$$

Maximum bending moment :

Considering any section XX at a distance x from the left support, the ordinate of the load w' at the section is :

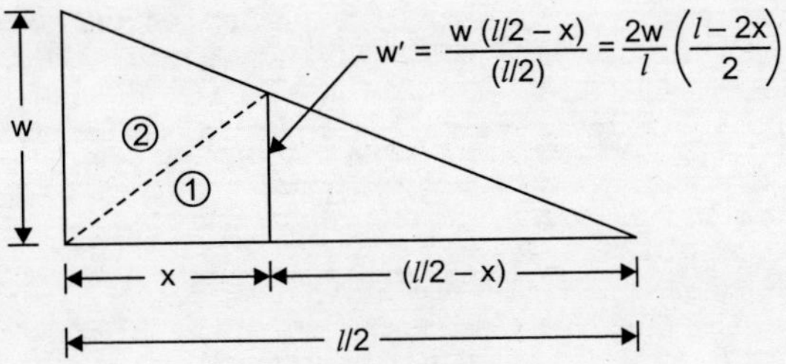

$$w' = \frac{2w}{l}\left(\frac{l - 2x}{2}\right) = \frac{w}{l}(l - 2x)$$

Bending moment at any section XX,

$$M_x = V_A \cdot x - \left(\frac{1}{2} \times x \times w'\right) \times \frac{1}{3}x - \left(\frac{1}{2} \times x \times w\right) \times \frac{2}{3}x - H \,.\, y$$

$$= \frac{wl}{4}x - \frac{1}{2} \times x \times \frac{w}{l}(l - 2x) \times \frac{1}{3}x - \frac{1}{2} \times x \times w \times \frac{2}{3}x$$

$$- \frac{wl^2}{24\, y_C} \times \frac{4y_C}{l^2}\, x(l - x)$$

$$= \frac{wl\,x}{4} - \frac{wx^2}{6l}(l - 2x) - \frac{wx^2}{3} - \frac{wx}{6}(l - x)$$

$$= \frac{wl\,x}{4} - \frac{wx^2 l}{6l} + \frac{2wx^3}{6l} - \frac{wx^2}{3} - \frac{wx\,l}{6} + \frac{wx^2}{6}$$

$$= \frac{wl\,x}{4} - \frac{wl\,x}{6} - \frac{wx^2}{3} + \frac{wx^2}{6} + \frac{wx^3}{3l} - \frac{wx^2}{6}$$

$$= \frac{wl\,x}{12}(3-2) - \frac{wx^2}{3} + \frac{wx^3}{3l}$$

$$M_x = \frac{wl\,x}{12} - \frac{wx^2}{3} + \frac{wx^3}{3l} \qquad \ldots(1)$$

For this bending moment to be maximum $\frac{dM_x}{dx} = 0$

Differentiating (1) w.r.t. x

$$\frac{dM_x}{dx} = \frac{wl}{12} - \frac{2\,wx}{3} + \frac{3\,wx^2}{3l}$$

$$w\left(\frac{l}{12} - \frac{2}{3}x + \frac{x^2}{l}\right) = 0$$

$$\frac{l^2 - 8lx + 12\,x^2}{12\,l} = 0$$

$$12x^2 - 8lx + l^2 = 0$$

$$x = \frac{-(-8l) \pm \sqrt{(-8l)^2 - 4 \times 12 \times l^2}}{2 \times 12}$$

$$= \frac{8l \pm \sqrt{64l^2 - 48l^2}}{24}$$

$$= \frac{8l \pm 4l}{24}$$

$$x = \frac{12l}{24} \quad \text{or} \quad \frac{4l}{24}$$

$$x = \frac{l}{2} \quad \text{or} \quad \frac{l}{6}$$

$x = \frac{l}{2}$ cannot be a point of maximum bending moment because of the hinge at the crown.

$$\therefore \qquad x = \frac{l}{6}$$

Substituting $x = \frac{l}{6}$ in eqn. (1)

Maximum bending moment,

$$M_{max} = \frac{wl}{12} \cdot \frac{l}{6} - \frac{w}{3}\left(\frac{l}{6}\right)^2 + \frac{w}{3l}\left(\frac{l}{6}\right)^3$$

$$= \frac{wl^2}{72} - \frac{w}{3} \times \frac{l^2}{36} + \frac{w}{3l} \times \frac{l^3}{216}$$

$$= \frac{wl^2}{648}[9 - 6 + 1] = \frac{4\,wl^2}{648} = \frac{wl^2}{162}$$

$$M_{max} = \frac{wl^2}{162} \text{ at } x = \frac{l}{6} \text{ from both ends.}$$

4.9 CIRCULAR ARCHES

Circular arches are more pleasing to look at. The important merit of the parabolic arch is that the calculations are easier. Circle geometry is a bit awkward when expressed in Cartesian co-ordinates.

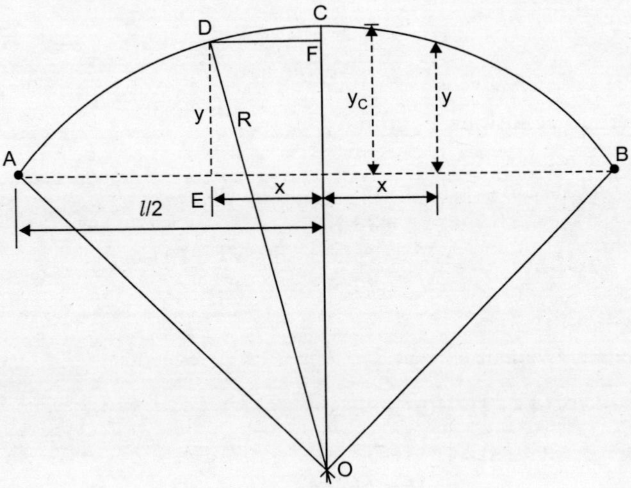

Fig. 4.15

In Fig. 4.15, $OD^2 = DF^2 + OF^2$

$$R^2 = x^2 + (R - y_C + y)^2$$

R can be determined from the relationship

$$(2R - y_C)\, y_C = \left(\frac{l}{2}\right)^2$$

(Actually there would be no difference in the geometry or action of a circular arch and a parabolic arch if the central rise is small compared to R.)

To get the slope of any given point on the arch it is easier to get the deviation of the normal (radius) from the vertical.

Example 4.8 *A 3 hinged arch is circular, 25 m in span with a central rise of 5 m. It is loaded with a concentrated load of 10 kN at 7.5 m from the left hand hinge. Find the*

(a) Horizontal thrust

(b) Reaction at each end hinge

(c) Bending moment under the load.

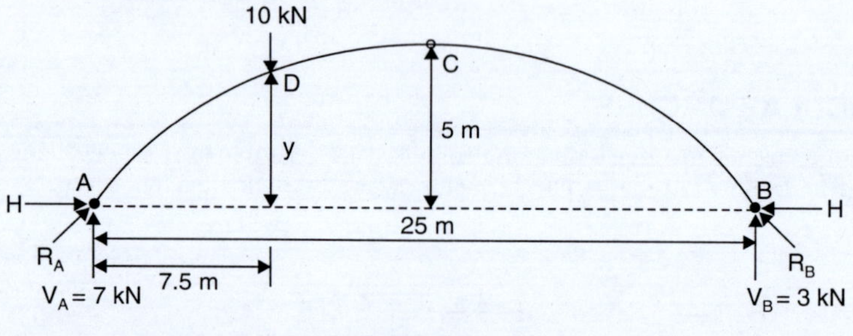

Fig. 4.16

Solution **Vertical reactions V_A and V_B**

Taking moments about A,

$$10(7.5) - V_B(25) = 0$$

$$V_B = 3 \text{ kN}$$

$$V_A = \text{Total load} - V_B = 10 - 3$$

$$V_A = 7 \text{ kN}$$

(*a*) **Horizontal thrust (H) :**

Taking moments about C,

$$V_B\left(\frac{25}{2}\right) - H(5) = 0$$

$$3(12.5) - H(5) = 0$$

$$H = 7.5 \text{ kN.}$$

(*b*) **Reactions R_A and R_B :**

$$R_A = \sqrt{V_A^2 + H^2} = \sqrt{7^2 + 7.5^2} = 10.26 \text{ kN}$$

$$R_B = \sqrt{V_B^2 + H^2} = \sqrt{3^2 + 7.5^2} = 8.08 \text{ kN}$$

(*c*) **Bending moment under the load :**

$$(BM)_D = V_A(7.50) - H.y = 7(7.5) - \mathbf{7.5y} \quad ...(1)$$

y = Rise of the arch at D.

To find y :

$$\frac{l}{2} \times \frac{l}{2} = 5(2R - 5)$$

$$12.5 \times 12.5 = 5(2R - 5)$$

$$R = 18.125 \text{ m}$$

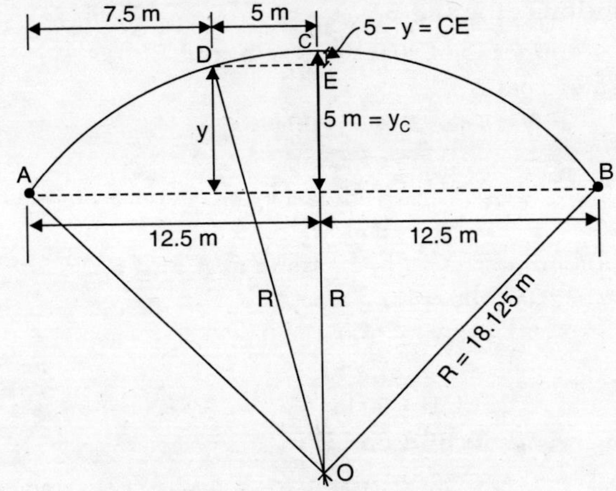

Fig. 4.17

In the above figure, ODE is a right angled triangle

$$OD^2 = OE^2 + ED^2 = (OC - CE)^2 + ED^2$$

$$R^2 = (R - (5 - y))^2 + 5^2$$

$$(18.125)^2 = (18.125 - 5 + y)^2 + 25$$

$$303.515 = (13.125 + y)^2$$

$$\sqrt{303.515} = 13.125 + y$$

$$17.421 = 13.125 + y$$

$$y = 4.3 \text{ m ; substitute in (4.9)}$$

$$(BM)_D = 7(7.5) - 7.5\,(4.3)$$

$$(BM)_D = 20.25 \text{ kNm.}$$

Example 4.9 *A three hinged circular arch of span 16 m and rise 4 m is subjected to two point loads of 100 kN and 80 kN at the left and right quarter span points respectively. Find the reactions at supports. Find also the bending moment, radial shear and normal thrust at 6 m from left support.*

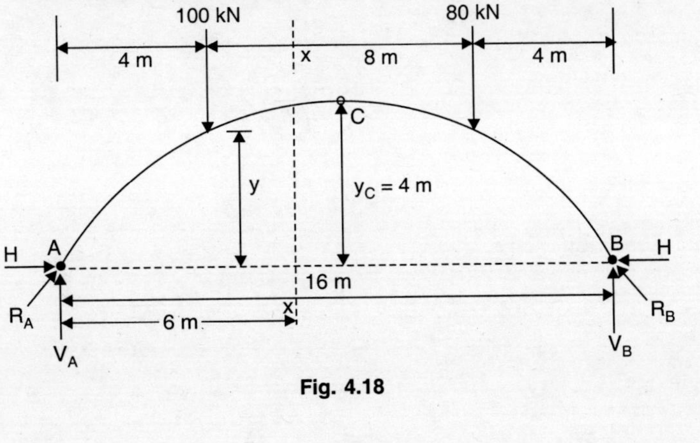

Fig. 4.18

Solution (*a*) **Reactions at A and B**

(*i*) **Vertical components of reactions at A and B :**

Taking moments about A,

$$100(4) + 80(12) - V_B(16) = 0$$
$$V_B = 85 \text{ kN}$$
$$V_A = \text{Total load} - V_B = (100 + 80) - 85$$
$$V_A = 95 \text{ kN}$$

(*ii*) **Horizontal components of reactions at A and B :**

Taking moments about the crown point C,

$$V_A(8) - H\,(y_C) - 100(4) = 0$$
$$95(8) - H(4) - 400 = 0$$
$$H = 90 \text{ kN}$$

(*iii*) **Resultant reactions at A and B :**

$$R_A = \sqrt{V_A{}^2 + H^2} = \sqrt{95^2 + 90^2} = 130.86 \text{ kN}$$
$$R_B = \sqrt{V_B{}^2 + H^2} = \sqrt{85^2 + 90^2} = 123.79 \text{ kN}$$

(*b*) **Bending moment at 6 m from left support**

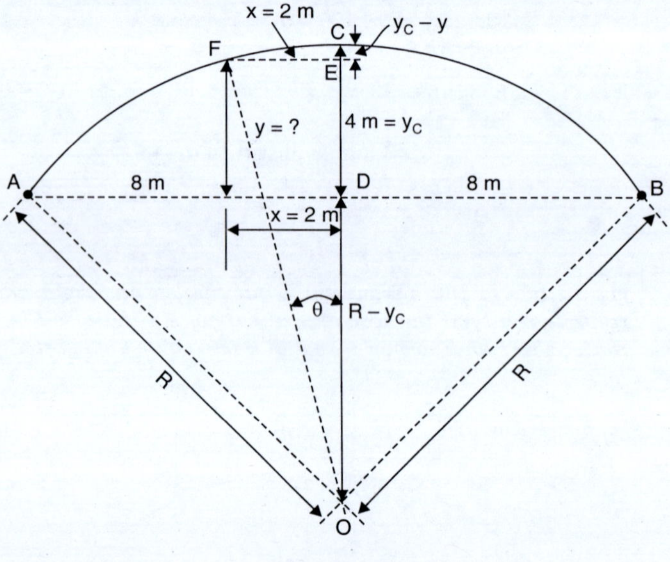

Fig. 4.19

To find the radius "R" :

$$\frac{l}{2} \times \frac{l}{2} = y_C\,(2R - y_C)$$

$y_C = 4$; $l = 16$

Therefore, $R = 10$ m

To find 'y' at $x = 2$ m from centre :

In Fig 4.19, OEF is a right angled triangle.

$$R^2 = (R - 4 + y)^2 + 2^2$$
$$10^2 = (10 - 4 + y)^2 + 2^2$$
$$96 = (6 + y)^2$$
$$9.8 = 6 + y$$
$$9.8 - 6 = y$$
$$y = 3.8 \text{ m at } x = 2 \text{ m from the centre}$$

Bending moment
$$= V_A(6) - H_A(y) - 100(2)$$
$$= 95(6) - 90y - 200$$
$$= 95(6) - 90(3.8) - 200$$
$$BM_{XX} = 28 \text{ kNm}$$

(*c*) **Radial shear force 'F'**

From Fig. 4.19,
$$\theta = \tan^{-1}\frac{2}{OE} = \tan^{-1}\frac{2}{6+y} = \tan^{-1}\frac{2}{6+3.8} = 11° \; 32'$$
$$F = V_X \cos\theta - H \sin\theta$$

where,
$$V_X = \text{net vertical shear force at } x = 6 \text{ m from A.}$$
$$= V_A - 100 = 95 - 100$$
$$V_X = -5 \text{ kN}$$
$$H = 90 \text{ kN}$$
$$F = -5 \cos(11° \; 32') - 90 \sin 11° \; 32' = -22.895$$
$$F = -22.9 \text{ kN}$$

(*d*) **Normal thrust** (at $x = 6$ m from 'A')
$$N = V_X \sin\theta + H \cos\theta$$
$$= -5 \sin 11° \; 32' + 90 \cos 11° \; 32' = -87.17 \text{ kN}$$

4.10 SPRINGINGS AT DIFFERENT LEVELS

While constructing arch bridges in mountainous terrain it may be necessary to bridge a gorge between a high mountain and a low mountain. In such cases it will be neither possible, nor necessary to locate the springings at the same level. In the case of parabolic arches we can fix the geometry easily by realizing that the equation to the arch with the crown as origin would be (see Fig. 4.20) the same for both the arches, namely, ACB and A′CB′

$$y = \left(\frac{4y_C}{l^2}\right)x^2 \qquad \ldots(4.5)$$

Since ACB and A′CB′ refer to the same arch, the constant $\left(\frac{4y_C}{l^2}\right) = \left(\frac{4y_C'}{(l')^2}\right)$

Then
$$\frac{y_C'}{y_C} = \frac{(l')^2}{l^2} \quad \text{and} \quad \frac{l'}{l} = \sqrt{\frac{y_C'}{y_C}} \qquad \ldots(4.6)$$

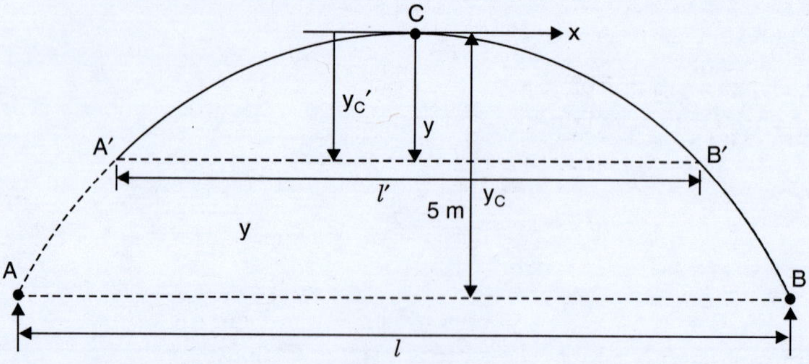

Fig. 4.20

Example 4.10 *A three hinged parabolic arch of 40 m span has abutments at unequal levels. The highest point of the arch is 4 m above left support and 9 m above the right abutment. The arch is subjected to an u.d.l. of 15 kN/m over its entire horizontal span. Find the horizontal thrust and bending moment at a point 8 m from the left support.*

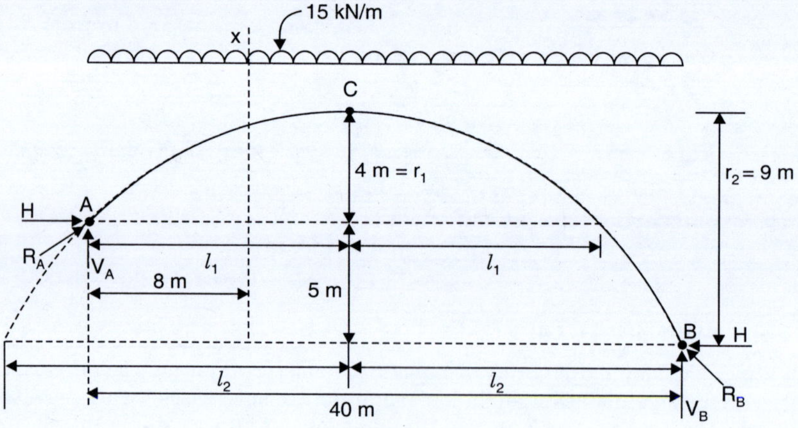

Fig. 4.21

Solution (*a*) **Reactions at A, B and H**

$$\frac{l_1}{l_2} = \sqrt{\frac{r_1}{r_2}}$$

$$\frac{l_1}{40 - l_1} = \sqrt{\frac{4}{9}}$$

$$l_1 = (40 - l_1) \times \frac{2}{3}$$

$$l_1 = 16 \text{ m}$$

$$l_2 = 40 - l_1 = 40 - 16 = 24 \text{ m}$$

Considering the equilibrium of the left side of C, $\Sigma M = 0$

$$V_A(16) - H(4) - 15 \times 16 \times \frac{16}{2} = 0$$

$$16V_A - 4H - 1920 = 0$$

$$4V_A - H - 480 = 0 \quad \ldots(1)$$

Considering the equilibrium of the right side of C,

$$-V_B(24) + H(9) + 15 \times 24 \times \frac{24}{2} = 0$$

$$-24\,V_B + 9H + 4320 = 0$$

$$-8\,V_B + 3H + 1440 = 0$$

$$V_A + V_B = 600 \quad \ldots(2)$$

$$V_B = 600 - V_A \quad \ldots(3)$$

Substitute the above in (2)

$$-8(600 - V_A) + 3H + 1440 = 0$$

$$-4800 + 8V_A + 3H + 1440 = 0$$

$$8V_A + 3H - 3360 = 0 \quad \ldots(4)$$

(1) × (2) $\quad 8V_A - 2H - 960 = 0 \quad \ldots(5)$

(4) – (5) $\quad 5H - 2400 = 0$

$$H = \frac{2400}{5} = 480 \text{ kN}$$

From (1) $\quad 4\,V_A - 480 - 480 = 0$

$$V_A = 240 \text{ kN}$$

From (3) $\quad V_B = 600 - V_A = 600 - 240$

$$V_B = 360 \text{ kN}$$

(*b*) **Bending moment at x = 8 m from 'A'**

$$BMx = V_A(8) - 15 \times 8 \times \frac{8}{2} - Hy$$

$$= 240 \times 8 - 15 \times 32 - 480y$$

$$BMx = 1440 - 480y \quad \ldots(6)$$

y = Rise of the arch at x = 8 m from 'A'

$$y = \frac{4r}{l^2}x\,(l - x)$$

$$= \frac{4 \times 4}{(2 \times 16)^2} \times 8((2 \times 16) - 8)$$

$y = 3$ m ; substitute in (6)

$$BM = 1440 - 480 \times 3$$

$$BM = 0$$

(*c*) **Radial shear F at x = 8 m from 'A'**

$$F = V_X \cos\theta - H \sin\theta$$

V_X = net vertical shear force = $V_A - 15 \times 8 = 240 - 15 \times 8$

$$V_X = 120 \text{ kN}$$

$$H = 480 \text{ kN}$$

$$\theta = \tan^{-1}\left[\frac{4r}{l^2}(l-2x)\right] \Rightarrow \theta = \tan^{-1}\left[\frac{4\times 4}{(32)^2}(32-2\times 8)\right]$$

$\theta = 14°\ 2'$

$F = 120 \cos 14°\ 2' - 480 \sin 14°\ 2'$

$F = 0$

(*d*) **Normal thrust N at *x* = 8 m from 'A'**

$N = V_X \sin\theta + H\cos\theta = 120 \sin 14°\ 2' + 480 \cos 14°\ 2'$

$N = 494.77$ kN.

Example 4.11 *A three hinged parabolic arch of span 30 m has its supports at depths of 4 m and 16 m below crown C. The arch carries a load of 80 kN at a distance of 5 m to the left of C and a second load of 100 kN at 10 m to the right of C. Determine the reactions at supports and bending moments under the loads.*

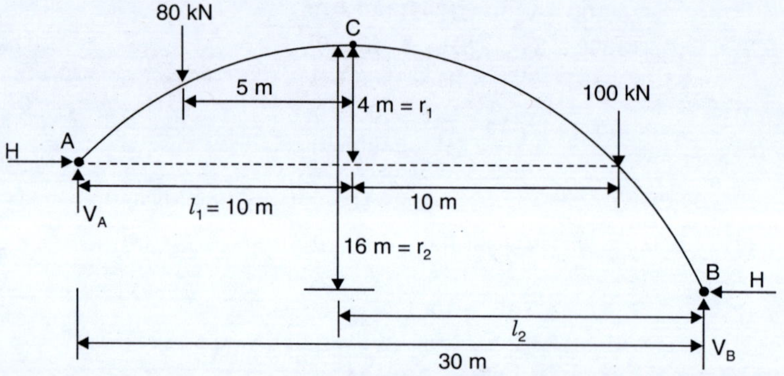

Fig. 4.22

Solution (*a*) **To find l_1 and l_2**

$$\frac{l_1}{l_2} = \sqrt{\frac{r_1}{r_2}}$$

$$\frac{l_1}{30 - l_1} = \sqrt{\frac{4}{16}} = \frac{1}{2}$$

$$2l_1 = 30 - l_1$$

$$l_1 = \frac{30}{3} = 10 \text{ m}$$

$$l_2 = 30 - 10 = 20 \text{ m}$$

(*b*) **To find H, V_A and V_B**

Considering the portion to the left of C and taking moments

$$V_A(10) - H(4) - 80(5) = 0$$

$$10V_A - 4H - 400 = 0$$

$$2.5V_A - H - 100 = 0 \qquad \text{...(1)}$$

Considering the portion to the right of C and taking moments

$$V_B(20) - H(16) - 100(10) = 0$$

$$20V_B - 16H - 1000 = 0$$

$$V_B - 0.8H - 50 = 0 \qquad \text{...(2)}$$

$$V_A + V_B = 80 + 100 = 180$$

$$V_B = 180 - V_A \qquad \text{...(3)}$$

Substitute in (2)

(2) ⇒ $$180 - V_A - 0.8H - 50 = 0$$

$$-V_A - 0.8H + 130 = 0$$

$$V_A + 0.8H - 130 = 0 \qquad \text{... (4)}$$

(4) × 2.5 $$2.5V_A + 2H - 325 = 0 \qquad \text{...(5)}$$

(1) ⇒ $$2.5V_A - H - 100 = 0$$

(5) – (1) $$3H - 225 = 0$$

$$H = 75 \text{ kN}$$

Substitute in (1) and (2)

(1) ⇒ $$2.5\,V_A - 75 - 100 = 0$$

$$V_A = \frac{175}{2.5}$$

$$V_A = 70 \text{ kN}$$

(2) ⇒ $$V_B - 0.8\,(75) - 50 = 0$$

$$V_B = 110 \text{ kN}$$

(*c*) **Resultant reactions R_A and R_B**

$$R_A = \sqrt{V_A^2 + H^2} = \sqrt{70^2 + 75^2} = 102.59 \text{ kN}$$

$$R_B = \sqrt{V_B^2 + H^2} = \sqrt{110^2 + 75^2} = 133.13 \text{ kN}$$

(*d*) **Bending moments under loads**

(*i*) ***BM under 80 kN :***

$$(BM)_{80} = V_A\,(5) - H(y_5)$$

$$y_5 = \frac{4r}{l^2} x\,(l - x) = \frac{4 \times 4}{(2 \times 10)^2} \times 5 \times [(2 \times 10) - 5]$$

$$y_5 = 3 \text{ m}$$

$$(BM)_{80} = 70(5) - 75 \times 3$$

$$(BM)_{80} = 125 \text{ kNm}$$

(*ii*) ***BM under 100 kN:***

$$(BM)_{100} = -V_B\,(10) + H(y_{10})$$

$$y_{10} = \frac{4r}{l^2} x\,(l - x) = \frac{4 \times 16}{(2 \times 20)^2} \times 10((2 \times 20) - 10)$$

$$y_{10} = 12 \text{ m}$$

∴ $$(BM)_{100} = +\,110\,(10) - 75\,(12)$$

$$(BM)_{100} = +\,200 \text{ kNm}$$

4.10.1 Three Hinged Parabolic Unsymmetrical Arches

Consider a three hinged parabolic arch of unsymmetrical shape whose dimensions are as shown in Fig. 4.23.

y_C = Rise of the arch at the crown section from the line joining the springings.

y_x = Rise of the arch at the section X measured from the line joining the springings.

$$y'_x = y_x \cos \phi$$

$$y'_C = y_C \cos \phi$$

A unit load is assumed to roll from left to right. For any position of load, x from A,

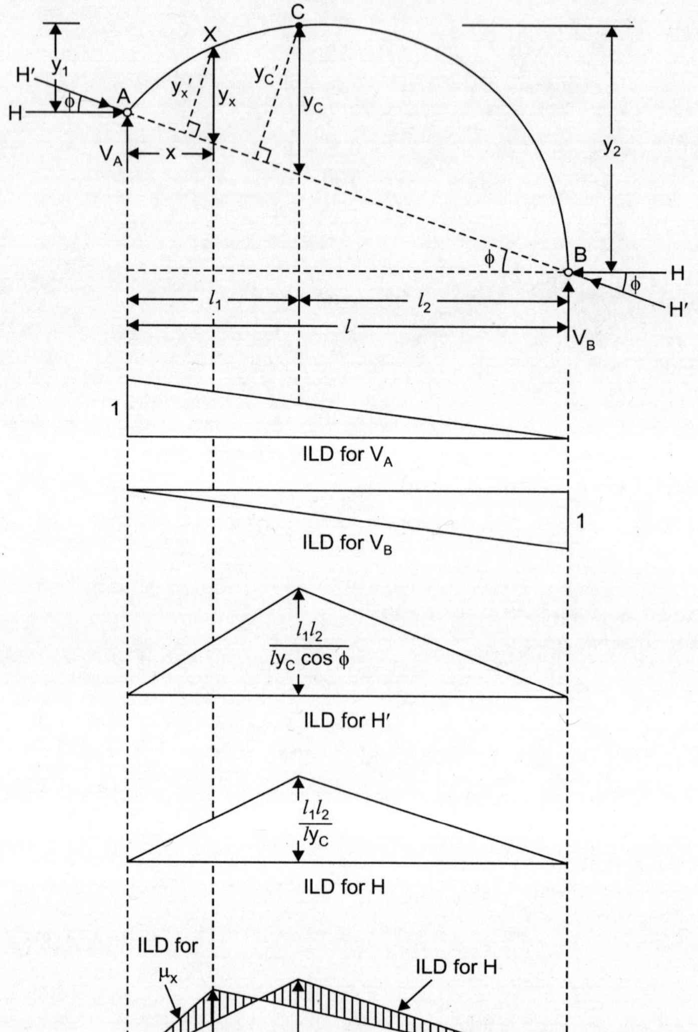

Fig. 4.23

$$V_A \cdot l - 1\,(l - x) = 0$$

$$V_A = \left(\frac{l - x}{l}\right) \qquad \text{...(1)}$$

$$V_B = 1 - \left(\frac{l - x}{l}\right) = \frac{l - l + x}{l}$$

$$V_B = \left(\frac{x}{l}\right) \qquad \text{...(2)}$$

At $x = 0$; $V_A = 1$; $V_B = 0$

At $x = l$; $V_A = 0$; $V_B = 1$

Load between C and B :

Assuming the load to the right of the crown C and taking moments about the crown C,

$$V_A \times l_1 - H' \times y_C \cos \phi = 0$$

$$V_A = \frac{H' y_C \cos \phi}{l_1}$$

H′ is thrust acting along the line joining the springings.

$$H' = \frac{V_A\, l_1}{y_C \cos \phi} \qquad \text{...(3)}$$

Substituting V_A in equation (3)

$$H' = \left(\frac{l - x}{l}\right) \frac{l_1}{y_C \cos \phi} \qquad \text{...(4)}$$

When the load is on the right hand support,

i.e., when $x = l$, $H' = 0$

when the load is on the crown C,

i.e., when $x = l_1$, $H' = \left(\frac{l - l_1}{l}\right) \frac{l_1}{y_C \cos \phi}$

$$H' = \frac{l_2\, l_1}{l\, y_C \cos \phi} \qquad \text{...(5)}$$

Load between A and C :

Assuming the load to be between A and C and taking moments about the crown C,

$$V_B \times l_2 - H' y_C \cos \phi = 0$$

$$V_B = \frac{H' y_C \cos \phi}{l_2}$$

$$H' = \frac{V_B\, l_2}{y_C \cos \phi} = \frac{x}{l} \cdot \frac{l_2}{y_C \cos \phi} \qquad \text{...(6)}$$

when $x = 0$, $H' = 0$

$$x = l_1,\ H' = \frac{l_1\, l_2}{l\, y_C \cos \phi} \qquad \text{...(5)}$$

∴ ILD for H′ will have a linear variation with zero ordinates at either springings and a maximum ordinate of $\frac{l_1 l_2}{l\, y_C \cos\phi}$ at the crown section.

Horizontal thrust, H = H′ cos ϕ

$$= \frac{l_1 l_2}{l\, y_C \cos\phi} \times \cos\phi$$

$$H = \frac{l_1 l_2}{l\, y_C}$$

∴ When the load is on the crown, the horizontal thrust

$$H = \frac{l_1 l_2}{l\, y_C} \qquad \text{...(7)}$$

ILD for bending moment at X :

ILD for B.M. at X in the arch = ILD for B.M. as a simply supported beam – ILD for H multiplied by y_x.

ILD for B.M. as a simply supported beam will be a triangle having a maximum ordinate of $\frac{x(l-x)}{l}$ at the section under consideration.

ILD for the reduction is B.M. due to H ($H'y'_x = \frac{l_1 l_2}{l\, y_C \cos\phi} \cdot y_x \cos\phi = \frac{l_1 l_2}{l\, y_C} \cdot y_x$, *i.e.*, H multiplied by y_x) will be triangle having a maximum ordinate of $\frac{l_1 l_2}{l\, y_C} \cdot y_x$ under the crown section. These two will be superimposed to get the ILD for B.M. at x.

Example 4.12 *A three hinged parabolic arch of span 30 m has its supports at depths of 4 m and 16 m below crown C. The arch carries a load of 80 kN at a distance of 5 m to the left of C and a second load of 100 kN at 10 m to the right of C. Determine the horizontal thrust and bending moments under the loads.*

Solution **To find l_1 and l_2 :**

$$\frac{l_1}{l_2} = \sqrt{\frac{y_1}{y_2}} = \sqrt{\frac{4}{16}} = \frac{1}{2}$$

$$2l_1 = l_2 \qquad \text{...(1)}$$

$$l_1 + l_2 = 30 \qquad \text{...(2)}$$

$$l_1 + 2l_1 = 30$$

$$3l_1 = 30$$

$$\therefore \quad l_1 = 10 \text{ m},\ l_2 = 20 \text{ m}$$

To find the horizontal thrust, H :

ILD for H is shown in Fig. 4.24 (*b*)

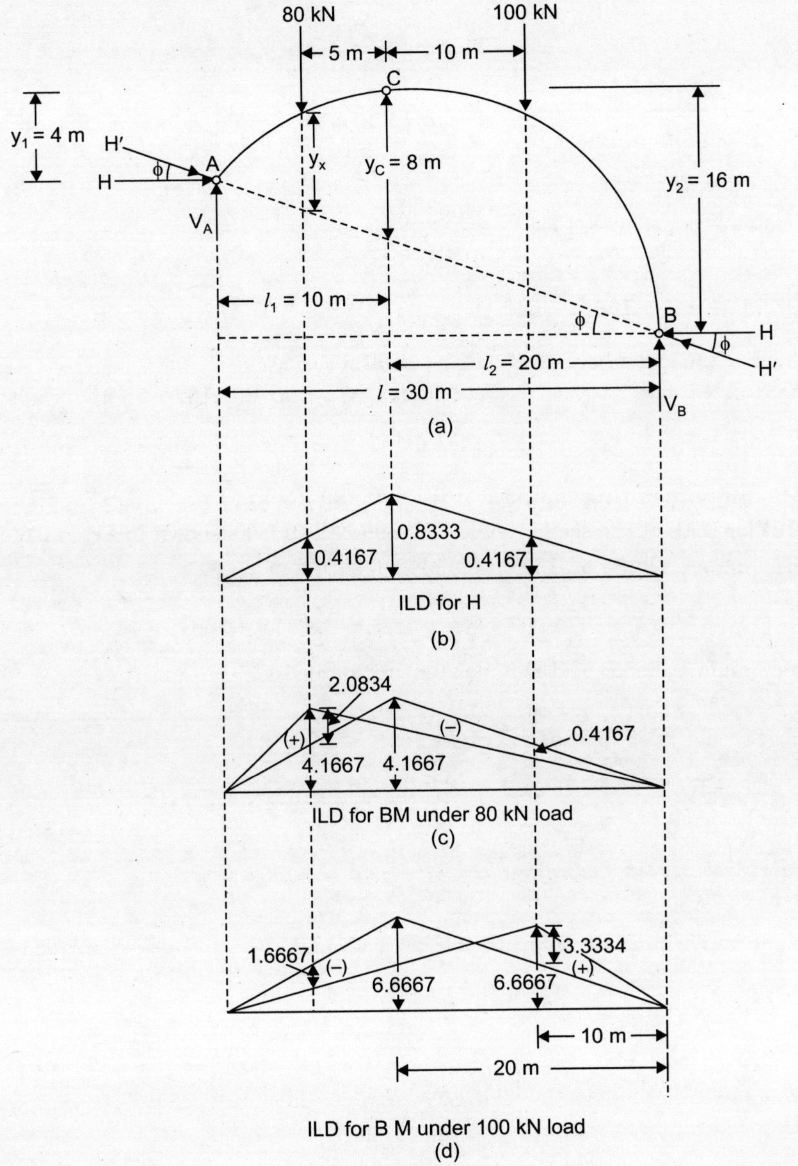

Fig. 4.24

Maximum ordinate of ILD under C = $\dfrac{l_1 l_2}{l\,y_C}$

$$y_C = 4 + \frac{(16-4)}{30} \times 10 = 4 + 4 = 8 \text{ m}$$

∴ Maximum ordinate $= \dfrac{l_1 l_2}{l\,y_C} = \dfrac{10 \times 20}{30 \times 8} = 0.8333$

Ordinate under 80 kN load $= \dfrac{0.8333}{10} \times 5 = 0.4167$

Ordinate under 100 kN load $= \dfrac{0.8333}{20} \times 10 = 0.4167$

∴ For the given load positions, horizontal thrust,

$$H = 80 \times 0.4167 + 100 \times 0.4167 = 75 \text{ kN}$$

$$\boxed{H = 75 \text{ kN}}$$

Maximum bending moment under the 80 kN load :

ILD for B.M. at the section under the 80 kN load is as shown in Fig. 4.24 (*c*)

Maximum ordinate of ILD for simply supported moment

$$= \frac{x(l-x)}{l} = \frac{5 \times 25}{30} = 4.1667$$

Maximum ordinate of ILD under the crown due to

$$H = \frac{l_1 l_2}{l\,y_C} \cdot y_x$$

$$y_x = \frac{3}{4} \times 4 + \frac{(16-4) \times 5}{30} = 3 + 2 = 5 \text{ m}$$

∴ Ordinate $= \dfrac{10 \times 20}{30 \times 8} \times 5 = 4.1667$

Net positive ordinate under the 80 kN load $= 4.1667 - \dfrac{4.1667 \times 5}{10} = 2.0834$

Net negative ordinate under the 100 kN load

$$= \frac{4.1667 \times 10}{20} - \frac{4.1667 \times 10}{25} = 0.4167 \text{ (–ve)}$$

Maximum B.M. under the 80 kN load = 80 × 2.0834 – 100 × 0.4167

$$\boxed{\text{B.M.} = 125 \text{ kNm}}$$

Maximum bending moment under the 100 kN load:

ILD for B.M. at the section under the 100 kN load is as shown in Fig. 4.24 (*d*)

Maximum ordinate of ILD for simply supported moment

$$= \frac{x\,(l-x)}{l} = \frac{20\,(30-20)}{30} = 6.6667$$

Maximum ordinate of ILD under the crown due to H

$$= \frac{l_1 l_2}{l\, y_C} \cdot y_x$$

$$y_x = \frac{3}{4} \times 16 - \frac{(16-4)}{30} \times 10 = 12 - 4 = 8 \text{ m}$$

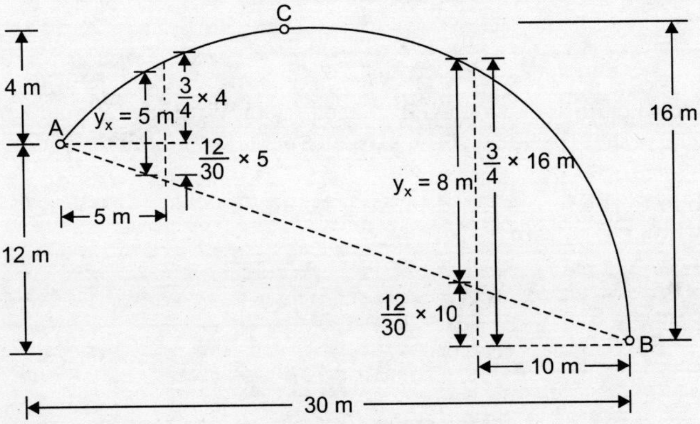

$$\therefore \text{ Ordinate} = \frac{10 \times 20}{30 \times 8} \times 8 = 6.6667$$

Net +ve ordinate under the 100 kN load

$$= 6.6667 - \frac{6.6667}{20} \times 10$$

$$= 3.3334$$

Net –ve ordinate under the 80 kN load

$$= \frac{6.6667 \times 5}{10} - \frac{6.6667 \times 5}{20}$$

$$= 1.6667 \text{ (–ve)}$$

Maximum bending moment under the 100 kN load

$$= 100 \times 3.3334 - 80 \times 1.6667$$

$$= 200 \text{ kNm}$$

B.M. = 200 kNm

4.11 INFLUENCE LINES FOR THREE HINGED ARCHES

Let us consider a point X on a 3 hinged parabolic arch at x from A [Fig. 4.25(a)]. We will investigate the following: V_A, V_B, vertical shear V_x at X, beam bending moment μ_x at X and the horizontal thrust H as a load W moves from A to B.

If 'a' be the horizontal distance of W from A,

$$V_A = \frac{W(l-a)}{l} \; ; \quad V_B = \frac{Wa}{l}$$

$$\mu_x = V_B \, . \, (l-x) = \frac{Wa}{l}(l-x) \qquad \text{...for } a < x$$

$$\mu_x = V_A \, . \, x = \frac{W(l-a) \, . \, x}{l} \qquad \text{...for } a > x$$

$$V_x = -V_B = \frac{-Wa}{l} \qquad \text{...for } a < x \qquad \text{...(4.7)}$$

$$V_x = V_A = \frac{W(l-a)}{l} \qquad \text{...for } a > x$$

(in the above equations we have replaced R_A by V_A since R_A might mean the resultant reaction at A. For arches, because of the presence of H, the reactions would be inclined and not vertical.)

In addition to the forces listed in equation (4.7), we have H as well. To find H, equate the moment at hinge C to zero.

$$V_B \, . \, \frac{l}{2} - H \, . \, y_C = 0 \; ; \quad V_B = -\frac{Wa}{l} \qquad \text{... for } a < \frac{l}{2}$$

$$H = V_B \, . \, \frac{l}{2y_C} = \frac{Wa}{2y_C} \qquad \text{.... for } a < \frac{l}{2} \qquad \text{...(4.8)}$$

$$H = \frac{W(l-a)}{2y_C} \qquad \text{... for } a > \frac{l}{2}$$

We can take W as unity in equations (4.7) and (4.8) and plot them as influence lines for V_A, V_B, V_x, μ_x and H. [Fig. 4.21 (b) to (f)]

4.11.1 Net Bending Moment

μ_x is the so called beam bending moment. This would be considerably mitigated by the arch action. At any section x, the bending moment is

$$M_x = \mu_x - Hy \qquad \text{...(4.9)}$$

$$M_x = \frac{Wa(l-x)}{l} - \frac{Wa \, . \, y}{2y_C} \qquad \text{...for } a < x \qquad \text{...[4.9 (}a\text{)]}$$

$$M_x = \frac{W(l-a)}{l} x - \frac{W(l-a)y}{2y_C} \qquad \text{...for } a > x \qquad \text{...[4.9 (}b\text{)]}$$

Equations [4.9 (a)] and [4.9 (b)] would yield the influence line for M_x if we take W as one unit. We shall remember that IL is for a given point X. So x and y would be constant and 'a' would be the only variable when we sketch the ILD for M_x. Figure 4.25 is the influence line for M_x. This comes in 2 parts, the positive part due to μ_x and the negative part due to Hy. We plot them on the same side of the x-axis so that the positive and negative parts cancel each other by and large, and we are left with the net moment diagram.

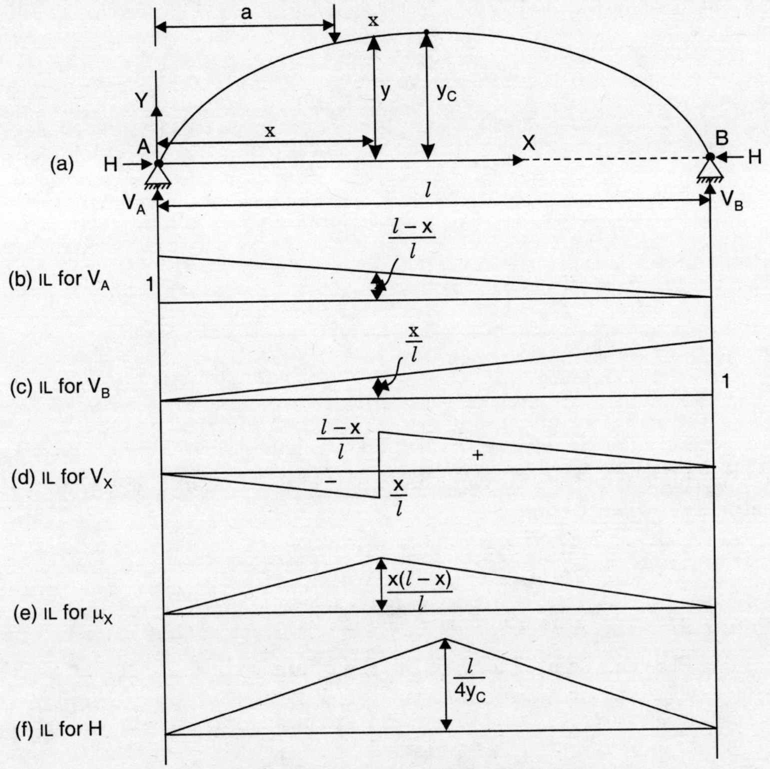

Fig. 4.25. IL for a three hinged arch.

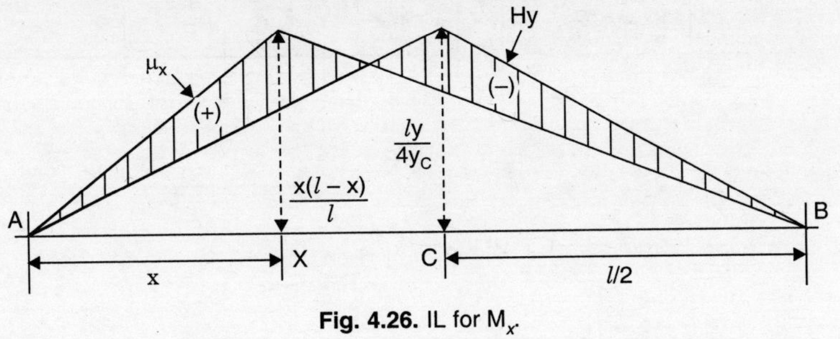

Fig. 4.26. IL for M_x.

In fact, 4.26 is the ILD for bending moment at any x. It becomes complete only when we assign a value for x.

4.11.2 Radial Shear and Normal Thrust

Since the arch centre line is curved, its initial slope is different at every other section. [equation (4.2)]. So the vertical shear V_x and horizontal thrust H combine to produce different values of normal thrust N_x and radial shear R_x or F_x at x. (Fig. 4.27), depending upon the initial slope at x.

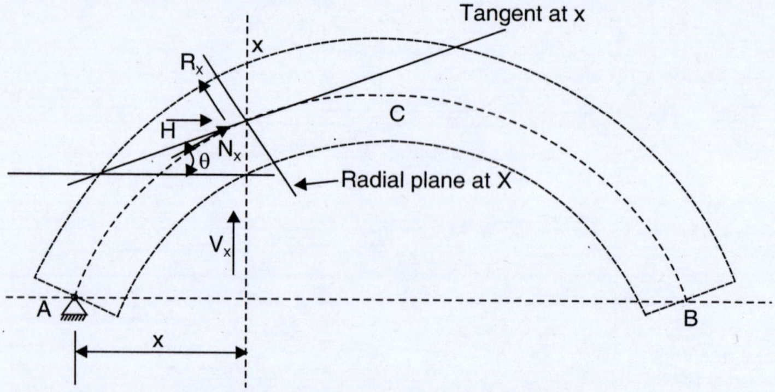

Fig. 4.27. Normal thrust and radial shear.

Thus, in the free body of the right hand part of the arch in Fig. 4.27, at a section X, θ is the inclination of the centre line.

$$P_x \text{ or } N_x = H\cos\theta + V_x \sin\theta$$

$$F_x \text{ or } R_x = V_x \cos\theta - H\sin\theta \quad \text{...(4.4)}$$

and

$$\tan\theta = \text{slope} = \frac{4y_C}{l^2}(l - 2x) \quad \text{...(4.2)}$$

Equation (4.4) can be used for plotting the influence lines for N_x and R_x (Fig. 4.28).

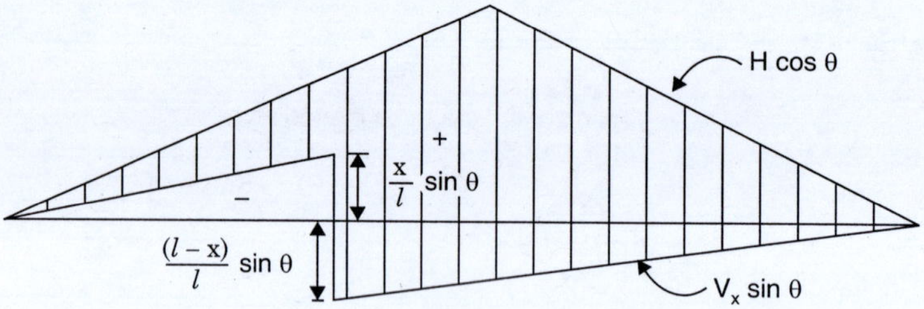

(a) N_x

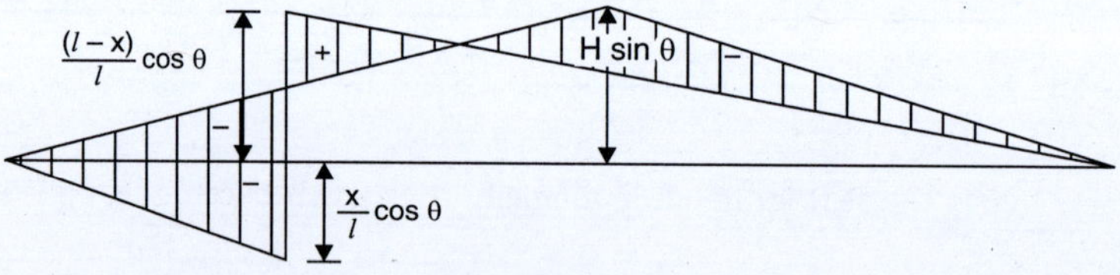

(b) R_x

Fig. 4.28. IL for N_x and R_x.

Example 4.13 *A three hinged parabolic arch has a span of 30 m and a central rise of 6 m. Five wheel loads of 4, 5, 5, 3, 3 kN spaced 3 m, 2 m, 3 m and 2 m in order, cross the arch from left to right with the 4 kN load leading. When the leading load is 20 m from the left hinge, calculate the horizontal thrust in the arch. Also, calculate the bending moment, normal thrust and shear force at the section under the tail load.*

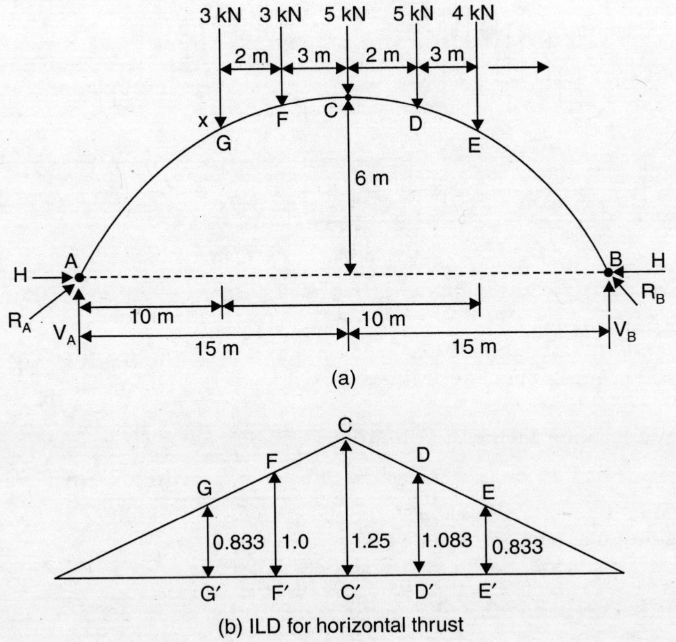

Fig. 4.29

Solution (*a*) **Horizontal thrust**

Ordinates of ILD:

$$CC' = \frac{l}{4r} = \frac{30}{4 \times 6} = 1.25$$

$$DD' = \frac{1.25}{15} \times (15 - 2) = 1.083$$

$$EE' = \frac{1.25}{15} \times (15 - (2 + 3)) = 0.833$$

$$FF' = \frac{1.25}{15} \times (15 - 3) = 1.0$$

$$GG' = \frac{1.25}{15} \times (15 - (2 + 3)) = 0.833$$

Horizontal thrust $= 4\,(0.833) + 5\,(1.083) + 5(1.25) + 3(1) + 3(0.833)$

$H = 20.496$ kN.

(*b*) **Bending moment at x = 10 m from A**

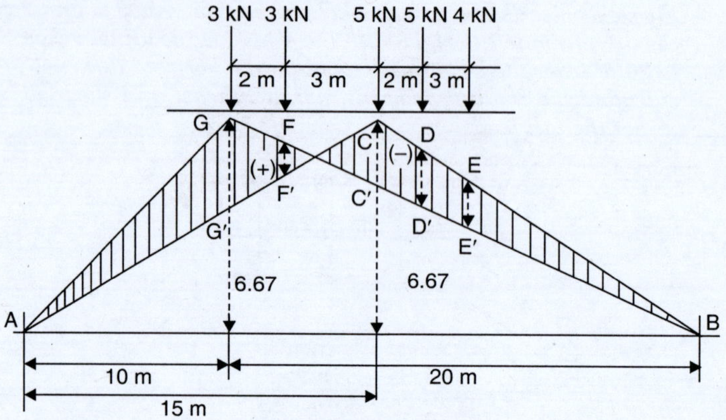

ILD for BM at x = 10 m

Fig. 4.30

Ordinates of ILD :

Maximum sagging moment ordinate $= \dfrac{x(l-x)}{l} = \dfrac{10(30-10)}{30} = 6.67$

Maximum hogging moment ordinate $= \dfrac{l}{4r} \times \dfrac{4r}{l^2} x\,(l-x)$

$$= \frac{x(l-x)}{l} = \frac{10(30-10)}{30} = 6.67$$

Net sagging moment ordinates

$$GG' = 6.67 - \frac{6.67}{15} \times 10$$

$$GG' = 2.22$$

$$FF' = \frac{6.67}{20} \times (20-2) - \frac{6.67}{15} \times (15-3) = 0.667$$

Net hogging moment ordinates

$$CC' = + \frac{6.67}{20} \times (20-5) - 6.67 = -1.67$$

$$DD' = + \frac{6.67}{20} \times (20-7) - \frac{6.67}{15} \times (15-2) = -1.45$$

$$EE' = + \frac{6.67}{15} \times (20-10) - \frac{6.67}{15} \times (15-5) = -1.11$$

Maximum bending moment at xx = 3 × 2.22 + 3 (0.667) – 5 (1.67) – 5 (1.45) – 4 (1.11)

Bending moment = – 11.379 kNm

(*c*) **Normal thrust at x = 10 m from 'A'**

Normal thrust = $V_x \sin\theta + H \cos\theta$

$$\theta = \tan^{-1}\left[\frac{4r}{l^2}(l-2x)\right] = \tan^{-1}\left[\frac{4\times 6}{(30)^2}(30-2(10))\right]$$

$\theta = 14° \, 55'$

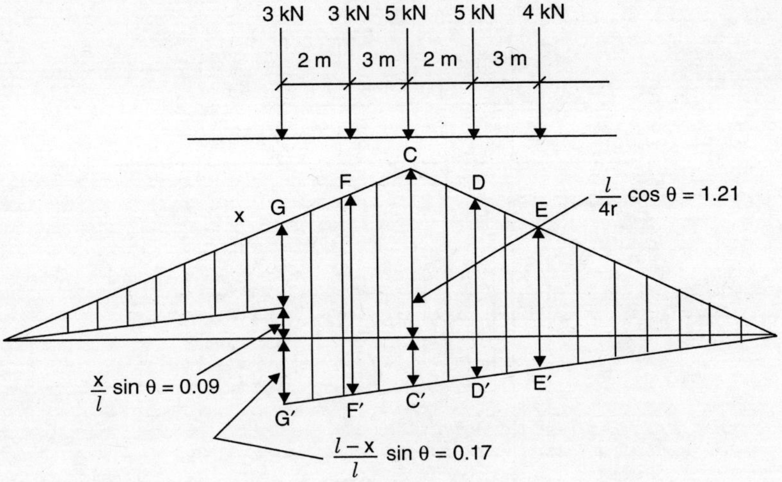

Fig. 4.31

Ordinates of ILD :

$$CC' = \frac{l}{4r}\cos\theta + \left[\frac{l-x}{l}\sin\theta\right] \times \frac{1}{20} \times 15$$

$$= \frac{30}{4 \times 6}\cos 14° \, 55' + \left[\frac{20}{30}\sin 14° \, 55'\right] \times \frac{15}{20}$$

$$CC' = 1.34$$

$$DD' = \frac{1.21}{15} \times (15-2) + \frac{0.17}{20} \times (20-7) = 1.16$$

$$EE' = \frac{1.21}{15} \times (15-5) + \frac{0.17}{20} \times (20-10) = 0.89$$

$$FF' = \frac{1.21}{15} \times (15-3) + \frac{0.17}{20} \times (20-2) = 1.121$$

$$GG' = \frac{1.21}{15} \times (15-5) + 0.17 = 0.98$$

Normal thrust = Σ(load × ordinate)

$$= 3 \times 0.98 + 3 \times 1.121 + 5 \times 1.34 + 5 \times 1.16 + 4 \times 0.89 = 22.36 \text{ kN}$$

(*d*) **Radial shear force at x = 10 m from 'A'**

Radial shear force, $F = V_x \cos\theta - H \sin\theta$

Ordinates of ILD :

$$CC' = \left[\frac{l-x}{l}\cos\theta\right] \times \frac{1}{20} \times (20-5) - \frac{l}{4r}\sin\theta$$

$$= \left[\frac{(30-10)}{30}\cos 14° \, 55'\right] \times \frac{1}{20} \times 15 - \frac{30}{4 \times 6}\sin 14° \, 55'$$

$$= 0.65 \times \frac{1}{20} \times 15 - 0.32$$

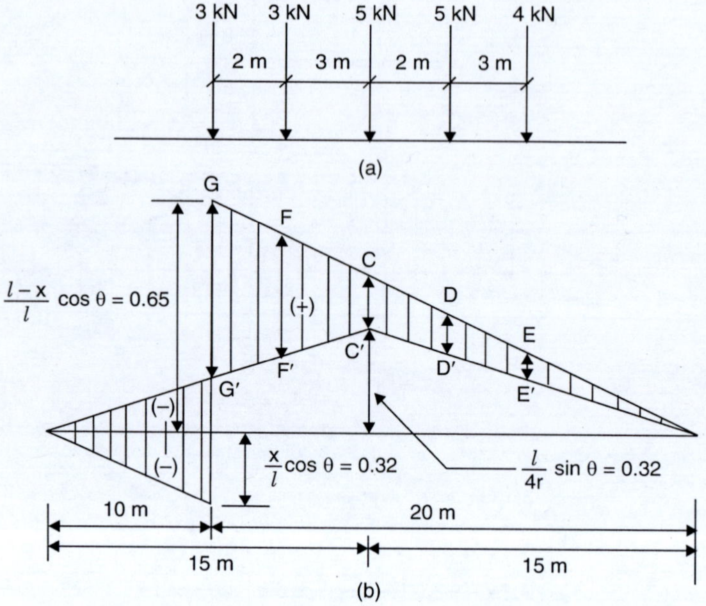

Fig. 4.32

$$CC' = 0.17$$

$$DD' = \frac{0.65}{20} \times (20 - 7) - \frac{0.32}{15} \times (15 - 2)$$

$$DD' = 0.145$$

$$EE' = \frac{0.65}{20} \times (20 - 10) - \frac{0.32}{15} \times (15 - 5) = 0.11$$

$$FF' = \frac{0.65}{20} \times (20 - 2) - \frac{0.32}{15} \times (15 - 3) = 0.33$$

$$GG' = 0.65 - \frac{0.32}{15} \times (15 - 5) = 0.44$$

Radial shear force under the tail load

$$= 3 \times GG' + 3 \times FF' + 5 \times CC' + 5 \times DD' + 4 \times EE'$$

$$= 3 \times 0.44 + 3 \times 0.33 + 5 \times 0.17 + 5 \times 0.15 + 4 \times 0.11 = 4.35 \text{ kN}$$

4.12 MAXIMUM BENDING MOMENT DIAGRAMS

To determine the maximum bending moments due to moving loads, it is convenient to start with the ILD for bending moments.

4.12.1 Positive Bending Moment

If we have a single concentrated load traversing a 3 hinged parabolic arch, Fig. 4.26 prompts us to conclude that the load shall be at X for M_{max} + to occur at X and x shall vary from 0 to $l/2$ and from $l/2$ to l.

Ordinate at $$X = \frac{x(l-x)}{l} - \frac{ly}{4y_C} \cdot \frac{x}{l/2}$$

$$M_x = \frac{x(l-x)}{l} - \frac{l}{4y_C} \frac{4y_C}{l^2} x(l-x) \cdot \frac{x}{l/2}$$

$$M_x = \frac{x(l-x)}{l}\left(1 - \frac{2x}{l}\right) \qquad \text{...(4.10)}$$

$$M_x = x(1 - 3k + 2k^2), \text{ where } k = x/l$$

At $x = 0$, $k = 0$, $M_x = 0$

At $x = l/2$, $k = 1/2$, $M_x = 0$

M_x is maximum when $\frac{dM_x}{dx} = 0$

$$1 - \frac{6x}{l} + \frac{6x^2}{l^2} = 0; \; x = 0.2113l, \; 0.7887l$$

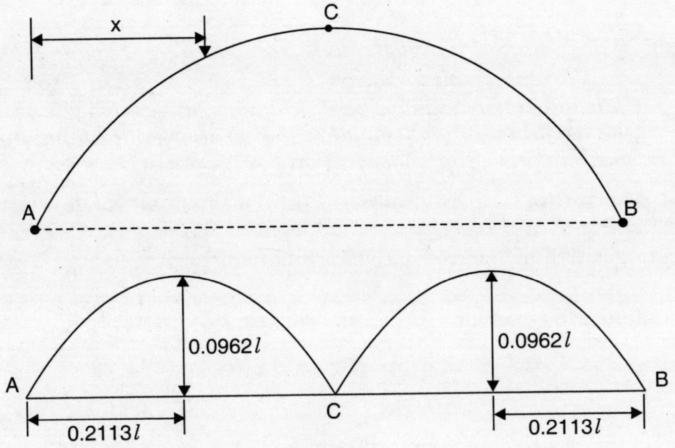

Fig. 4.33. Maximum positive BM due to unit load.

$M_{max\,max}$ is got by substituting 0.2113 for K

$M_{max\,max} = x(1 - 3 \times 0.2113 + 2 \times 0.2113^2) = 0.2113 \times l(1 - 0.6339 + 0.0893)$

$M_{max\,max} + = 0.0962l$ ...(4.9)

4.12.2 Negative Bending Moment

From Fig. 4.26 it is clear that the maximum negative bending moment occurs at x, when the load is at C.

$$M_x - = -\frac{ly}{4y_C} + \frac{x(l-x)}{(l-x) \times l} \times \frac{l}{2}$$

$$= -\frac{l\,4y_C}{l^2}\frac{(x)(l-x)}{4y_C} + \frac{x}{2} = -\frac{l \times (l-x)}{l^2} + \frac{x}{2}$$

$$M_x - = \left(\frac{x^2}{l} - \frac{x}{2}\right) \qquad \text{...(4.11)}$$

M_x – has the max value when $\frac{dM_x}{dx} = 0$

$$+\frac{2x}{l} - \frac{1}{2} = 0\ ;\ \frac{x}{l} = +\frac{1}{4}\ ;\ x = \frac{l}{4}$$

This must be valid even when x is measured from B towards C.

$$M_{\text{max max}} - = \frac{l^2}{16l} - \frac{l}{8} = -\frac{l}{16} \qquad \text{...(4.12)}$$

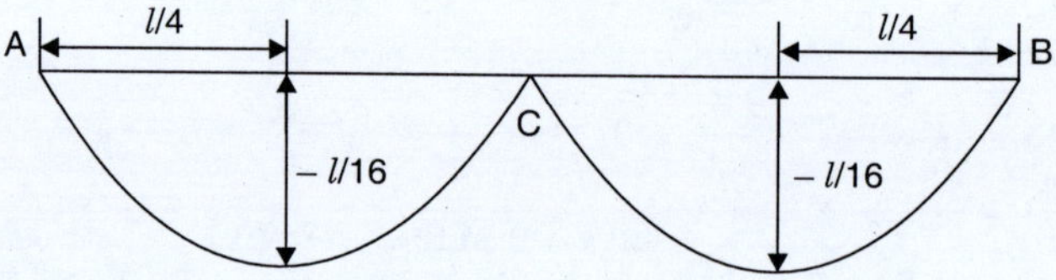

Fig. 4.34. Maximum negative BM due to unit load.

Example 4.14 *A three hinged parabolic arch has a horizontal span of 30 m with a central rise of 5 m. A point load of 10 kN moves across the span from left to right. Calculate the maximum positive and negative moments at a section 8 m from the left hinge. Also, calculate the position and magnitude of the absolute maximum bending moment that may occur in the arch.*

Solution (*a*) **Maximum bending moment** (at 8 m from left hinge)

Maximum positive moment (sagging) ordinate $= \frac{x(l-x)}{l} = \frac{8 \times 22}{30} = 5.87$

Maximum negative moment (hogging) ordinate $= \left[\frac{l}{4r}\right] y$

$$= \frac{l}{4r} \times \frac{4r}{l^2} x\,(l-x) = \frac{x(l-x)}{l} = \frac{8 \times 22}{30} = 5.87$$

Net maximum positive moment ordinate $= 5.87 - \frac{5.87}{15} \times 8 = 2.74$

Net maximum negative moment ordinate $= 5.87 - \frac{5.87}{22} \times 15 = 1.87$

To get maximum positive moment at the section 8 m from 'A', the load should be placed on that section itself

Then maximum positive moment = Load × Net max. positive ordinate
= 10 × 2.74
= 27.4 kNm

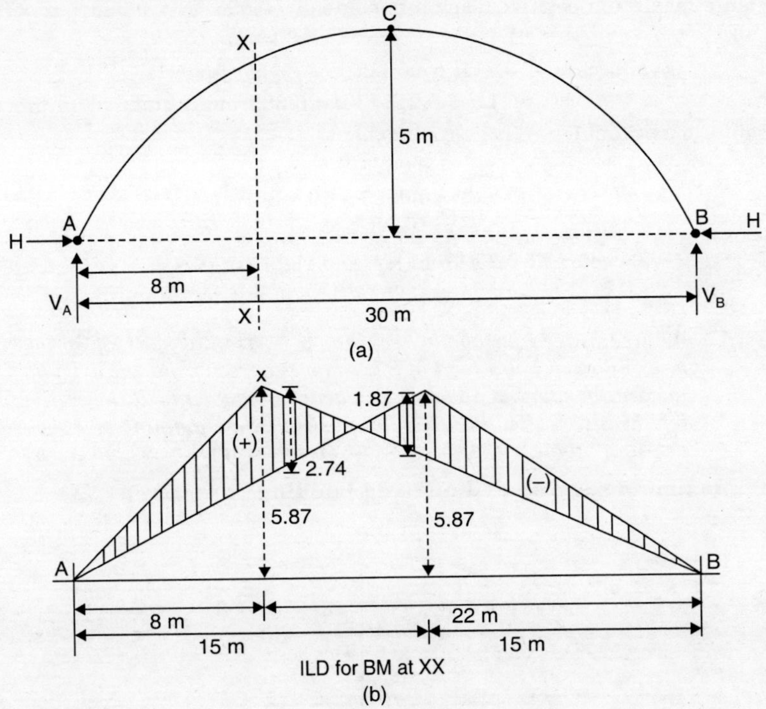

Fig. 4.35

To get max. negative moment, the load should be placed at midspan

Maximum negative moment = Load × Net max. negative ordinate

= 10 × 1.87

= 18.7 kNm

(*b*) **Absolute maximum bending moment**

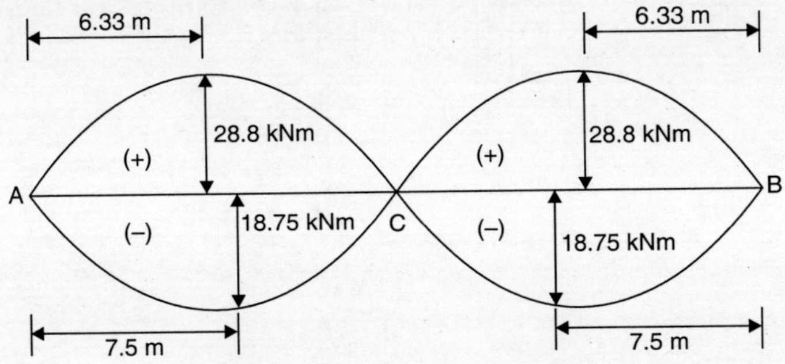

Fig. 4.36. Absolute max. BM diagram.

Absolute maximum positive (sagging) moment = 0.096 Wl = 0.096 × 10 × 30

$$BM_{max\ max}\ (+ve) = 28.8\ kNm$$

$BM_{max\ max}$ (+ve) occurs at 0.211l from left and right supports.

$$0.211l = 0.211\ (30) = 6.33 \text{ from A and } 6.33 \text{ m from B}$$

Absolute maximum (–ve) (hogging) moment

$$= 0.0625\ Wl$$
$$= 0.0625 \times 10 \times 30$$
$$BM_{max\ max}\ (-ve) = 18.75\ kNm$$

$BM_{max\ max}$ (–ve) occurs at 0.25l from left and right supports.

$$0.25l = 0.25\ (30) = 7.5 \text{ m from A and B.}$$

Example 4.15 *A three hinged parabolic arch has a horizontal span of 36 m with a central rise of 6 m. A point load of 8 kN moves across from left to right. Calculate the maximum sagging and hogging bending moments at the section 9 m from the left hinge. Calculate also the position and amount of absolute maximum bending moment that may occur on the arch.*

Solution **Maximum sagging and hogging bending moments at XX**

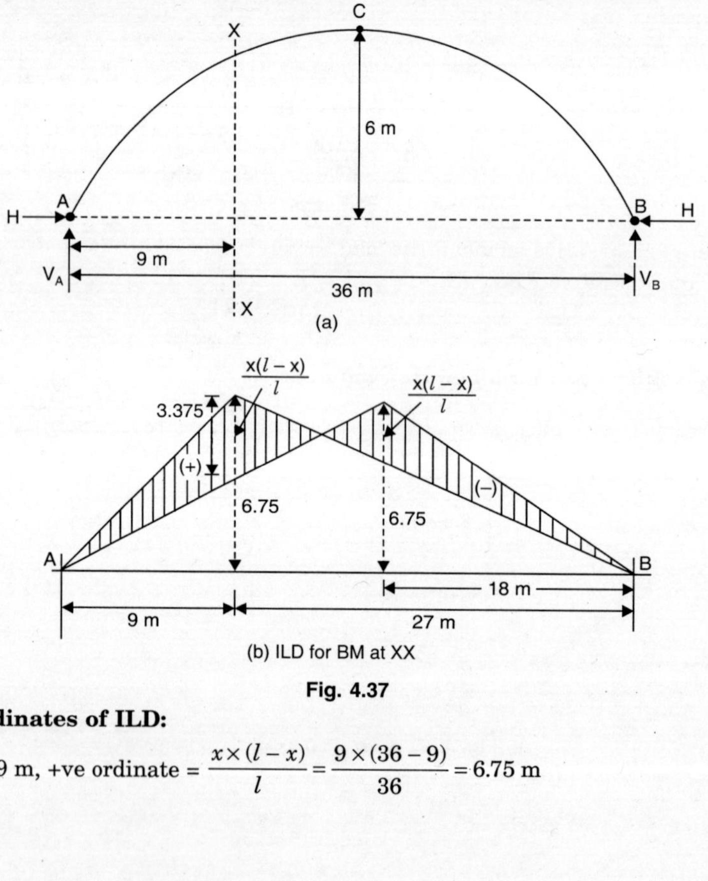

Fig. 4.37

(*a*) **Ordinates of ILD:**

At x = 9 m, +ve ordinate $= \dfrac{x \times (l - x)}{l} = \dfrac{9 \times (36 - 9)}{36} = 6.75$ m

At $x = 18$ m –ve ordinate $= \text{H.y} = \frac{l}{4r} \times \frac{4r}{l^2} x\,(l - x)$

$= \frac{x(l-x)}{l} = \frac{9(36-9)}{36}$

Negative ordinate $= 6.75$ m.

(*b*) **Net positive moment at $x = 9$ m from 'A' :**

Net positive ordinate of ILD $= 6.75 - \frac{6.75}{18} \times 9 = 3.375$ m

For maximum positive bending moment at x, the load has to be placed on the section x.

Maximum net sagging moment = Load × Net positive ordinate of ILD

$= 8 \times 3.375 = 27$ kNm

(*c*) **Net negative moment at $x = 9$ m from 'A' :**

Net negative ordinate $= 6.75 - \frac{6.75}{27} \times 18 = 2.25$ m

For maximum negative bending moment at x, the load has to be placed at the centre of span.

Maximum net hogging moment = Load × Net negative ordinate of ILD

$= 8 \times 2.25 = 18$ kNm

(*d*) **Absolute maximum bending moment :**

Absolute maximum positive bending moment

$= 0.096\ \text{W}l$

$= 0.096 \times 8 \times 36$

$= 27.648$ kNm. Occurs at $0.211l$

$0.211\,(36) = 7.6$ m from left and right supports.

Absolute maximum negative bending moment

$= 0.0625\ \text{W}l$

$= 0.0625 \times 8 \times 36 = 18.09$ kNm, occurs

at $0.25l = 0.25(36) = 9$ m from left and right supports.

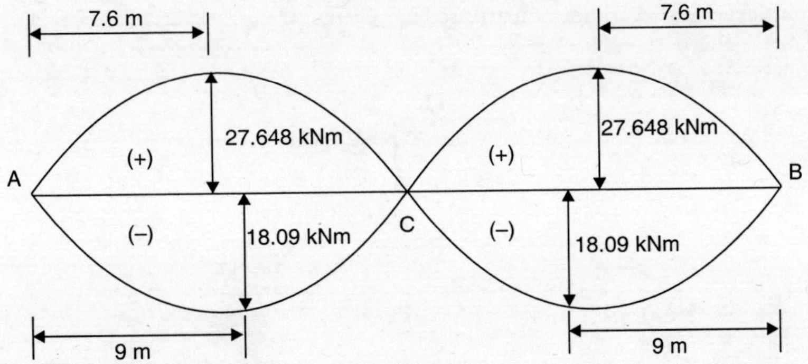

Fig. 4.38

4.13 TWO HINGED ARCHES

Three hinged arches have a hinge at the crown, not to make the structure more efficient, but to make the computations simple. Two hinged arches are more practical. But they are statically indeterminate and we have to invoke Castigliano's II theorem to sort out the support reactions.

4.13.1 Solving Two Hinged Arches

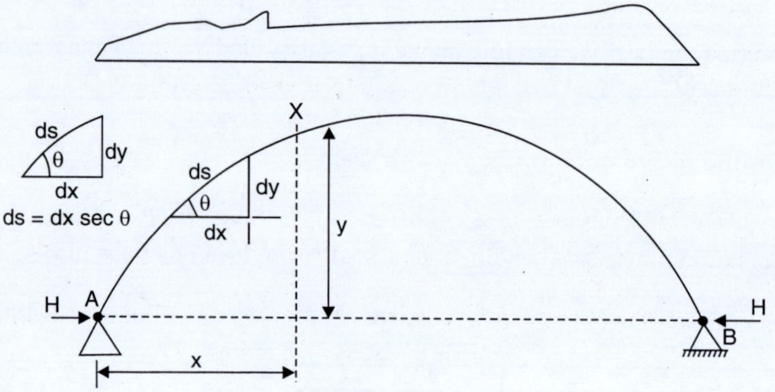

Fig. 4.39

In the arch in Fig. 4.39, let us treat the horizontal force H at B as a redundant. B is actually hinged. But we state that it is on rollers and has a reaction H (the unknown here).

We will assume that the bending moment is the major force creating strain energy in the arch and willfully ignore strain energies due to shear and normal thrust.

$$M_x = \mu_x - Hy$$

$$U = \int_A^B \frac{M_x^2\, ds}{2EI},$$

the integration being performed along the curve of the arch from A to B. Since B is hinged, there is no horizontal displacement there.

Hence, $\dfrac{\partial U}{\partial H} = 0$

$$U = \int_A^B \frac{M_x^2\, ds}{2EI} = \int_A^B \frac{(\mu - Hy)^2\, ds}{2EI}$$

$$\frac{\partial U}{\partial H} = 0 \qquad \text{...(4.13)}$$

$$\int_A^B \frac{(\mu - Hy).(-y)\, ds}{EI} = 0$$

$$-\int \frac{\mu y\, ds}{EI} + \int \frac{Hy^2\, ds}{EI} = 0 \qquad \text{...(4.14)}$$

$$H = \frac{\int_A^B \mu y \, ds/EI}{\int_A^B y^2 \, ds/EI} \qquad ...(4.15)$$

This integration is usually difficult to perform since $ds = \sqrt{dx^2 + dy^2}$

To make things somewhat simpler for calculations, it is customary to assume that I varies as

$$I = I_0 \sec \theta \qquad ...(4.16)$$

where I_0 is the M.I at the crown.

Since $ds = dx \sec \theta$, equation (4.15) gets simplified as,

$$H = \frac{\int_0^l \mu y \, dx}{\int_0^l y^2 \, dx} \qquad ...(4.17)$$

This is applicable to all shapes of arches. But y will vary according to the geometry of the arch. If the simplification in eqn. (4.16) is not available, it is possible to carryout the integrations in 4.13 by a suitable numerical or graphical method, to evaluate H.

4.13.2 Two Hinged Parabolic Arch

(*a*) If the arch profile is parabolic, the loading uniform, and if $I = I_0 \sec \theta$, we know that

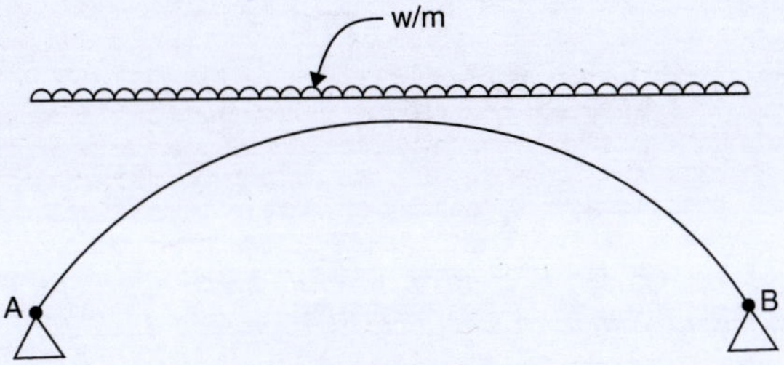

Fig. 4.40 (*a*)

$$y = \frac{4y_C}{l^2} x (l - x)$$

$$\mu = \frac{wl}{2} \cdot x - w.x. \frac{x}{2} = \frac{wx}{2} (l - x)$$

$$\int_0^l \mu y \, dx = \frac{4y_C w}{2l^2} \int_0^l x^2 (l - x)^2 \, dx$$

$$= \frac{2y_C w}{l^2} \int_0^l (x^2 l^2 - 2x^3 l + x^4) \, dx$$

$$= \frac{2y_C w}{l^2} l^5 \left[\frac{1}{3} - \frac{2}{4} + \frac{1}{5}\right] = \frac{y_C w l^3}{15}$$

$$\int_0^l y^2\, dx = \frac{16{y_C}^2}{l^4} \frac{l^5}{30} = \frac{8{y_C}^2 l}{15} \qquad \text{...(4.18)}$$

$$H = \frac{\int_0^l \mu y\, dx}{\int_0^l y^2\, dx} = \frac{y_C w l^3}{15} \cdot \frac{15}{8{y_C}^2 l} = \frac{wl^2}{8y_C}$$

Thus, in a parabolic two hinged arch uniformly loaded, with $I = I_0 \sec \theta$.

The horizontal thrust is $H = \dfrac{wl^2}{8y_C}$ ...(4.19)

(*b*) Parabolic arch with a single concentrated, load at centre, if $I = I_0 \sec \theta$

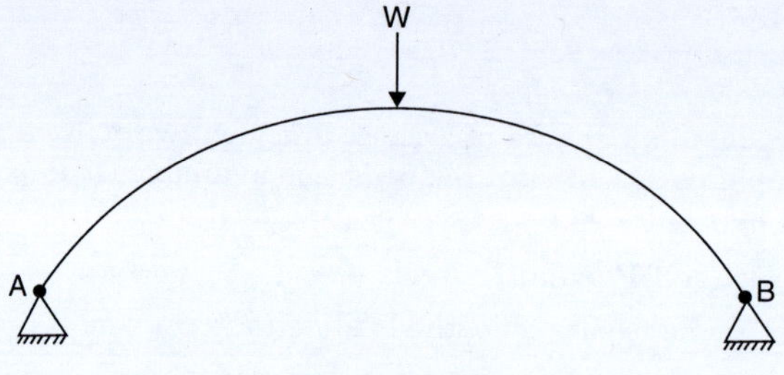

Fig. 4.40 (*b*)

$$y = \frac{4y_C}{l^2} . x(l - x)$$

$$\mu = \frac{Wx}{2}$$

Example 4.16 *A parabolic arch hinged at the ends has a span of 60 m and a rise of 12 m. A concentrated load of 8 kN acts at 15 m from the left hinge. The second moment of area varies as the secant of the inclination of the arch axis. Calculate the horizontal thrust and the reactions at the hinge. Also calculate the net bending moment at the section.*

Solution (*a*) **Vertical reactions at A and B**

Taking moments about A,

$$8(15) - V_B (60) = 0$$

$$V_B = 2 \text{ kN}$$

$$V_A = \text{Total load} - V_B$$

$$V_A = 8 - 2 = 6 \text{ kN.}$$

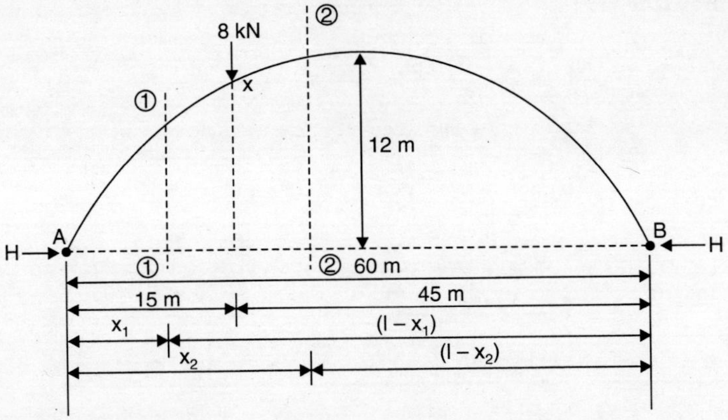

Fig. 4.41

(*b*) **Horizontal thrust (H)**

Horizontal thrust,
$$H = \frac{\int_0^l \mu y\, dx}{\int_0^l y^2\, dx}$$

$$H = \frac{\int_0^{15} \mu_1 y\, dx + \int_{15}^{60} \mu_2 y\, dx}{\int_0^l y^2\, dx} = \frac{\text{Nr}(1) + \text{Nr}(2)}{\text{Dr}}$$

where μ_1 = Beam bending moment in AX

μ_2 = Beam bending moment in XB

$$y = \frac{4r}{l^2} x\,(l - x)$$

(*i*) **Denominator :**

$$\text{Dr} = \int_0^l y^2\, dx = \int_0^l \left(\frac{4r\,x}{l^2}(l - x)\right)^2 dx$$

$$= \int_0^{60} \left(\frac{4 \times 12}{(60)^2} x\,(60 - x)\right)^2 dx$$

$$= \frac{1}{5625}\left[\frac{3600x^3}{3} - \frac{120x^4}{4} + \frac{x^5}{5}\right]_0^{60}$$

$$= \frac{1}{5625}\left[1200\,(60)^3 - 30\,(60)^4 + \frac{(60)^5}{5} - 0\right]$$

$$\text{Dr} = 4608$$

(*ii*) **Numerator (1) :**

$$\text{Nr }(1) = \int_0^{15} \mu_1 \, y \, dx$$

$$\mu_1 = V_A \, x_1 = 6x_1 = 6x$$

$$\text{Nr }(1) = \int_0^{15} \left[6x \, . \frac{4r}{l^2} x \, (l - x) \right] dx$$

$$= \int_0^{15} \left[6x \, . \frac{4 \times 12}{(60)^2} x \, (60 - x) \right] dx = \int_0^{15} \frac{1}{12.5} \, (x^2 \, (60 - x)) \, dx$$

$$\text{Nr }(1) = \frac{1}{12.5} \int_0^{15} (60x^2 - x^3) \, dx = \frac{1}{12.5} \left[\frac{60x^3}{3} - \frac{x^4}{4} \right]_0^{15}$$

$$= \frac{1}{12.5} \left[20(15)^3 - \frac{(15)^4}{4} \right]$$

$$\text{Nr }(1) = 4387.5$$

(*iii*) **Numerator (2) :**

$$\text{Nr }(2) = \int_{15}^{60} \mu_2 \, y \, dx$$

$$\mu_2 = V_A \, (x_2) - 8 \, (x_2 - 15) = V_A \, . \, x_2 - 8x_2 + 120$$

$$= 6x_2 - 8x_2 + 120 = 120 - 2x_2$$

$$\mu_2 = 120 - 2x$$

$$\text{Nr }(2) = \int_{15}^{60} (120 - 2x) \left[\frac{4r}{l^2} (lx - x^2) \right] dx$$

$$= \int_{15}^{60} (120 - 2x) \left[\frac{1}{75} (60x - x^2) \right] dx$$

$$\text{Nr }(2) = \frac{1}{75} \int_{15}^{60} x(120 - 2x)(60 - x) \, dx$$

$$= \frac{1}{75} \int_{15}^{60} x \, [7200 - 120x - 120x + 2x^2] \, dx$$

$$= \frac{1}{75} \int_{15}^{60} [7200x - 240x^2 + 2x^3] \, dx$$

$$= \frac{1}{75} \left[7200 \frac{x^2}{2} - 240 \frac{x^3}{3} + \frac{2x^4}{4} \right]_{15}^{60}$$

$$= \frac{1}{75} \left[3600 \, (60^2 - 15^2) - 80(60^3 - 15^3) + \frac{(60^4 - 15^4)}{2} \right]$$

$$\text{Nr }(2) = 21262.5$$

Horizontal thrust, $$H = \frac{Nr(1) + Nr(2)}{Dr(1)} = \frac{4387.5 + 21262.5}{4608}$$

$$H = 5.566 = 5.57 \text{ kN}$$

(*c*) **Reactions at 'A' and 'B'**

$$R_A = \sqrt{V_A^2 + H^2} = \sqrt{6^2 + 5.57^2} = 8.18 \text{ kN}$$

$$R_B = \sqrt{V_B^2 + H^2} = \sqrt{2^2 + 5.57^2} = 5.91 \text{ kN}$$

(*d*) **Maximum bending moment**

$$M_x = + V_A (15) - H(y)$$

$$y = \frac{4 \times 12}{(60)^2} \times 15 \times (60 - 15) = 9 \text{ m}$$

$$M_{max} = 6(15) - 5.57 (9) = 39.87 \text{ kNm.}$$

Example 4.17 *A two hinged parabolic arch of span l and rise r carries a u.d.l. of w/metre run over the left hand half of the span. The moment of inertia of the arch rib varies as the secant of the slope of the rib axis .*

(a) Obtain the expression for the horizontal thrust H.

(b) Calculate the horizontal thrust and bending moment at quarter span point on the right half of the span if l = 20 m, r = 4 m and w = 20 kN/m.

Solution (*a*) **Reactions**

To find V_A, taking moments about B

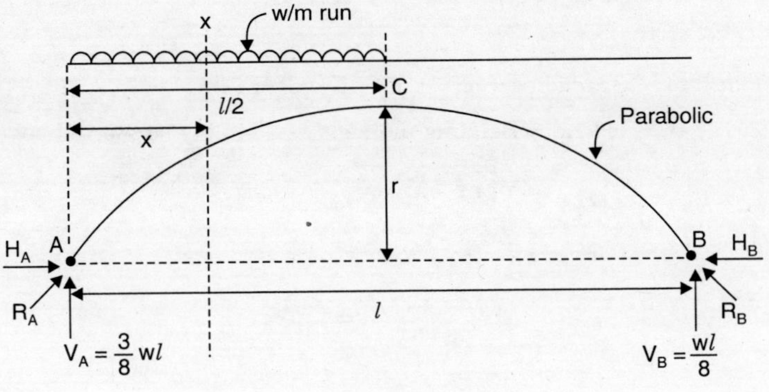

Fig. 4.42

$$V_A \times l - w \times \frac{l}{2}\left(\frac{l}{2} + \frac{1}{2} \times \frac{l}{2}\right) = 0$$

$$V_A \times l = \frac{3}{8} wl^2$$

$$V_A = \frac{3}{8} wl$$

$$V_B = \text{Total load} - V_A = w\frac{l}{2} - \frac{3}{8}wl$$

$$V_B = \frac{wl}{8}$$

Horizontal thrust H is given by, $H = \dfrac{\int \mu\, y\, dx}{\int y^2\, dx}$

Denominator:

$$\int_0^l y^2\, dx = \int_0^l \left(\frac{4r}{l^2}x\,(l-x)\right)^2 dx = \int_0^l \frac{16r^2}{l^4}\, x^2\,(l^2 + x^2 - 2lx)\, dx$$

$$= \frac{16r^2}{l^4}\int_0^l (l^2x^2 + x^4 - 2lx^3)\, dx = \frac{16r^2}{l^4}\left[\frac{l^2x^3}{3} + \frac{x^5}{5} - \frac{2lx^4}{4}\right]_0^l$$

$$= \frac{16r^2}{l^4}\left[\left(\frac{l^2\,l^3}{3} + \frac{l^5}{5} - \frac{2l\,l^4}{4}\right) - (0+0-0)\right]$$

$$= \frac{16r^2}{l^4}\left[\frac{l^5}{3} + \frac{l^5}{5} - \frac{l^5}{2}\right] = \frac{16r^2}{l^4}\left[\frac{10l^5 + 6l^5 - 15l^5}{30}\right]$$

$$= \frac{16r^2}{l^4} \times \frac{l^5}{30} = \frac{8r^2l}{15}$$

Therefore, $\displaystyle\int y^2\, dx = \frac{8r^2l}{15}$ ...(1)

Numerator = Nr (1) + Nr (2)

Nr (1) for 0 to $\dfrac{l}{2}$ (loaded portion)

μ_1 = Bending moment as a simply supported beam

$$= V_A \,.\, x - w \,.\, x \,.\, \frac{x}{2} = \frac{3}{8}wlx - \frac{wx^2}{2}$$

$$\text{Nr}\,(1) = \int_0^{l/2} \mu_1\, y\, dx$$

$$= \int_0^{l/2}\left(\frac{3}{8}wlx - \frac{wx^2}{2}\right)\left(\frac{4r}{l^2}x\,(l-x)\right) dx$$

$$= \frac{4r}{l^2}\int_0^{l/2}\left(\frac{3}{8}wl^2x^2 - \frac{3}{8}wl\,x^3 - \frac{wlx^3}{2} + \frac{wx^4}{2}\right) dx$$

$$= \frac{4r}{l^2}\left[\frac{3}{8}wl^2\,\frac{x^3}{3} - \frac{3}{8}wl\,\frac{x^4}{4} - \frac{wl}{2}\,\frac{x^4}{4} + \frac{w}{2}\,\frac{x^5}{5}\right]_0^{l/2}$$

$$= \frac{4r}{l^2}\left[\frac{wl^2}{8}\left(\frac{l}{2}\right)^3 - \frac{3}{32}wl\left(\frac{l}{2}\right)^4 - \frac{wl}{8}\left(\frac{l}{2}\right)^4 + \frac{w}{10}\left(\frac{l}{2}\right)^5\right]$$

$$= \frac{4r}{l^2}\left[\frac{wl^2}{8}\times\frac{l^3}{8}-\frac{3}{32}wl\times\frac{l^4}{16}-\frac{wl}{8}\times\frac{l^4}{16}+\frac{w}{10}\times\frac{l^5}{32}\right]$$

$$= \frac{4r}{l^2}\times wl^5\left[\frac{1}{64}-\frac{3}{512}-\frac{1}{128}+\frac{1}{320}\right]$$

$$= 4wrl^3\left[\frac{20-7.5-10+4}{1280}\right]$$

$$\text{Nr (1)} = \frac{6.5\,wl^3r}{320} \quad \text{...(2)}$$

Nr (2) for $\frac{l}{2}$ to l (unloaded portion)

$$\mu_2 = \text{Bending moment as a simply supported beam}$$

$$= V_B(l-x) = \frac{wl}{8}(l-x)$$

$$\text{Nr (2)} = \int_{l/2}^{l}\mu_2\, y\, dx = \int_{l/2}^{l}\frac{wl}{8}(l-x)\frac{4r}{l^2}x(l-x)\,dx$$

$$= \frac{wl}{8}\times\frac{4r}{l^2}\int_{l/2}^{l}x(l-x)^2\,dx = \frac{wr}{2l}\int_{l/2}^{l}x(l^2+x^2-2lx)\,dx$$

$$= \frac{wr}{2l}\int_{l/2}^{l}(l^2x+x^3-2lx^2)\,dx = \frac{wr}{2l}\left[l^2\frac{x^2}{2}+\frac{x^4}{4}-2l\frac{x^3}{3}\right]_{l/2}^{l}$$

$$= \frac{wr}{2l}\left[\frac{l^2}{2}x^2+\frac{1}{4}x^4-\frac{2l}{3}x^3\right]_{l/2}^{l}$$

$$= \frac{wr}{2l}\left[\frac{l^2}{2}\left(l^2-\frac{l^2}{4}\right)+\frac{1}{4}\left(l^4-\frac{l^4}{16}\right)-\frac{2l}{3}\left(l^3-\frac{l^3}{8}\right)\right]$$

$$= \frac{wr}{2l}\left[\frac{l^2}{2}\times\frac{3}{4}l^2+\frac{1}{4}\times\frac{15l^4}{16}-\frac{2l}{3}\times\frac{7l^3}{8}\right]$$

$$= \frac{wrl^4}{2l}\left[\frac{3}{8}+\frac{15}{64}-\frac{14}{24}\right] = \frac{wrl^3}{2}\left[\frac{72+45-112}{192}\right]$$

$$\text{Nr (2)} = \frac{5wrl^3}{384} \quad \text{...(3)}$$

$$\text{Nr} = \text{Nr (1)} + \text{Nr (2)}$$

$$= \frac{6.5wrl^3}{320}+\frac{5wrl^3}{384} = \frac{(39+25)\,wrl^3}{1920} = \frac{64wrl^3}{1920}$$

$$\text{Nr} = \frac{wrl^3}{30}$$

Therefore,

$$H = \frac{\text{Nr}}{\text{Dr}} = \frac{\left(\frac{wrl^3}{30}\right)}{\left(\frac{8}{15}r^2l\right)} = \frac{wrl^3}{30}\times\frac{15}{8r^2l} = \frac{wl^2}{16r}$$

Therefore, horizontal thrust, $H = \frac{wl^2}{16r}$*

(*b*) **When l = 20 m, r = 4 m and w = 20 kN/m**

$$H = \frac{20 \times 20^2}{16 \times 4} = 125 \text{ kN}$$

$$V_A = \frac{3}{8}wl = \frac{3}{8} \times 20 \times 20 = 150 \text{ kN}$$

$$V_B = \frac{wl}{8} = \frac{20 \times 20}{8} = 50 \text{ kN}$$

Rise of the arch at right quarter span point

$$y = \frac{4r}{l^2}x(l-x)$$

$$= \frac{4 \times 4}{20^2} \times \frac{3}{4} \times 20 \times \frac{20}{4} = 3 \text{ m}$$

$$x = \frac{3}{4}l = \frac{3}{4} \times 20$$

$$(l - x) = \left(l - \frac{3}{4}l\right) = \frac{l}{4} = \frac{20}{4}$$

Bending moment at right quarter span point

$$= V_B \times 5 - H \times y_5 = 50 \times 5 - 125 \times 3 = -125 \text{ kNm}$$

B.M. = – 125 kNm.

Example 4.18 *Derive an expression for the horizontal thrust in a two hinged semicircular arch carrying a point load at the crown.*

Solution

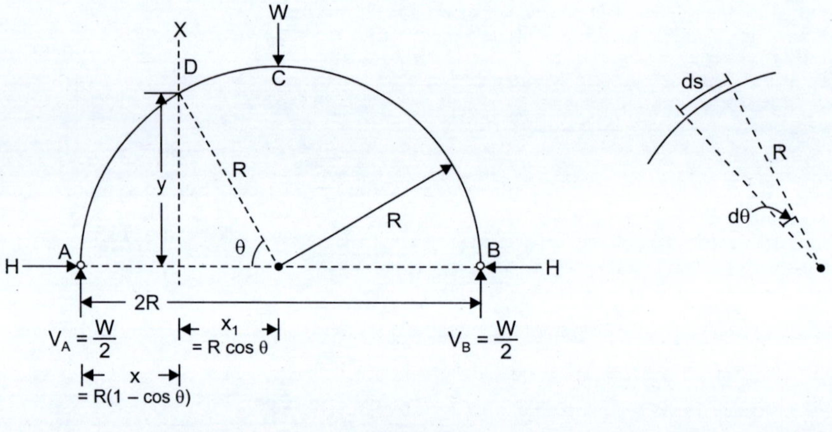

Fig. 4.43

* This result can be obtained by considering the fact that if the right side were also loaded, H would be $\frac{wl^2}{8r}$ [equation (4.19)]. For one half alone loaded $H = \frac{wl^2}{16r}$.

Figure 4.43 shows the two hinged semi-circular arch of radius R with a point load W at the crown C. Considering a point D on the arch at a section XX at x_1 from the centre of its span the coordinates of the point D are,

$$\frac{x_1}{R} = \cos\theta \qquad x_1 = R\cos\theta$$

$$\frac{y}{R} = \sin\theta \qquad y = R\sin\theta$$

Distance of section XX from the left hand hinge

$$x = R - x_1 = R - R\cos\theta = R(1-\cos\theta)$$

Bending moment at section XX, *i.e.*, at D = $M_x = \mu - Hy$

$$\mu = V_A \cdot x = \frac{W}{2} \cdot R(1-\cos\theta)$$

$$Hy = H \cdot R\sin\theta$$

$$\therefore \qquad M_x = \frac{W}{2} R(1-\cos\theta) - H \cdot R\sin\theta$$

Partially differentiating w.r.t. to H

$$\frac{\partial M}{\partial H} = -R\sin\theta$$

$$ds = R \cdot d\theta$$

Strain energy due to bending, $U = \int \frac{M_x^2\, ds}{2\,EI}$

$$U = \frac{1}{2\,EI}\int (\mu - Hy)^2 \cdot R \cdot d\theta \qquad \left[\begin{aligned} M &= \mu - Hy \\ \frac{\partial M}{\partial H} &= -y \\ &= (-R\sin\theta) \end{aligned}\right.$$

$$\frac{\partial U}{\partial H} = 0 \qquad (\because \text{ A and B are hinged})$$

$$\frac{\partial U}{\partial H} = 0 = \frac{2}{2\,EI}\int_0^{\pi/2} 2(\mu - Hy) \cdot \frac{\partial M}{\partial H} \cdot R \cdot d\theta$$

$$0 = \frac{2}{EI}\int_0^{\pi/2}\left[\frac{WR}{2}(1-\cos\theta) - HR\sin\theta\right][-R\sin\theta]\, R \cdot d\theta$$

$$0 = \frac{2}{EI}\int_0^{\pi/2} R\left[\frac{W}{2}(1-\cos\theta) - H\sin\theta\right][-R\sin\theta]\, R \times d\theta$$

$$0 = \frac{2}{EI}\int_0^{\pi/2} -R^3\left[\frac{W}{2}(1-\cos\theta)\sin\theta - H\sin^2\theta\right] d\theta$$

$$0 = \frac{-2R^3}{EI}\left[\int_0^{\pi/2}\frac{W}{2}(1-\cos\theta)\sin\theta\, d\theta - \int_0^{\pi/2} H\sin^2\theta\, d\theta\right]$$

$$0 = \int_0^{\pi/2} \frac{W}{2}[1 - \cos\theta]\sin\theta\, d\theta - \int_0^{\pi/2} H \times \sin^2\theta\, d\theta$$

$$0 = \frac{W}{2}\int_0^1 u\, du - \int_0^{\pi/2} H \times \sin^2\theta\, d\theta$$

$$1 - \cos\theta = u$$

$$-(-\sin\theta)\, d\theta = du$$

$$\sin\theta\, d\theta = du$$

$$0 = \frac{W}{2}\left[\frac{u^2}{2}\right]_0^1 - H \times \left[\frac{1}{2} \times \frac{\pi}{2}\right]$$

$$0 = \frac{W}{2} \times \left[\frac{1}{2}\right] - H \times \left[\frac{\pi}{4}\right]$$

θ	0	$\frac{\pi}{2}$
u	0	1

$$\int_0^{\pi/2} \sin^2\theta = \frac{1}{2} \times \frac{\pi}{2}$$

$$\frac{H\pi}{4} = \frac{W}{4}$$

∴ Horizontal thrust, $\mathbf{H = \frac{W}{\pi}}$.

Example 4.19 *The equation to the axis of a two hinged arch is* $y = y_C \sin\frac{\pi x}{l}$ *where* y_C *is the central rise of the arch and* l *is the span. The moment of inertia of any section of the arch is equal to* $I_0 \sec\theta$, *where* I_0 *is the moment of inertia at the crown and* θ *is the slope of the arch axis at the section. A point load W is acting vertically at the left quarter point. Show that the horizontal thrust in the arch is given by*

$$H = \frac{\sqrt{2}Wl}{\pi^2 y_C}$$

Solution

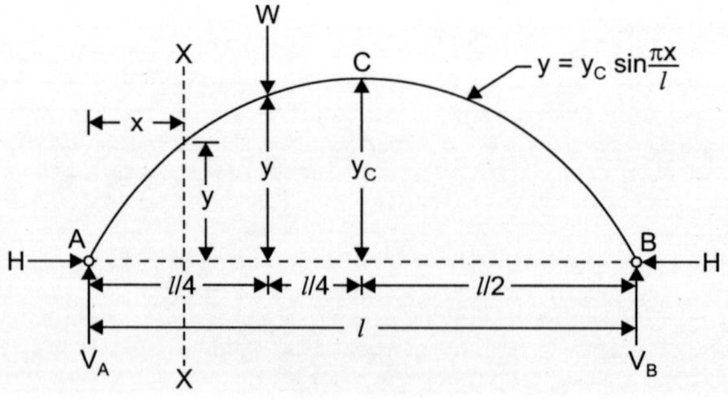

Fig. 4.44

The two hinged arch with load is shown in Fig. 4.44

Taking moments about B,

$$V_A \times l - W \times \frac{3}{4} l = 0$$

$$V_A = \frac{3}{4} W$$

$$V_B = \text{Total load} - V_A$$

$$= W - \frac{3}{4} W = \frac{W}{4}$$

In the case of a two hinged arch, the horizontal thrust is given by,

$$H = \frac{\int_0^l \mu y \, dx}{\int_0^l y^2 \, dx}$$

Numerator :

$$\int_0^l \mu y \, dx = \int_0^{l/4} V_A \,.\, x \,.\, y_C \sin \frac{\pi x}{l} \,.\, dx + \int_{l/4}^l V_B \,.\, (l - x) \,.\, y_C \sin \frac{\pi x}{l} \,.\, dx$$

$$= \int_0^{l/4} \frac{3}{4} W \,.\, x \,.\, y_C \sin \frac{\pi x}{l} \,.\, dx + \int_{l/4}^l \frac{W}{4} \,.\, (l - x) \,.\, y_C \sin \frac{\pi x}{l} \,.\, dx$$

$$= \frac{3}{4} W y_C \left[\int_0^{l/4} x \sin \frac{\pi x}{l} dx \right] + \frac{W}{4} y_C \left[\int_{l/4}^l (l - x) \sin \frac{\pi x}{l} dx \right]$$

$$\boxed{\int u dv = uv - \int v \, du}$$

$$= \frac{3}{4} W y_C \left[\left[x \times \frac{-\cos \frac{\pi x}{l}}{\frac{\pi}{l}} \right] - \left[1 \times \frac{-\sin \frac{\pi x}{l}}{\frac{\pi^2}{l^2}} \right] \right]_0^{l/4}$$

$$+ \frac{W}{4} y_C \left[(l - x) \left[\frac{-\cos \pi \frac{\pi x}{l}}{\frac{\pi}{l}} \right] - \left[-1 \times \frac{-\sin \frac{\pi x}{l}}{\frac{\pi^2}{l^2}} \right] \right]_{l/4}^l$$

$$= \frac{3}{4} W y_C \left[-\frac{l}{4} \times \frac{\cos \frac{\pi}{4}}{\frac{\pi}{l}} + \frac{\sin \frac{\pi}{4}}{\frac{\pi^2}{l^2}} \right]$$

$$+ \frac{W}{4} y_C \left[0 - 0 - \left[\left(l - \frac{l}{4} \right) \left[\frac{-\cos \frac{\pi}{4}}{\frac{\pi}{l}} \right] - \frac{\sin \frac{\pi}{4}}{\frac{\pi^2}{l^2}} \right] \right]$$

$$= \frac{3}{4} W y_C \left[-\frac{l}{4} \times \frac{\sqrt{3}}{2} \times \frac{l}{\pi} + \frac{1}{\sqrt{2}} \times \frac{l^2}{\pi^2} \right]$$

$$+ \frac{W}{4} y_C \left[\frac{3l}{4} \times \frac{\sqrt{3}}{2} \times \frac{l}{\pi} + \frac{1}{\sqrt{2}} \times \frac{l^2}{\pi^2} \right]$$

$$= -\frac{3}{4} W \times y_C \times \frac{\sqrt{3} \times l^2}{8\pi} + \frac{3}{4} W y_C \times \frac{l^2}{\sqrt{2}\pi^2}$$

$$+ \frac{W}{4} y_C \left[\frac{3\sqrt{3} l^2}{8\pi} \right] + \frac{W}{4} y_C \times \frac{l^2}{\sqrt{2}\pi^2}$$

$$= \frac{4 W y_C l^2}{4 \times \sqrt{2} \times \pi^2}$$

$$\boxed{\int \mu \, y \, dx = W y_C \cdot \frac{l^2}{\pi^2} \cdot \frac{1}{\sqrt{2}}} \qquad \ldots(1)$$

Denominator :

$$\int_0^l y^2 \, . \, dx = \int_0^l \left[y_C \, . \, \sin \frac{\pi x}{l} \right]^2 . \, dx = \int_0^l y_C^2 \sin^2 \frac{\pi x}{l} \, . \, dx$$

$$= y_C^2 \times \int_0^l \frac{1 - \cos \frac{2\pi x}{l}}{2} \, dx \qquad \left| \sin^2 \theta = \frac{1 - \cos 2\theta}{2} \right.$$

$$= y_C^2 \times \int_0^l \left[\frac{1}{2} - \frac{\cos \frac{2\pi x}{l}}{2} \right] dx = y_C^2 \times \left[\frac{x}{2} - \frac{\sin \frac{2\pi x}{l}}{2 \times \frac{2\pi}{l}} \right]_0^l$$

$$= y_C^2 \left[\frac{l}{2} - 0 - [0 - 0] \right] = y_C^2 \times \frac{l}{2}$$

$$\boxed{\int_0^l y^2 \, dx = y_C^2 \times \frac{l}{2}} \qquad \ldots(2)$$

$$H = \frac{\text{Nr}}{\text{Dr}} = \frac{(1)}{(2)} = \frac{W \, . \, y_C \, \frac{l^2}{\pi^2} \times \frac{1}{\sqrt{2}}}{y_C^2 \times \frac{l}{2}} = \frac{W}{y_C} \cdot \frac{l}{\pi^2} \times \frac{2}{\sqrt{2}}$$

$$H = \frac{\sqrt{2} W l}{\pi^2 y_C}$$

Hence the proof.

4.14 INFLUENCE LINES FOR A PARABOLIC TWO HINGED ARCH WITH $I = I_0 \sec\theta$

(*a*) **Influence line for μ_x :**

Let X be the section for which we want the influence line. Let X be at a distance x from A.

Figure 4.45 (*a*) and (*b*) show the free bms (μ_x) in the arch for two load positions one to the left of X and another to the right of X.

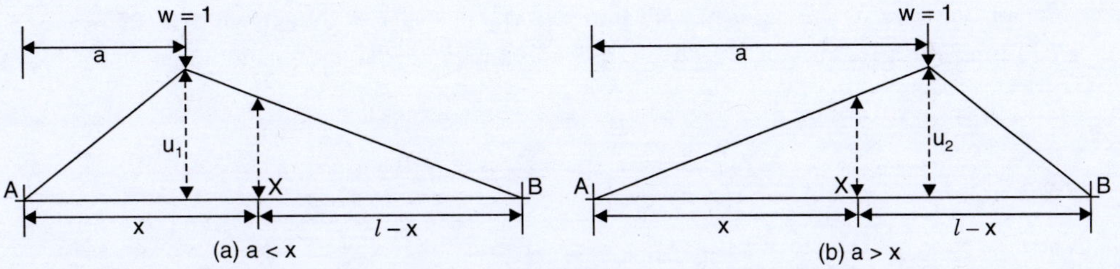

Fig. 4.45

For $a < x$, $$\mu_x = \frac{\mu_1 (l - x)}{l - a} = \frac{a\,(l - a)}{l} \times \frac{(l - x)}{(l - a)}$$

$$\mu_x = \frac{a\,(l - x)}{l} \qquad \ldots[4.20(a)]$$

For $a > x$ $$\mu_x = \frac{\mu_2\, x}{a} = \frac{a\,(l - a)}{l} \times \frac{x}{a}$$

$$\mu_x = \frac{(l - a)\,x}{l} \qquad \ldots[4.20(b)]$$

Figure 4.46 shows the ILD for μ_x.

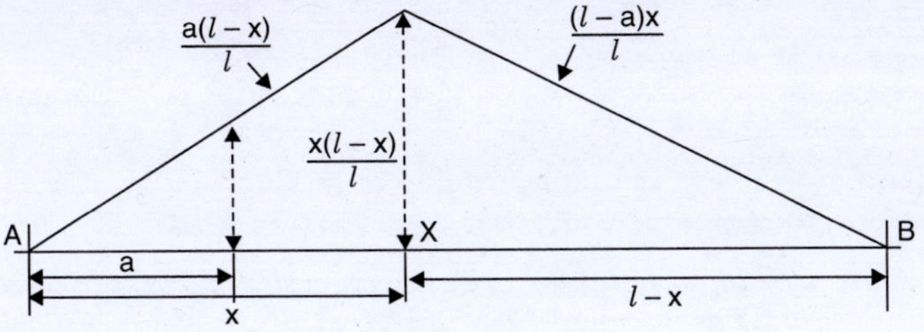

Fig. 4.46. ILD for μ_x.

(*b*) **Influence line for H :**

$$H = \frac{\int \mu y \, dx}{\int y^2 \, dx}$$

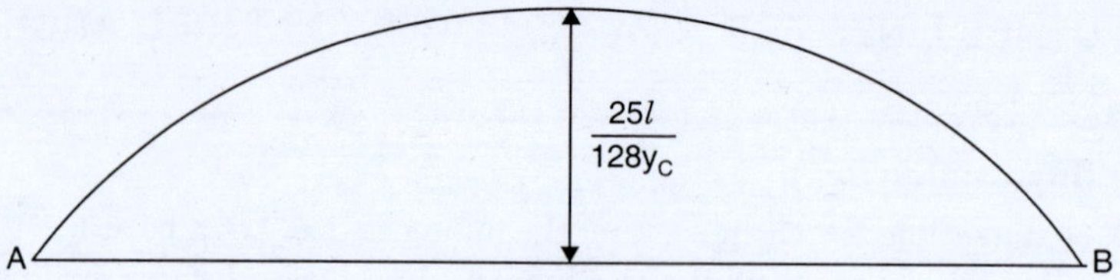

Fig. 4.47. ILD for H.

Figure 4.47 shows the influence line (For the derivation you can consult (Reference 5, 6 or 7 Punmia Jain & Jain) for Hy, the arch bending moment. The net bending moment diagram is shown in Fig. 4.48.

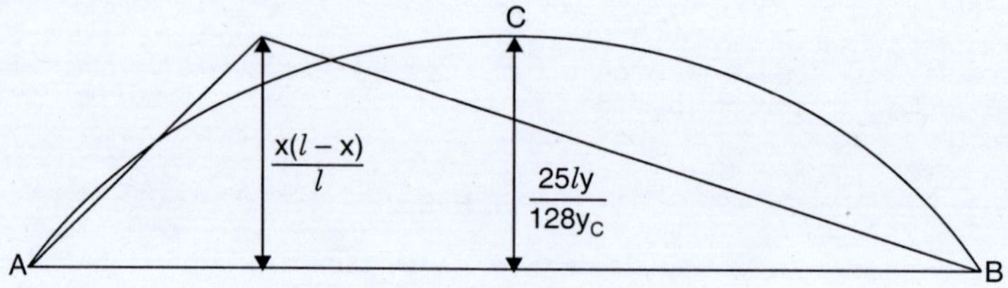

Fig. 4.48. ILD for bm in a two hinged arch.

In Fig. 4.48, the equation to the curve is given by

$$\text{ILO for H} = \frac{5}{8}\frac{l}{y_C} k\,(1-k)\,(1+k-k^2) \qquad \ldots[4.20(c)]$$

where $k = \frac{x}{l}$, representing the load position. Please notice that it is not a parabola, but a 4th degree curve.

when $k = 0.5$, as for C, $H = \frac{25}{128}\frac{l}{y_C}$.

Example 4.20 employs the influence line to tackle the case of a moving load on a 2 hinged parabolic arch.

Example 4.20 *A two hinged parabolic arch has a varying moment of inertia given by $I = I_0$ sec θ. It has a span of 40 m and a central rise of 8 m. Calculate the maximum positive and negative bending moments at a section D 12 m from the left support, due to a moving point load of 6 kN.*

Solution (*a*) **ILD for bending moment**

Figure 4.49 shows the ILD for bending moment at D.

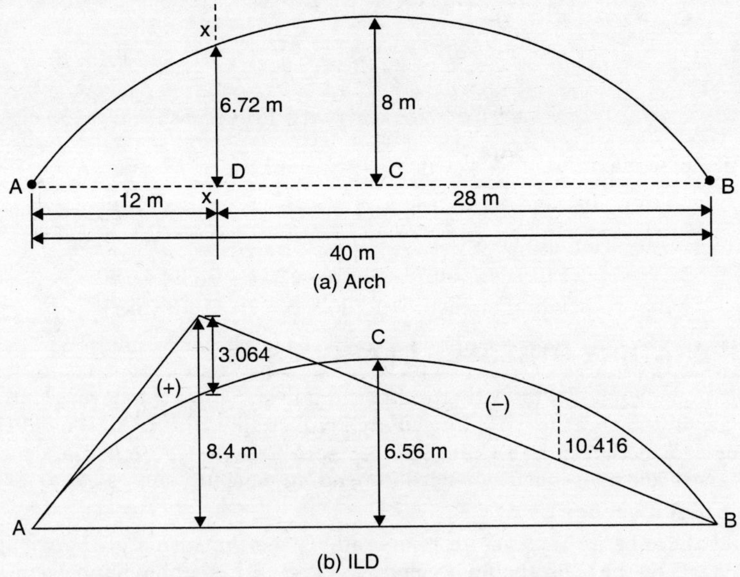

Fig. 4.49

Maximum positive bending moment = $\dfrac{x\,(l-x)}{l}$ = 12 × 28/40 = 8.4 m

Arch rise at D, $y = \dfrac{4y_C}{l^2}x\,(l-x) = \dfrac{4\times 8\times 12\times 28}{40^2} = 6.72$ m

Maximum negative bending moment ordinate = $\dfrac{25ly}{128y_C} = \dfrac{25\times 40\times 6.72}{128\times 8} = 6.56$ m

(*b*) **Maximum positive bending moment at D**

It is clear from the ILD that the load shall be at D to cause maximum positive bending moment at D.

$$k = \frac{x}{l} = \frac{12}{40} = 0.3$$

ILO at D = 8.4 – OD

$$\text{OD} = \frac{5l}{8r}k\,(1-k)\,(1+k-k^2)\,y$$

$$= \frac{5\times 40}{8\times 8}\ 0.3\,(0.7)\,(1+0.3-(0.3)^2)\times 6.72 = 5.336 \text{ m}$$

ILO = 8.4 – 5.336 = 3.064 m

Bending moment at D = 3.064 × 6 = 18.38 kNm.

(*c*) **Maximum negative bending moment**

The maximum negative ILO will occur between C and B. At any x given by kl, bending moment due to unit load (ILO) M_U is given by

$$M_U = \frac{5l}{8y_C} k(1-k)(1+k-k^2)y - \frac{8.4\,(1-k)\,40}{28}$$

$$= \frac{5 \times 40}{8 \times 8} \times 6.72k\,(1-k)(1+k-k^2) - \frac{8.4 \times 40}{28}(1-k)$$

$$M_U = 21(k - 2k^3 + k^4) - 12(1-k)$$

For M_U to be maximum, $\frac{dM_U}{dk} = 0$

$$33 + k^2\,(84k - 126) = 0$$

Solving this by trial and error, $k = 0.7$

$$M_{U(max)} = 21\,(0.7 - 2 \times 0.7^3 + 0.7^4) - 12\,(1 - 0.7) = 1.736$$

Max –ve bending moment = 1.736 × 6 = 10.416 kNm.

4.15 TEMPERATURE STRESSES IN ARCHES

A two hinged arch like any other indeterminate structure can develop internal stresses due to changes in temperature. A 3 hinged arch will simply change its central rise if there is expansion or contraction due to change in temperature. A two hinged arch will develop a horizontal thrust and consequent changes in bending moment, shears etc.

(*a*) **3 hinged arches :**

Should there be an increase in temperate T, the dimension of the whole arch would change by $1 + \alpha T$. So the span should become $l + l\,\alpha t$. But this cannot happen since A and B are position fixed. Hence lengths AC and BC would increase to AC′ and BC′ such that

$$BC' = BC(1 + \alpha T) \text{ and } AC' = AC(1 + \alpha T)$$

increase in central rise= $CC' = \Delta/\sin\theta$

$$\Delta = BC'\,\alpha t = BC\,\alpha t = \frac{l}{2}\sec\theta\,\alpha T$$

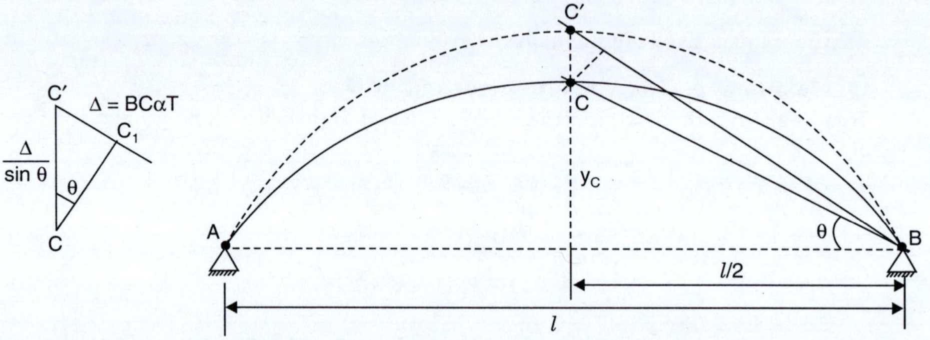

Fig. 4.50

$$CC' = \frac{l}{2}\sec\theta\,\alpha T / \sin\theta = \frac{l}{2}\alpha T \frac{AC}{l/2}\cdot\frac{AC}{y_C} = \frac{\alpha T}{y_C} AC^2$$

Therefore, $$CC' = \frac{l^2 + 4y_C^2}{4y_C}\alpha T \quad \text{...(4.21)}$$

CC′ is the increase in y_C when the temperature change is T°. There would be no consequent increase in H or other forces.

(*b*) **2 hinged arches :**

Let us assume that an increase in temperature of T° increases the span by $l\,\alpha T$. Going back to article 4.13.1 wherein we derived equation (4.17), we will make 2 modifications.

(*i*) H is made up of 2 parts H_0 and H_1, H_0 when there is no change of temperature and H_1 due to the changes of temperature. H_1 is the force that brings the arch into its natural position from an expansion of $l\alpha T$.

(*ii*) So $\dfrac{\partial U}{\partial H} = l\alpha T$ instead of zero.

$$U = \int \frac{(\mu - Hy)^2\,ds}{2EI} = \int \frac{(\mu - (H_0 + H_1)y)^2\,ds}{2EI}$$

$$\frac{\partial U}{\partial H} = l\alpha T \qquad ...(4.22)$$

$$\int \frac{(\mu - Hy)(-y)\,ds}{EI} = l\alpha T$$

$$\int \frac{-\mu\,y\,ds}{EI} + \int \frac{H_0\,y^2\,ds}{EI} + \int \frac{H_1\,y^2\,ds}{EI} = l\alpha T$$

Here the first two terms are equal to zero [equation (4.14)]

So
$$\int \frac{H_1\,y^2\,ds}{EI} = l\alpha T$$

$$H_1 = \frac{l\alpha T\,EI}{\int y^2\,ds} \qquad ...(4.23)$$

H_1 is the additional horizontal thrust due to temperature change.

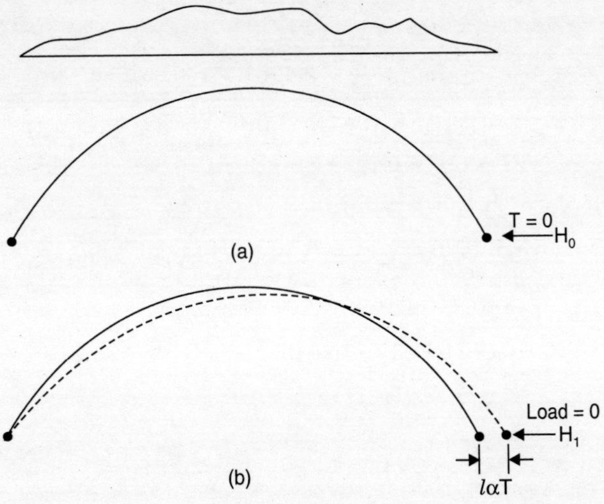

Fig. 4.51

Example 4.21 *A 3 hinged arch has a span of 24 m and a central rise of 8 m. The body of the arch is fabricated from rolled steel sections. Find the change in central rise.*

(i) due to an increase in temperature of 30°C.

(ii) due to a drop in temperature of 20°C.

Solution $l = 24$; $y_C = 8$; $a = 12 \times 10^{-6}/°C$

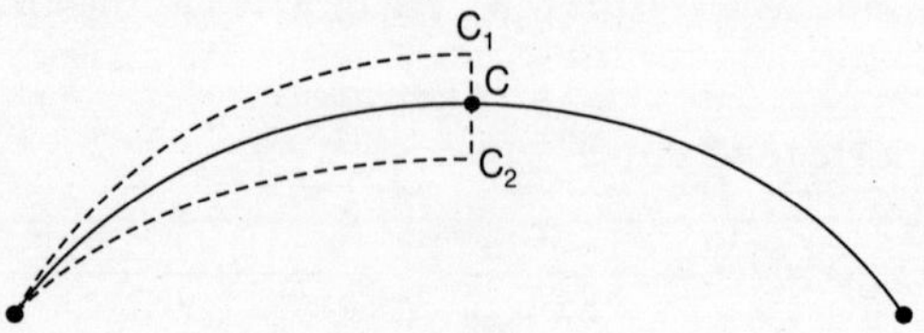

Fig. 4.52

From equation (4.21),

$$(i)\ \frac{l^2 + 4y_C^2}{4y_C}\alpha T = \frac{24^2 + 4 \times 8^2}{4 \times 8} \times 12 \times 10^{-6} \times 30$$

$$= 9360 \times 10^{-6} \text{ m} = 9.36 \text{ mm up}$$

$$(ii)\ CC_2 = \frac{l^2 + 4y_C^2}{4y_C} \times 12 \times 10^{-6} \times 20 = 6.24 \text{ mm down.}$$

Example 4.22 *A reinforced concrete arch is hinged between haunches 42 m apart. It has a central rise of 7 m and a parabolic profile. Find the increment in horizontal thrust in the arch for a temperature increase of 24°C. The section is 120 cm deep and 40 cm wide. E for concrete can be taken as 3×10^4 MPa. α for concrete $= 11.2 \times 10^{-6}/°C$.*

Solution

$$I = \frac{1}{12}bd^3 = \frac{1}{12} \times 0.4 \times 1.2^3 = 0.0576 \text{ m}^4$$

$$E = 3 \times 10^4 \text{ MPa} = 3 \times 10^4 \times 10^6 \text{ N/m}^2 = 3 \times 10^{10} \text{ N/m}^2$$

$$EI = 10^{10} \times 0.1728 \text{ Nm}^2$$

Equation to the arch centre line is

$$y = \frac{4y_C}{l^2}x\,(l - x) = \frac{4 \times 7}{42^2}x\,(42 - x) = \frac{1}{63}(42x - x^2)$$

From equation (4.23),

$$H_1 = \frac{l\,\alpha T\,EI}{\int_0^l y^2\,ds} \approx \frac{l\,\alpha T\,EI}{\int_0^l y^2\,dx}$$

$$\int_0^l y^2\,dx = \int_0^l \frac{1}{63^2}(42^2 x^2 + x^4 - 2 \times 42x^3)\,dx$$

$$= \frac{1}{63^2}\left[42^2 \times \frac{42^3}{3} + \frac{42^5}{5} - 2 \times 42 \times \frac{42^4}{4}\right] = 1097.6$$

$$H_1 = \frac{0.1728 \times 10^{10} \times 42 \times 11.2 \times 10^{-6} \times 24}{1097.6} = 1.777 \times 10^4 \text{ N}$$

$$= 17.77 \text{ kN}$$

This is the horizontal thrust due to temperature rise.

4.16 TWO HINGED CIRCULAR ARCHES

For shallow arches ($y_C < .1l$) circular arches tend to behave like parabolic ones. Semicircular arches are much two deep for convenient use in bridges. They can be used in buildings as decorative arches.

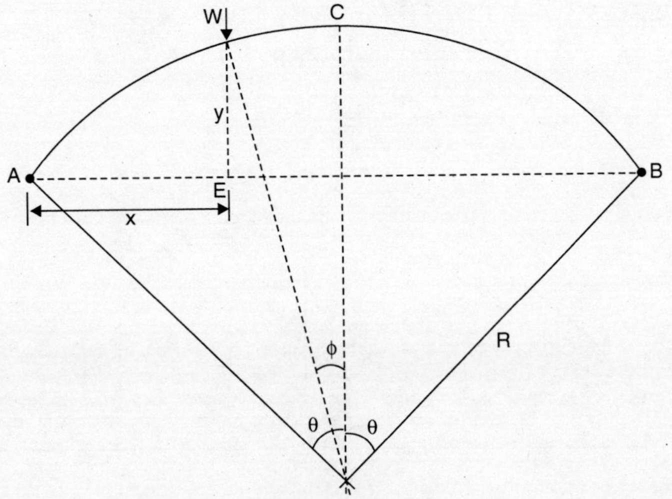

Fig. 4.53

In the mid range between shallow and deep (semicircular), circular arches are distinctive in their behaviour.

It is convenient to consider points on the arch with reference to the angle made by their radius vector from the vertical (Fig. 4.53).

In a circular arch with semicentral angle θ, it can be proved * that the horizontal thrust is given by

$$H = \frac{W \sin^2 \theta - \sin^2 \phi - 2 \cos \theta (\cos \theta - \cos \phi + \theta \sin \theta - \phi \sin \phi)}{(4\, \theta \cos^2 \theta + 2\theta - 3 \sin 2\theta)} \qquad \text{...(4.24)}$$

When we extend this to semicircular arches,

We take $\theta = \frac{\pi}{2}$, then $H = \frac{W \cos^2 \phi}{\pi}$ ...(4.25)

(*Vide reference 1. Punmia, Jain & Jain)

Example 4.23 *Sketch the influence line for bending moment at the crown of a 2 hinged semicircular arch.*

Solution $\theta = 90°$; $y_C = R$; $l = 2R$

In Fig. 4.53 $x = R(1 - \sin \phi)$

$$y = R \cos \phi$$

From eqn. (4.25), $H = \dfrac{W \cos^2 \phi}{\pi}$

When W = 1 $H_U = \dfrac{\cos^2 \phi}{\pi}$

$$M = \mu - H.y = \mu - HR$$

$$M = \mu - \frac{R \cos^2 \phi}{\pi}.$$

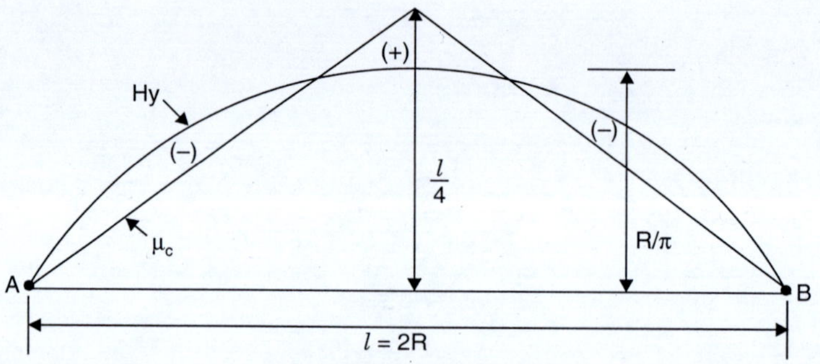

Fig. 4.54. IL for BM at C.

4.17 FIXED ARCHES

Fixed arches are more common than hinged arches. All the masonry arches of the previous centuries should really be considered as fixed arches.

Fixed arches are a little more difficult to analyse than hinged arches. In fact this aspect is the only justification for constructing hinged arches. Actually it takes a lot more effort and expenditure to introduce hinges, than to realize fixity at both supports.

4.17.1 Analysis of Fixed Arches

Fixed arches are these days used for long spans.

In the case of fixed arches, there are six reaction components, 3 at each fixed support. The equilibrium equations available are only 3 hence the structure is statically indeterminate to the 3rd degree. Fixed arches can be analysed using

1. Castigliano's theorem
2. Elastic centre method, or
3. Column analogy method.

This can also be done by the use of indirect model analysis.

Herein the analysis will be done using Castigliano's theorem. The structure will first have to be reduced to a statically determinate structure by identifying any three redundants. This can be done in several ways.

Usually any one fixed support will be removed and the same will be replaced by the reactions at the support. Now, considering the bending moment equation due to the given loading and the redundant, the strain energy (due to bending alone) will be obtained. This will be in terms of the 3 redundants.

We know that, according to Castigliano's theorem 1, the partial derivative of the total strain energy with respect to a force or moment gives the displacement (or rotation). Differentiating the total strain energy with respect to each of the redundant forces, one at a time, and equating to zero (since all displacement at a fixed end will be zero), three simultaneous equations involving the 3 unknown redundants will be obtained. Solving them, the redundants will be obtained. Once these 3 values are known, the forces / moments at any section of the arch can be computed easily.

Example 4.24 *A parabolic arch fixed at both ends has a span of 42 m and a central rise of 8.5 m. It is subjected to concentrated loads of 75 kN and 100 kN at 8 m and 16 m respectively from the left end. The moment of inertia of the arch rib varies as the secant of the inclination of the rib axis. Analyse the arch and find the bending moments at either support and at the crown.*

Solution There are 6 unknown reaction components, 3 at each end. There are only 3 equations of static equilibrium available for the structure. Hence the structure is statically indeterminate to 3rd degree. Treat the 3 reactions *i.e.*, V_B, H_B and M_B as redundants. The statically determinate structure (primary structure) will be as shown in Fig. 4.56.

(*a*) **Equation to the arch**

$$y = \frac{4r}{l^2} \cdot x \cdot (l - x) = \frac{4 \times 8.5}{42^2} \cdot x \cdot (l - x)$$

$$y = 0.0193x\,(l - x) \qquad \text{...(1)}$$

(*b*) **Strain energy (due to bending)**

$$U = \int_0^l \frac{M^2\,ds}{2EI} = \int_0^l \frac{M^2\,dx \sec\theta}{2EI_0 \sec\theta} = \int_0^l \frac{M^2\,dx}{2EI_0}$$

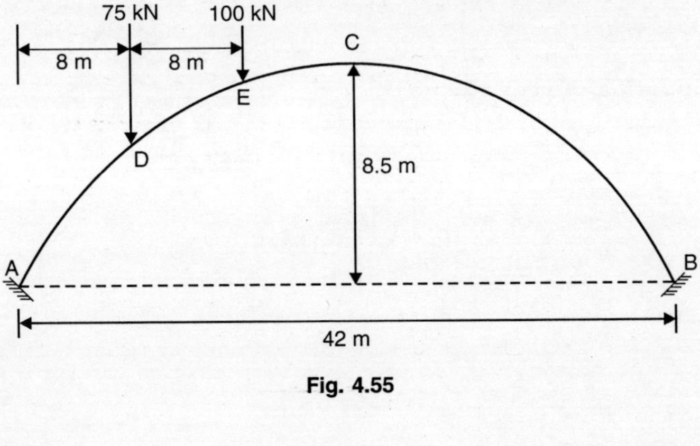

Fig. 4.55

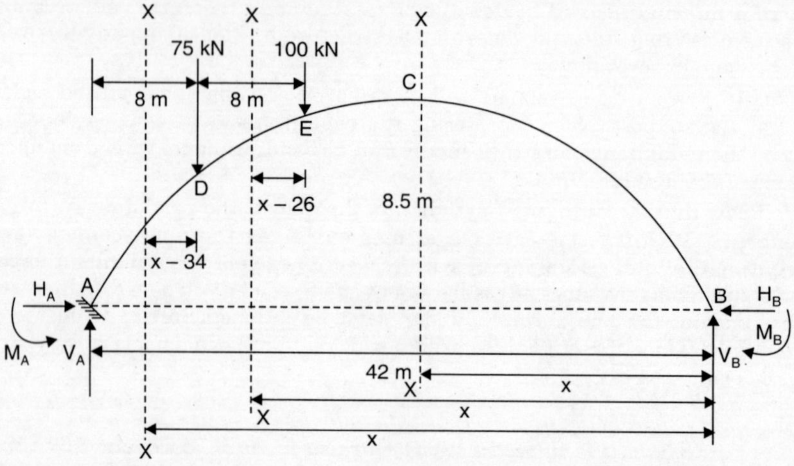

Fig. 4.56

(c) **Equations for M**

These will be different for portions BE, ED and DA and are listed in Table 4.1 below. Since differentiation of the strain energy with respect to the 3 redundant V_B, H_B and M_B are involved in the analysis, these differentials are also listed below the table.

Table 4.1

Portion	*Origin*	*Limits*	M_x
BE	B	0 – 26 m	$V_B \cdot x - H_B y - M_B$
ED	B	26 – 34 m	$V_B \cdot x - H_B y - M_B - 100\,(x - 26)$
DA	B	34 – 42 m	$V_B \cdot x - H_B y - M_B - 100\,(x - 26) - 75\,(x - 34)$

$$\frac{\partial M}{\partial V_B} = x; \quad \frac{\partial M}{\partial H_B} = -y; \quad \frac{\partial M}{\partial M_B} = -1 \qquad \text{...(2)}$$

$$y = \frac{4r}{l^2} \cdot x(l - x) = 0.0193x\,(42 - x) \qquad \text{...(3)}$$

(d) **Evaluation of redundants**

$$\frac{\partial U}{\partial V_B} = \int_0^l \frac{2Mx}{2EI_0} \times \frac{\partial\, Mx}{\partial\, V_B} . dx = \frac{1}{EI_0}\int_0^l Mx . \frac{\partial\, Mx}{\partial\, V_B} . dx$$

$$= \frac{1}{EI_0}\int_0^l Mx . \frac{\partial Mx}{\partial V_B} . dx$$

$$\frac{\partial U}{\partial V_B} = \frac{1}{EI_0}$$

$$\left[\int_0^{26} (V_B\, x - H_B\, y - M_B)\,(x)\,.\,dx + \int_{26}^{34} (V_B\, x - H_B\, y - M_B - 100\,(x-26))\,(x)\,.\,dx + \right.$$

$$\left.\int_{34}^{42} V_B\,.\,x - H_B\,.\,y - M_B - 100(x-26) - 75\,(x-34)\,(x)\,.\,dx\right]$$

$$= \frac{1}{EI_0}\left[\int_0^{42} (V_B\, x^2 - H_B\, 0.0193\, x\,(42-x)\,.\,(x) - M_B\, x)\, dx + \int_{26}^{42} - 100\,(x-26)\,(x)\,.\,dx\right.$$

$$\left. + \int_{34}^{42} - 75\,(x-34)\,(x)\,.\,dx\right]$$

$$= \frac{1}{EI_0}\left[\int_0^{42} (V_B\, x^2 - (0.8106x^2 - 0.0193x^3)\, H_B - M_B\,(x))\, dx + \int_{26}^{42} - 100x^2 + 2600x)\,.\,dx\right.$$

$$\left. + \int_{34}^{42} (-75x^2 + 2550x)\,.\,dx\right]$$

$$= \frac{1}{EI_0}\left[\left(\left\{\frac{V_B\, x^3}{3} - 0.8106\,\frac{x^3}{3} - 0.0193\,\frac{x^4}{4}\right\} H_B - \frac{M_B\, x^2}{2}\right)_0^{42} + \left(\frac{-100x^3}{3} + \frac{2600x^2}{2}\right)_{26}^{42}\right.$$

$$\left. + \left(\frac{-75x^3}{3} + \frac{2550x^2}{2}\right)_{34}^{42}\right]$$

$$= \frac{1}{EI_0}\,[2.47 \times 10^4\, V_B - 5 \times 10^3\, H_B - 882\, M_B$$

$$+ (-176400 - 292933) + (396900 - 491300)]$$

$$= \frac{1}{EI_0}\,[2.47 \times 10^4\, V_B - 5 \times 10^3\, H_B - 882\, M_B - 4.69 \times 10^5 - 9.44 \times 10^4]$$

$$= \frac{1}{EI_0}\,[2.47 \times 10^4\, V_B - 5 \times 10^3\, H_B - 882\, M_B - 5.637 \times 10^5]$$

$$\frac{\partial U}{\partial V_B} = \frac{1}{EI_0}\,[2.47 \times 10^4\, V_B - 5 \times 10^3\, H_B - 882\, M_B - 5.637 \times 10^5] \qquad \ldots(4)$$

$$\frac{\partial U}{\partial H_B} = \frac{1}{EI_0}\int_0^{l} Mx \times \frac{\partial\, Mx}{\partial H_B}\,.\,dx,\quad \frac{\partial U}{\partial H_B} = \frac{1}{EI_0}$$

$$\left[\int_0^{26} (V_B\, x - H_B\, y - M_B)\,(-y)\,.\,dx + \int_{26}^{34} (V_B\, x - H_B\, y - M_B - 100(x-26))\,(-y)\,.\,dx + \right.$$

$$\left.\int_{34}^{42} (V_B\, x - H_B\, y - M_B - 100\,(x-26) - 75\,(x-34)\,(-y)\,.\,dx\right]$$

$$= \frac{1}{EI_0}\left[\int_0^{42} (V_B\, x - H_B\, y - M_B)\,(-y)\, dx + \int_{26}^{42} - 100\,(x-26)\,(-y)\,.\,dx\right.$$

$$\left. + \int_{34}^{42} - 75\,(x-34)\,(-y)\,.\,dx\right]$$

$$= \frac{1}{EI_0}\left[\int_0^{42} (V_B\, x - H_B\, y - M_B)x - 0.0193\, x\,(42 - x)\, dx + \int_{26}^{42} - 100\,(x - 26)\, x \right.$$

$$\left. (- 0.0193x\,(42 - x))\, dx + \int_{34}^{42} - 75x\,(x - 34)\,(- 0.0193x\,(42 - x))\,.\, dx\right]$$

$$= \frac{1}{EI_0}\left[\int_0^{42} \{(- 0.8103x^2 + 0.0193x^3)\, V_B + (0.0193\, x\,(42 - x))^2\, H_B + (0.8106x - 0.0193x^2)\, M_B\}\right.$$

$$\left. + \int_{26}^{42} (1.93x\,(42x - x^2 - 1092 + 26x)\,.\, dx + \int_{34}^{42} (1.45x\,(42x - x^2 - 1428 + 34x)\, dx\right]$$

$$= \frac{1}{EI_0}\left[\int_0^{42} \{(- 0.8103x^2 + 0.0193x^3)\, V_B + (3.72 \times 10^{-4} \times x^2\,(1764 - 84x + x^2)\, H_B\right.$$

$$+ (0.8106 - 0.0193x^2)\, M_B\}\, dx + \int_{26}^{42} (81.06x^2 - 1.93x^3 - 2107.56x$$

$$\left. + 50.18x^2)\,.\, dx + \int_{34}^{42} (60.9x^2 - 1.45x^3 - 2070.6x + 49.3x^2)\,.\, dx\right]$$

$$= \frac{1}{EI_0}\left[\int_0^{42} (- 81.06x^2 + 0.0193x^3)\, V_B + (0.656x^2 - 0.031x^3 + 3.72x \times 10^{-4}\, x^4)\, H_B\right.$$

$$+ (0.8106x - 0.0193x^2)\, M_B\}\, dx$$

$$\left. + \int_{26}^{42} (131.24x^2 - 1.93x^3 - 2107.56x)\,.\, dx + \int_{34}^{42} (110.2x^2 - 1.45x^3 - 2070.6x)\,.\, dx\right]$$

$$= \frac{1}{EI_0}\left[\left[\left(- 0.8106\, \frac{x^3}{3} + 0.0193\, \frac{x^4}{4}\right) V_B + \left(0.656\, \frac{x^3}{3} - 0.031 \frac{x^4}{4} + 3.72 \times 10^{-4}\, \frac{x^5}{5}\right)\right.\right.$$

$$\left. H_B + \left(0.8106\, \frac{x^2}{2} - 0.0193\, \frac{x^3}{3}\right) M_B\right]_0^{42}$$

$$\left. + \left[\left(131.24\, \frac{x^3}{3} - 1.93\, \frac{x^4}{4} - 2107.56\, \frac{x^2}{2}\right)\right]_{26}^{42} + \left[\left(110.2\, \frac{x^3}{3} - 1.45\, \frac{x^4}{4} - 2070.6\, \frac{x^2}{2}\right)\right]_{34}^{42}\right]$$

$$= \frac{1}{EI_0}\ [- 5 \times 10^{-3}\, V_B + 1.81 \times 10^3\, H_B + 2.38 \times 10^2\, M_B + 4.98 \times 10^4]$$

$$\frac{\partial U}{\partial H_B} = \frac{1}{EI_0}$$

$$[- 5 \times 10^{-3}\, V_B + 1.81 \times 10^3\, H_B + 2.38 \times 10^2\, M_B + 4.98 \times 10^4] \quad \ldots.(5)$$

$$\frac{\partial U}{\partial M_B} = \frac{1}{EI_0}\int_0^l M_x\, \frac{\partial M}{\partial M_B}\, dx = \frac{1}{EI_0}\int_0^{42} M_x\, \frac{\partial M}{\partial M_B}\, dx$$

$$= \frac{1}{EI_0}\left[\int_0^{42} (V_B\, x - H_B y - M_B)\,(- 1)\, dx \int_{26}^{42} (- 100\,(x - 26)\,(- 1))\, dx\right.$$

$$\left. + \int_{34}^{42} (- 75\,(x - 34)\,(- 1))\,.\, dx\right]$$

$$= \frac{1}{EI_0}\left[\int_0^{42} (-V_B\, x + H_B\, 0.0193x\,(42 - x) + M_B)\, dx + \int_{26}^{42} (100x - 2600)\, dx + \int_{34}^{42} (-75x - 2550)\, dx\right]$$

$$= \frac{1}{EI_0}\left[\int_0^{42} (-V_B\, x + (0.8106x - 0.0193x^2)\, H_B + M_B)\, dx + \int_{26}^{42} (100x - 2600)\, dx + \int_{34}^{42} (75x - 2550)\,.\, dx\right]$$

$$= \frac{1}{EI_0}\left[\left[\left(-V_B\frac{x^2}{2} + \left(0.8106\frac{x^2}{2} - 0.0193\frac{x^3}{3}\right)H_B + M_B\,.\,x\right)\right]_0^{42} + \left[\left(100\frac{x^2}{2} - 2600x\right)\right]_{26}^{42} + \left[\left(75\frac{x^2}{2} - 2550x\right)\right]_{34}^{42}\right]$$

$$= \frac{1}{EI_0}\,[-882\, V_B + 2.38 \times 10^2\, H_B + 42\, M_B + 1.51 \times 10^4]$$

$$\frac{\partial U}{\partial M_B} = \frac{1}{EI_0} = [-882\, V_B + 2.38 \times 10^2\, H_B + 42\, M_B + 1.51 \times 10^4] \qquad ...(6)$$

$$\frac{\partial U}{\partial V_B} = 0\,; \quad \frac{\partial U}{\partial H_B} = 0\,; \quad \frac{\partial U}{\partial M_B} = 0$$

Therefore, $2.47 \times 10^4\, V_B - 5 \times 10^3\, H_B - 882\, M_B - 5.637 \times 10^5 = 0$ ...(7)

$-5 \times 10^3\, V_B + 1.81 \times 10^3\, H_B + 2.38 \times 10^2\, M_B + 4.98 \times 10^4 = 0$ ...(8)

$-882\, V_B + 2.38 \times 10^2\, H_B + 42\, M_B + 1.51 \times 10^4 = 0$ ...(9)

Solving equations (7), (8), and (9),

$$V_B = 39.6 \text{ kN};\quad H_B = 78.47 \text{ kN};\ M_B = 25 \text{ kNm}$$

(*e*) **Evaluation of other reactions and moments**

Equating moments about A to zero

$$V_B \times 42 - H_B \times 0 - M_B - 100 \times 16 - 75 \times 8 + M_A = 0$$

$$M_A = -42\, V_B + M_B + 1600 + 75 \times 8$$

$$= -42 \times 40.48 + 34.064 + 1600 + 600$$

$$M_A = 561.8 \text{ kNm}$$

$$M_C = V_B \times 21 - M_B - H_B \times 8.5$$

$$= 40.48 \times 21 - 34.064 - 76.89 \times 8.5$$

$$M_C = 139.61 \text{ kNm}$$

Since $\Sigma Y = 0$

$$V_A + V_B = 175 \text{ kN}$$

$$V_A = 175 - 39.60$$

$$V_A = 135.4 \text{ kN}$$

$\Sigma H = 0$ yields

$$H_A = 78.47 \text{ kN.}$$

PROBLEMS

1. A three hinged parabolic symmetric arch has a span of 20 m and central rise of 5 m. It carries a concentrated load of 64 kN at a distance of 5 m from the left support. Determine the maximum bending moment and sketch the bending moment diagram.

[**Ans.** M_{max} (+ve) = 120 kNm under the load
M_{max} (–ve) = –40 kNm at 15 m from left hand support]

2. A three hinged parabolic arch, hinged at the crown and springings, has a horizontal span of 15 metres with a central rise of 3 metres. It carries a uniformly distributed load of 32 kN/m (of horizontal span) over the left hand half of the span. Calculate the normal thrust, radial shear and bending moment at 5 metres from the left hand hinge.

[**Ans.** Normal thrust P_5 = 150.03 kN; Radial shear
F_5 = – 20.93 kN and bending moment M_5 = 100 kNm]

3. A three hinged symmetric parabolic arch is of span 24 m with a central rise of 4 m. It carries a distributed load which varies uniformly from 30 kN/m at the crown to 60 kN/m at the springings. Calculate the normal thrust, shear force and bending moment at sections 3 metres and 6 metres from the left end hinge. [**Ans.** P_3 = 810 kN; F_3 = 10.2 kN; M_3 = 101.3 kNm; P_6 = 754 kN;
F_6 = – 14.25 kN; M_6 = 90 kNm]

4. A three hinged circular arch of span 40 m and central rise 5 m is carrying a u.d.l. of 15 kN/m over the entire span. Point loads of 100 kN, 150 kN and 200 kN are acting at horizontal distances of 15, 20 and 25 m respectively from the left hand hinge. Calculate the horizontal thrust and reaction at the hinges. Also find the bending moment at a section 8 m from the left hand support.

[**Ans.** H = 1350 kN ; R_A = 1444 kN ; R_B = 1453.06 kN ; B.M. = 794.50 kNm]

5. Figure 4.57 shows a three hinged parabolic arched rib ACB with supports at different levels. It has a horizontal span of 15 m, the hinge at the crown C being at 6 m horizontally from A and 2 m vertically above it. Calculate the difference between the levels of the supports A and B. The arch carries an u.d.l. of 20 kN/m from A to C. Calculate the horizontal thrust and the negative bending moment. [**Ans.** B is lower than A by 2.5 m, H = 72 kN; M_{max} = – 81 kNm
at x = 4.5 m from left end A]

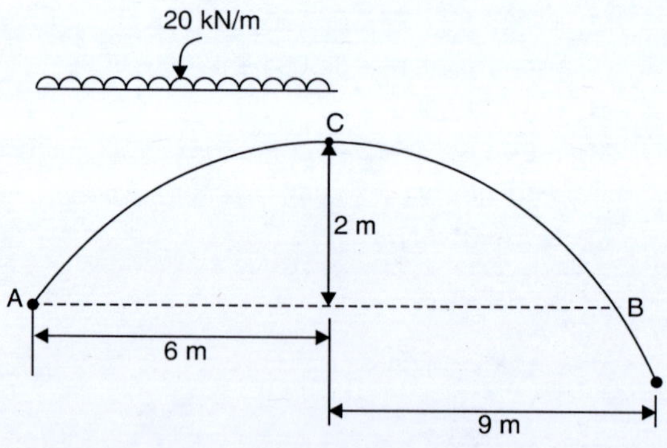

Fig. 4.57

6. A parabolic unsymmetrical three hinged arch is as shown in Fig. 4.58 sketch the influence lines for the horizontal thrust and bending moment at X, 10 m from the left hand support, due to a moving load of 1800 N.

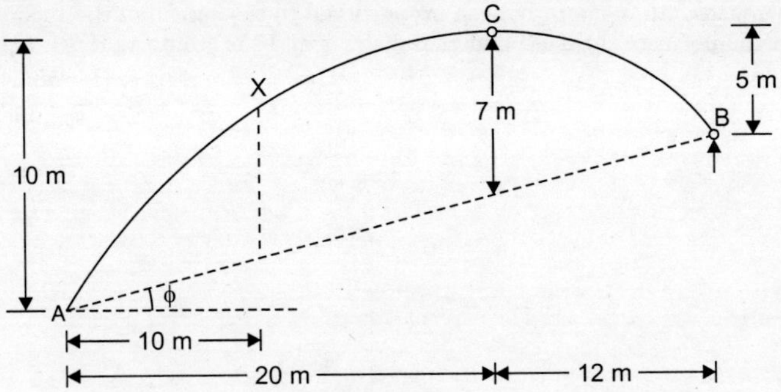

Fig. 4.58

[**Ans.** Max. ILO for H′ = 1.071 sec ϕ ; Max. ILO for H = 1.071 ; H = 1927.8 kN ; Max +ve B.M. = 7252.2 Nm ; Max –ve B.M. = 3495.6 Nm]

7. A three hinged symmetric circular arch is loaded as shown in Fig. 4.59. Determine the bending moment, normal thrust and radial shear at 9 m from the left support.

[**Ans.** M = 347.4 kNm; P = 379.3 kN; F = – 11.5 kN]

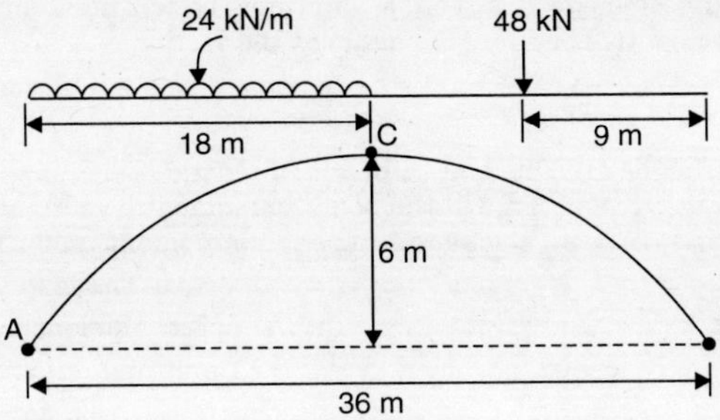

Fig. 4.59

8. A three hinged parabolic arch of span 30 m has a central rise of 5 m. Find the increase in rise of the arch crown if the temperature rises through 30°C. Take co-efficient of linear expansion for the arch material as 12×10^{-6}/°C. [**Ans.** 18 mm]

9. A two hinged parabolic arch of span 25 m and rise 6 m suffers a rise of temperature of 20°C. Find the horizontal thrust developed in the arch. Take $E = 2 \times 10^5$ N/mm² and $\alpha = 11 \times 10^{-6}$ per °C. Moment of inertia of the arch at the crown is 100×10^8 mm⁴. [**Ans.** 28.65 kN]

10. A two hinged parabolic arch of span 36 m and central rise 8 m carries a uniformly distributed load of 32 kN/m over the left half of the span. Determine the position and value of maximum bending moment. Also find the normal thrust and radial shear at the section. Assume that the moment of inertia at a section varies as secant of the inclination at the section.

[**Ans.** x = 9 m from left end and M_{max} = 648 kNm at x = 9 m from right end and M_{max} = –648 kNm; P_9 = 354.6 kN; F_9 = 0]

11. A two hinged semi-circular arch of span 12 m has its hinged supports on the same horizontal level. The arch is loaded at the crown with a concentrated load of 24 kN acting vertically. Calculate the horizontal thrust at the hinged supports. Assume EI is constant. [**Ans.** H = 7.68 kN]

12. A two hinged parabolic arch has a span 28 m and central rise of 4.5 m as shown in Fig. 4.60. It has a varying second moment of area proportional to the secant of the inclination of the neutral axis. Find the horizontal thrust and radial shear at 10 m from the right support.

[**Ans.** H = 58.24 kN; M = 39.76 kN;
P = 59.54 kN; F = 1.93 kN]

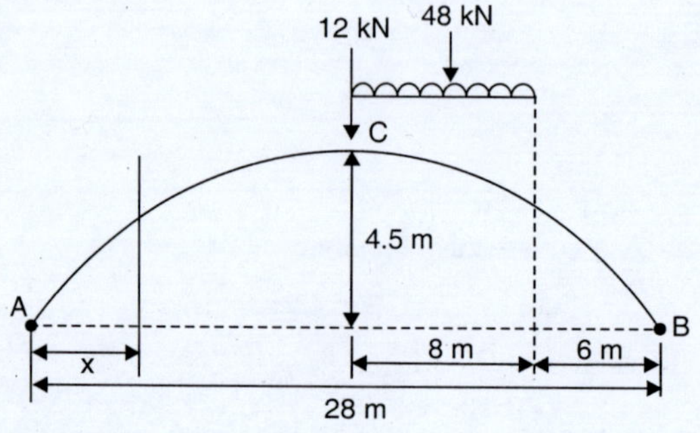

Fig. 4.60

13. A three hinged circular arch has a span of 20 m and a central rise of 5 m. Draw the influence line diagram for bending moment at a section 5 m from the left hand support. Hence find the maximum positive and negative bending moments at the section due to a moving point load of 72 kN.

[**Ans.** M_{max} (+ve) = 127.58 kNm;
M_{max} (–ve) = 104.83 kNm]

14. A three hinged parabolic arch of span 24 m and of central rise 5 m, carries a moving point load of 120 kN. At what position of load will the bending moment be zero at a section 8 m from the left hinge ? For this load position, calculate the bending moment under the load.

[**Ans.** The load would be 10.286 m from left
support. Moment under the load = 100.8 kNm]

15. Analyse the fixed arch, loaded as shown in Fig. 4.61 and find out the support moments.

[**Ans.** M_A = 75 kNm; M_B = – 75 kNm]

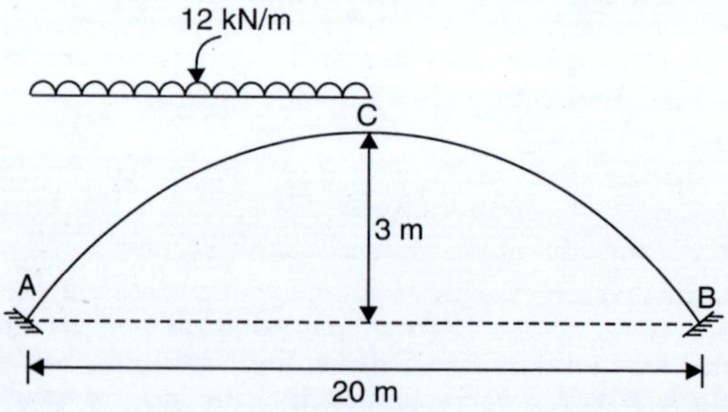

Fig. 4.61

CHAPTER 5

CABLES AND SUSPENSION BRIDGES

5.1 INTRODUCTION

During the 20th century a large number of remarkable suspension bridges were constructed in North America and Australia. They have become landmarks of certain cities. For example, the Golden Gate bridge in San Francisco (Fig. 5.1) California is considered by Civil Engineers the world over as the pride of their profession. Spanning across a seaway nearly 2 miles wide, it has a central span of 1280 m between its pylons and supports 2 tiers of traffic. Hundreds of thousands of visitors visit San Francisco, just to look at the bridge.

However in the latter part of the century the accent has shifted from suspension bridges to cable stayed bridges. One of the reasons stated for this waning of interest in suspension bridges is that they are vulnerable to the phenomenon of aerodynamic instability (or flutter) during certain directions and velocities of cross winds.

Structurally, suspension bridge remains the most efficient mode of transferring span loads to the supports. It is more efficient than arches since arches are predominantly in compression and cables, in tension. Since compression implies buckling also, compression members have to be made really stiff, thus making them expensive.

5.2 CABLES

Cables in the form of wire ropes or parallel wires joined with clips are used in suspension bridges and transmission towers. Suspension bridges are used for roadways for spans more than 200 m and for light traffic ways of spans more than 300 m.

The traffic load on the deck of the bridge is transferred by 2 sets of vertical suspenders to the two suspension cables, one on either side of the width of the bridge. The suspension cables are the main load bearing members. The suspension cables are flexible throughout. They can take only direct tension and cannot take any bending moment.

The maximum sag or dip of the cable varies from $\frac{l}{10}$ to $\frac{l}{15}$, where l is the horizontal span. The greater the dip, the less will be the cable tension due to a given loading. The suspension cable after passing over the supporting towers will be anchored by back stays to a huge mass of concrete. The concrete mass shall be capable of resisting the tension in the cable.

Since the cable is flexible throughout, the cable shape will vary according to the nature and position of loading, if unstiffened. The initial shape of the cable can be retained without change, by providing a stiffening girder, three hinged or two hinged. The components of a suspension bridge and their functions without and with stiffening girders are discussed in the subsequent articles of this chapter.

Fig. 5.1

5.3 ROPEWAYS AND BRIDGES

A suspended cable as a structure spanning distances has a lot of advantages and drawbacks.

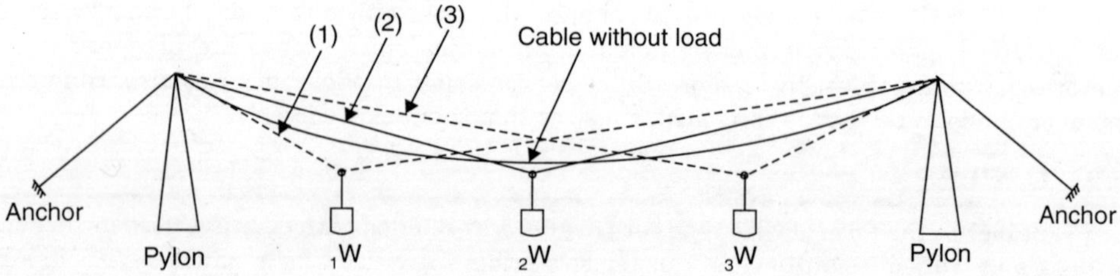

Fig. 5.2

A cable is in tension. Tension members are much more efficient than compression or bending members. You just have to provide sufficient area of cross-section to take the tensile force. That way a cable is very efficient, economical.

Suppose a cable has to carry a load W at variable positions. Depending upon the load position, the cable would change its profile. So $_1W$, $_2W$ and $_3W$ will create profiles 1, 2 and 3. This means, there would be enormous deflections in the cable.

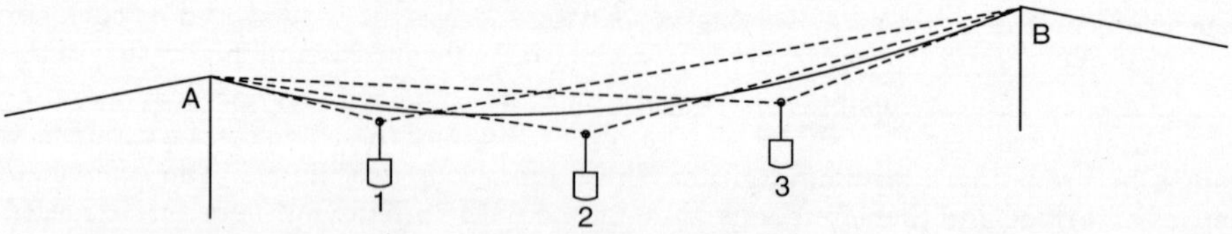

Fig. 5.3

If the cable is to be used as a rope way to support a cable car, these large deflections would not matter.

A single cradle suspended from a cable way would then travel as shown in Fig. 5.3. In the figures, 1, 2 and 3 are 3 positions of the same cradle, as it moves from A to B. If we compute the vertical displacements of the cable, it can be really astonishing. Suppose the span AB is 100 m and the initial central dip 10 m. Depending upon the load position, the change in vertical position of the cable can be 4 to 5 metres at midspan.

This kind of large displacements are unsuitable for bridges. For a bridge, we need a level deck that would keep its position. The vertical displacements, if any, can be a few centimetres but definitely not approaching metres. So we need a **stiffening girder**. This girder is usually suspended from the cable by a number of suspenders (or hangers) and carries the bridge deck. It is capable of taking bending moment.

A stiffening girder, because of its stiffness, would not undergo large deflections. That way it will prevent the cable from suffering large deflections. So the cable would more or less keep its original, parabolic shape, even when loads act at different points along its span.

5.4 COMPONENTS AND THEIR FUNCTIONS

Figure 5.4 shows a suspension bridge with a stiffening girder.

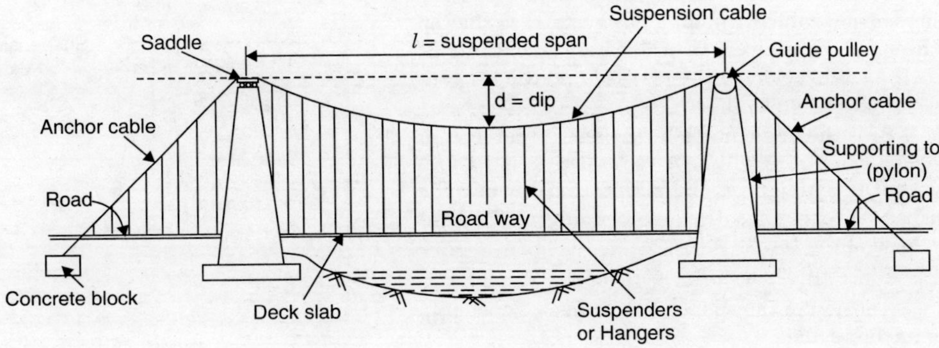

Fig. 5.4

(*a*) **Suspension cable :**

This is the main load bearing member. Suspension cable will have a central dip of 1/10 to 1/15 of the horizontal span. Suspension cables are flexible and hence can change their shape under the load systems since cables cannot take any bending moment and can take only direct tension.

(*b*) **Suspenders :**

The stiffening girder with deck slab of roadway is suspended from the suspension cables by means of suspenders or hangers. Suspenders are closely spaced and transfer a part of the traffic load on the deck slab to the suspension cable as a uniformly distributed load.

(*c*) **Supporting towers :**

Each suspension cable is supported on 2 towers or pylons on either side. A guide pulley or saddle placed on rollers is usually provided on the towers for passing the suspension cable on them. The forces on supporting towers are discussed subsequently in this chapter. Pylons can be as tall as 20 to 200 m and usually provide substance and majesty to the bridge. Since they dominate the landscape, they have to be designed not only for strength and stability but also for their architectural value.

(*d*) **Anchor cables :**

After passing over the tall pylons, suspension cables have to be anchored to the bed rock. Anchoring the cables can be a stupendous task since the cable tensions can be of the order of several hundred tonnes. Each anchor span can be about 25 to 30 percent of the suspended span. In most bridges anchor spans are also above water, the pylons being founded inside the water way. In view of the need for firm anchoring of cables, suspension bridge can be built only in regions where bed rock is accessible.

5.5 PASSING THE CABLE OVER PYLONS

The suspension cable is supported by passing it on towers (or piers) on either side and anchoring it into a huge mass of concrete or on firm rock. There are two types of arrangements for passing the cable on a supporting tower.

(*i*) Guide pulley support (*ii*) Roller or saddle support.

(*i*) **Guide pulley support**

Figure 5.5 shows the arrangement of passing a suspension cable through a pulley fixed to the top of the supporting tower. The cable here is continuous. θ_S is the angle between the vertical and the tangent to the suspension cable.

θ_A is the angle made by the anchor cable with the vertical. There will be a vertical reaction and a horizontal reaction on the supporting tower. The horizontal reaction will cause bending moment at the base of the tower.

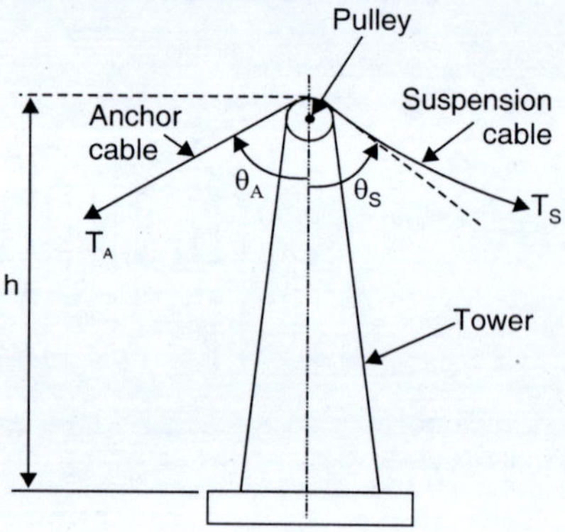

Fig. 5.5. Guide pulley support.

Since this is a pulley support,

Tension in the suspension cable = Tension in the anchor cable

$$T_S = T_A = T$$

Vertical reaction on the top of the tower,

$$V_T = T \cos \theta_S + T \cos \theta_A$$

$$V_T = V + T \cos \theta_A \quad \ldots(5.1)$$

Horizontal reaction on the top of the tower

$$H_T = T \sin \theta_S - T \sin \theta_A$$

$$H_T = T (\sin \theta_S - \sin \theta_A) \quad \ldots(5.2)$$

If h is the height of the tower,

Bending moment at the base of the tower = $H_T \times h$

Bending moment = $T (\sin \theta_S - \sin \theta_A) \times h$ ...(5.3)

(*ii*) **Roller or saddle support**

Figure 5.6 shows the arrangement of passing a suspension cable through a saddle mounted on rollers. Here the cable is non continuous. Since the saddle is placed on rollers, the net horizontal reaction on the top of the tower must be zero. *i.e.,* The horizontal components of the tension in the suspension cable and the anchor cable must be equal.

$$T_S \sin \theta_S = T_A \sin \theta_A = H$$

The vertical pressure on the top of the tower

$$V_T = T_S \cos \theta_S + T_A \cos \theta_A \qquad ...(5.4)$$

Since the net horizontal force on the top of this tower is zero, the bending moment at the base of the tower due to the cable reactions is zero. Thus the tower is rendered free of bending moments due to the net horizontal cable pull.

However this nullification of horizontal pull is achieved by an appropriate horizontal movement of the saddle. This movement can create an eccentricity in the vertical component and the consequent bending moment on the pylon.

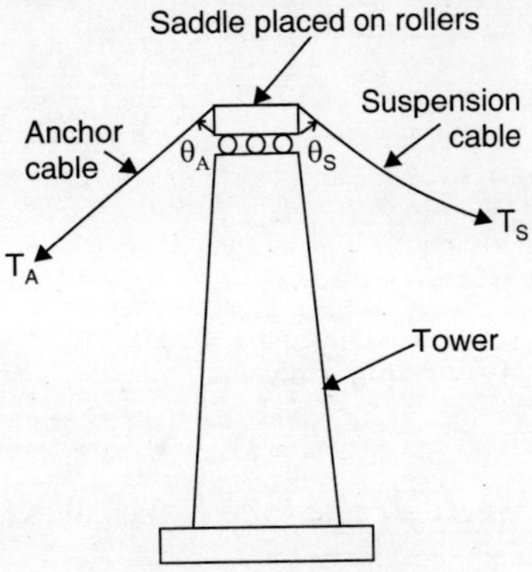

Fig. 5.6. Roller support.

5.6 REACTIONS, TENSION AND LENGTH OF SUSPENSION CABLE

(*a*) **Supports at the Same Level**

(*i*) *Profile :*

Consider a suspension cable of horizontal span l supported at A and B at the same level. Let it be subjected to a udl of intensity p/unit length of its horizontal span. Due to symmetry in loading and supports at the same level the maximum dip, d will be at the centre C. Since there are no horizontal forces, the horizontal tensions at A and B are equal *i.e.,* $H_A = H_B = H$

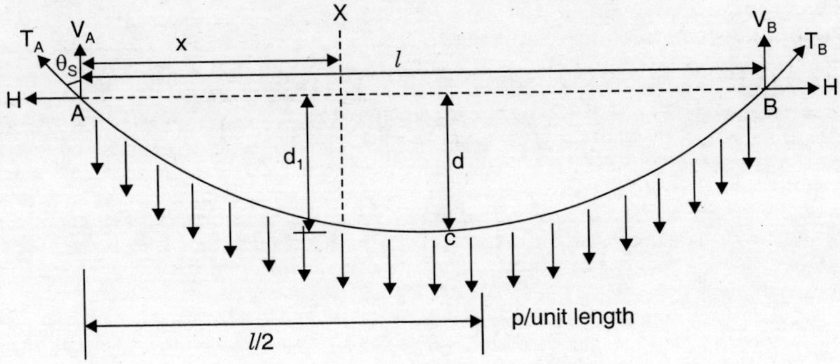

Fig. 5.7

Due to symmetry, vertical reactions

$$V_A = V_B = \frac{pl}{2}$$

Since the cable can take direct tension only, bending moment at any section of the cable shall be zero. Considering the left half of the cable and taking moments about C,

$$V_A \times \frac{l}{2} - p\frac{l}{2}.\frac{l}{4} - H \times d = 0$$

$$\frac{pl}{2} \times \frac{l}{2} - \frac{pl^2}{8} = \text{H}\, d$$

$$\text{H} = \frac{1}{d}\left[\frac{pl^2}{4} - \frac{pl^2}{8}\right]$$

$$\text{H} = \frac{pl^2}{8d} \qquad \ldots[5.4(a)]$$

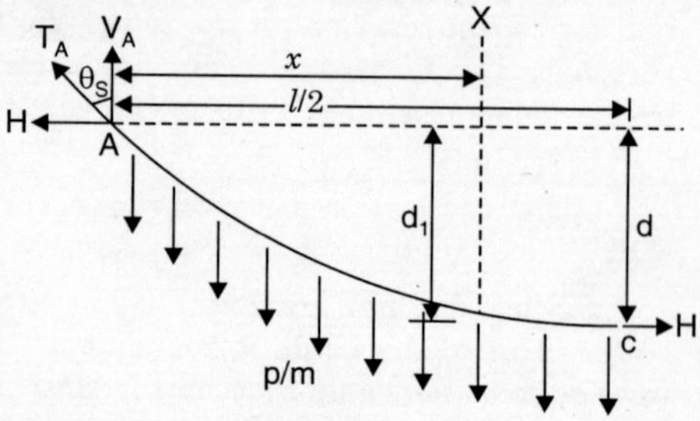

Fig. 5.8

Considering a section X at distance x from A, bending moment,

$$\text{M}_x = \text{V}_\text{A}\, x - p\,.\,x\,.\,\frac{x}{2} - \text{H}\,.\,d_1$$

where d_1 is the dip of the cable at the section X.

$$\text{M}_x = \frac{plx}{2} - \frac{px^2}{2} - \frac{pl^2}{8d} \times d_1 = 0$$

$$-\frac{pl^2}{8d} \times d_1 = -\frac{plx}{2} + \frac{px^2}{2}$$

$$d_1 = -\frac{8d}{pl^2}\left[-\frac{plx}{2} + \frac{px^2}{2}\right]$$

$$d_1 = \frac{8d}{pl^2} \times \frac{p}{2}(lx - x^2)$$

$$d_1 = \frac{4d}{l^2} x(l - x) \qquad \ldots(5.5)$$

This is the equation of a parabola. Hence the cable subjected to a udl takes the shape of a parabola.

(ii) Cable tensions :

Since the supports are at the same level,

$$\text{T}_\text{A} = \text{T}_\text{B} = \sqrt{\text{V}^2 + \text{H}^2} = \sqrt{\left(\frac{pl}{2}\right)^2 + \left(\frac{pl^2}{8d}\right)^2}$$

$$= \sqrt{\left(\frac{pl}{2}\right)^2 + \left(\frac{pl}{2}\right)^2\left(\frac{l}{4d}\right)^2} = \frac{pl}{2}\sqrt{1 + \frac{l^2}{16d^2}}$$

$$\text{T} = \text{V}\sqrt{1 + \frac{l^2}{16d^2}} \qquad \ldots(5.6)$$

Also,

$$\text{T} = \sqrt{\left(\frac{pl}{2}\right)^2 + \left(\frac{pl^2}{8d}\right)^2}$$

$$= \sqrt{\left(\frac{pl}{2} \times \frac{l}{4d}\right)^2 \times \left(\frac{4d}{l}\right)^2 + \left(\frac{pl^2}{8d}\right)^2} = \frac{pl^2}{8d}\sqrt{\frac{16d^2}{l^2} + 1}$$

$$\text{T} = \text{H}\sqrt{\frac{16d^2}{l^2} + 1} \qquad \ldots(5.7)$$

(iii) Inclination of cable with the vertical (Fig. 5.8) :

$$\tan \theta_S = \frac{H}{V}$$

From (5.4) and (5.5), $$\tan \theta_S = \frac{pl^2}{8d} \times \frac{2}{pl} = \frac{l}{4d}$$

$$\theta_S = \tan^{-1}\left(\frac{l}{4d}\right) \qquad ...(5.8)$$

(iv) Length of the cable :

Consider an elementary length ds of the cable

Length of this cable = ds

$$ds = \sqrt{dx^2 + dy^2} = dx\sqrt{1 + \left(\frac{dy}{dx}\right)^2}$$

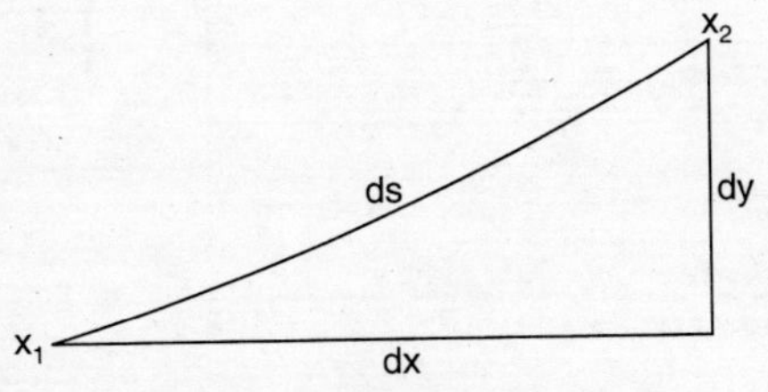

Fig. 5.9

For the part CB of the cable (Fig. 5.7), the equation for the parabola, with C as origin can be taken as $y = k \cdot x^2$

where k is a constant,

At $x = \frac{l}{2}$; $y = d$; $d = k \cdot \frac{l^2}{4}$; $k = \frac{4d}{l^2}$

Therefore, $$y = \frac{4d}{l^2} x^2$$

$$\frac{dy}{dx} = \frac{4d}{l^2} \times 2x = \frac{8dx}{l^2}$$

Substituting for $\frac{dy}{dx}$ in the equation for ds

$$ds = dx\sqrt{1 + \left(\frac{8dx}{l^2}\right)^2} = dx\sqrt{1 + \frac{64\, d^2\, x^2}{l^4}}$$

Total length of the cable, $$S = 2\int_0^{l/2} ds$$

$$S = 2\int_0^{l/2} dx \left[1 + \frac{64d^2\, x^2}{l^4}\right]^{1/2}$$

Expanding $\left[1 + \frac{64d^2}{l^4} x^2\right]^{1/2}$ by Binomial theorem and ignoring higher powers of $\left(\frac{dx}{l}\right)$

$$S = 2\int_0^{l/2} \left(1 + \frac{1}{2}\frac{64d^2}{l^4} x^2\right) dx = 2\left(x + \frac{1}{2}\frac{64d^2}{l^4}\frac{x^3}{3}\right)_0^{l/2}$$

$$= 2\left(\frac{l}{2} + \frac{32d^2}{l^4}\frac{1}{3}\frac{l^3}{8}\right) = \left(\frac{2l}{2} + \frac{2}{3}\frac{32d^2\, l^3}{8l^4}\right)$$

$$S = l + \frac{8}{3}\frac{d^2}{l} \text{ is the length of the cable.} \qquad ...(5.9)$$

(*b*) **Supports at Different Levels :**

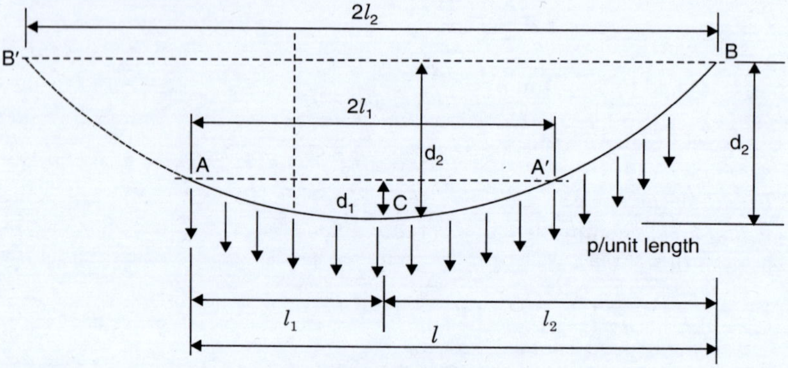

Fig. 5.10

Here the cable AB is supported at different levels at A and B. C is the lowest point of the cable.

d_1 is the dip and $2l_1$ the length of the portion ACA′.

d_2 is the dip and $2l_2$ the length of the extended cable B′CB.

Consider the portion ACA′ of the cable.

For the free body in Fig. 5.11, $V_A = pl_1$

Using (5.4),

$$H = \frac{p\,(2l_1)^2}{8d_1}; \quad H = \frac{pl_1^2}{2d_1}$$

Tension at A, $T_A = \sqrt{V_A^2 + H^2}$

$$T_A = \sqrt{(pl_1)^2 + \left(\frac{pl_1^2}{2d_1}\right)^2}$$

Tension at C, $H = \dfrac{pl_1^2}{2d_1}$

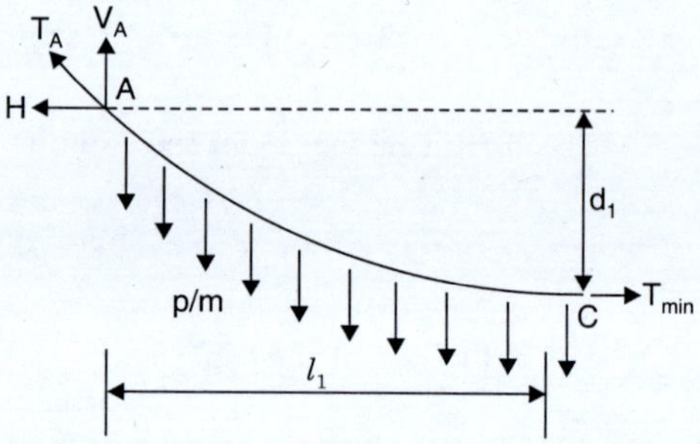

Fig. 5.11

Likewise, in the free body in Fig. 5.12,

$$V_B = pl_2$$

$$H = \frac{pl^2}{8d} = \frac{p\,(2l_2)^2}{8d_2}$$

$$H = \frac{pl_2^2}{2d_2}$$

$$T_B = \sqrt{V_B^2 + H^2}$$

$$T_B = \sqrt{(pl_2)^2 + \left(\frac{pl_2^2}{2d_2}\right)^2}$$

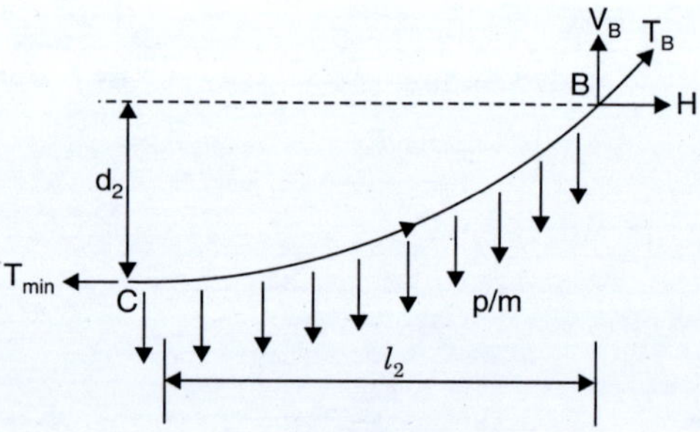

Fig. 5.12

Tension at C, $H = \dfrac{pl_2^2}{2d_2}$

The tension at C in both parts AC and CB must be equal. So,

$$\frac{pl_1^2}{2d_1} = \frac{pl_2^2}{2d_2}$$

$$\frac{l_1^2}{d_1} = \frac{l_2^2}{d_2}$$

$$\frac{l_1^2}{l_2^2} = \frac{d_1}{d_2} \quad \text{or} \quad \frac{l_1}{l_2} = \sqrt{\frac{d_1}{d_2}} \qquad \text{...(5.10)}$$

$$l_1 + l_2 = l$$

l_1 and l_2 can be obtained from the above two equations.

The length of the cable $\quad S = S_{AC} + S_{CB}$

Therefore, $\quad ACA' = 2l_1 \quad$ and $\quad B'CB = 2l_2$

$$S = \frac{1}{2}\left[2l_1 + \frac{8}{3}\frac{d_1^2}{2l_1} + 2l_2 + \frac{8}{3}\frac{d_2^2}{2l_2}\right]$$

$$= l_1 + \frac{2d_1^2}{3l_1} + l_2 + \frac{2d_2^2}{3l_2} = l + \frac{2d_1^2}{3l_1} + \frac{2d_2^2}{3l_2}$$

Total length of the cable, $\quad S = l + \dfrac{2d_1^2}{3l_1} + \dfrac{2d_2^2}{3l_2}. \qquad$ ...(5.11)

Example 5.1 *A suspension cable having supports at the same level, has a span of 30 m and a maximum dip of 3 m. The cable is loaded with a UDL of 10 kN/m throughout its length. Find from the first principles, the maximum tension in the cable.*

Solution

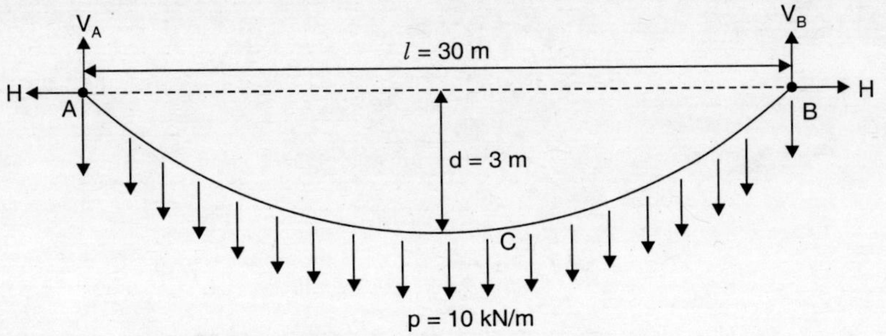

Fig. 5.13

Vertical reaction at the supports

$$V_A = V_B = \frac{pl}{2} = \frac{10 \times 30}{2} = 150 \text{ kN}$$

Horizontal pull in the cable

$$H = \frac{pl^2}{8d} = \frac{10 \times 30^2}{8 \times 3} = 375 \text{ kN}$$

Maximum tension in the cable

$$T_{max} = \sqrt{V_A^2 + H^2} = \sqrt{150^2 + 375^2} = 403.9 \text{ kN}$$

$$T_{max} = 403.9 \text{ kN.}$$

Example 5.2 *A suspension cable is supported at 2 points 25 m apart. The left support is 2.5 m above the right support. The cable is loaded with a uniformly distributed load of 10 kN/m throughout the span. The maximum dip in the cable from the left support is 4 m. Find the maximum and minimum tensions in the cable.*

Solution

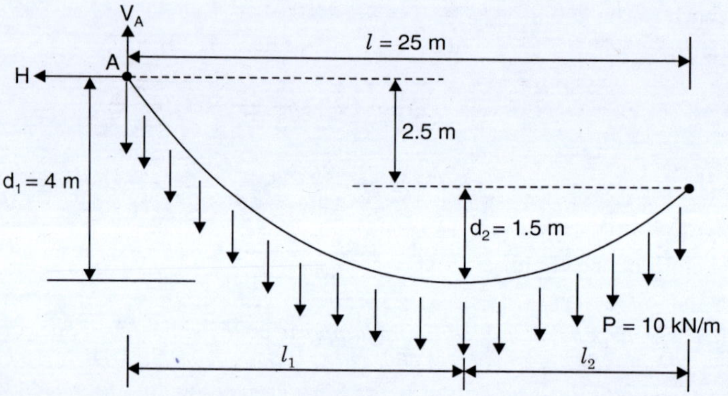

Fig. 5.14

$$\frac{l_1}{l_2} = \sqrt{\frac{d_1}{d_2}}$$

$$l_1 = \sqrt{\frac{4}{1.5}}\, l_2 = 1.6329\, l_2$$

$$l_1 + l_2 = l$$

$$1.6329\, l_2 + l_2 = 25$$

$$2.6329\, l_2 = 25$$

$$l_2 = 9.495$$

$$l_1 = 15.505$$

$$V_A = pl_1 = 10 \times 15.505 = 155.05 \text{ kN}$$

$$V_B = 10 \times 9.495 = 94.95 \text{ kN}$$

$$H = \frac{pl_1^2}{2d_1} = \frac{10 \times 15.505^2}{2 \times 4} = 300.51 \text{ kN}$$

Also, $$H = \frac{pl_2^2}{2d_2} = \frac{10 \times 9.495^2}{2 \times 1.5} = 300.51 \text{ kN}$$

$$T_A = \sqrt{V_A^2 + H^2} = \sqrt{155.05^2 + 300.51^2} = 338.15 \text{ kN}$$

$$T_A = 338.15 \text{ kN}$$

$$T_B = \sqrt{V_B^2 + H^2} = \sqrt{94.95^2 + 300.51^2} = 315.15 \text{ kN}$$

Therefore maximum tension in the cable = 338.15 kN

Minimum tension will occur at the lowest point in the cable and its value will be equal to H itself.

Minimum tension, T_{min} = H = 300.51 kN.

Example 5.3 *A suspension cable of 130 m horizontal span is supported at the same level. It is subjected to a uniformly distributed load of 28.5 kN per horizontal metre. If the maximum tension in the cable is limited to 5000 kN, calculate the minimum central dip needed.*

Solution

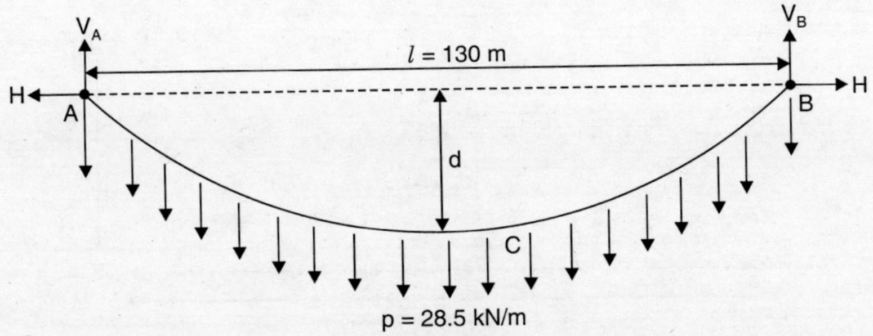

Fig. 5.15

$$V_A = V_B = \frac{\text{Total load}}{2} = \frac{28.5 \times 130}{2} = 1852.5 \text{ kN}.$$

Horizontal tension,

$$H = \frac{pl^2}{8d} = \frac{28.5 \times 130^2}{8d} = \frac{60206.25}{d} \text{ kN}$$

Maximum tension, T_{max} = 5000 kN

$$T_{max} = \sqrt{V^2 + H^2}$$

$$5000 = \sqrt{(1852.5)^2 + \left(\frac{60206.25}{d}\right)^2}$$

Squaring both sides $$5000^2 = (1852.5)^2 + \frac{(60206.25)^2}{d^2}$$

$$\frac{5000^2 - (1852.5)^2}{(60206.25)^2} = \frac{1}{d^2}$$

$$d = \sqrt{\frac{(60206.25)^2}{5000^2 - (1852.5)^2}} = 12.96 \text{ m}$$

$$d = 12.96 \text{ m}$$

The minimum dip required is 12.96 m.

Example 5.4 *A suspension cable of horizontal span 95 m is supported at two different levels. The right support is higher than the left support by 4 m. The dip to the lowest point of the cable below the left support is 5 m. The cross-sectional area of the cable is 3500 mm². Find the uniformly distributed load that can be carried by the cable if the maximum stress is limited to 600 N/mm².*

Solution Cross-sectional area of the cable = 3500 mm² ; σ in cable = 600 N/mm²

$$\frac{l_1}{l_2} = \sqrt{\frac{d_1}{d_2}}; \quad \frac{l_1}{l_2} = \sqrt{\frac{5}{9}}$$

$$l_1 = 0.745 l_2 \quad \text{...(1)}$$

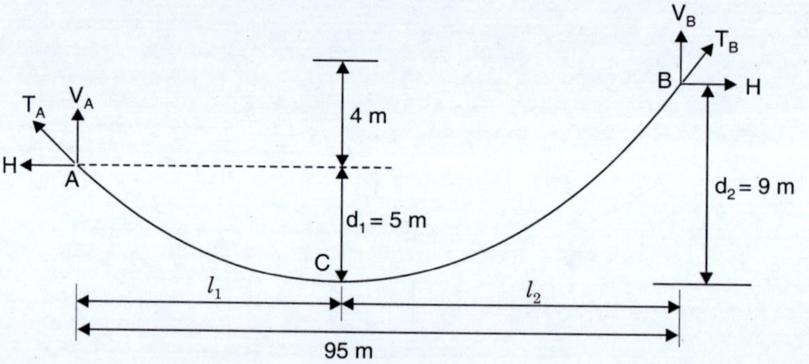

Fig. 5.16

$$l_1 + l_2 = 95$$

i.e.,

$$0.745\, l_2 + l_2 = 95$$

$$l_2 = 54.44 \text{ m}$$

$$l_1 = 40.56 \text{ m}$$

$$V_A = pl_1 = 40.56\, p$$

$$V_B = pl_2 = 54.44\, p$$

$$H = \frac{pl_2^2}{2d_2} = \frac{p\,(54.44)^2}{2 \times 9} = 164.65\, p$$

Maximum tension will occur at the right support (since $V_B > V_A$ and hence $T_B > T_A$)

Max. tension,

$$T_{max} = T_B = \sqrt{V_B^2 + H^2} = \sqrt{(54.44p)^2 + (164.65p)^2}$$

$$= p\sqrt{2963.71 + 27109.62}$$

$$T_{max} = 173.417p \text{ N}$$

$$T_{max} = \sigma \,.\, A.$$

$$173.417p = 600 \times 3500$$

$$p = \frac{600 \times 3500}{173.417} \text{ N/m}$$

$$p = 12109 \text{ N/m}$$

Therefore, the udl including self weight, that can be safely carried by the cable is

$$p = 12.109 \text{ kN/m}.$$

Example 5.5 *A suspension bridge is of 50 m span with a 16 m wide roadway. It is subjected to a load of 25 kN/m² including dead loads. The bridge is supported by a pair of cables having a central dip of 4.2 m. Find the cross-sectional area of the cable necessary if the maximum permissible stress in the cable material is not to exceed 600 N/mm².*

Solution Considering 1 m length of the bridge, load taken by two cables = 1 × 16 × 25

= 400 kN/m length

Load in each cable = p = 400/2 = 200 kN/m

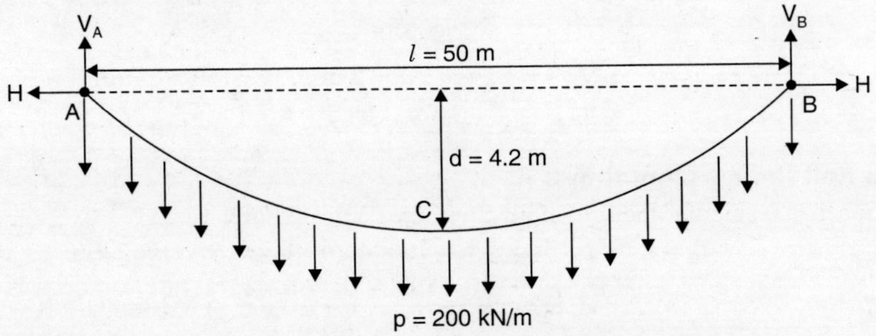

Fig. 5.17

$$V_A = V_B = \frac{pl}{2} = \frac{200 \times 50}{2} = 5000 \text{ kN}$$

Horizontal pull in the cable, $H = \frac{pl^2}{8d} = \frac{200 \times 50^2}{8 \times 4.2} = 14881 \text{ kN}$

Max. tension in the cable $= T = \sqrt{V^2 + H^2}$

$$= \sqrt{(5000)^2 + (14881)^2} = 15698.5 \text{ kN}$$

$$T = 15698.5 \times 1000 \text{ N}$$

$$T_{max} = \sigma A$$

$$15698.5 \times 1000 = 600 \text{ A}$$

$$A = 26164 \text{ mm}^2$$

The required minimum cross-sectional area of the cable = 26164 mm².

Example 5.6 *A cable of horizontal span 21 m is to be used to support six equal loads of 40 kN each at 3 m spacing. The central dip of the cable is limited to 2.0 m. Find the length of the cable required and also its sectional area if the safe tensile stress is 750 N/mm².*

Solution

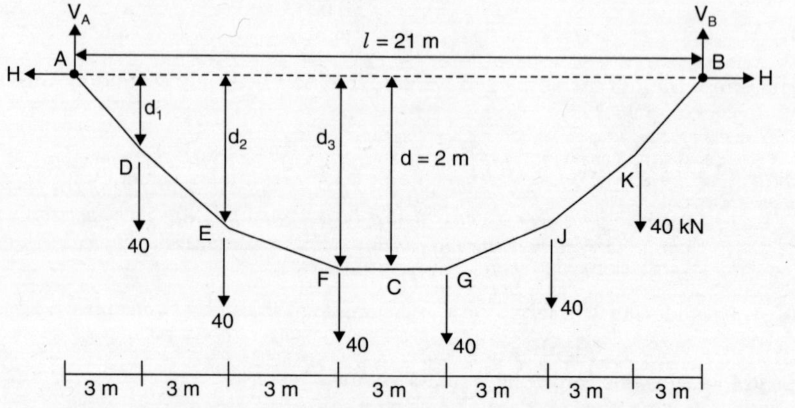

Fig. 5.18

$$l = 21 \text{ m}; \quad d = 2 \text{ m}; \quad \sigma = 750 \text{ N/mm}^2.$$

$$V_A = V_B = \frac{\text{Total load}}{2} = \frac{6 \times 40}{2} = 120 \text{ kN}.$$

To find the horizontal pull :

Equating moments about C of all the forces one side of C to zero,

$$V_A \times 10.5 - 40 \times 7.5 - 40 \times 4.5 - 40 \times 1.5 - H \times 2 = 0$$

$$H = \frac{120 \times 10.5 - 40\,(7.5 + 4.5 + 1.5)}{2} = 360 \text{ kN}$$

Equating moments about D to zero,

$$120 \times 3 - 360 \times d_1 = 0$$

$$d_1 = \frac{120 \times 3}{360} = 1 \text{ m}$$

$$d_1 = 1 \text{ m}$$

$$AD = \sqrt{3^2 + 1^2} = 3.162 \text{ m}$$

Taking moments about E

$$120 \times 6 - 40 \times 3 - 360 \times d_2 = 0$$

$$\frac{120 \times 6 - 40 \times 3}{360} = d_2$$

$$d_2 = 1.667 \text{ m}$$

$$DE = \sqrt{3^2 + (1.667 - 1.0)^2} = 3.073 \text{ m}$$

Taking moments about F

$$120 \times 9 - 360 \times d_3 - 40 \times 6 - 40 \times 3 = 0$$

$$d_3 = \frac{120 \times 9 - 40 \times 6 - 40 \times 3}{360} = 2 \text{ m (agrees with the given values)}$$

$$EF = \sqrt{3^2 + (2 - 1.667)^2} = 3.018 \text{ m}$$

$$FC = 1.5 \text{ m}$$

Therefore, length of cable,

$$S = 2\,(AD + DE + EF + FC) = 2\,(3.162 + 3.073 + 3.018 + 1.5)$$

$$S = 21.506 \text{ m}$$

Maximum tension in the cable,

$$T_{max} = \sqrt{V^2 + H^2} = \sqrt{120^2 + 360^2} = 379.47 \text{ kN}$$

$$T_{max} = \sigma \,.\, A$$

$$A = \frac{T_{max}}{\sigma} = \frac{379.47 \times 1000}{750} = 505.96 \text{ mm}^2$$

The required sectional area of the cable, $A = 506 \text{ mm}^2$.

Example 5.7 *A suspension cable having supports at the same level has a horizontal span of 42 m and maximum dip of 4 m. The cable is loaded with a UDL of 16 kN/m throughout its length and concentrated loads of 50 kN and 100 kN at 1/3 points. Find the maximum tension in the cable.*

Solution

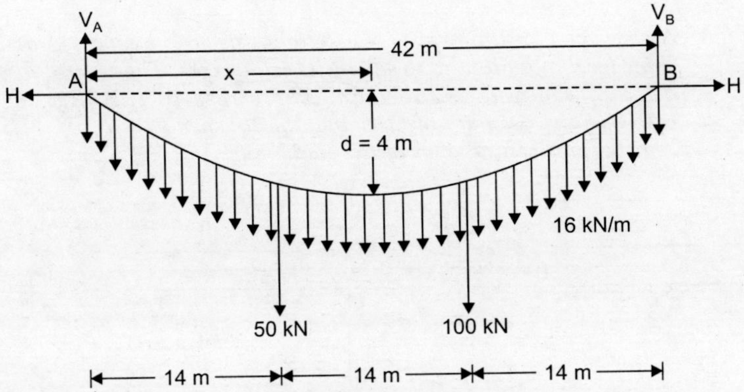

Fig. 5.19

Vertical reactions :

Taking moments about the right end B,

$$V_A \times 42 - 16 \times 42 \times \frac{42}{2} - 50 \times 28 - 100 \times 14 = 0$$

$$42\,V_A = 16 \times 42 \times 21 + 50 \times 28 + 100 \times 14$$

$$V_A = 402.67 \text{ kN}$$

$$V_B = \text{Total load} - V_A = 16 \times 42 + 50 + 100 - 402.67$$

$$V_B = 419.33 \text{ kN}$$

Location of maximum dip :

Because of the concentrated loads, the shape of the flexible cable will not be symmetrical. Let us assume that the maximum dip of the cable is between the point loads and at a distance x m from A. At this section, the tension in the cable will be horizontal and vertical component of the tension in the cable will be zero.

Considering the equilibrium of the portion of the cable to the left of section X,

$$\Sigma V = 0 \text{ gives,}$$

$$V_A - p.x - 50 = 0$$

$$402.67 - 16x - 50 = 0$$

$$16x = 402.67 - 50$$

$$x = 22.04 \text{ m}$$

Horizontal pull in the cable :

Taking moments about C,

$$V_A \times 22.04 - 16 \times 22.04 \times \frac{22.04}{2} - 50 \times 8.04 - H_A \times 4 = 0$$

$$402.67 \times 22.04 - 8 \times 22.04 \times 22.04 - 50 \times 8.04 = H_A \times 4$$

Solving, $H_A = 1146.69 \text{ kN} = H_B$

Maximum tension in the cable :

Since V_B is more, tension at B will be maximum

$$T_B = \sqrt{V_B^2 + H^2} = \sqrt{(419.33)^2 + (1146.69)^2} = 1220.96 \text{ kN}$$

$\therefore$ Maximum tension in the cable = 1220.96 kN.

Example 5.8 *A steel wire of uniform cross-sectional area is supported at its ends at the same level and allowed to sag by its self weight. The central dip is restricted to 1/15 of the horizontal span. The stress in the wire is not to exceed 640 N/mm². Density of steel may be taken as 78 × 10⁻⁶ N/mm³. Find the limiting horizontal span of the cable.*

Solution

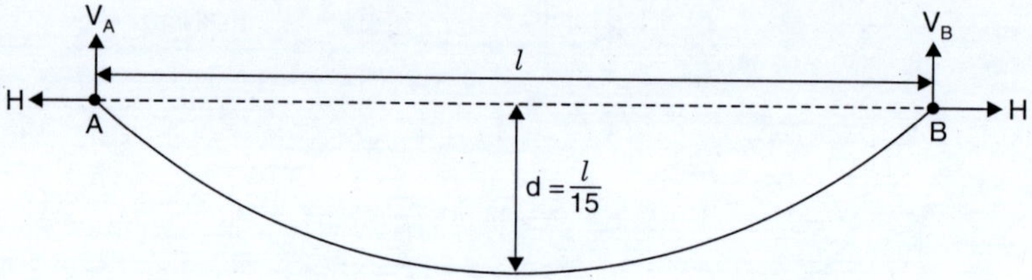

Fig. 5.20

(*a*) **Approximate solution :**

Here the profile of the cable will be assumed to be a parabola. Let l be the horizontal span and $d\left(=\frac{l}{15}\right)$ the central dip.

Length of cable, $$S = l + \frac{8}{3}\frac{d^2}{l} = l + \frac{8}{3}\frac{(l/15)^2}{l} = l + \frac{8}{3}\frac{l^2}{225\,l} = l + \frac{8l}{675}$$

$$S = \frac{675l + 8l}{675} = \frac{683l}{675} = 1.0119l$$

Total weight of wire, $$W = A \times S \times 78 \times 10^{-6} = A \times 1.0119l \times 78 \times 10^{-6}$$

$$= 78.9282 \times 10^{-6}\,Al$$

Horizontal tension, $H = \frac{pl^2}{8d} = \frac{(pl)\,l}{8d} = \frac{Wl}{8d}$

pl = Total load W = 78.9282×10^{-6} Al

Therefore, $$H = \frac{78.9282 \times 10^{-6}\ Al \,.\, l}{8 \times \left(\frac{l}{15}\right)} = \frac{15 \times 78.9282 \times 10^{-6}}{8} Al$$

$$H = 147.99\ Al \times 10^{-6}$$

$$V = \frac{pl}{2} = \frac{W}{2} \qquad (\because\ pl = W)$$

$$= \frac{78.9282 \times 10^{-6}\ Al}{2} = 39.464 \times 10^{-6}\ Al$$

$$V = 39.464 \times 10^{-6}\ Al$$

$$T_{max} = \sqrt{V^2 + H^2} = \sqrt{(39.4641)^2 + (147.99)^2} \times Al \times 10^{-6}$$

$$= 153.1615 \times 10^{-6}\ Al \qquad ...(1)$$

$$T_{max} = \sigma \,.\, A = 640 \times A \qquad ...(2)$$

Equations (1) and (2)

$$153.1615 \times 10^{-6}\ Al = 640\ A$$

$$l = \frac{640 \times 10^6}{153.1615} = 4.179 \times 10^6 \text{ mm}$$

Therefore, horizontal span of the cable, l = 4,179 m.

This is limiting span of the cable. Beyond this span, it cannot safely support its self weight. It can be noticed that the limiting span is independent of the sectional area of the cable. It is determined by the strength to unit weight ratio of the cable material.

(*b*) **Rigorous solution :**

A cable carrying a loading that is constant over **unit horizontal length** takes the form of a parabola. When it sags due to its self weight, its loading is uniform over **unit cable length**. Its profile is not a parabola but a **catenary**.

The equation to a catenary can be written as *(For the derivation the student is referred to "Mechanics of Structures" Vol. II by S.B. Junnarkar, Charotar Publishing House, 1996)*

$$y = c \cosh \frac{x}{c} \qquad ...(5.12)$$

Here $c = H/w$

where H is the horizontal component of the cable tension and w is the weight per unit length of the cable.

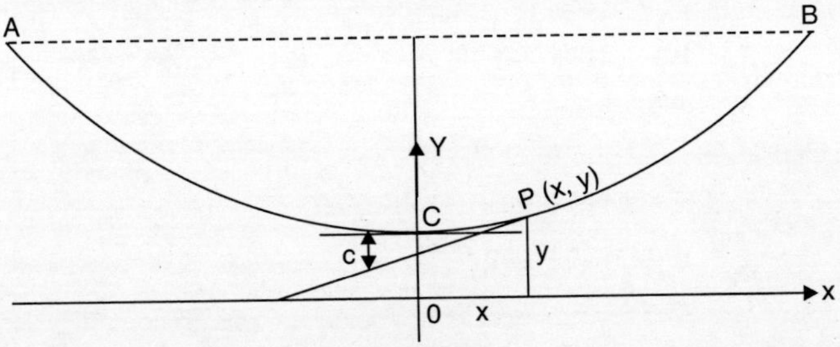

Fig. 5.21

c also satifies the relation $$y^2 = c^2 + s^2 \quad ...(5.13)$$
where s is the curved length CP at any point P on the cable.

Using the relations (5.12) and (5.13) the problem can be solved (with a lot of difficulty).

Example 5.9 *A suspension cable of span 100 m and dip 10 m carries a uniformly distributed load of 8 kN/m of horizontal span over the full span. Find the vertical and horizontal forces transmitted to the supporting pylons,*

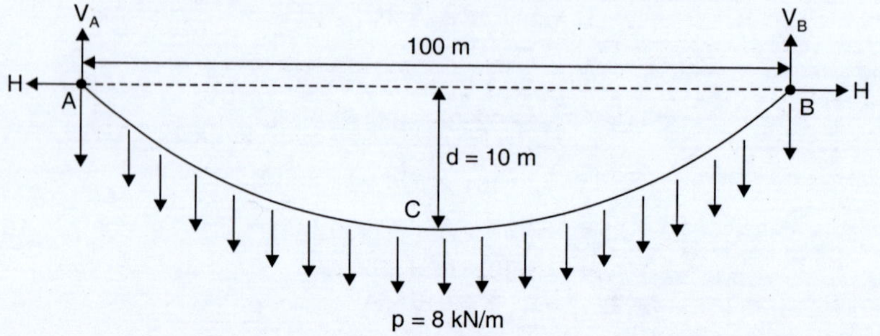

Fig. 5.22

(*a*) *If the cable is passed over a smooth pulley.*

(*b*) *If the cable is clamped to a saddle with rollers on the top of the piers.*

The anchor cable makes 30° to the horizontal at the pylons.

Solution Vertical reactions,

$$V_A = V_B = \frac{pl}{2} = \frac{8 \times 100}{2} = 400 \text{ kN}$$

Horizontal tension, $$H = \frac{pl^2}{8d} = \frac{8 \times 100^2}{8 \times 10} = 1000 \text{ kN}$$

Tension in the cable at A,

$$T = \sqrt{V^2 + H^2} = \sqrt{(400)^2 + (1000)^2}$$

$$T = 1077.03 \text{ kN}$$

(*a*) **Anchor cable passing over pulley :**

In case of pulley support,

Vertical pressure on the pier

$$= V_A + T \cos 60°$$

$$= 400 + 1077 \cos 60°$$

$$V = 938.5 \text{ kN}$$

Net horizontal force at the top of pylon

$$= H - T \sin 60°$$

$$= 1000 - 1077 \sin 60°$$

$$= 67.29 \text{ kN}$$

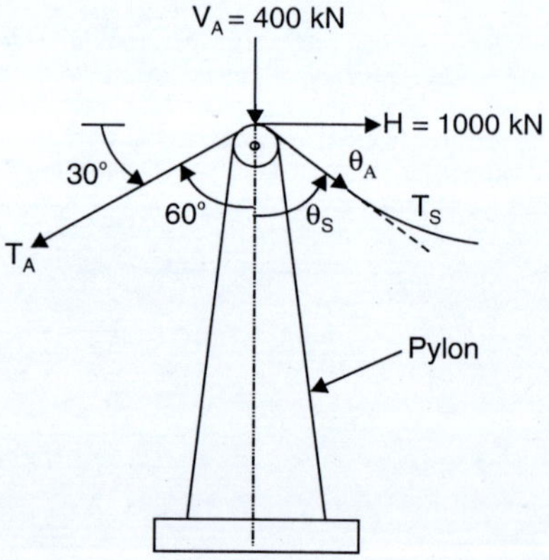

Fig. 5.23

(*b*) **Cable passing over saddle support :**

Since the saddle is placed on rollers, for equilibrium, horizontal component H of tension in the suspension cable = Horizontal component of tension in the anchor cable.

$$1000 = T_1 \sin 60°$$

$$T_1 = 1000/\sin 60°$$

$$= 1154.70 \text{ kN}$$

Net vertical pressure on the pier

$$= V + T_1 \cos 60°$$

$$= 400 + 1154.7 \cos 60°$$

$$= 977.35 \text{ kN}$$

Net horizontal force at the top of the pier = 0.

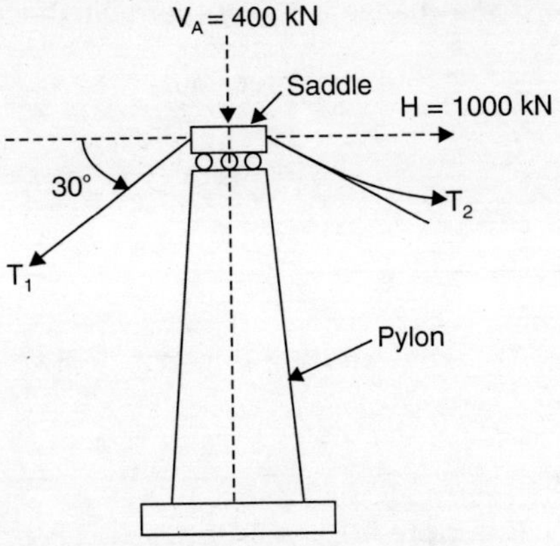

Fig. 5.24

5.7 EFFECT OF CHANGE IN TEMPERATURE IN SUSPENSION CABLES

Consider a suspension cable of horizontal span l supported at the same level and having a maximum central dip of **d**. Let it be subjected to a change in temperature of **t**°C. Let $\boldsymbol{\alpha}$ be the coefficient of linear expansion of the material of the cable. When the cable is subjected to a change of temperature, there will be a change in the length of the cable. The supporting towers cannot undergo any change of span and hence there will be a change in the dip of the cable.

We know that the length of the cable, [From eqn. (5.9)]

$$S = l + \frac{8}{3}\frac{d^2}{l}$$

Differentiating with respect to d

$$\frac{\partial S}{\partial d} = \frac{8}{3} \times \frac{2d}{l}$$

$$\partial S = \frac{16d}{3l}\,\partial d$$

$$\partial d = \frac{3l}{16d}\,\partial S$$

Change in length of the cable, $ds = S\alpha t = \left(l + \frac{8}{3}\frac{d^2}{l}\right)\alpha t = l\,\alpha\,t + \frac{8}{3}\frac{d^2}{l}\alpha\,t$

$\frac{8}{3}\frac{d^2}{l}\alpha\,t$ can be ignored since this term is small compared to $l\,\alpha\,t$

Therefore, $\partial S = l\,\alpha\,t$

Substituting ∂S in the expression for ∂d

$$\partial d = \frac{3l}{16d} \times l\,\alpha\,t$$

$$\partial d = \frac{3l^2}{16d}\alpha\,t \qquad ...(5.14)$$

is the change in dip of the cable due to the change in temperature.

Horizontal tension, $$H = \frac{pl^2}{8d}$$

$$\frac{\partial H}{\partial d} = \frac{pl^2}{8}\frac{\partial}{\partial d}d^{-1} = \frac{pl^2}{8}\left(-\frac{1}{d^2}\right)$$

$$\partial H = -\frac{pl^2}{8d^2}\partial d$$

Substituting from eqn. (5.14), $$\partial H = -\frac{pl^2}{8d^2}\frac{3l^2\,\alpha\,t}{16d} = -\frac{3l^2\,\alpha\,t}{16d^2}H$$

$$\partial H = -\frac{3l^2\,\alpha\,t}{16d^2}H. \qquad ...(5.15)$$

Example 5.10 *A suspension cable of horizontal span 210 m is supported at the same level and has a central dip of 20 m. Find the increase in dip of the cable if the cable is subjected to a rise in temperature of 28°C. Take $\alpha = 12 \times 10^{-6}$ per °C.*

Solution $l = 210$ m ; $d = 20$ m ; $t = 28$°C ; $\alpha = 12 \times 10^{-6}$ per °C

Change in dip, $\delta d = \frac{3l^2}{16d}\alpha\,t = \frac{3 \times 210^2}{16 \times 20} \times 12 \times 10^{-6} \times 28 = 0.1389 \text{ m} = 138.9 \text{ mm}$

Increase in the central dip = 138.9 mm.

Example 5.11 *A cable supported at the same level on either end is of 140 m horizontal span with a central dip of 14 m. It carries a load of 15 kN/m on its horizontal span. Calculate the change in the horizontal tension when the temperature rises through 28°C. Coefficient of linear expansion of the cable materials, $\alpha = 4 \times 10^{-6}$/°C.*

Solution $l = 140$ m ; $d = 14$ m ; $p = 15$ kN/m ; $t = 28$°C ; $\alpha = 4 \times 10^{-6}$ per °C

$$H = \frac{pl^2}{8d} = \frac{15 \times 140^2}{8 \times 14} = 2625 \text{ kN}$$

Change in horizontal tension,

$$\Delta H = -\frac{3l^2}{16d^2}\alpha\,t\,H$$

$$\therefore \quad \Delta H = -\frac{3 \times 140^2}{16 \times 14^2} \times 4 \times 10^{-6} \times 28 \times 2625 = -5.513 \text{ kN}$$

Reduction in horizontal tension = 5.513 kN.

5.8 STIFFENING GIRDERS—NECESSITY AND TYPES

We have seen that in bridges suspension cables are the main load bearing members and because they are flexible they would change their shape depending upon the nature and position of the moving live load on the deck slab, if unstiffened. Stiffening girders are usually suspended from the cable through closely spaced hangers or suspenders, one on each side of the width of the bridges and they serve the following functions :

1. Stiffening girders enable the suspension bridge decks to remain fairly level.

2. Whatever be the live load on the deck slab, the stiffening girders will convert and transmit the load on the deck slab as a uniformly distributed load and thereby help the cable retain the parabolic shape during the passage of loads.

3. The dead load of the girders which is a udl is directly transmitted to the cables and is taken up entirely by the tension in the cables.

4. Thus the uniformly distributed dead load will not cause any shear force or bending moment in the stiffening girder.

5. The stiffening girders will have to resist the shear force and bending moment due to live loads.

Stiffening girders may be either three hinged or two hinged.

5.8.1 Suspension Bridges with Three Hinged Stiffening Girders

Figure 5.25 shows a suspension cable stiffened by a three hinged stiffening girder hinged at supports E and F and at centre G.

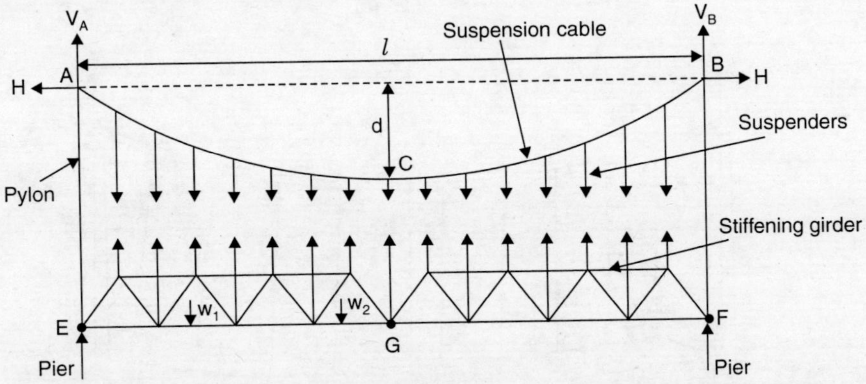

Fig. 5.25

The girder and the cable complement each other. The girder by itself will not be enough to span the enormous distances. The cable by pulling the girder up via its closely spaced suspenders relieves the girder of a huge part of the load, by developing a uniform tension in its suspenders. In other words the suspenders apply on the girder an upward acting udl. This udl produces on the girder a negative bending moment parabolically distributed. This hogging bending moment is very similar to the hogging bending moment developed in an arch due to the horizontal thrust. Simple calculations would show us that the horizontal pull H in the cable plays the same role on the stiffening girder as the horizontal thrust plays on the arch.

To visualize this, let us have a three hinged stiffening girder inadequately supported by the cable which (let us say) has a very small tension [Fig. 5.26 (*a*)]

Now let us activate the suspenders by applying at A and B outward pulls T,T.

T will pull the cable up. The cable would pull the suspenders up which in turn would restore the girders to their functional position [Fig. 5.26 (*b*)].

Thus, the cable strengthens the girder by developing an appropriate tension. But for the huge uniformly distributed upward force given by the cable, the stiffening girder can not function as a bridge girder, (particularly so, if it is 3 hinged, since a 3 hinged girder is statically deficient).

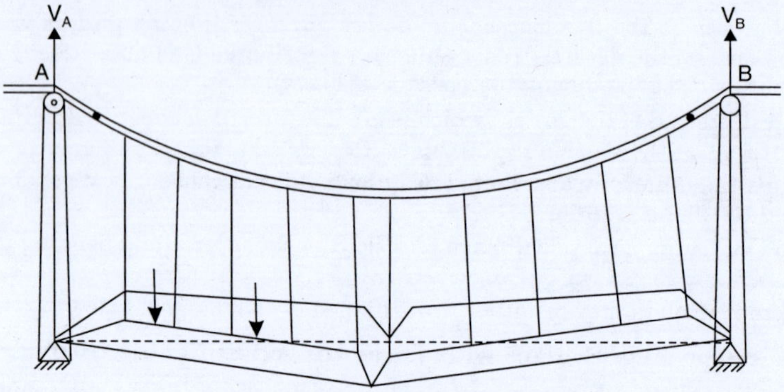

Insufficient tension in cable; girder sags

Fig. 5.26 (*a*)

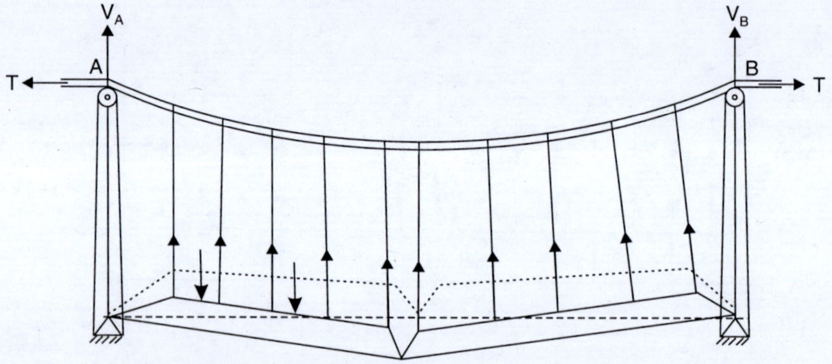

The cable being pulled tight ; girder back in place

Fig. 5.26 (*b*)

On the other hand, the girder is an essential part of the structural combine. The girder makes sure that the parabolic shape of the cable does not change (considerably). Of course, there would be small elastic displacements, but nothing large enough to shift the shape of the parabolic cable. If the cable remains parabolic then we can infer that the load acting on it is uniform over the horizontal length.

Thus, the girder ensures that the cable does not deform unduly (vide Figs. 5.2, 5.3). In other words, the girder stiffens the cable and the cable strengthens the girder.

The corollary is that whatever pattern of loads act on the girder, the loading transferred to the cable is a udl and by the same count the loading that the cable applies on the girder is uniform and upward (vide Fig. 5.25).

5.8.1 (*a*) Bending Moment and Shear Force in the Girder

Let us go back to Fig. 5.25. If the loading on the bridge is made up of 2 loads $\mathbf{W_1}$ and $\mathbf{W_2}$, then the system is going to produce an upward udl of **w/m** on the girder and a downward udl of **w/m** on the cable.

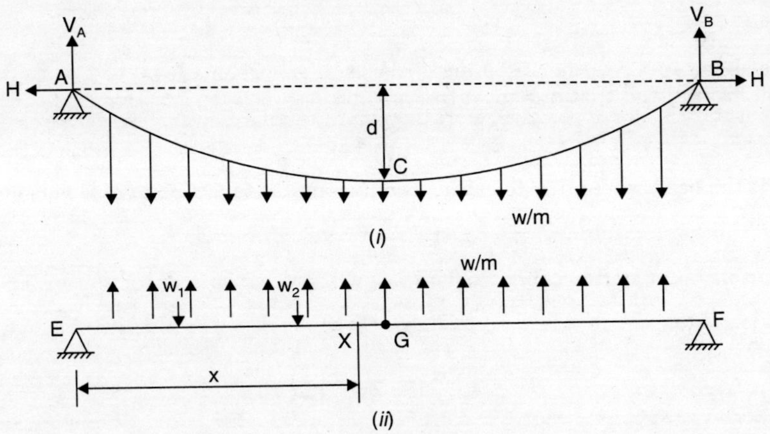

Fig. 5.26 (c)

The girder load is thus made up of 2 systems.

(*i*) W_1 and W_2 downwards producing positive bending moment

(*ii*) **w**/m upwards producing negative bending moment

At any section X on the girder, let us denote the bending moment due to external loads as μ_x, and the bending moment due to **w** would be $-\mathbf{w}x\,(l-x)/2$.

Hence, $$M_x = \mu_x - \frac{\mathbf{w}}{2}x\,(l-x) \qquad \text{...(5.16)}$$

What does w produce on the cable ? Obviously, the reactions,

$$V_A = V_B = \frac{\mathbf{w}l}{2} \quad \text{and} \quad H = \frac{\mathbf{w}l^2}{8d} \qquad \text{...[5.16 (a)]}$$

Now we will make an attempt to identify the behaviour of the arch with that of the suspension bridge.

We can recall that the hogging bending moment in arches (which is responsible for reducing the net bending moment in the arch section) is H.y.

For the parabolic cable, $$y = \frac{4d}{l^2}x\,(l-x)$$

Therefore, $$Hy = \frac{\mathbf{w}l^2}{8d}\cdot\frac{4d}{l^2}x\,(l-x)$$

$$Hy = \frac{\mathbf{w}}{2}x\,(l-x)$$

So we are in a position to modify the equation (5.16) as

$$M_x = \mu_x - Hy \qquad \text{...(5.17)}$$

Quite surprisingly, we find that, what modifies (or reduces) the beam bending moment μ_x is the horizontal cable tension. The cable tension achieves this effect via **w**, the suspender pulls.

We were able to convert equations (5.16) to (5.17) because the cable profile and the BMD due to udl, both have the same parabolic equation.

Since, $$\mathrm{H} = \frac{\mathbf{w}l^2}{8d}, \quad \mathbf{w} = \frac{8\mathrm{H}d}{l^2} \qquad \text{...[5.17 (a)]}$$

The **shear force** in the stiffening girder is, like wise, made up of 2 parts, the part due to external loading and that due to w. In Fig. 5.26,

$$\mathrm{S}_x = \mathrm{S}_b - \frac{\mathbf{w}l}{2} + \mathbf{w}x$$

where S_b is the beam shear due to ext loads only (similar to μ_x due to ext loads only)

$$\mathrm{S}_x = \mathrm{S}_b - \frac{\mathbf{w}}{2}(l - 2x) \qquad \text{...(5.18)}$$

Can this be related to cable tension? We will try.

For the cable, $$y = \frac{4d}{l^2}x(l - x)$$

$$\frac{dy}{dx} = \frac{4d}{l^2}(l - 2x) = \tan\theta$$

$$\mathrm{H}\tan\theta = \frac{\mathbf{w}l^2}{8d}\frac{4d}{l^2}(l - 2x) = \frac{\mathbf{w}}{2}(l - 2x)$$

The second part of the RHS in eqn. (5.18) happens to be, after all, **H tan θ**

So, $$\mathrm{S}_x = \mathrm{S}_b - \mathrm{H}\tan\theta \qquad \text{...(5.19)}$$

In conclusion, the bending moment and shear force at every section in the stiffening girder is mitigated by the cable tension indirectly, since cable tension produces w and reduces M_x and S_x. Equations (5.17) and (5.19) clearly reveal the beneficial effect of H on M_x and S_x.

5.8.1 (*b*) Evaluating H and w from the Girder Loading

In a 3 hinged stiffening girder, the net bending moment at the central hinge would be zero. This condition would help us evaluate H (or **w**) from this statics condition. In fact, this ease of calculation is the only justification for providing a third hinge at the centre.

In equation (5.17), when $x = l/2$, $y = d$ and $\mathrm{M}_x = 0 = \mu_C - \mathrm{H}d$

$$\mathrm{H} = \frac{\mu_C}{d} \qquad \text{...(5.20)}$$

$$\mathbf{w} = \frac{8\mathrm{H}d}{l^2} = \frac{\mu_C}{d}\frac{8d}{l^2} = \frac{8\mu_C}{l^2}$$

$$\mathbf{w} = \frac{8\mu_C}{l^2}. \qquad \text{...(5.21)}$$

Example 5.12 *A suspension bridge of 250 m span has two nos. of three hinged stiffening girders supported by cables with a central dip of 25 m. If 4 point loads of 300 kN each are placed at the centre line of the roadway at 20, 30, 40 and 50 m from the left hand hinge, find the shear force and bending moment in each girder at 62.5 m from each end. Calculate also the maximum tension in the cable.*

Solution The load system is shared equally by the two girders and cables. We will therefore take the loads as 150 kN each.

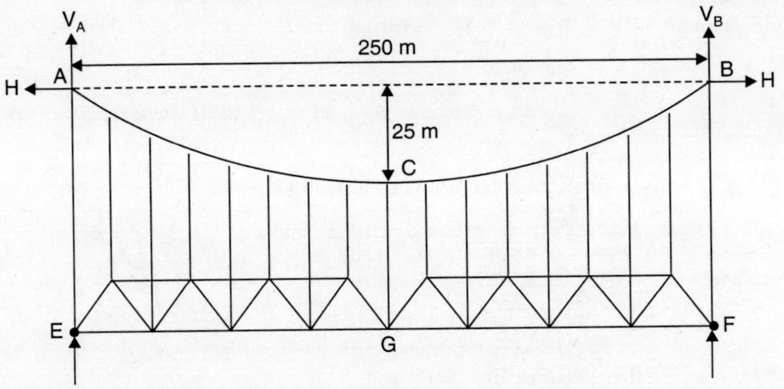

Fig. 5.27

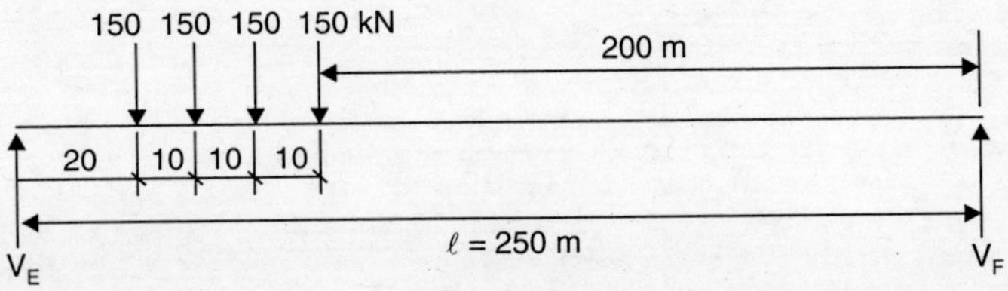

Fig. 5.27 (*a*)

Taking moments about F

$$V_E \times 250 - 150 \times 200 - 150 \times 210 - 150 \times 220 - 150 \times 230 = 0$$

$$V_E = 516 \text{ kN}$$

$$V_F = 84 \text{ kN}$$

From equation (5.20),

Horizontal tension, $\quad H = \dfrac{\mu_C}{d} = \dfrac{V_F \times 125}{25} = \dfrac{84 \times 125}{25} = 420 \text{ kN}$

$$H = 420 \text{ kN}$$

(*a*) **Bending moment :**

Bending moment at 62.5 m from left hand hinge

$$= V_F \times 187.5 - H \times y$$

y at 62.5 m from each end $= \dfrac{4d}{l^2}\, x\,(l - x) = \dfrac{4 \times 25}{(250)^2} \times 62.5 \times (250 - 62.5) = 18.75$ m

Therefore bending moment = 84 × 187.5 – 420 × 18.75 = 7875 kNm

Bending moment at 62.5 m from left hinge = 7875 kNm

Bending moment 62.5 m from right hand hinge

$$= V_F \times 62.5 - H \times y = 84 \times 62.5 - 420 \times 18.75$$

$$= -2625 \text{ kNm}$$

Bending moment 62.5 m from right hand hinge = – 2625 kNm.

(*b*) **Shear force :**

Shear force at 62.5 m from left hand hinge

$$V_{62.5} = V_b - H \tan\theta$$

$$\tan\theta = \frac{4d}{l^2}(l - 2x) = \frac{4 \times 25}{(250)^2}(250 - 2 \times 62.5) = 0.2$$

$$\tan\theta = 0.2$$

Net vertical shear at 62.5 m from left hand hinge

$$V_b = V_E - 4 \times \frac{300}{2} = 516 - 600 = -84 \text{ kN}$$

Therefore, $V_{62.5} = -84 - 420 \times 0.2 = -168$ kN

Shear force at 62.5 m from right hand hinge (or) 187.5 m from left hand hinge

$$V_{187.5} = -V + H \tan\theta = -84 + 420 \times 0.2 = 0$$

(*c*) **Vertical pull on the cable (w) :**

$$Hd = \frac{\mathbf{w}l^2}{8}$$

$$\mathbf{w} = \frac{H \times d \times 8}{l^2}$$

$$\mathbf{w} = \frac{420 \times 25 \times 8}{(250)^2} = 1.344 \text{ kN/m}$$

$$V_A = V_B = \frac{\mathbf{w}l}{2} = \frac{1.344 \times 250}{2} = 168 \text{ kN}$$

$$T_{max} = \sqrt{V_A^2 + H^2} = \sqrt{(168)^2 + (420)^2} = 452.35 \text{ kN}$$

Maximum tension in the cable = 452.35 kN.

Example 5.13 *A suspension bridge of 100 m span has two numbers of three hinged stiffening girders supported by two cables with a central dip of 10 m. If three point loads of 20 kN each are placed along the centre line of the roadway at 10, 15 and 20 m from left hand hinge, find the shear force and bending moment in each girder at 30 m from each end. Calculate the maximum tension in the cable.*

Solution The load system is shared equally by the 2 girders. Each point load carried by each girder = 20/2 = 10 kN.

(*a*) **To find V_D and V_F :**

Taking moments about the hinge

$$V_F \times 100 = 10 \times 10 + 10 \times 15 + 10 \times 20$$

$$V_F = 4.5 \text{ kN}$$

$$V_D = \text{Total load} - V_F = 30 - 4.5$$

$$V_D = 25.5 \text{ kN}$$

(*b*) **To find horizontal reaction (H) :**

Taking moments about the central hinge E,

$$M_E = V_F \times 50 - H \times 10 = 0$$

$$4.5 \times 50 - H \times 10 = 0$$

$$H = 22.5 \text{ kN}$$

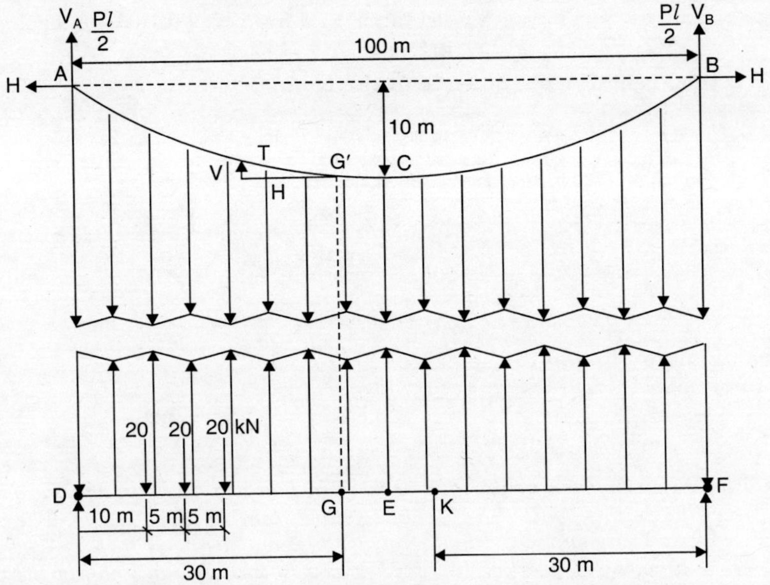

Fig. 5.28

Equation of the parabola, $$y = \frac{4d}{l^2} x(l - x)$$

At $x = 30$ m, $$y = \frac{4 \times 10}{100^2} \times 30 \times 70 = 8.4 \text{ m}$$

Therefore, at point G and K, $y = 8.4$ m

(*c*) **At *x* = 30 m :**

$$\frac{dy}{dx} = \frac{4d}{l^2}(l - 2x)$$

$$\tan \theta = \frac{4 \times 10}{100^2}(100 - 2 \times 30) = 0.16$$

$$M_G = V_F \times 70 - H \times 8.4$$

$$= 4.5 \times 70 - 22.5 \times 8.4 = 126 \text{ kNm}$$

F_G = shear force due to simply supported action of the girder-shear force due to the effect of horizontal reaction.

$$= -V_F - H \tan \theta = -4.5 - 22.5 \times 0.16 = -8.10 \text{ kN}$$

$$F_A = -8.10 \text{ kN}$$

(*d*) **At *x* = 70 m :**

$$\frac{dy}{dx} = \frac{4d}{l^2}(l - 2x)$$

$$\tan \theta = \frac{4 \times 10}{100^2}(100 - 2 \times 70) = -0.16$$

$$M_K = V_F \times 30 - H \times 8.4 = 4.5 \times 30 - 22.5 \times 8.4 = -54 \text{ kNm}$$
$$F_K = -V_F - H \tan\theta = -4.5 - 22.5 \times 0.16 = -0.9 \text{ kN}$$
$$F_K = -0.9 \text{ kN}$$

(*e*) **Vertical reaction at the ends of the cable (V) :**

$$V = \frac{pl}{2}$$

But,
$$H = \frac{pl^2}{8d}; \quad p = \frac{8dH}{l^2}$$
$$p = \frac{8 \times 10 \times 22.5}{100^2} = 0.18 \text{ kN/m}$$
$$V = \frac{0.18 \times 100}{2} = 9 \text{ kN}$$

(*f*) **Maximum tension in the cable (T) :**

$$T_{max} = \sqrt{V^2 + H^2} = \sqrt{9^2 + 22.5^2} = 24.23 \text{ kN}$$
$$T_{max} = 24.23 \text{ kN.}$$

Example 5.14 *A suspension bridge of 140 m span has two numbers of 3 hinged stiffening girders supported by 2 cables with a central dip of 12 m. The width of the roadway supported by the girder is 5 m. The dead load is 7 kN/m² of floor area. A live load of 12.5 kN/m² covers the left hand half of the bridge. Find the shear force and bending moment at the loaded quarter span point. Determine also the maximum tension in the cable.*

Solution

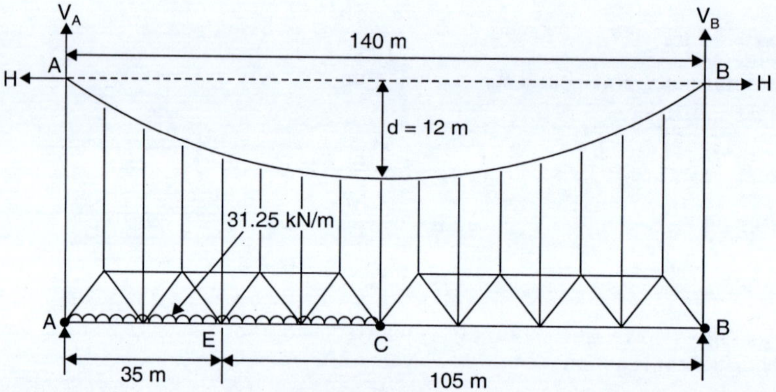

Fig. 5.29

Load per girder and cable

$$\text{Dead load} = \frac{5 \times 7}{2} = 17.5 \text{ kN/m}$$

$$\text{Live load} = \frac{5 \times 12.5}{2} = 31.25 \text{ kN/m over the left hand half of the span only.}$$

(*a*) **For the stiffening girder :**

The girder will not be subjected to shear force or bending moment anywhere due to the dead load but there will be shear force and bending moment due to live load.

Vertical reactions :

Taking moments about A, (omitting dead load)

$$V_B \times 140 - 31.25 \times 70 \times 35 = 0$$

$$V_B = 546.875 \text{ kN}$$

$$V_A = 1640.625 \text{ kN}$$

(*b*) **Horizontal pull H_1 (due to live load) :**

Equating the moments about the hinge C to zero,

$$M_C = V_B \times 70 - H_1 \times 12 = 0 = 546.875 \times 70 - H_1 \times 12 = 0$$

$$H_1 = 3190.1 \text{ kN}$$

At the loaded quarter span point (x = 35 m from A)

Dip of the cable, $$y = \frac{4d}{l^2} x(l-x) = \frac{4 \times 12}{140^2} \times 35 \times 105 = 9 \text{ m}$$

$$\tan \theta = \frac{dy}{dx} = \frac{4d}{l^2}(l - 2x) = \frac{4 \times 12}{140^2}(140 - 2 \times 35) = 0.1714$$

$$M_E = V_A \times 35 - 31.25 \times 35 \times 17.5 - Hy$$

$$= 1640.625 \times 35 - 31.25 \times 35 \times 17.5 - 3190.1 \times 9$$

$$M_E = 9570.35 \text{ kNm}$$

$$F_E = V_A - 31.25 \times 35 - H \tan \theta$$

$$= 1640.625 - 31.25 \times 35 - 3190.1 \times 0.1714$$

$$F_E = 0$$

(*c*) **For the suspension cable :**

Total equivalent uniformly distributed live load transmitted to the cable:

$$\frac{W_1 l}{8d} = H_1;$$

$$W_1 = \frac{8dH_1}{l} = \frac{8 \times 12 \times 3190.1}{140} = 2187.5 \text{ kN}$$

Total dead load transmitted to the cable, $W_d = 17.5 \times 140 = 2450$ kN

Total load, $W = W_2 + W_1 = 4637.5$ kN.

Vertical reaction at each end of cable,

$$V = \frac{W}{2} = 2318.75 \text{ kN}$$

Horizontal pull due to dead load, $$H_2 = \frac{W_2 l^2}{8d} = \frac{17.5 \times 140^2}{8 \times 12} = 3572.92 \text{ kN}$$

Total horizontal pull, $H = H_2 + H_1 = 3190.1 + 3572.92 = 6763.02$ kN

Therefore, maximum tension in the cable,

$$T = \sqrt{V^2 + H^2} = \sqrt{(2318.75)^2 + (6763.02)^2}$$

$$T = 7149.48 \text{ kN}.$$

5.9 INFLUENCE LINE DIAGRAM FOR CABLES WITH THREE HINGED STIFFENING GIRDERS

(*a*) **ILD for Horizontal Tension/Pull**

Consider a suspension bridge, the cable of which is stiffened by a three hinged stiffening girder [Fig. 5.30 (*a*)]. Let a unit load move from left to right of the girder and let it be at a distance **'a'** from the left end.

Taking moments about B

$$V_A \times l - 1\,(l - a) = 0$$

$$V_A = \frac{(l-a)}{l}; \quad V_B = \frac{a}{l}$$

Equating moments about C of the forces to the right of C to zero,

$$V_B \times \frac{l}{2} - H \times d = 0$$

$$\frac{a}{l} \times \frac{l}{2} = H\,d$$

$$H = \frac{a}{2d}$$

This is the equation for H, H varies linearly

When $a = 0,\ H = 0,$

When $a = \frac{l}{2},\quad H = \frac{l}{4d}$ ...(5.22)

By symmetry, when the unit load is at B, H will be zero. Hence the influence line diagram for H is a triangle with the maximum ordinate of $\frac{l}{4d}$ at the centre of span [Fig. 5.27 (*b*)].

(*b*) **ILD for w the Intensity of Load Transmitted to the Cable**

A part of the live load on the girder is converted as uniformly distributed load *w* by the stiffening girder, and the same is transmitted to the suspension cable. The variation of **w** transmitted to the cable with the load position '*a*' will be studied here.

Considering the unit load position in previous subdivision,

Horizontal reaction, $H = \frac{a}{2d}$

The uniformly distributed load **w** transmitted to the cable is related to H [eqns. (5.16) (*a*) and (5.17) (*a*)].

Horizontal reaction, $H = \frac{\mathbf{w}l^2}{8d}$

$$\frac{\mathbf{w}l^2}{8d} = \frac{a}{2d}; \quad \mathbf{w} = \frac{a}{2d} \times \frac{8d}{l^2} = \frac{4a}{l^2}$$

This is a linear variation and will be symmetrical about C.

At $a = 0,\quad \mathbf{w} = 0$

At $a = \frac{l}{2},\quad \mathbf{w} = \frac{2}{l}$ ...(5.23)

(*c*) **ILD for Bending Moment at a Given Section X**

We know that for the stiffening girder, the bending moment at any section is given by

$Mx = \mu_x - Hy$, where μ_x = Bending moment at the section considering the girder as a simply supported beam.

For the section considered, the bending moment due to unit

$$\text{load} = \frac{x(l-x)}{l}$$

The ILD for μ_x will be a triangle with the maximum ordinate of $\frac{x(l-x)}{l}$ at the section X.

From equation (5.22), $H = \frac{l}{4d}$ at centre

$$y = \frac{4d}{l^2} x(l-x) \text{ is a fixed quantity for a given section.}$$

Therefore, the IL for H*y* will be a triangle, with maximum ordinate at C.

$$\text{Ordinate at C} = \frac{l}{4d} \times \frac{4d}{l^2} \times (l-x)$$

$$\text{Ordinate at C} = \frac{x(l-x)}{l} \qquad \text{...(5.24)}$$

Super imposing the IL for μ_x and H*y*, the ILD for bending moment at a given section is as shown in Fig. 5.30(*d*).

You may notice that this is very similar to the IL for M_x in arches (Fig. 4.26).

(*d*) **Maximum Bending Moment Diagram Due to a Single Point Load *w* Moving from One End to the Other**

(*i*) *Maximum positive bending moment :*

From the influence line for bending moment at X, it is obvious that the maximum positive bending moment at a section due to a single point load occurs when the load is on the section itself.

At the section X, $M_{x\ max}$ positive = Load × Net ordinate

$$= W \times \left[\frac{x(l-x)}{l} - \frac{\left[\frac{x(l-x)}{l}\right] x}{\left(\frac{l}{2}\right)}\right] = W\left[\frac{x(l-x)}{l} - \frac{2}{l} \times \frac{x(l-x)}{l} x\right]$$

$$= \frac{Wx(l-x)}{l}\left[1 - \frac{2x}{l}\right] = \frac{Wx(l-x)}{l}\left(\frac{l-2x}{l}\right)$$

$$M_{max}(+ve) = \frac{W}{l^2}(lx - x^2)(l-2x) \qquad \text{...(5.25)}$$

This moment will be maximum when $= \frac{dM_{max}}{dx} = 0$

$$\frac{d}{dx}\left[\frac{W}{l^2}(lx - x^2)(l-2x)\right] = 0$$

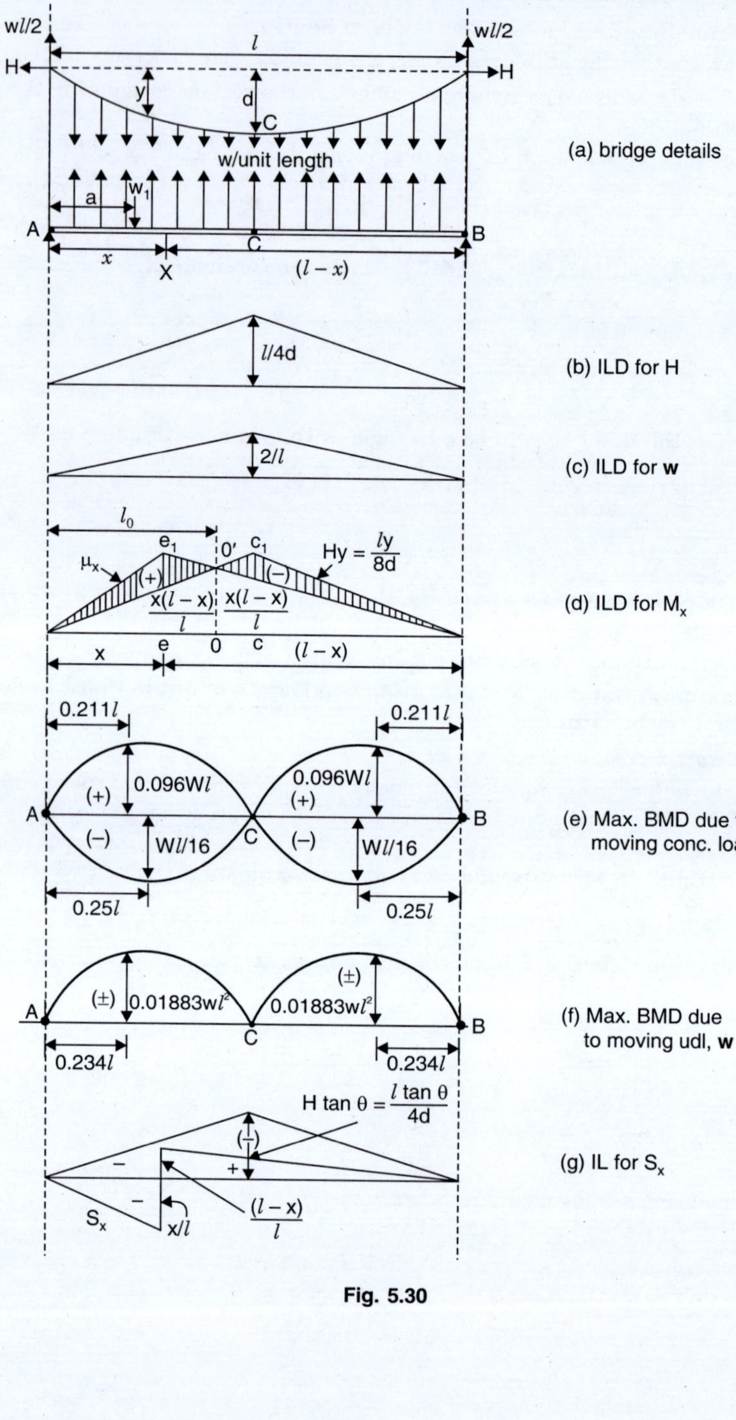
wl/2
wl/2
l
H
H
y
d
C
w/unit length
a
w1
A
B
C
x
X
(l – x)
(a) bridge details
l/4d
(b) ILD for H
2/l
(c) ILD for w
l0
e1
0′
c1
μx
(+)
(–)
Hy = ly/8d
x(l – x)/l
x(l – x)/l
e
0
c
x
(l – x)
(d) ILD for Mx
0.211l
0.211l
0.096Wl
0.096Wl
(+)
(+)
A
B
(–)
Wl/16
C
(–)
Wl/16
0.25l
0.25l
(e) Max. BMD due to moving conc. load W
(±)
(±)
0.01883wl²
0.01883wl²
A
C
B
0.234l
0.234l
(f) Max. BMD due to moving udl, w
H tan θ = l tan θ/4d
(+)
–
+
–
Sx
x/l
(l – x)/l
(g) IL for Sx

Fig. 5.30

$$\frac{W}{l^2}[(lx - x^2)(-2) + (l - 2x)(l - 2x)] = 0$$

$$-2lx + 2x^2 + l^2 - 2lx - 2lx + 4x^2 = 0$$

$$6x^2 - 6lx + l^2 = 0$$

$$x = \frac{-(-6l) \pm \sqrt{(-6l)^2 - 4 \times 6 \times l^2}}{2 \times 6}$$

$$x_{cr} = \frac{6l \pm 3.464\, l}{12} = 0.789\, l,\ 0.211\, l \qquad \text{...(5.26)}$$

i.e., the absolute maximum positive bending moment occurs at $0.211l$ from either end

Substituting $x = 0.211l$ in eqn. (5.25),

$$M_{\text{max max}} \text{ positive} = \frac{W}{l^2}[l \times 0.211\, l - (0.211\, l)^2]\,[l - 2 \times 0.211\, l]$$

$$= \frac{W}{l^2} \times 0.166479\, l^2 \times 0.578\, l = 0.096\, Wl$$

$$M_{\text{max max}} (+\text{ve}) = 0.096\, Wl \qquad \text{...(5.27)}$$

Again, from the ILD for M_x, it is clear that the maximum negative bending moment at a section due to a single point load occur when the load is on the central hinge.

For the section X,

$$M_{\text{max}} (-\text{ve}) = \text{Load} \times \text{Net ordinate}$$

$$= W \times \left[\frac{x(l - x)}{l} - \frac{\left(\frac{x(l - x)}{l}\right)\frac{l}{2}}{(l - x)}\right] = W\frac{x(l - x)}{l}\left[1 - \frac{l}{2(l - x)}\right]$$

$$= W\frac{x(l - x)}{l}\left[\frac{2(l - x) - l}{2(l - x)}\right] = \frac{Wx(l - x)}{l} \times \frac{(2l - 2x - l)}{2(l - x)}$$

$$M_{\text{max}} (-\text{ve}) = \frac{Wx(l - 2x)}{2l} = \frac{W}{2l}(lx - 2x^2) \qquad \text{...(5.28)}$$

For the moment to be the absolute maximum

$$\frac{dM_{\text{max}}}{dx} = 0; \quad \frac{d}{dx}\left[\frac{W}{2l}(lx - 2x^2)\right] = 0$$

$$\frac{W}{2l}(l - 4x) = 0$$

$$l - 4x = 0; \quad l = 4x; \quad x = \frac{l}{4}$$

i.e., the absolute maximum negative bending moment occurs at $0.25l$ from either end.

Substituting this value of x in equation (5.28)

$$M_{\text{max max}} \text{ negative} = \frac{W}{2l}[l \times 0.25\, l - 2(0.25\, l)^2]$$

$$M_{\text{max max}} \text{ negative} = 0.0625\, Wl \qquad \text{...(5.29)}$$

The maximum positive and negative bending moment diagrams due to a single concentrated moving load are shown in Fig. 5.30 (*e*).

(*e*) **Maximum Bending Moment for a Long udl**

We proceed from the ILD for M_x [Fig. 5.30 (*d*)]. Here the live load is considered to be of unlimited length.

Obviously the maximum positive bending moment occurs at X, if the load covers AO and maximum negative bending moment occurs at X when the load covers OC. When the entire span is covered, we would find that M_x is zero since the net area of the IL diagram is zero (area of A, e_1, B = area of A, c_1, B). In the figure, the ordinate at O is zero. Its position l_0 can be obtained by computation of OO′ in 2 ways and equating them.

In triangle Ae_1B, $OO' = \frac{x(l-x)(l-l_0)}{l(l-x)} = \frac{x(l-l_0)}{l}$

In triangle AC_1B, $OO' = \frac{x(l-x)}{l} \frac{l_0}{l/2} = \frac{2x(l-x)l_0}{l^2}$

Equating the above two values of OO′, we get

$$\frac{l-l_0}{l} = \frac{2(l-x)l_0}{l^2}$$

$$l_0 = \frac{l^2}{(3l-2x)}$$

$$OO' = \frac{2x(l-x)}{(3l-2x)}$$

Area under positive ILD $= \frac{x(l-x)(l-2x)}{2(3l-2x)}$

The same would be the area under negative ILD also

Therefore, $$M_{max} = \frac{wx(l-x)(l-2x)}{2(3l-2x)} \quad ...(5.30)$$

Equation (5.31) also gives the maximum bending moment at X due to a moving udl, **w**. This is plotted in Fig. 5.30 (*f*). There is a critical value of x that would maximise M_{max}. For M_{max} to be maximum.

$$\frac{d}{dx} M_{max} = 0 = \frac{w}{2}\left[(3l-2x)\frac{[(lx-x^2)-(-2)+(l-2x)^2]-(lx-x^2)(l-2x)(-2)}{(3l-2x)^2}\right]$$

Solving this, we get

$$x_{cr} = 0.234l \quad ...(5.31)$$

Substituting this value of x_{cr} in equation (5.31),

$$M_{max\ max} = \frac{w}{2} \frac{[(l \times 0.234\ l) - (0.234\ l)^2]\ (l - 2 \times 0.234\ l)}{(3l - 2 \times 0.234\ l)}$$

$$= \frac{w}{2} \times \frac{0.1792\ l^2 \times 0.532\ l}{2.532\ l} = 0.01883\ wl^2$$

$$M_{max\ max} = 0.01883\ wl^2 \quad ...(5.32)$$

Load length, $$l_0 = \frac{l^2}{(3l-2x)} = \frac{l^2}{(3l - 2 \times 0.234\ l)} = 0.395\ l$$

For max. negative bending moment at X, the portion AO will be empty and portion OB will be loaded. Since the intensity udl and areas of ILD for net positive and negative b.m. are the same, the absolute max. negative bending moment will also be the same as absolute max. positive bending moment.

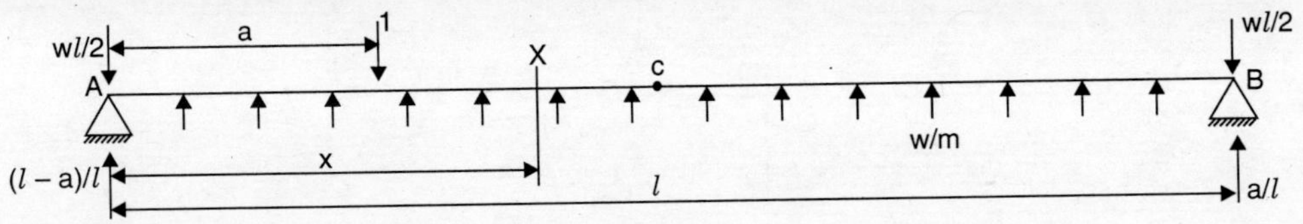

Fig. 5.31

(*f*) **ILD for Shear Force**

In Fig. 5.31, at a section X,

$$\text{ILO } F_x = -\frac{a}{l} - \frac{w(l-a)}{l} + \frac{wl}{2} = -\frac{a}{l} - w\left(l - a - \frac{l}{2}\right)$$

$$\text{ILO } F_x = -\frac{a}{l} - w\left(\frac{l}{2} - a\right) \quad \text{...(1)}$$

Here $- a/l$ is the ILO for a S.S beam. We will call this S_x,

So, $$\text{ILO } F_x = S_x - w\left(\frac{l}{2} - a\right)$$

The ordinate at X, $(a = x)$ is

$$\text{ILO } F_x = S_x - w\,(l/2 - x) \quad \text{...(2)}$$

We can relate **w** the cable pull to the cable tension. For the parabolic cable,

$$y = \frac{4d}{l^2}x\,(l-x)\,;\quad \frac{dy}{dx} = \frac{8d}{l^2}\left(\frac{l}{2} - x\right) = \tan\theta.$$

From equation [5.16(*a*)], $H = \dfrac{wl^2}{8d}$; $\dfrac{8d^2}{l^2} - \dfrac{w}{H}$

Therefore, $$\tan\theta = \frac{w}{H}\left(\frac{l}{2} - x\right)$$

$$\mathbf{w}\left(\frac{l}{2} - x\right) = H\tan\theta.$$

From eqn. (2), $$\text{ILO } F_x = S_x - H\tan\theta \quad \text{...(5.33)}$$

θ is the inclination of the cable at X. Figure 5.30 (*g*) gives the plot of ILD for F_x. Here, one part is the beam shear and the other part is the reduction (or change) in shear due to H. For any section X, θ is a constant and H is the only variable. The shape of this part, H tan θ will be similar to H-plot in Fig. 5.30 (*b*).

Depending upon the position of X, Fig. 5.30 (*g*) can have different forms as given in Figs. 5.30 (*a*) to (*d*).

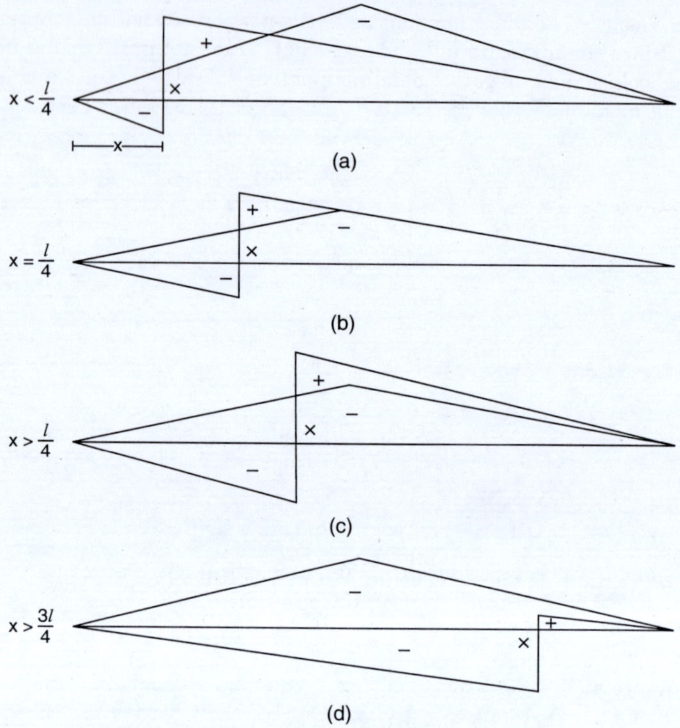

Fig. 5.32. Various possible ILDs for shear at X.

Example 5.15 *A suspension bridge is of 160 m span. The cable of the bridge has a dip of 12 m. The cable is stiffened by a three hinged girder with hinges at either end and at centre. The dead load of the girder is 15 kN/m. Find the greatest positive and negative bending moments in the girder when a single concentrated load of 340 kN passes through it. Also find the maximum tension in the cable.*

Solution l = 160 m ; d = 12 m ; w_d = 15 kN/m ; w_l = 340 kN

(*a*) **Bending moment :**

The uniformly distributed dead load will not cause any bending moment in the stiffening girder. The live load is a single concentrated moving load. The maximum bending moment diagram due to a single concentrated load is given below as in Fig. 5.30 (*e*).

Maximum +ve bending moment,

$$M_{max\ max} \text{ +ve} = 0.096Wl = 0.096 \times 340 \times 160 = 5222.4 \text{ kNm}$$

and this will occur at $0.211l = 0.211 \times 160 = 33.76$ m from either end

Maximum –ve bending moment,

$$M_{max\ max} \text{ –ve} = -\frac{Wl}{16} = -\frac{340 \times 160}{16} = -3400 \text{ kNm}$$

and this will occur at $0.25l = 0.25 \times 160 = 40$ m from either end.

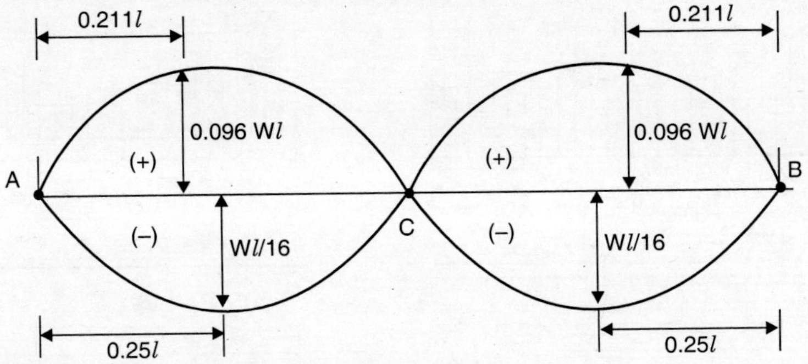

Fig. 5.33

(*b*) **Maximum tension in the cable :**

Dead load of the girder (transmitted to the cable directly)

$$p_d = 15 \text{ kN/m}$$

Equivalent udl transmitted to the cable due to the

moving concentrated load, $p_l = \dfrac{2 \times 340}{160} = \dfrac{680}{160} = 4.25$ kN/m

Total load transmitted to the cable, $p = p_d + p_l = 15 + 4.25 = 19.25$ kN/m

Vertical reaction, $V = \dfrac{pl}{2} = \dfrac{19.25 \times 160}{2} = 1540$ kN

Horizontal pull, $H = \dfrac{pl^2}{8d} = \dfrac{19.25 \times 160^2}{8 \times 12} = 5133.2$ kN

Maximum tension, $T_{max} = \sqrt{V^2 + H^2} = \sqrt{(1540)^2 + (5133.3)^2}$

$T_{max} = 5359.3$ kN.

5.10 TWO HINGED STIFFENING GIRDERS

Like in arches, it is almost inevitable that a 3 hinged girder is followed by a 2 hinged girder. It might have been guessed that a 2 hinged stiffening girder is statically indeterminate. This is true, but we are going to make a simplification that makes 2 hinged girder problems easier than its predecessor.

The assumption made with 2 hinged girders is that they transfer a udl to the cable and the value of the udl (**w**) is the same wherever the load is and what ever the loading. Figure 5.34 will show the situation better.

In a 3 hinged girder the value of **w** depends on W and the load position a. In a two hinged girder, **w** does not depend upon **a**, the load position. The girder (it is assumed) transfers a uniform load equal to the **total load ÷ span**. This however is a gross assumption. Any error in the assumption can have the effect of

1. Over estimating the negative bending moment in the girder when the loads are near supports and

2. Under estimating the positive bending moment in the girder, when the loads are near midspan.

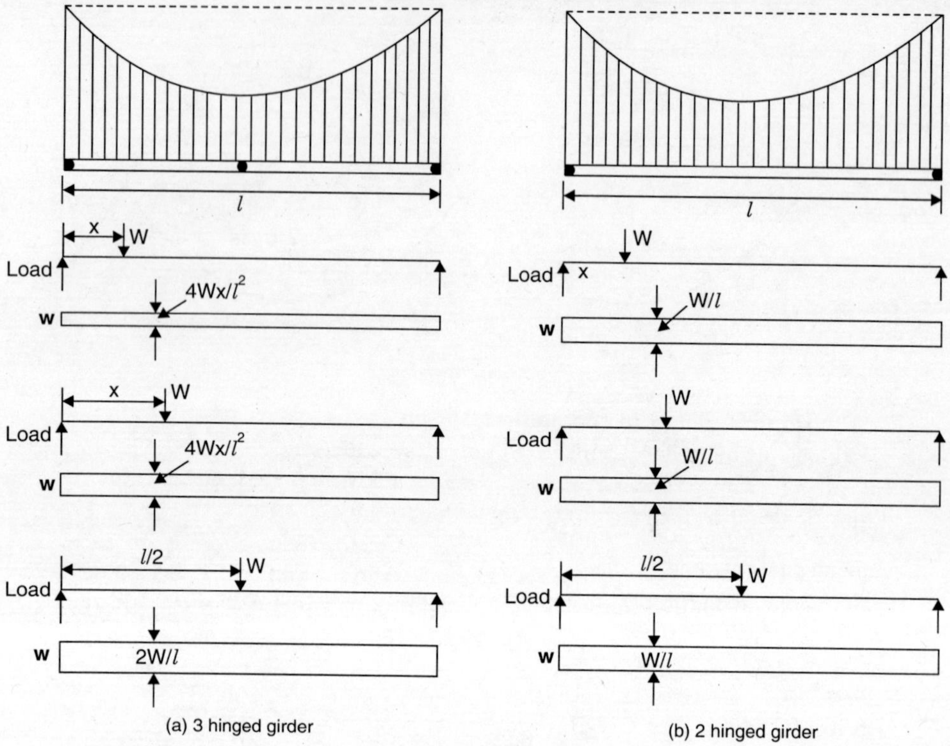

Fig. 5.34. How much upward udl is developed when a load W moves on the deck ?

5.11 INFLUENCE LINES FOR H AND w

Influence line for **w** and H are ridiculously simple (due to the simplifying assumption made)

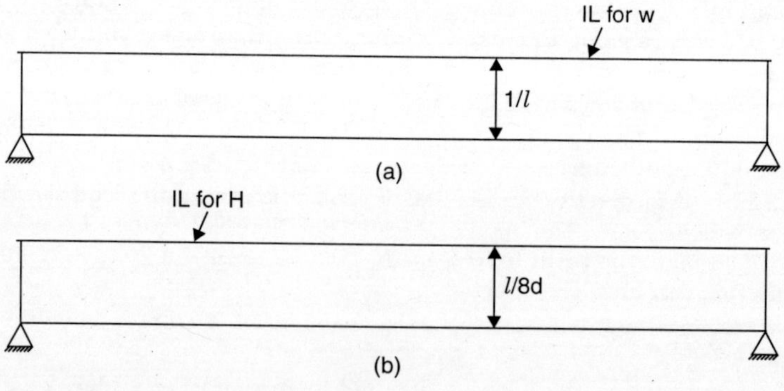

Fig. 5.35

Since $w = \dfrac{W}{l}$, the ILO for **w** is just $1/l$ (when W = 1) and the ILD is a level line [Fig. 5.35 (*a*)].

Likewise $H = \dfrac{wl^2}{8d} = \dfrac{Wl}{8d}$ independent of **'a'**, Hence ILO = $l/8d$ uniformly.

Here is it worthwhile comparing these with their counterparts for 3 hinged girders (Fig. 5.30)

Table 5.1 Comparison of influence lines.

IL for	*3 hinged girder*	*2 hinged girder*
H	$l/4d$	$l/8d$
w	$2/l$	$1/l$

5.12 INFLUENCE LINES FOR M_x AND S_x

H and **w** are not for any particular point on the girder. They are for the structure as a whole. Only the load position (*a*) varies. For the 2 hinged girders even **'a'** is vanishing from the ILO expression.

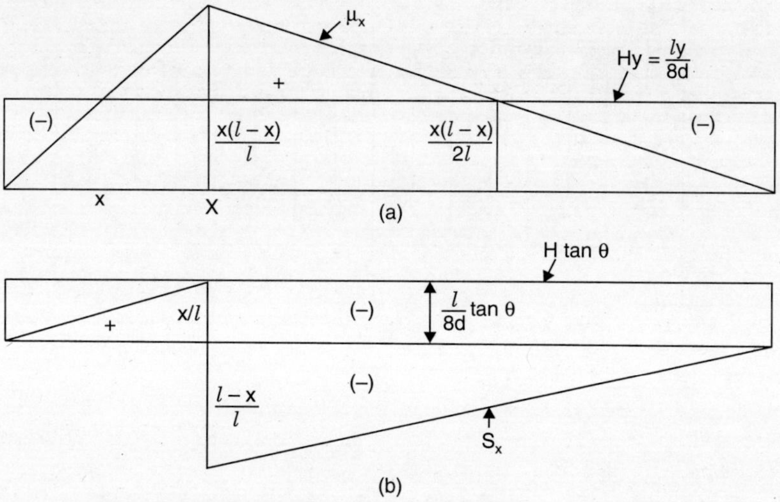

Fig. 5.36. IL for M_x and F_x.

We can see from Figs. 5.30 (*d*) and (*g*) that the ILD is made up of 2 parts (*i*) The beam ILD and (*ii*) the effect of H (on M_x and S_x).

In Fig. 5.30 (*d*), $M_x = \mu_x - Hy$. Here, since the ILD is for any given point, *y* becomes a constant and Hy takes the form of H only. Using the same logic, the ILD for M_x in a 2 hinged arch is made up of

(*i*) ILD for μ_x, a triangle

(*ii*) ILD for Hy, a rectangle, with the ordinate = $Hy = \frac{l}{8d}y$. Figure 5.36 (*a*) shows the influence line. In the same way IL for S_x is drawn by modifying Fig. 5.30 (*g*). In the new diagram [Fig. 5.36 (*b*)] H tan θ diagram would be horizontal.

Example 5.16 *A suspension cable of 75 metres horizontal span and central dip 6 metres has a stiffening girder hinged at both ends. The dead load transmitted to the cable including its own weight is 1500 kN. The girder carries a live load of 30 kN/m uniformly distributed over the left half of the span. Assuming the girder to be rigid, calculate the shear force and bending moment in the girder at 20 m from the left support. Also calculate the maximum tension in the cable.*

Solution

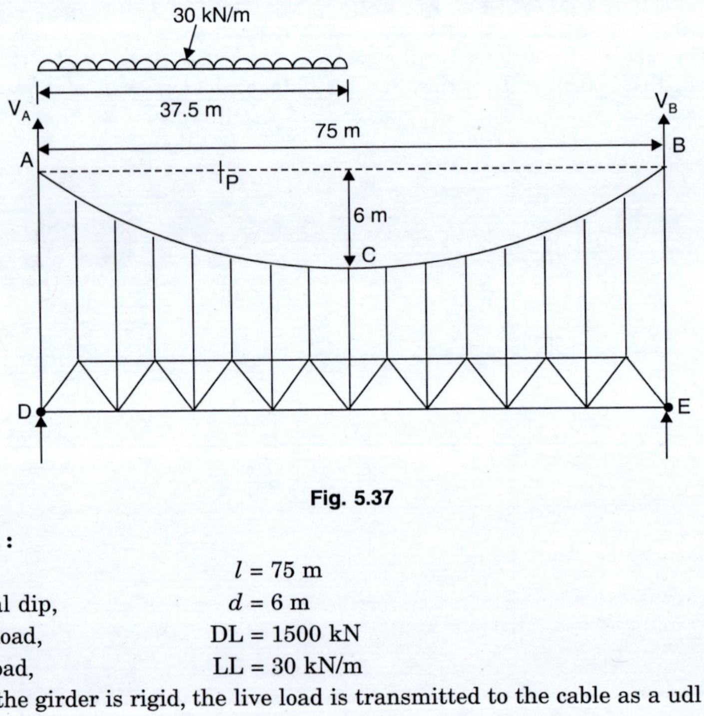

Fig. 5.37

Given :

Span, l = 75 m

Central dip, d = 6 m

Dead load, DL = 1500 kN

Live load, LL = 30 kN/m

Since the girder is rigid, the live load is transmitted to the cable as a udl whatever the position of the load.

Horizontal force due to live load, $H_l = \frac{Pl}{8d} = \frac{(30 \times 37.5) \times 75}{8 \times 6} = 1757.8$ kN

Horizontal force due to dead load $H_d = \frac{Pl}{8d} = \frac{1500 \times 75}{8 \times 6} = 2343.8$ kN

Total horizontal force, $H = H_l + H_d$

$$H = 1757.8 + 2343.8$$

$$H = 4101.6 \text{ kN}$$

$$V_A = V_B = \text{Total load}/2$$

$$= \frac{W_d + W_l}{2} = \frac{(30 \times 37.5) + 1500}{2} = 1312.5 \text{ kN}$$

Maximum tension in the cable

$$T_{max} = \sqrt{H^2 + V^2} = \sqrt{4101.6^2 + 1312.5^2} = 4306.5 \text{ kN}$$

$$T_{max} = 4306.5 \text{ kN}$$

Dip at x = 20 m : $y = \frac{4d}{l^2} x (l - x) = \frac{4 \times 6}{75^2} \times 20 \,(75 - 20) = 4.69$ m

$$\tan\theta = \frac{4d}{l^2}(l - 2x) = \frac{4 \times 6}{75^2}(75 - 2 \times 20) = 0.149$$

To find V_A and V_B :

Equating moments about A to zero

$$V_B \times 75 - 30 \times 37.5 \times 37.5/2 = 0$$

$$V_B = 281.25 \text{ kN}$$

$$V_A = 30 \times 37.5 - 281.25$$

$$V_A = 843.75 \text{ kN}$$

Bending moment at P :

$$BM = M_{20\,m} = V_A \times 20 - H_l \times y - w \,.\, l\,\frac{l}{2}$$

$$= 843.75 \times 20 - 1757.8 \times 4.69 - 30 \times 20 \times 20/2$$

$$BM = 2630.92 \text{ kNm}$$

Shear force at P :

$$S.F = V_A - H_l \tan\theta - wl = 843.75 - 1757.8 \times 0.149 - 30 \times 20$$

$$S.F = -18.16 \text{ kN.}$$

Example 5.17 *A suspension cable has a span of 120 m and a central dip of 10 m and is suspended from the same level at both towers. The bridge is stiffened by a stiffening girder hinged at the end supports. The girder carries a single concentrated load of 100 kN at a point 30 m from left end. Assuming equal tension in the suspension hangers.*

Calculate : (*i*) *The horizontal tension in the cable.*

(*ii*) *The maximum positive bending moment.*

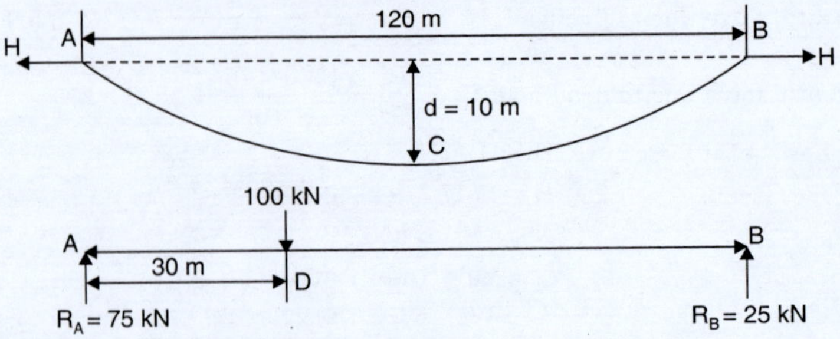

Fig. 5.38

Solution 1. **Vertical reactions R_A and R_B :**

Taking moments about A,

$$100(30) - R_B(120) = 0$$

$$R_B = 25 \text{ kN} ; \quad R_A = 75 \text{ kN}$$

2. **Horizontal pull (H) :**

Taking moments about C, considering the right side of C,

$$-R_B(60) + H(10) = 0$$

$$-25(60) + 10H = 0$$

$$H = 150 \text{ kN}$$

3. **Maximum tension in the cable :**

(*i*) *To find equivalent UDL :*

We know, $\mathbf{w} = \dfrac{100}{l} = \dfrac{100}{120} = 0.83 \text{ kN/m}$

$$\mathbf{w} = 0.83 \text{ kN/m}$$

(*ii*) *Reactions for equivalent UDL :*

$$V_A = V_B = \frac{\mathbf{w} \times l}{2} = \frac{0.83 \times 120}{2} = 50 \text{ kN}$$

(*iii*) *Maximum tension (T_{max}) :*

$$T_{max} = \sqrt{V_A^2 + H^2} = \sqrt{50^2 + 150^2} = 158.1 \text{ kN}$$

4. **Maximum positive bending moment in the girder :**

Maximum positive bending moment will occur under the point of application of load.

$$\text{BM at D} = V_A(30) - Hy$$

$$y = \frac{4y_C}{l^2} \times (l - x) = \frac{4 \times 10}{120^2} \times 30 \times 90 = 7.5 \text{ m}$$

$$M_D = 75 \times 30 - 150 \times 7.5 = 1125 \text{ kNm}$$

PROBLEMS

1. A suspension cable of span 45 m and central dip 3 m carries a uniformly distributed load of 60 kN/m over the entire span. Find the following :
   (*a*) Maximum tension in the cable (*b*) Minimum tension in the cable
   (*c*) The length of the cable. [**Ans.** 5239.41 kN; 5062.5 kN; 45.53 m]
2. A suspension cable of length 174.53 m is supported at the two ends at the same levels. The supports are 170 m a part. The cable is subjected to a uniformly distributed load of 20 kN/m of horizontal length over its entire span. Determine the reactions developed at the supports and their inclinations to the horizontal. [**Ans.** 4577.39 kN; 21° 48′]
3. A cable is hung between two points at a horizontal distance of 120 m. The vertical distance between the tops of supports is 24 m as shown in Fig. 5.39. It carries three vertical loads of magnitude 60, 75 and 30 kN at horizontal distances of 30, 60 and 90 m from the left support respectively. Determine the cable profile and the length of the cable. [**Ans.** 148.44 mm]

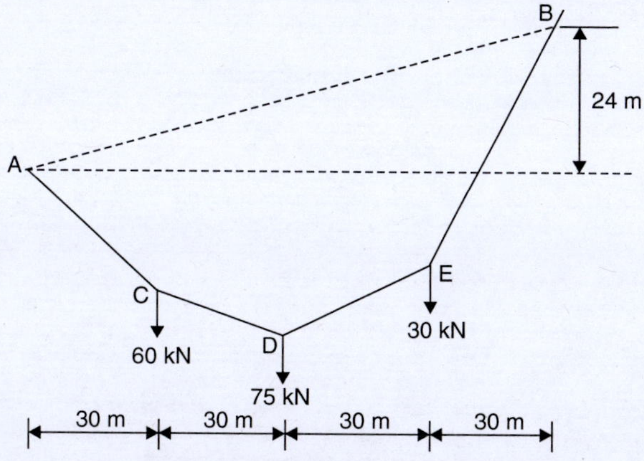

Fig. 5.39

4. A suspension cable of span 180 m hangs between two supports at the same level. It carries a uniformly distributed load of 30 kN/m over the entire span. The dip of the cable is 16.5 m and the slope of the anchor cable at the top of the tower is 35° with the horizontal. Determine the horizontal and vertical reactions on the top of the towers if
   (*a*) the cable passes over a frictionless pulley
   (*b*) the cable passes over a saddle supported on frictionless rollers.
   [**Ans.** 939.03 kN; 7198.6 kN; 0; 7856 kN]
5. A suspension cable of uniform material is hung between two points A and B separated by a horizontal distance of 92 m and carries a uniformly distributed load of 25 kN/m. The maximum dip of the cable measured from points A and B are 8 m and 4 m respectively. Determine the horizontal component of tension in the cable. Also determine the length of the cable required. Neglect the self weight of the cable. [**Ans.** H = 4538.04 kN; s = 93.072 m]
6. A suspension cable of span 110 m and central dip 10 m is subjected to a rise in temperature of 28°C. Find the increase in the dip of the cable. Take the co-efficient of linear expansion of the material as 12×10^{-6} per °C. Also find the change in the horizontal tension, if the cable carries a uniform load of 30 kN/m. [**Ans.** Δd = 0.076 m ; ΔH = – 34.59 kN]

7. The three hinged stiffening girder (with hinges at the ends and at the midspan) of a suspension bridge of span 200 metres is subjected to two point loads of 450 kN and 280 kN at distances of 40 m and 150 m respectively from the left end. Find the shear forces and bending moment in the girder at a section 75 m from the left end. The central dip of the supporting cable is 16 metres. Find also the maximum tension in the cable and draw the bending moment diagram for the girder. [**Ans.** F_{75} = – 100 kN; BM_{75} = 1500 kNm; T_{max} = 1049.95 kN; BM_{40} = 6960 kNm; BM_{150} = 3000 kNm]

8. A suspension cable of span 50 m and dip 5 m is stiffened by a three hinged girder. The dead load is 7.5 kN/m. Determine the maximum bending moment anywhere in the girder when a load of 100 kN rolls from left to right. Determine the maximum tension in the cable.

[**Ans.** $M_{max\ max}$ (+ve) = 480 kNm at 10.55 m from either end; $M_{max\ max}$ (–ve) = 312.5 kNm at 12.5 m from either end; T_{max} = 774.11 kN]

CHAPTER 6

BEAMS CURVED IN PLAN

6.1 INTRODUCTION

Arches are in fact beams with an initial curvature. The curvature is visible only in elevation. In plan they would appear straight. But there are also beams which are curved in plan. Bridges that negotiate a curve have to have curved (in plan) beams. The other cases of curved beams are, ring beams supporting water tanks, silos etc., beam supporting corner lintels and curved balconies etc. Ramps in traffic interchanges invariably have curved-in-plan beams.

Curved beams, in addition to bending moments and shears would also develop torsional moments (or twisting moments)

Figure 6.1 (*a*) to (*g*) show some cases of beams curved in plan.

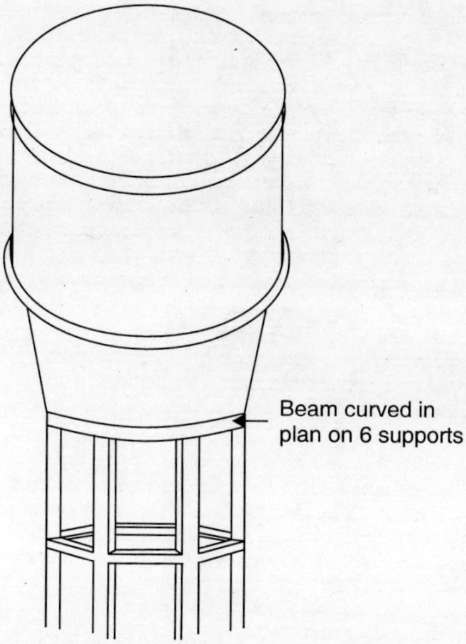

(a)

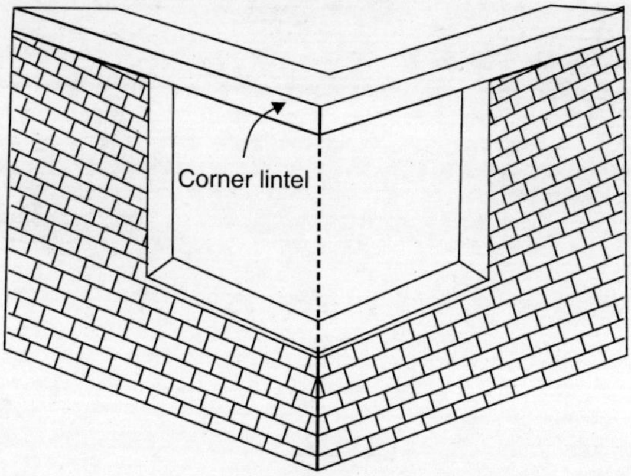

(b)

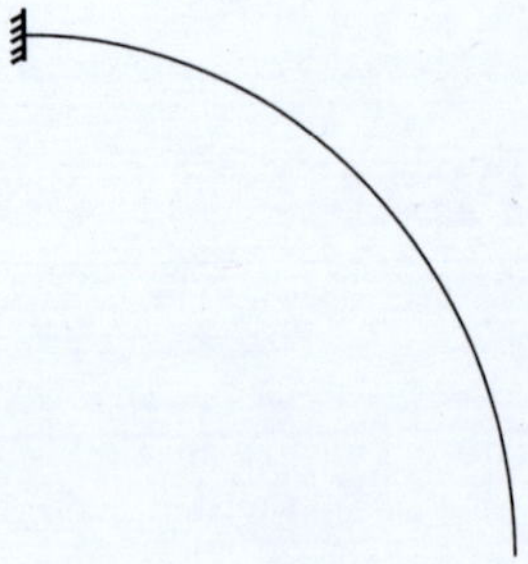
(c)

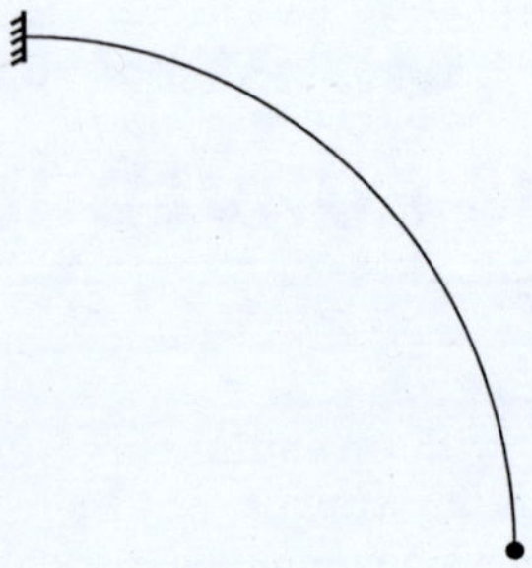
(d)

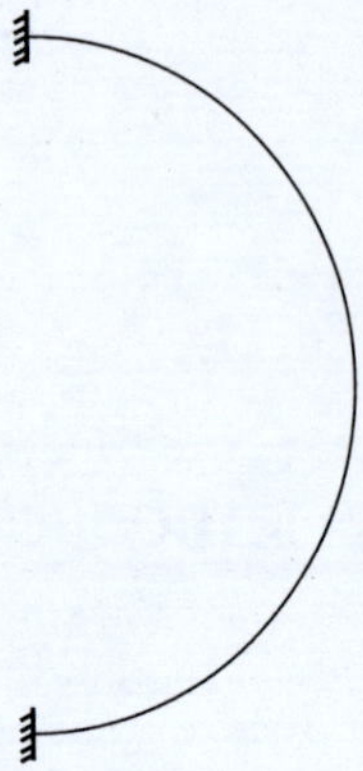
(e)

(f)

(g)

Fig. 6.1

6.2 SOLVING FOR MOMENT, SHEAR AND TORSION

The three diverse force components have one thing in common – the strain energy stored in a beam due to each type of force. Among the 3, we normally ignore the strain energy due to shear forces as negligible. That leaves us with moment (M) and torque (T). The strain energy due to both is given by

$$U = \int \frac{M^2\,ds}{2EI} + \int \frac{T^2\,ds}{2GJ} \qquad \text{...(6.1)}$$

integrated over the relevant length of the beam. Then, displacements, if desired under any real or fictitious load P can be obtained using Castigliano's first theorem. Thus,

$$\frac{\partial U}{\partial P} = \delta_P \qquad \text{...(6.2)}$$

If the beam is indeterminate, the redundant forces/moments, say R_1, R_2, ..., R_i, ... can be solved using Castigliano's second theorem. Thus,

$$\frac{\partial U}{\partial R_i} = 0 \qquad \text{...(6.3)}$$

Equation (6.3) involving R_1, R_2, ..., R_i can be solved to determine R_1, R_2, etc.

Equation (6.3) would involve both flexural rigidity EI and torsional rigidity GJ. It might simplify matters if we assume that EI/GJ is a constant, say, α.

6.3 SIGN CONVENTION

The following sign conventions would be adopted.

(*i*) Downward load ○ (Open circle)

(*ii*) Upward load ● (Solid circle)

(*iii*) Positive shear 

(*iv*) Positive moment 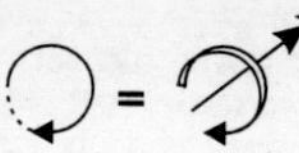

(*v*) Positive bending moment 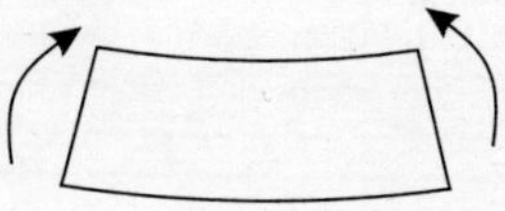

(*vi*) Positive torsional moment 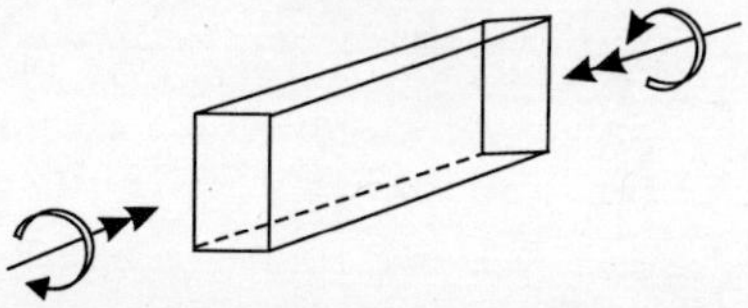

(For positive torque, the twist is clockwise on the right face). The vector representation of a moment is done in this book, using 2 arrows. The arrow points to the advance of a right handed screw as it is turned in the (clockwise or anti-clockwise) sense indicated. For actual solution of problems involving resolution of moments, the vector notation would be employed.

Example 6.1 *A curved beam in the form of a quadrant of a circle of radius R and having a uniform cross-section is in a horizontal plane. It is fixed at A and free at B as shown in Fig. 6.2. It carries a vertical concentrated load W at the free end B. Compute the shear force, bending moment and twisting moment values and sketch variations of the above quantities. Also determine the vertical deflection of the free end B.*

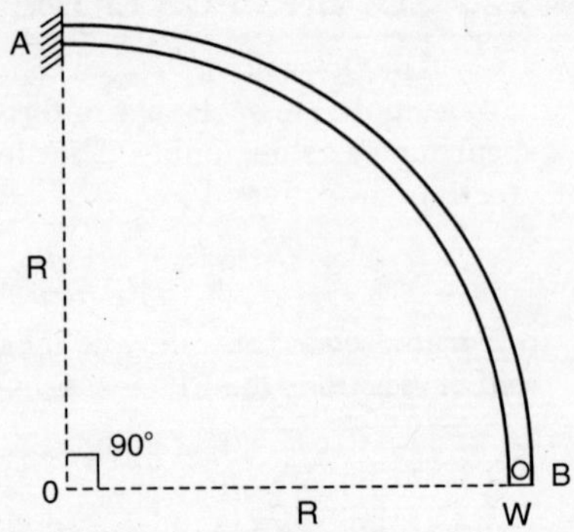

Fig. 6.2

Solution The given cantilever is a statically determinate structure. Consider any point X on the beam at an angle θ from OB (Fig. 6.3)

$$CX = R(1 - \cos\theta)$$

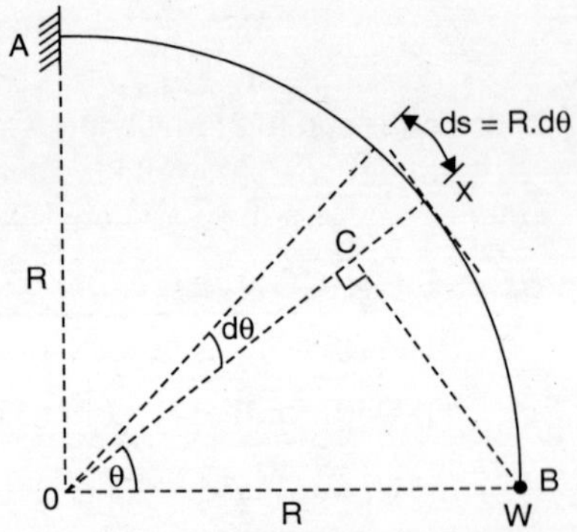

Fig. 6.3

(*i*) Shear force at the section X, $F_\theta = W$:

F_θ is independent of θ, and uniform throughout.

(*ii*) Bending moment at the section X, $M_\theta = -W(CB)$:

$$M_\theta = -W.\, R\sin\theta$$

$$\text{At } \theta = 0, \qquad M_B = 0$$

$$\text{At } \theta = \frac{\pi}{2}, \qquad M_A = -WR$$

(*iii*) Twisting moment at the section X, $T_\theta = -W(CX)$:

$$T_\theta = -WR\,(1 - \cos\theta)$$

$$\text{at } \theta = 0, \qquad T_B = -WR\,(1 - \cos\theta) = 0$$

$$\text{at } \theta = \frac{\pi}{2}, \qquad T_A = -WR\left(1 - \cos\frac{\pi}{2}\right) = -WR$$

The shear force, bending moment and T.M.s are sketched in Fig. 6.4

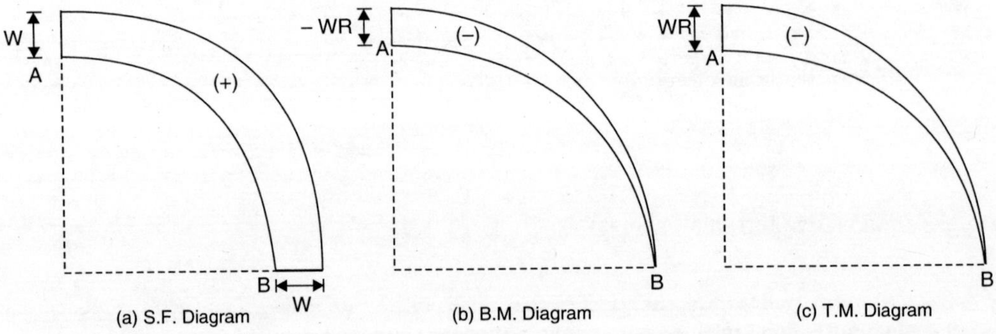

Fig. 6.4

(*iv*) Deflection at the free end B

Method of strain energy is used to find the deflection at the free end B.

$$\text{Strain energy, U} = \int \frac{M_\theta^2\, ds}{2EI} + \int \frac{T_\theta^2\, ds}{2GJ}$$

$$= \frac{1}{2EI}\int_0^{\pi/2} (-\,WR\sin\theta)^2\, Rd\theta + \frac{1}{2GJ}\int_0^{\pi/2} [-\,WR(1-\cos\theta)]^2\, Rd\theta$$

$$= \frac{1}{2EI}\int_0^{\pi/2} (W^2R^2\sin^2\theta\, R.d\theta + \frac{1}{2GJ}\int_0^{\pi/2} W^2R^2\,(1+\cos^2\theta - 2\cos\theta)\, Rd\theta$$

$$= \frac{1}{2EI} W^2R^3 \int_0^{\pi/2} \left(\frac{1-\cos 2\theta}{2}\right) d\theta + \frac{1}{2GJ} \times W^2R^3 \int_0^{\pi/2} \left(1 + \frac{(1+\cos 2\theta)}{2} - 2\cos\theta\right) d\theta$$

$$= \frac{1}{2EI}\frac{W^2R^3}{2}\int_0^{\pi/2}(1-\cos 2\theta)\, d\theta + \frac{W^2R^3}{4GJ}\int_0^{\pi/2}(2+1+\cos 2\theta - 4\cos\theta)\, d\theta$$

$$= \frac{W^2R^3}{4EI}\left[\theta - \frac{\sin 2\theta}{2}\right]_0^{\pi/2} + \frac{W^2R^3}{4GJ}\left[3\theta + \frac{\sin 2\theta}{2} - 4\sin\theta\right]_0^{\pi/2}$$

$$= \frac{W^2R^3}{4EI}[\pi/2] + \frac{W^2R^3}{4GJ}\left[\frac{3\pi}{2} - 4\right]$$

$$U = \frac{\pi\, W^2R^3}{8EI} + \frac{W^2R^3}{8GJ}[3\pi - 8]$$

$$\delta_B = \frac{dU}{dW}$$

$$\delta_B = \frac{\pi WR^3}{4EI} + \frac{WR^3}{4GJ}(3\pi - 8) \qquad \text{...(6.4)}$$

Example 6.2 *A curved beam AB of uniform cross-section is horizontal in plan and in the form of a quadrant of a circle of radius R. The beam is fixed at A and free at B. It carries a uniformly distributed load of w/unit run over the entire length of the beam, as shown in Fig. 6.5. Calculate the shear force, bending moment and T.M values at A and B and sketch the variations of the same. Also determine the deflection at the free end B.*

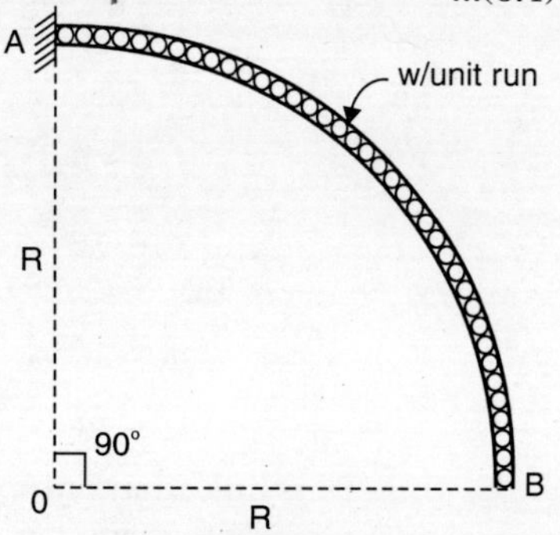

Fig. 6.5

Solution The given cantilever is a statically determinate structure. Consider any point X at an angle θ from B as shown in Fig. 6.6.

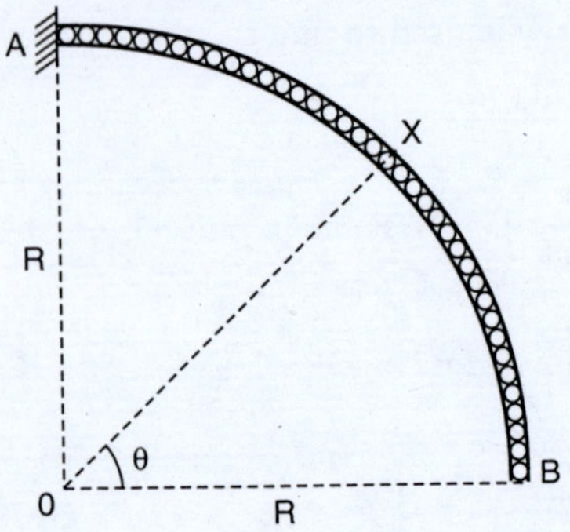

Fig. 6.6

(*i*) **Shear force at the section X, $F_\theta = w \cdot R\theta$:**

when $\theta = 0, \quad F_B = 0$

when $\theta = \frac{\pi}{2}, \quad F_A = w \cdot R \cdot \frac{\pi}{2} = \frac{\pi w R}{2}$

(*ii*) **Bending moment at the section X, M_θ :**

The bending moment due to the udl will be negative since tension will occur at the top.

Let us consider the bending moment M_θ at X, at θ from the free end. The bending moment due to load on an element $Rd\theta$ at an angle ϕ from OX is given by

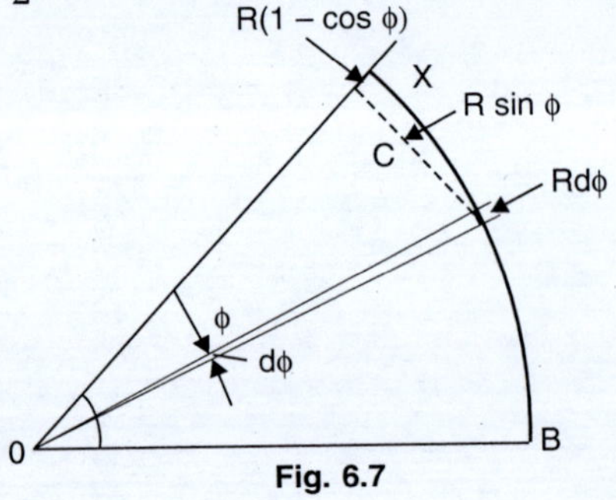

Fig. 6.7

$$dM_\theta = -w\, R\, d\phi \,.\, R \sin\phi$$

$$M_\theta = -\int_0^\theta w \,.\, Rd\phi\, R\sin\phi$$

$$= -wR^2 \int_0^\theta \sin\phi\, d\phi$$

$$= -wR^2 \Big[-\cos\phi\Big]_0^\theta$$

$$= -wR^2\, [-\cos\theta + (\cos 0)]$$

$$M_\theta = -wR^2\,(1 - \cos\theta) \qquad \text{...(6.5)}$$

at $\theta = 0, \quad M_B = -wR^2\,(1 - \cos 0) = 0$ (free end)

at $\theta = \frac{\pi}{2}, \quad M_A = -wR^2\left(1 - \cos\frac{\pi}{2}\right) = -wR^2$ (fixed end)

(*iii*) **Twisting moment at section X, T_θ :**

The twisting moment due to the udl will be negative since the twist would be anti-clockwise on the right face.

$$T_\theta = -\int_0^\theta wR\, d\phi\; R(1 - \cos\phi) = -\int_0^\theta wR^2\,(1 - \cos\phi)\, d\phi$$

$$T_\theta = -wR^2\,(\theta - \sin\theta) \qquad \text{...(6.6)}$$

at $\theta = 0$, $T_B = -wR^2 (0 - \sin 0) = 0$

at $\theta = \frac{\pi}{2}$, $T_A = -wR^2 \left(\frac{\pi}{2} - \sin\frac{\pi}{2}\right)$

at the fixed end, $T_A = -wR^2 \left(\frac{\pi}{2} - 1\right)$ ...(6.7)

The shear force, bending moment and T.M. diagrams are sketched in Fig. 6.8 (*a*) to (*c*)

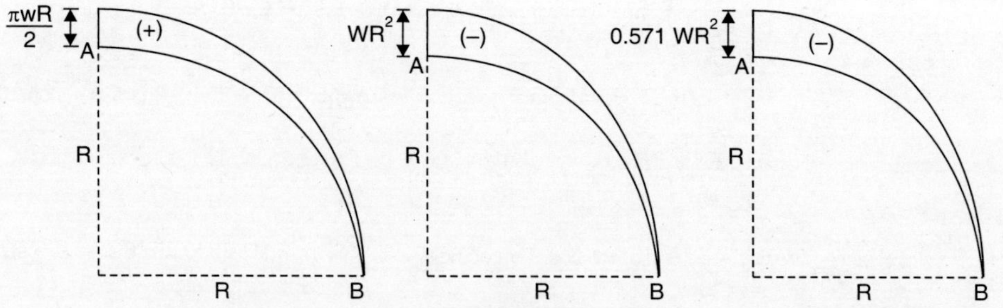

(a) S.F. Diagram (b) B.M. Diagram (c) T.M. Diagram

Fig. 6.8

(*iv*) **Deflection at the free end B :**

The method of strain energy will be used to find the deflection at the free end B. A ficticious load P will be applied at the end B. The deflection at B is then

$$\delta_B = \left.\frac{dU}{dP}\right|_{p=0}$$

Bending moment at the section X due to udl [from equation (6.5)] and due to the point load P,

$$M_\theta = -wR^2 (1 - \cos\theta) - PR\sin\theta$$

$$\frac{dM_\theta}{dP} = -R\sin\theta$$

T.M. at the section X due to udl [from eqn. (6.6)] and due to the point load P

$$T_\theta = -wR^2 (\theta - \sin\theta) - PR(1 - \cos\theta)$$

$$\frac{dT_\theta}{dP} = -R(1 - \cos\theta)$$

$$\frac{dU}{dP} = \frac{1}{EI}\int M_\theta \frac{dM_\theta}{dP} R\,d\theta + \frac{1}{GJ}\int T_\theta \frac{dT_\theta}{dP} R\,d\theta$$

$$= \frac{1}{EI}\int_0^{\pi/2} [-wR^2 (1 - \cos\theta) - PR\sin\theta]\,(-R\sin\theta)\,R\,d\theta$$

$$+ \frac{1}{GJ}\int_0^{\pi/2} [-wR^2 (\theta - \sin\theta) - PR(1 - \cos\theta)]\,[-R(1 - \cos\theta)]\,R\,d\theta$$

$$= \frac{1}{EI}\int_0^{\pi/2} [wR^4 (1 - \cos\theta)\sin\theta + PR^3 \sin^2\theta]\, d\theta$$

$$+ \frac{1}{GJ}\int_0^{\pi/2} [wR(\theta - \sin\theta)(1 - \cos\theta) + PR^3 (1 - \cos\theta)^2]\, d\theta$$

$$= \frac{R^3}{EI}\int_0^{\pi/2} [wR(\sin\theta - \sin\theta\cos\theta) + P\sin^2\theta] d\theta$$

$$+ \frac{R^3}{GJ}\int_0^{\pi/2} [wR(\theta - \theta\cos\theta - \sin\theta + \sin\theta\cos\theta] + P(1 - 2\cos\theta + \cos^2\theta)\, d\theta.$$

$$= \frac{R^3}{EI}\left[wR\left(-\cos\theta + \frac{1}{2}\frac{\cos 2\theta}{2}\right) + P\left(\frac{\theta}{2} - \frac{\sin 2\theta}{4}\right)\right]$$

$$+ \frac{R^3}{GJ}\left[wR\left(\frac{\theta^2}{2} - \theta\sin\theta - \cos\theta + \cos\theta - \frac{1}{2}\frac{\cos 2\theta}{2}\right)\right] + P\left(\theta - 2\sin\theta + \frac{\theta}{2} + \frac{1}{4}\sin 2\theta\right)$$

$$= \frac{R^3}{EI}\left[wR\left(1 + \frac{1}{4}(-2)\right) + P\left(\frac{\pi}{4} - \frac{1}{4}(0)\right)\right]$$

$$+ \frac{R^3}{GJ}\left[wR\left(\frac{\pi^2}{8} - \frac{\pi}{2} - \frac{1}{4}(-2)\right)\right] + P\left[\left(\frac{\pi}{2} - 2 + \frac{\pi}{4} + \frac{1}{4}(0)\right)\right]$$

$$= \frac{R^3}{EI}\left[\frac{wR}{2} + \frac{P\pi}{4}\right] + \frac{R^3}{GJ}\left[wR\left(\frac{\pi^2}{8} - \frac{\pi}{2} + \frac{1}{2}\right) + P\left(\frac{3\pi}{4} - 2\right)\right]$$

$$\frac{dU}{dP} = \frac{wR^4}{2EI} + \frac{\pi PR^3}{4EI} + \frac{wR^4}{8GJ}(\pi - 2)^2 + \frac{PR^3}{4GJ}(3\pi - 8)$$

If $P = 0$

$$\delta_B = \frac{wR^4}{2EI} + \frac{wR^4}{8GJ}(\pi - 2)^2. \qquad ...(6.8)$$

Example 6.3 *Figure 6.9 shows a curved beam, semi-circular in plan and supported on three equally spaced supports. The beam carries a uniformly distributed load of w / unit of the circular length. Analyse the beam and sketch the bending moment and twisting moment diagrams.*

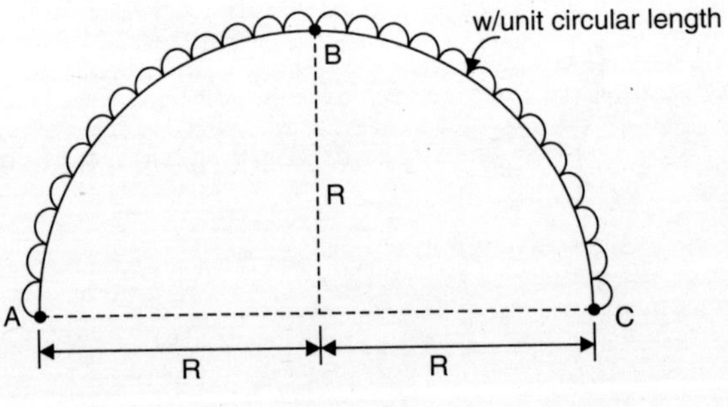

Fig. 6.9

Solution The curved beam given is shown in Fig. 6.10. The XX and YY-axes are as shown and ZZ is the vertical axis. The unknown reactions are V_A, V_B and V_C. The end supported do not exert any moment reaction. There are three equations of static equilibrium, *i.e.*, $\Sigma M_{xx} = 0$, $\Sigma M_{yy} = 0$ and $\Sigma F_{zz} = 0$. Hence the structure is externally determinate.

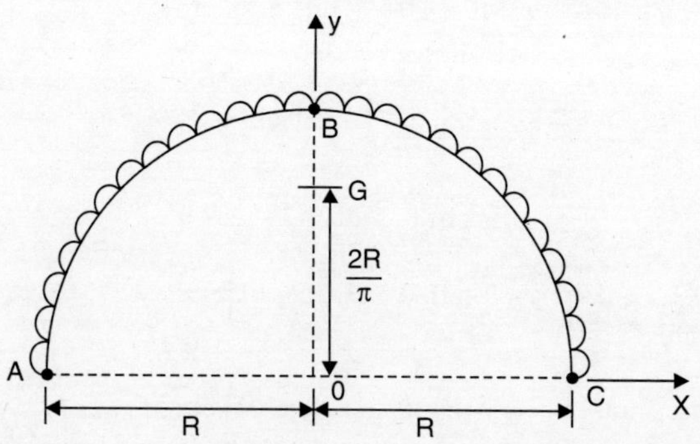

Fig. 6.10

Total load on the beam = $w \times \pi R$

Vertical support reactions :

The structure, loading and support conditions are symmetric with respect to the YY- axis

Hence, $V_A = V_C$

Taking moments of all forces about a tangent at point B,

$$2(V_A \times R) - \pi wR\left(R - \frac{2R}{\pi}\right) = 0$$

$$V_A = \frac{1}{2R} \times \pi wR\left(\frac{\pi R - 2R}{\pi}\right)$$

$$= \frac{w}{2} R(\pi - 2)$$

$$V_A = \frac{(\pi - 2)}{2} wR \qquad ...(6.9)$$

$$V_B = \text{Total load} - 2V_A$$

$$= \pi wR - 2x\frac{(\pi - 2)}{2} wR$$

$$= wR[\pi - (\pi - 2)] \qquad ...(6.10)$$

$$V_B = 2wR$$

Bending moment and twisting moment :

Consider a section X located at an angle θ with OA

Take a segment R$d\phi$ at an angle ϕ from x :

Bending moment, M_θ at x :

$$M_\theta = V_A \times (AN) - \int_0^\theta wR\,d\phi\,R\sin\phi$$

Using equation (6.9)

$$M_\theta = \frac{(\pi-2)}{2} wR\,R\sin\theta - \int_0^\theta wR\,d\phi\,R\sin\phi$$

$$= \frac{(\pi-2)}{2} wR^2 \sin\theta - \int_0^\theta wR^2 \sin\phi\,d\phi$$

$$= \frac{(\pi-2)}{2} wR^2 \sin\theta - wR^2 (1-\cos\theta)$$

$$M_\theta = wR^2\left[\frac{(\pi-2)}{2}\sin\theta - (1-\cos\theta)\right] \quad ...(6.11)$$

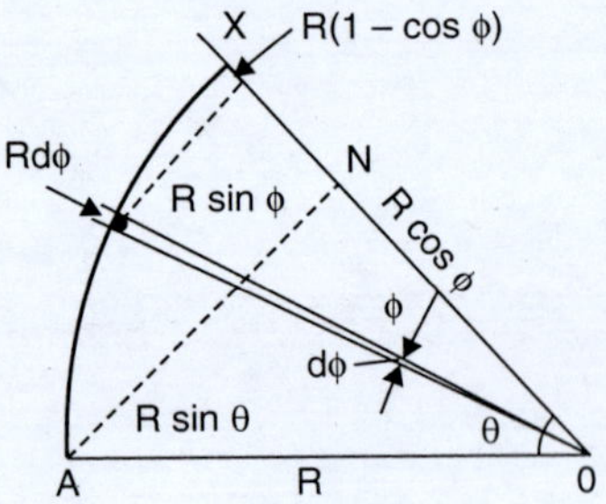

Fig. 6.11

at $\theta = 0$, $$M_A = wR^2\left[\frac{(\pi-2)}{2}\sin 0 - (1-\cos 0)\right]$$

$M_A = 0$ (hinged end)

at $\theta = \frac{\pi}{2}$, $$M_B = wR^2\left[\frac{(\pi-2)}{2}\sin\frac{\pi}{2} - \left(1-\cos\frac{\pi}{2}\right)\right]$$

$$= wR^2\left[\frac{(\pi-2)}{2}\times 1 - (1-0)\right]$$

$$= wR^2\left[\frac{(\pi-2)}{2} - 1\right]$$

$$= -0.429\,wR^2$$

$$M_B = -0.429\,wR^2 \quad ...(6.12)$$

Therefore moment at B = $0.429\,wR^2$ (Hogging)

Maximum bending moment might occur between A and B (and also between B and C by symmetry)

For M_θ to be maximum

$$\frac{dM_\theta}{d\theta} = 0$$

$$\frac{d}{d\theta}\left(wR^2\left[\frac{(\pi-2)}{2}\sin\theta - (1-\cos\theta)\right]\right) = 0$$

$$\frac{(\pi-2)}{2}\cos\theta - (0+\sin\theta) = 0$$

$$\frac{(\pi-2)}{2}\cos\theta - \sin\theta = 0$$

$$\frac{(\pi-2)}{2} = \frac{\sin\theta}{\cos\theta}$$

$$\tan\theta = \frac{(\pi-2)}{2}$$

For M_{max}, $$\theta = 29°43' \quad ...(6.13)$$

Therefore, M_{max} $= wR^2 \left[\frac{(\pi - 2)}{2} \sin 29° 43' - (1 - \cos 29° 43')\right]$

$= wR^2 \, [0.2830 - 0.1315]$

$M_{max} = 0.1515 \, wR^2$ sagging ...(6.14)

Maximum sagging moment = $0.1515 \, wR^2$ and this will occur at $\theta = 29° \, 43'$ from OA (and OC)

Point of contraflexure :

At the point of contraflexure, $M_\theta = 0$

$$wR^2 \left[\frac{(\pi - 2)}{2} \sin \theta - (1 - \cos \theta)\right] = 0$$

$$\frac{(\pi - 2)}{2} \sin \theta - 1 + \cos = 0$$

$$\frac{(1 - \cos \theta)}{\sin \theta} = \frac{(\pi - 2)}{2}$$

$\theta = 59° \, 27'$, by trial and error

Therefore the point of contraflexure occurs at 59° 27′ from OA (and from OC)

The bending moment diagram is as shown in Fig. 6.12

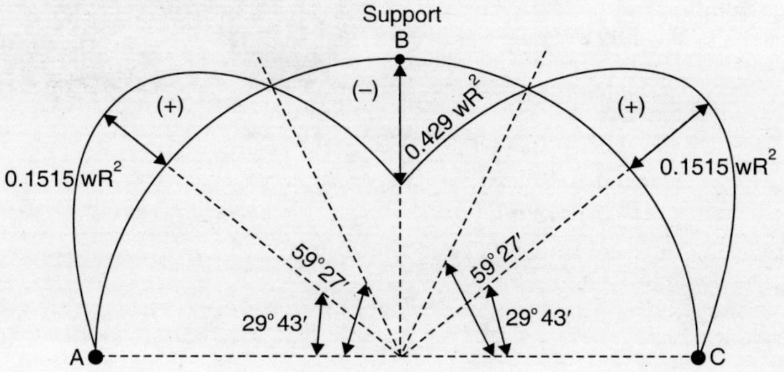

Fig. 6.12 B.M. Diagram

Twisting moment, T_θ at X :

$$T_\theta = -V_A \, (XN) + \int_0^\theta wR \, d\phi \, (1 - \cos \phi)$$

$$= -\frac{(\pi - 2)}{2} wR \, R \, (1 - \cos \theta) + \int_0^\theta wR^2 \, (1 - \cos \phi) \, d\phi$$

$$= -\frac{(\pi - 2)}{2} wR^2 \, (1 - \cos \theta) + wR^2 \, (\theta - \sin \theta)$$

$$T_\theta = wR^2 \left[-\frac{(\pi - 2)}{2}(1 - \cos \theta) + \theta - \sin \theta\right]$$

when $\theta = 0$, $T_A = wR^2\left[-\frac{(\pi-2)}{2}(1-1)+0-0\right]$

when $\theta = \frac{\pi}{2}$, $T_B = wR^2\left[-\frac{(\pi-2)}{2}(1-0)+\frac{\pi}{2}-1\right]$

$$= wR^2\left[-\frac{(\pi-2)}{2}+\frac{(\pi-2)}{2}\right] = 0$$

For maximum value of torsional moment,

$$\frac{dT_\theta}{d\theta} = 0$$

i.e.,

$$\frac{d}{d\theta}\left[wR^2\left(-\frac{(\pi-2)}{2}(1-\cos\theta)+\theta-\sin\theta\right)\right] = 0$$

$$wR^2\left[-\frac{(\pi-2)}{2}(\sin\theta)+1-\cos\theta\right] = 0$$

$$1-\cos\theta = \frac{(\pi-2)}{2}\sin\theta$$

$$\frac{1-\cos\theta}{\sin\theta} = \frac{(\pi-2)}{2}$$

$$\theta = 59°\ 27'$$

$$T_{max} = wR^2\left[-\frac{(\pi-2)}{2}(1-\cos 59°\ 27')+59°\ 27'\times\frac{\pi}{180}-\sin 59°\ 27'\right]$$

$$= wR^2\ [-\ 0.2807 + 1.0376 - 0.8612]$$

$$= -\ 0.1043\ wR^2$$

$$T_{max} = -\ 0.1043\ wR^2 \qquad \text{...(6.15)}$$

The twisting moment diagram is as shown in Fig. 6.13.

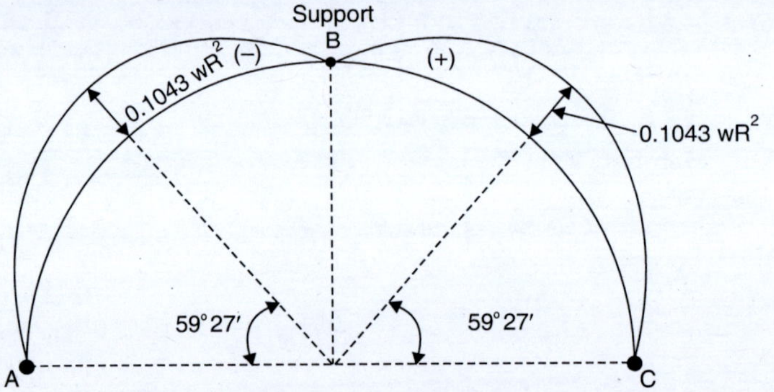

Fig. 6.13 T.M. Diagram.

It is observed that the twisting moment is maximum at the section where the bending moment is zero.

Example 6.4 *A rigid bent ABC, is of uniform cross-section and is in a horizontal plane. It is fixed at A and is supported on a spherical bearing at C, as shown in Fig. 6.14.*

It carries a vertical concentrated load P at B. Analyse the bent and determine the reaction at the supports of the beams.

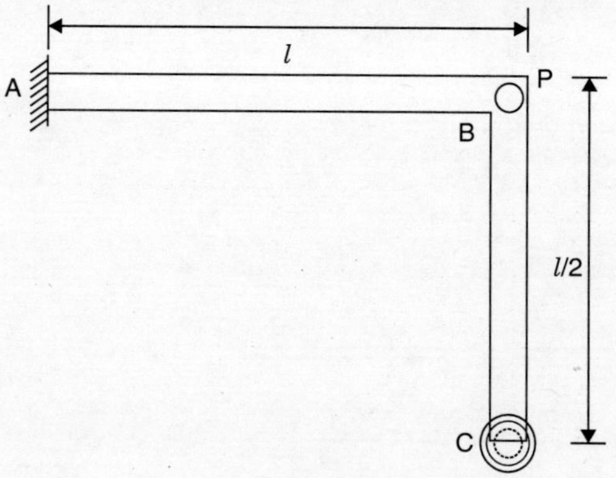

Fig. 6.14

Solution The structure is statically indeterminate to the first degree. The vertical reaction V_C at C is treated as the redundant.

From Castigliano's Theorem 2,

$$\frac{dU}{dV_C} = \frac{d}{dV_C}(U_{CB} + U_{BA})$$

For CB,
$$\frac{dU_{CB}}{dV_C} = \frac{1}{EI}\int_0^{l/2} Mx \frac{dMx}{dV_C} dx + \frac{1}{GJ}\int_0^{l/2} Tx \frac{dTx}{dV_C}$$

Table 6.1 Energy integral table.

Portion	*Origin*	*Limits*	M_x	T_x	$\frac{dM_x}{dV_C}$	$\frac{dT_x}{dV_C}$
CB	C	0 to $l/2$	$V_C \times x$	0	x	0
BA	B	0 to l	$V_C . x - Px$	$V_C \times l/2$	x	$l/2$

$$= \frac{1}{EI}\int_0^{l/2} (V_C . x)(x)\, dx + 0$$

$$= \frac{1}{EI}\int_0^{l/2} V_C\, x^2\, dx = \frac{1}{EI}\left[V_C . \frac{x^3}{3}\right]_0^{l/2} = \frac{1}{EI}\left[\frac{V_C}{3}\left(\frac{l}{2}\right)^3\right] = \frac{V_C l^3}{24 EI}$$

$$\frac{dU_{CB}}{dV_C} = \frac{V_C l^3}{24 EI} \quad ...(1)$$

For segment BA :

$$\frac{dU_{CB}}{dV_C} = \frac{1}{EI}\int_0^l M_x \frac{dM_x}{dV_C}\,dx + \frac{1}{GJ}\int_0^l T_x \frac{dT_x}{dV_C}\,dx$$

$$= \frac{1}{EI}\int_0^l (V_C\,x - Px)(x)\,dx + \frac{1}{GJ}\int_0^l V_C \frac{l}{2}\frac{l}{2}\,dx$$

$$= \frac{1}{EI}\int_0^l (V_C\,x^2 - Px^2)\,dx + \frac{1}{GJ}\int_0^l V_C \frac{l}{2}\frac{l}{2}\,dx$$

$$= \frac{1}{EI}\left[V_C\frac{x^3}{3} - P\frac{x^3}{3}\right]_0^l + \frac{1}{GJ}\frac{V_C\,l^2}{4}(x)_0^l$$

$$= \frac{1}{EI}\left[V_C\frac{l^3}{3} - P\frac{l^3}{3}\right] + \frac{1}{GJ}\frac{V_C\,l^3}{4}$$

$$\frac{dU_{BA}}{dV_C} = \frac{V_C\,l^3}{3EI} - \frac{Pl^3}{3EI} + \frac{V_C\,l^3}{4GJ} \qquad ...(2)$$

$$\frac{dU}{dV_C} = \frac{dU_{CB}}{dV_C} + \frac{dU_{BA}}{dV_C} = 0$$

$$\frac{V_C}{24EI} + \frac{V_C}{3EI} - \frac{P}{3EI} + \frac{V_C}{4GJ} = 0$$

$$= \frac{V_C}{24EI} + \frac{V_C}{3EI} + \frac{V_C}{4GJ} \times \frac{EI}{EI} = \frac{P}{3EI}$$

$$\frac{V_C}{EI}\left[\frac{1}{24} + \frac{1}{3} + \frac{EI}{4GJ}\right] = \frac{P}{3EI}$$

$$V_C\left[\frac{1+8+6\alpha}{24}\right] = \frac{P}{3}$$

$$V_C = \frac{P}{3} \times \frac{24}{(9\times 6\alpha)} = \frac{8P}{3(3+2\alpha)}$$

$$V_C = \frac{8P}{3(3+2\alpha)} \qquad ...(3)$$

$$V_A = P - V_C = P - \frac{8P}{3(2+2\alpha)}$$

$$V_A = P\,\frac{(1+6\alpha)}{(9+6\alpha)} \qquad ...(4)$$

Bending moment on CB :

$$M_x = V_C\,.\,x \qquad ...(5)$$

At $x = 0$, $M_C = 0$

At $x = \frac{l}{2}$, $M_B = V_C\,\frac{l}{2} = \frac{8Pl}{2(9+6\alpha)}$

$$M_B = \frac{4Pl}{9+6\alpha} \qquad ...(6)$$

Bending moment on BA :

$$M_x = V_C x - P_x = \left(\frac{8P}{9+6\alpha} - P\right) x$$

At $x = 0$ $\qquad M_B = 0$

At $x = l$ $\qquad M_A = \left(\frac{8P}{9+6\alpha} - P\right) l$

$$M_A = -\left(\frac{1+6\alpha}{9+6\alpha} Pl\right) \qquad ...(7)$$

6.4 CIRCULAR BEAM ON SEVERAL SUPPORTS

Circular beams are commonly used to support circular water tanks, silos and other storage structures. Depending upon the size of the storage tank it is customary to use 4, 6 or 8 columns. These beams experience both bending and torsion.

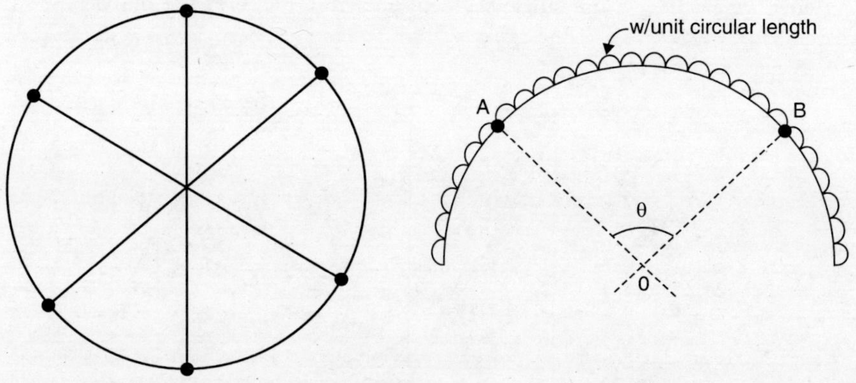

(a) (b)

Fig. 6.15

We will investigate the case of a circular beam of uniform cross-section and compute the critical bending moments and torsional moments in the same.

The given beam is statically indeterminate. However we can take advantage of symmetry of supports and loading. Since every column support is identically loaded on either side, we can make the following inferences.

(*i*) Since the frame and loading are symmetrical about a diametral line through any column, there would be no slope to either side. Hence the supports would have zero slopes and can be treated as fixed supports.

(*ii*) The total load would be equally shared and thus the support reactions are readily known.

(*iii*) Half this support load would come from the span to the left of the support and the other half from the right.

Hence reaction at any support = $wR\theta$

$$V_{AB} = V_{BA} = \frac{wR\theta}{2}$$

(*iv*) Torsional moment would be zero at every support. The reason is, any torque at a support would make it unsymmetrical. See Fig. 6.16 (*a*) and (*b*).

Here Fig. 6.16 (*a*) represents torsion. But it is unsymmetrical about the column and is therefore ruled out. Figure 6.16 (*b*) is a symmetrical axial moment case. But this does not represent a state of torsion.

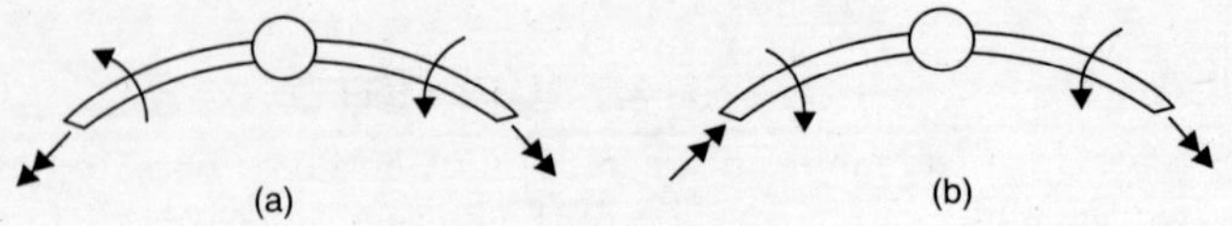

Fig. 6.16

Hence we are left with a single segment shown in Fig. 6.17 for analysis.

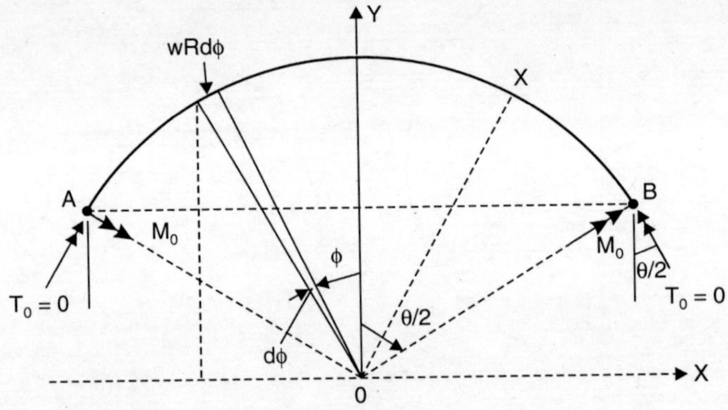

Fig. 6.17

Let M_0 be the support moments at A and B. The moment vectors M_0 acting at A and at B would be as shown. M_0 can be determined by considering the moment equilibrium about AB.

$$2M_0 \sin\frac{\theta}{2} - 2\int_0^{\theta/2} wRd\phi\left(R\cos\phi - R\cos\frac{\theta}{2}\right) = 0$$

$$M_0 \sin\frac{\theta}{2} = wR^2\left(\sin\phi - \phi\cos\frac{\theta}{2}\right)_0^{\theta/2}$$

$$= wR^2\left(\sin\frac{\theta}{2} - \frac{\theta}{2}\cos\frac{\theta}{2}\right)$$

$$M_0 = wR^2\left(1 - \frac{\theta}{2}\cot\frac{\theta}{2}\right) \quad ...(6.16)$$

To compute the bending moment and torsion at any given section in the beam, let us consider a section X at an angle α from OA (Fig. 6.18).

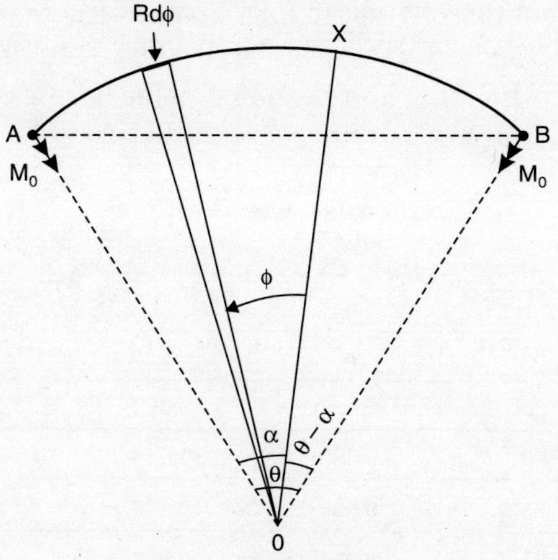

Fig. 6.18

$$M_0 = wR^2 \left(1 - \frac{\theta}{2} \cot \frac{\theta}{2}\right)$$

$$V_A = V_B = \frac{wR\theta}{2}$$

$$T_\alpha = M_0 \sin \alpha - V_A (1 - \cos \alpha) R + \int_0^\alpha wRd\phi\, R(1 - \cos \phi)$$

$$= wR^2 \left[\left(1 - \frac{\theta}{2} \cot \frac{\theta}{2}\right) s + \frac{\theta}{2}(1 - C) + (\alpha + s)\right]$$

where, $s = \sin \alpha$; $C = \cos \alpha$

$$T_\alpha = wR^2 \left[s - \frac{\theta}{2} \cot \frac{\theta}{2} s - \frac{\theta}{2} + \frac{\theta}{2} C + \alpha - s\right]$$

$$T_\alpha = wR^2 \left[\alpha - \frac{\theta}{2} + \frac{\theta}{2} C + \frac{\theta}{2} \cot \frac{\theta}{2} s\right] \qquad ...(6.17)$$

$$M_\alpha = -M_0 \cos \alpha - \int_0^\alpha wRd\phi \,.\, R \sin \phi + V_A R \sin \alpha$$

$$= -wR^2 \left(1 - \frac{\theta}{2} \cot \frac{\theta}{2}\right) C + wR \frac{\theta}{2} Rs - wR^2 \left(-\cos \phi\right)_0^\alpha$$

$$= wR^2 \left[-\left(1 - \frac{\theta}{2} \cot \frac{\theta}{2}\right) C + \frac{\theta}{2} s + (c - 1)\right]$$

$$M_\alpha = wR^2 \left[\frac{\theta}{2} s + \frac{\theta}{2} \cot \frac{\theta}{2} C - 1\right] \qquad ...(6.18)$$

Using equations (6.16), (6.17) and (6.18), we can construct a table (Table 6.2) giving the critical values of bending and torsional moments and their positions.

Table 6.2 Bending and torsional moments in circular beams on N number of supports

$$\theta = (360/N)^0$$
$$\phi_C = \text{angle from support}$$
$$M_0 = -a\,wR^2\,\theta$$
$$M_{max} = b\,wR^2\,\theta$$
$$T_{max} = cw\,R^2\theta$$

N	θ^0	*a*	*b*	*c*	ϕ_C
4	90	0.137	0.070	0.021	19° 15′
6	60	0.089	0.045	0.009	12° 45′
8	45	0.066	0.030	0.005	9° 20′
10	36	0.054	0.023	0.003	7° 30′

PROBLEMS

1. Using Table 6.2 find the number of columns needed to support a circular beam, 8 m diameter meant to carry a combined water and tank load of 1.24×10^6 N total. The maximum bending moment is limited to 40 kNm and max torque of 4 kNm. [**Ans.** 8 columns]
2. Find the bending moment at midspan of a semicircular beam loaded at the midspan with a concentrated load of 80 kN. The beam is fixed at both supports. Find the maximum bending moment and maximum torque in the beam.

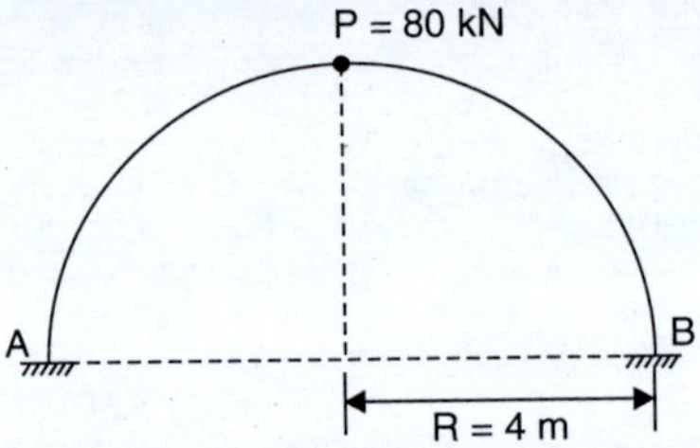

Fig. 6.19

$$\left[\textbf{Ans. } M_C = \frac{PR}{\pi} = 101.9 \text{ kNm}; M_{max} = \frac{PR}{2} = 160 \text{ kNm}; \quad T_{max} = PR\left(\frac{\pi - 2}{2\pi}\right) = 465.1 \text{ kNm.}\right]$$

3. In a circular beam supported on 6 columns, determine the position of maximum torsional moment. (**Hint.** Differentiate equation (6.17) w.r.t. α) [**Ans.** 12° 45′]

CHAPTER 7
SPACE TRUSSES

7.1 INTRODUCTION

Space trusses are more common in practice than we care to think. In a village fire wood depot, you might have seen a over sized balance hanging from a trestle. This tripod is a space truss. These days most of the station yard structures in railway junctions are space trusses. And of course there are all those transmission towers, antenna towers, microwave towers and the rest. These are typical space frames.

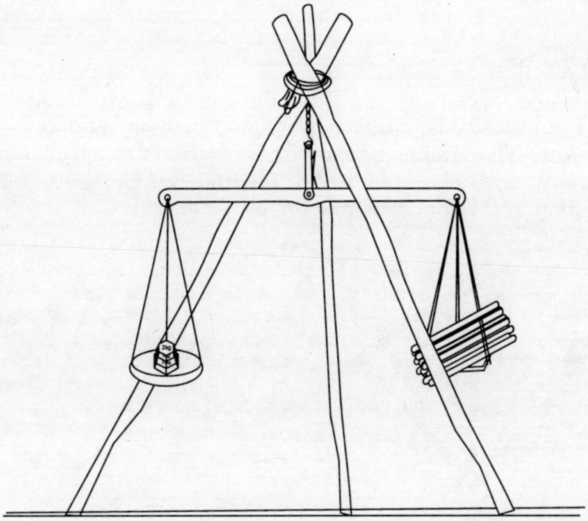

Fig. 7.1. A trestle.

Whereas in a plane truss all the joints are assumed to be hinged with freedom to swivel in the X-Y plane, all joints in a space frame are assumed to be ball and socket jointed. Such joints would enable each member to swivel in X,Y and Z directions. Such hinged and ball and socket joints are needed to ensure that the truss members can accommodate small displacements of joints without bending.

Hence hinges in plane frames and ball and socket joints in space frames are needed to ensure that the members develop only direct forces (tension or compression) and no bending forces (including bending shear).

A plane truss in the X-Y plane is inherently unstable in the Z direction. That is why plane trusses supporting roofs have to be connected to each other. There is no such stability problem with space trusses.

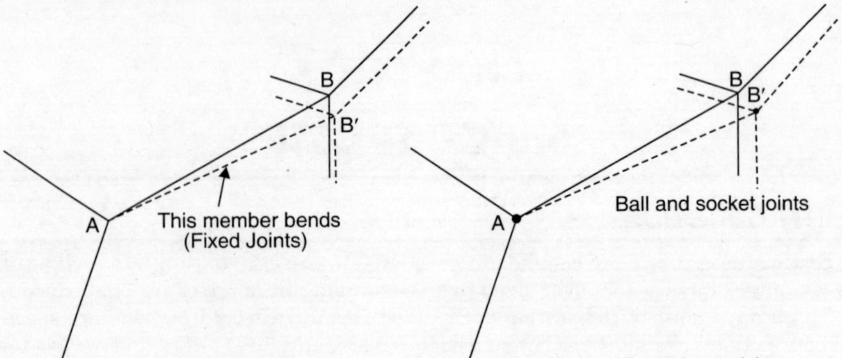

(a) Member AB bends (b) Member AB remains straight

Fig. 7.2

7.2 BASIC FORMS

A space frame or space truss is a three dimensional assemblage of line members, each member being joined at its ends, either to the foundation or to other members by frictionless ball-and-socket joints. The simplest space frame consists of six members joined to form a tetrahedron. Beginning with a six member tetrahedron, a stable space frame can be constructed by successive addition of three new members and a new joint.

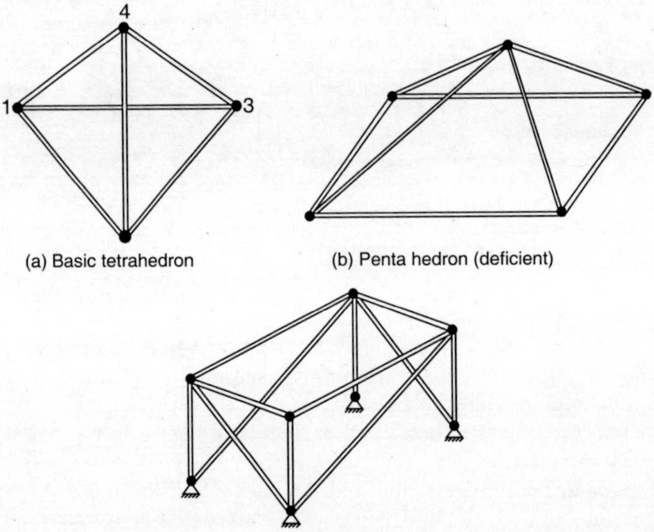

(a) Basic tetrahedron (b) Penta hedron (deficient)

(c) A utility space frame

Fig. 7.3. Typical space frames.

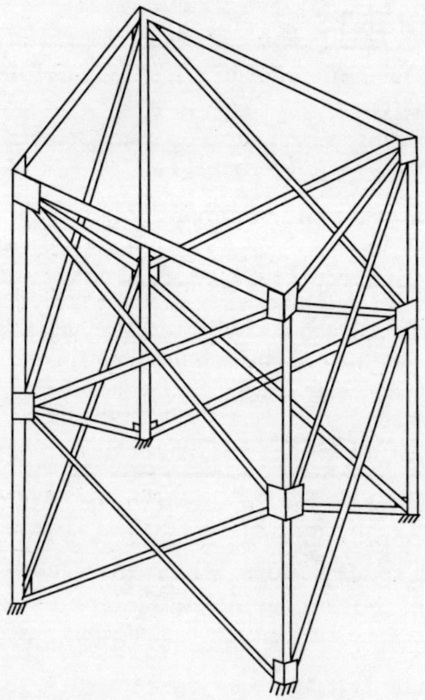

Fig. 7.4. A typical panel of a transmission tower.

In space trusses, the relationship between the number of members (m) and number of joints (j) is given by

$$m = (3j - 6) \qquad ...(7.1)$$

m gives the minimum number of members required to build a stable truss or frame.

$$m = 3j \qquad ...(7.2)$$

if the frame starts from a rigid base.

If a truss or frame has less number of members than what is given by the above equations the truss or frame will be unstable. If it has more members, the frame will be internally statically indeterminate.

The geometry of space structures makes their analysis slightly more complex compared to two-dimensional structures. This needs a systematic approach.

A systematic approach to analyse pin-jointed trusses was introduced by R.V. Southwell in 1920. The method based on the equilibrium conditions of joints, develops equations for the components of member forces along the coordinate axes at each joint. Southwell noted a relationship between the components of member forces along coordinate axes and the member coordinates at the connecting nodes. This relationship enables the use of force per unit length of the members in the analysis. The force per unit length of a member is usually known as the tension coefficient of the member. The concepts of analysis and their application to plane and space trusses are discussed below.

7.3 TENSION COEFFICIENTS

The tension coefficient for a member of a frame is defined as the pull or tension in that member divided by its length. Thus

$$t = \frac{T}{l}$$

where t is the tension coefficient for the member. This is +ve if the forces in the member is tensile.

T is the pull in the member, and l is its length.

Tension coefficients are initially assumed to be positive (tensile forces in the member) for all the members. Negative values of the tension coefficient, obtained in the analysis would indicate compressive forces in the members.

7.4 METHOD OF TENSION COEFFICIENTS

This method is equally applicable for the analysis of plane and space frames. For plane frames, this gives a more general approach, compared to the method of joints and method of sections. For the analysis of space frames, which involves tedious computations because of their geometry, the method of tension coefficients offers a systematic process.

To start with, let us learn the procedure for the analysis of plane frames and extend it for space frames.

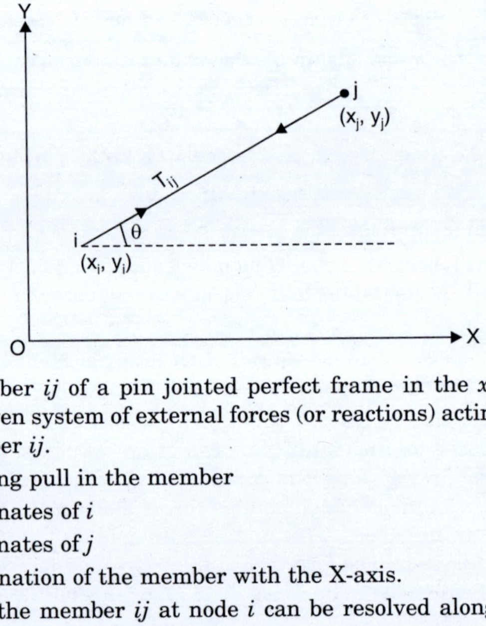

Consider a member ij of a pin jointed perfect frame in the $x - y$ plane, which is in equilibrium under a given system of external forces (or reactions) acting at the joints. Let l_{ij} be the length of the member ij.

T_{ij} be the resulting pull in the member

(x_i, y_i) the coordinates of i

(x_j, y_j) the coordinates of j

and θ the angle of inclination of the member with the X-axis.

The force T_{ij} in the member ij at node i can be resolved along the X and Y-axes and simplified as below :

$$T_x = T_{ij} \cos \theta$$

$$= T_{ij} \cdot \frac{x_{ij}}{l_{ij}}$$

$$T_x = \left(\frac{T_{ij}}{l_{ij}}\right) x_{ij}$$

$$x_{ij} = x_j - x_i$$
$$y_{ij} = y_j - y_i$$
$$\cos\theta = \frac{(x_j - x_i)}{l} = \frac{x_{ij}}{l}$$

$$T_x = t_{ij} \cdot x_{ij} \qquad \text{...(7.3)} \quad \left(\text{Since } t_{ij} = \frac{T_{ij}}{l_{ij}}\right)$$

This is the relation between the X component of the force and the member coordinates x_i, x_j of the member ij. That is the X component of the force in the member is expressed in terms of the tension coefficient and the coordinates of the ends of the member with reference to the chosen reference axis.

Similarly, the Y component of T_{ij} in the y-direction

$$T_y = T_{ij} \sin\theta = T_{ij} \frac{y_{ij}}{l_{ij}} = \left(\frac{T_{ij}}{l_{ij}}\right) y_{ij}$$

$$T_y = t_{ij}\, y_{ij} \; ... \qquad \text{...(7.4)}$$

t_{ij} in equations (7.3) and (7.4) is of course the tension coefficient of the member ij.

The length of the member ij,

$$l_{ij} = \sqrt{{x_{ij}}^2 + {y_{ij}}^2} \qquad \text{...(7.5)}$$

x_{ij} and y_{ij} represent the projected lengths of the member ij along the X and Y-axes respectively. It is to be noted that the projected length is always obtained by subtracting the coordinate of the near end from that of the far end.

Applying the above to formulate the equilibrium conditions of a typically plane truss joint i where members, say, ij, ik and im meet, the equilibrium equations in the X and Y-directions are

$$\Sigma F_x = t_{ij}\, x_{ij} + t_{ik}\, x_{ik} + t_{im}\, x_{im} + X_i$$
$$\Sigma F_y = t_{ij}\, y_{ij} + t_{ik}\, y_{ik} + t_{im}\, y_{im} + Y_i$$

The above equations can be expressed in a compact form as

$$\Sigma\, t_{ij}\, x_{ij} + X_i = 0 \qquad \text{...(7.6)}$$
$$\Sigma\, t_{ij}\, y_{ij} + Y_i = 0 \qquad \text{...(7.7)}$$

where X_i and Y_i are the X and Y components of the applied force at joint i.

Similar equations are formulated at each of the joints and are solved for the tension coefficients. Member forces are completed by multiplying the tension coefficients by the corresponding member length.

In a plane frame, we obtain two equations at each joint in which tension coefficients are the only unknowns. If the frame has j number of joints, there will be $2j$ number of equations. If there are m number of members in a stable and perfect frame,

We have, $\qquad m = 2j - 3$

and there will be m tension coefficients. To solve, we require only $(2j - 3)$ equations, while the available equations are $2j$. Thus, three excess equations are available and can be used for applying checks.

7.5 EXTENSION OF TENSION COEFFICIENT METHOD FOR SPACE TRUSSES

The principle and procedure discussed above can be extended to three dimensional structures by incorporating the force components and coordinates in the Z-direction also. The equilibrium equation in the Z-direction is,

$$\Sigma t_{ij} z_{ij} + Z_i = 0 \qquad \text{...(7.8)}$$

where $(z_{ij} = z_j - z_i)$. z_{ij} represents the projected length of the member along the Z-axis and Z_i are the Z components of the applied forces at joint i.

Length of the member in a three dimensional truss can be expressed as

$$l_{ij} = \sqrt{x_{ij}^2 + y_{ij}^2 + z_{ij}^2} \qquad \text{...(7.9)}$$

7.6 ANALYSIS PROCEDURE USING TENSION COEFFICIENTS—2D FRAMES

The method of tension coefficients for the analysis of plane trusses is summarized below:

1. List the coordinates of each joint (node) of the truss.
2. Determine the projected lengths x_{ij} and y_{ij} of each member of the truss. Determine the lengths l_{ij} of each member using equation (7.5).
3. Resolve the applied forces at each joint in the X and Y directions. Determine the support reactions and their X and Y-components.
4. Identify a node with only two unknown member forces and apply the equations of equilibrium. The solution yields the tension coefficients for the members at the node.
5. Select the next joint with only two unknown member forces and apply the equations of equilibrium and obtain the tension coefficients.
6. Repeat step 5 till the tension coefficients of all the members are obtained.
7. Compute the member forces from the tension coefficients obtained as above, using

$$T_{ij} = t_{ij} \times l_{ij}$$

and tabulate/sketch the same.

7.7 ANALYSIS PROCEDURE FOR SPACE FRAMES

1. In step 2 above, the projected lengths z_{ij} in the Z-direction are also to be computed. Length of any member ij, is obtained from equation (7.9).
2. In step 3 above, the components of forces and reactions in the Z-directions are also to be determined.
3. In steps 4 and 5, each time, nodes with not more than three unknown member forces are to be considered.

But for these 3 deviations, the general procedure of analysis is the same as for 2D frames.

Note. Due care must be exercised in assigning the signs for x_{ij}, y_{ij} etc. Let us consider a member AB, where A is (3,0) and B is (6,0).

For joint A, i is A and j is B and x_{ij} is X_{AB} and is equal to $X_B - X_A = 6 - 3 = +3$; $x_{ij} = 3$.

For joint B, i is B and j is A. Hence $x_{ij} = X_{BA} = X_A - X_B = 3 - 6 = -3$; $x_{ij} = -3$.

Example 7.1 *Using the method of tension coefficients, analyse the cantilever, plane truss shown in Fig. 7.5 and find the member forces.*

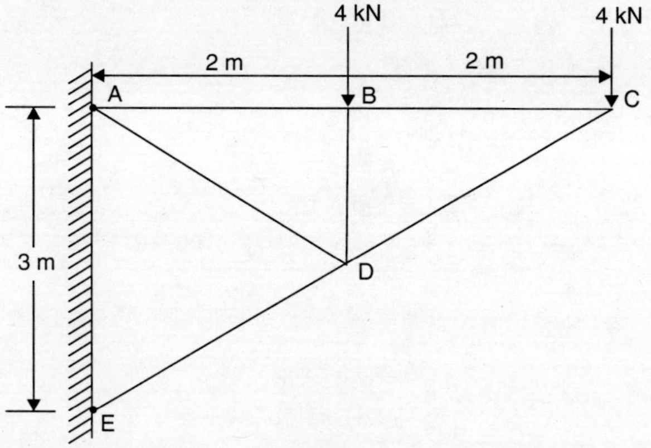

Fig. 7.5

Solution Taking E as origin, the X and Y-axes and the nodal coordinates are marked in Fig. 7.6.

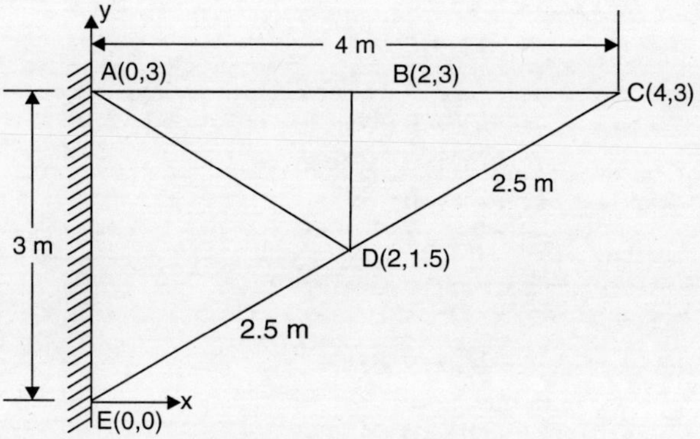

Fig. 7.6. Nodal coordinates.

Table 7.1 Member parameters of the truss.

S. No.	Member	x_i (m)	x_j (m)	$x_{ij} = (x_j - x_i)$ (m)	y_i (m)	y_j (m)	$y_{ij} = (y_j - y_i)$ (m)	$l_{ij} = \sqrt{x_{ij}^2 + y_{ij}^2}$ (m)
1.	AB	0	2	2	3	3	0	2
2.	BC	2	4	2	3	3	0	2
3.	CD	4	2	– 2	3	1.5	– 1.5	2.5
4.	DE	2	0	– 2	1.5	0	– 1.5	2.5
5.	AD	0	2	2	3	1.5	– 1.5	2.5
6.	BD	2	2	0	3	1.5	– 1.5	1.5

Calculation of tension coefficients:

Joint C :

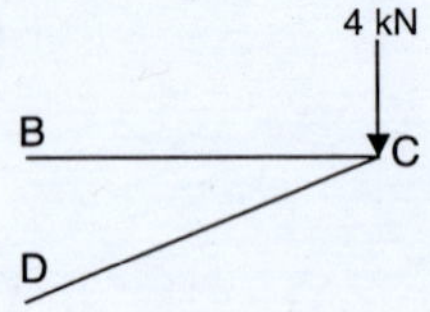

$$\Sigma H = 0 \text{ gives}$$

$$x_{cb} \cdot t_{cb} + x_{cd} \cdot t_{cd} = 0$$

$$(-2)\, t_{cb} + (-2)\, t_{cd} = 0$$

$$-2\, t_{cb} - 2\, t_{cd} = 0$$

$$t_{cb} = -t_{cd} \quad \text{...(1)}$$

$$\Sigma V = 0 \text{ gives}$$

$$y_{cb} \cdot t_{cb} + y_{cd}\, t_{cd} - 4 = 0 \quad \text{(the given 4 kN load is in the –ve Y-direction)}$$

$$0 \,.\, t_{cb} + (-1.5)\, t_{cd} = 4$$

$$t_{cd} = -\frac{4}{1.5}$$

$$t_{cd} = -2.67 \text{ kN/m}$$

Therefore, $t_{cb} = 2.67$ kN/m [From (1)]

Joint B :

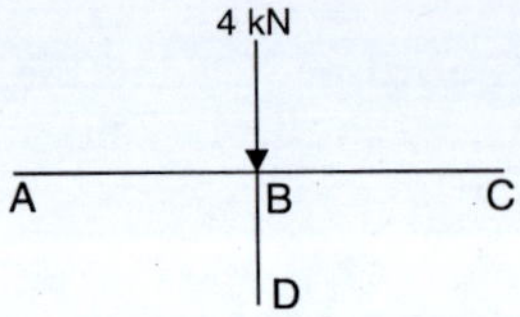

$$\Sigma H = 0 \text{ gives}$$

$$x_{ba} \cdot t_{ba} + x_{bc} \cdot t_{bc} = 0$$

$$(-2)\, t_{ba} + 2\, t_{bc} = 0$$

$$t_{ba} = t_{bc}$$

$$t_{ba} = 2.67 \text{ kN/m}$$

$$\Sigma V = 0 \text{ gives}$$

$$y_{bd} \cdot t_{bd} + Y_i = 0 \quad \text{(the given 4 kN load is in the negative Y-direction)}$$

$$(-1.5)\, t_{bd} - 4 = 0$$

$$-1.5\, t_{bd} = 4$$

$$t_{bd} = -2.67 \text{ kN/m.}$$

Joint D :

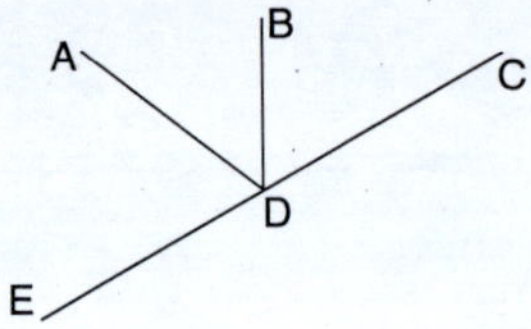

$$\Sigma H = 0 \text{ gives}$$

$$x_{da} \cdot t_{da} + x_{de}\, t_{de} + x_{dc}\, t_{dc} = 0$$

$$-2\, t_{da} - 2\, t_{de} + 2\,(-2.67) = 0$$

$$t_{da} + t_{de} + 2.67 = 0$$

$$t_{da} + t_{de} = -2.67 \quad \text{...(2)}$$

$$\Sigma V = 0 \text{ gives}$$

$$y_{da}\, t_{da} + y_{db}\, t_{db} + y_{dc}\, t_{dc} + y_{de}\, t_{de} = 0$$

$$1.5\, t_{da} + 1.5\,(-2.67) + 1.50\,(-2.67) - 1.5\, t_{de} = 0$$

$$t_{da} - t_{de} = 5.34 \quad \text{...(3)}$$

From eqns. (2) and (3)

$$t_{da} = 1.335 \text{ kN/m}$$

$$t_{de} = -4 \text{ kN/m}$$

Table 7.2 gives the final forces.

Table 7.2 Final forces in members.

S. No.	*Member*	t_{ij} *(kN/m)*	l_{ij} *(m)*	$T_{ij} = t_{ij} \times l_{ij}$ *(kN)*	*Nature of force*
1.	AB	2.67	2	5.34	Tensile
2.	BC	2.67	2	5.34	Tensile
3.	CD	– 2.67	2.5	– 6.675	Compressive
4.	DE	– 4.00	2.5	– 10.00	Compressive
5.	AD	1.335	2.5	3.33	Tensile
6.	BD	– 2.67	1.5	– 4.00	Compressive

Example 7.2 *Using the method of tension coefficients, analyse the plane truss shown in Fig. 7.7 and find the forces in the members.*

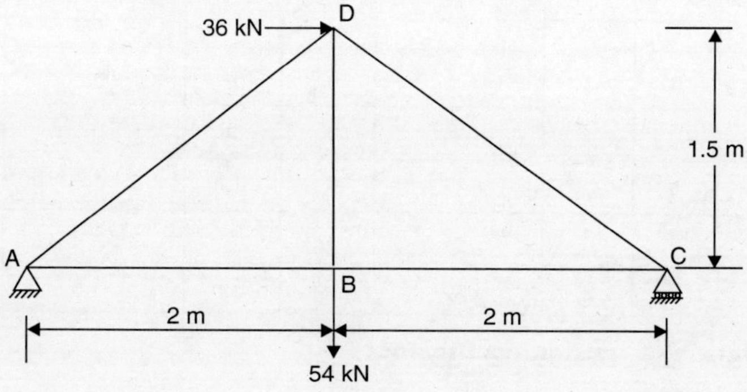

Fig. 7.7

Solution Taking moments about A,

$$-V_c \times 4 + 36 \times 1.5 + 54 \times 2 = 0$$

$$V_c = 40.5 \text{ kN}$$

$$V_A = 54 - 40.5 = 13.5 \text{ kN}$$

$$\Sigma H = 0 \text{ gives}$$

$$H_A = 36 \text{ kN} (\leftarrow)$$

For the given truss, loads, reactions and nodal coordinates (choosing A as the origin) are shown in Fig. 7.8.

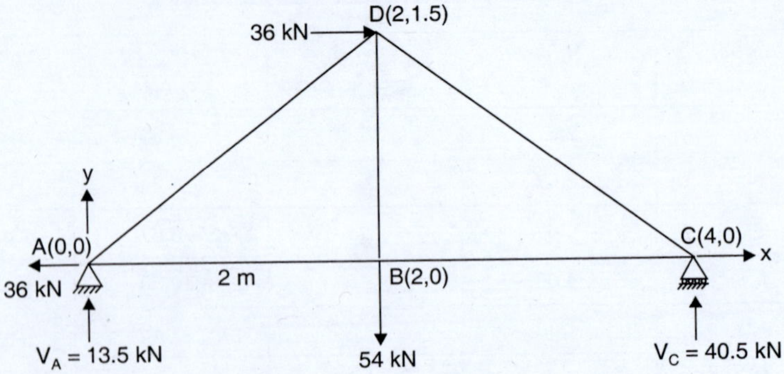

Fig. 7.8. Nodal coordinates.

Table 7.3 Member parameters of the truss.

S. No.	Member	$x_i(m)$	$x_j(m)$	$x_{ij} = (x_j - x_i)$ (m)	$y_i(m)$	$y_j(m)$	$y_{ij} = (y_j - y_i)$ (m)	$l_{ij} = \sqrt{x_{ij}^2 + y_{ij}^2}$ (m)
1.	AB	0	2	2	0	0	0	2
2.	BC	2	4	2	0	0	0	2
3.	CD	4	2	–2	0	1.5	1.5	2.5
4.	DA	2	0	–2	1.5	0	– 1.5	2.5
5.	BD	2	2	0	0	1.5	1.5	1.5

Calculation of tension coefficients :

Joint A :

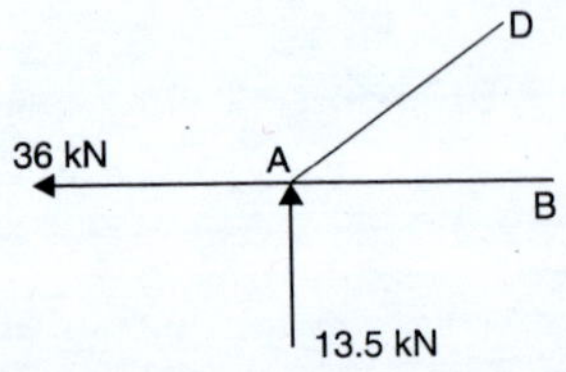

$$\Sigma H = 0 \text{ gives}$$

$$x_{ad} \cdot t_{ad} + x_{ab} \cdot t_{ab} - 36 = 0$$

$$(\rightarrow) \quad (\rightarrow) \quad (\leftarrow)$$

$$2\, t_{ad} + 2\, t_{ab} = 36$$

$$t_{ad} + t_{ab} = 18 \qquad \ldots(1)$$

$$\Sigma V = 0 \text{ gives}$$

$$y_{ad} \cdot t_{ad} + y_{ab} \cdot t_{ab} + 13.5 = 0$$

$$y_{ad} = -\, y_{da} = +\, 1.5$$

$$1.5\, t_{ad} + 0 = -\, 13.5$$

$$t_{ad} = \frac{-\,13.5}{1.5} = -\,9$$

$$t_{ad} = -\,9 \text{ kN/m}$$

From eqn. (1) $$t_{ab} = 18 - t_{ad} = 18 - (-\,9)$$

$$t_{ab} = 27 \text{ kN/m}$$

Joint B :

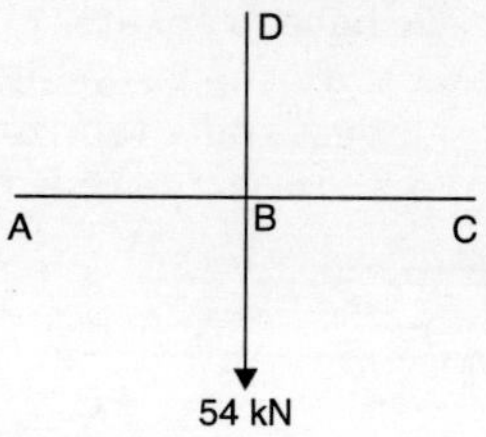

$$\Sigma H = 0 \text{ gives}$$
$$x_{ba} \cdot t_{ba} + x_{bc} \cdot t_{bc} = 0$$
$$-2 \cdot t_{ba} + 2\, t_{bc} = 0$$
$$t_{ba} = t_{bc}$$
$$t_{bc} = 27 \text{ kN/m}$$
$$\Sigma V = 0 \text{ gives}$$
$$y_{bd} \cdot t_{bd} - 54 = 0$$
$$1.5\, t_{bd} = 54$$
$$t_{bd} = 36 \text{ kN/m}$$

Joint C :

$$\Sigma H = 0 \text{ gives}$$
$$x_{cd}\, t_{cd} + x_{cb}\, t_{cb} = 0$$
$$-2\, t_{cd} - 2 \times 27 = 0$$
$$t_{cd} = -27 \text{ kN/m}$$

Table 7.4 Final forces in members.

S. No.	*Member*	t_{ij} *(kN/m)*	l_{ij}*(m)*	$T_{ij} = t_{ij} \times l_{ij}$ *(kN)*	*Nature of force*
1.	AB	27	2	54	Tensile
2.	BC	27	2	54	Tensile
3.	CD	– 27	2.5	– 67.5	Compressive
4.	DA	– 9	2.5	– 22.5	Compressive
5.	BD	36	1.5	54	Tensile

Each joint in space frames provides 3 conditions of equilibrium. Corresponding to hinged joints in plane frames, space frames shall (theoretically) have ball and socket joints at every node, which would allow rotation of all members in X, Y and Z planes.

The ground supports need considerable attention if external indeterminacy is to be avoided. We know that a plane truss hinged at both ends [Fig. 7.8(*a*)] has 4 possible ground reactions and only 3 conditions of overall equilibrium, namely, $\Sigma H = 0$, $\Sigma V = 0$ and $\Sigma M = 0$. So a hinged-hinged plane truss is one degree externally redundant.

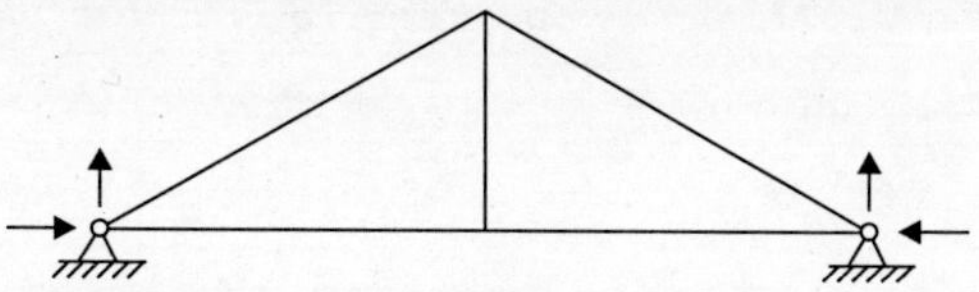

Fig. 7.8 (*a*)

Extending this reasoning to a space truss as in Fig. 7.8(*b*), AB, BC, CA, AD, BD and CD are the members. A, B and C are the ground points, and *can* offer reactions Z_A, Y_A, X_A, Z_B, Y_B, X_B and Z_C, Y_C, X_C.

Since there can be only 6 conditions of (overall) equilibrium, namely,

$$\Sigma X = \Sigma Y = \Sigma Z = \Sigma M_X = \Sigma M_Y = \Sigma M_Z = 0$$

we have to ensure that 3 of the reactions vanish, by the way we provide freedom of movements at the 3 supports [Fig. 7.8(*b*)]. This would mean that the ball joint at C would be allowed to roll along X and Y (horizontal) directions [Fig. 7.8(*c*)] and at joint B the movement would be limited to a groove in the X direction only [Fig. 7.8(*d*)]

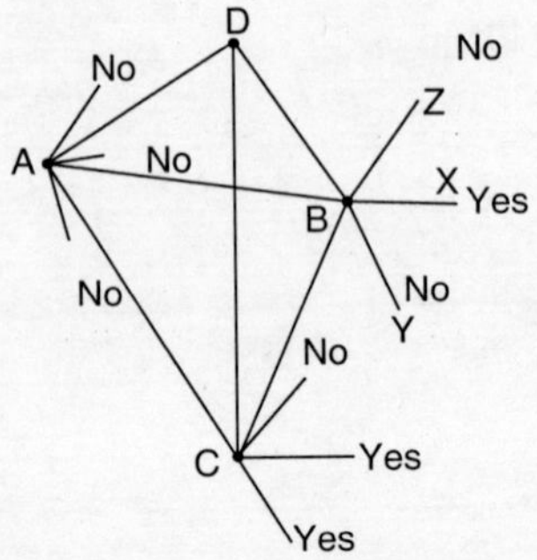

Fig. 7.8 (*b*) (Freedom of movement at the 3 supports)

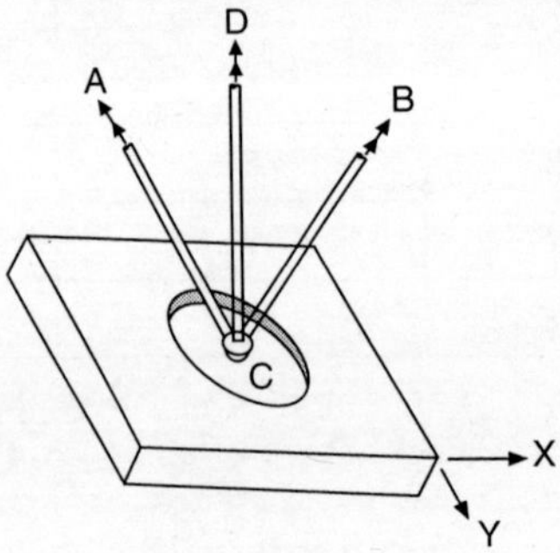

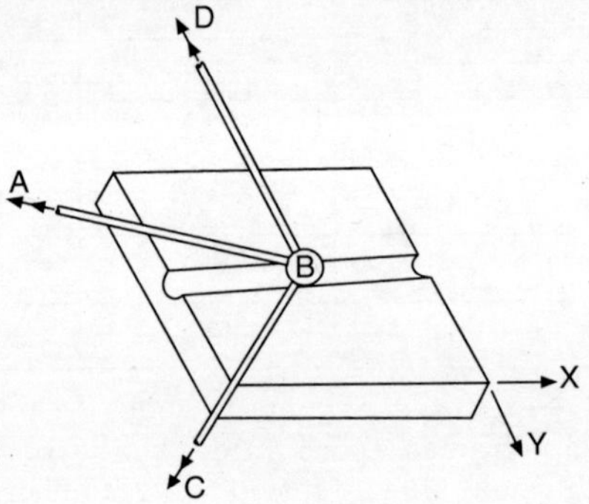

Fig. 7.8 (*c*) Support at C

Fig. 7.8 (*d*) Support at B

Example 7.3 *Determine the forces in the members of the space frame, the plan view of which is shown in Fig. 7.9 by the tension coefficient method.*

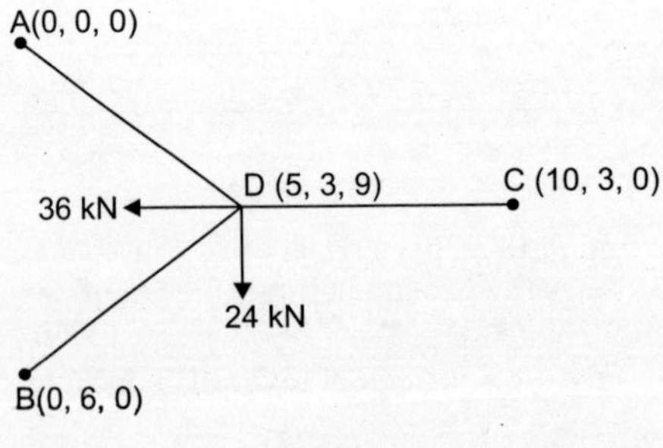

Fig. 7.9

Solution **Nodal coordinates:**

S. No.	*Node*	*Coordinates (m)*		
		X	*Y*	*Z*
1.	A	0	0	0
2.	B	0	6	0
3.	C	10	3	0
4.	D	5	3	9

Member parameters :

Member	x_{ij} *(m)*	y_{ij} *(m)*	z_{ij} *(m)*	$l_{ij} = \sqrt{x_{ij}^2 + y_{ij}^2 + z_{ij}^2}$ *(m)*	t_{ij} *(kN/m)*	$F_{ij} = t_{ij} \times l_{ij}$ *(kN)*	*Nature of force*
DA	– 5	– 3	– 9	10.724	2.2	23.59	Tensile
DB	– 5	3	– 9	10.724	– 5.8	– 62.20	Compressive
DC	5	0	– 9	10.296	3.6	37.06	Tensile

Determination of tension coefficients :

Joint D : (DA, DB, DC)

$$\Sigma F_X = 0,$$

$$-5\,t_{DA} - 5\,t_{DB} + 5\,t_{DC} - 36 = 0$$

$$-5\,t_{DA} - 5\,t_{DB} + 5\,t_{DC} = 36$$

$$t_{DA} + t_{DB} - t_{DC} = -7.2 \qquad \text{...(1)}$$

$$\Sigma F_Y = 0,$$

$$-3\,t_{DA} + 3\,t_{DB} + 0.\,t_{DC} + 24 = 0$$

$$-3\,t_{DA} + 3\,t_{DB} = -24$$

$$t_{DA} - t_{DB} = 8 \qquad \text{...(2)}$$

$$\Sigma F_Z = 0,$$

$$-9\,t_{DA} - 9\,t_{DB} - 9\,t_{DC} = 0$$

$$t_{DA} + t_{DB} + t_{DC} = 0 \qquad \text{...(3)}$$

(3) – (1) $\Rightarrow$ $2\,t_{DC} = 7.2 \quad \therefore \quad t_{DC} = 3.6$

(2) + (3) $\Rightarrow$ $2\,t_{DA} + t_{DC} = 8$

$2\,t_{DA} + 3.6 = 8 \quad \therefore \quad t_{DA} = 2.2$

Substituting $t_{DA} = 2.2$ in eqn. (2)

$2.2 - t_{DB} = 8 \quad \therefore \quad t_{DB} = -5.8$

Example 7.4 *Using the method of tension coefficients, analyse the space truss shown in Fig. 7.10 and find the forces in the members of the truss. Assume that supports A and C provide freedom of movement in the X direction only. Support B is free to move in Y direction only.*

Solution Joint A is considered as the origin and coordinates axes X,Y and Z are as shown. The coordinates of all nodes/joints marked in Fig. 7.10 are also tabulated. Suports A and C provide freedom of movement in the X direction only. Suport B is free to move in Y direction only.

Table 7.5 Joint coordinates.

S. No.	*Joint*	*Coordinates (m)*		
		X	*Y*	*Z*
1.	A	0	0	0
2.	B	3	4	0
3.	C	6	0	0
4.	D	3	2	5

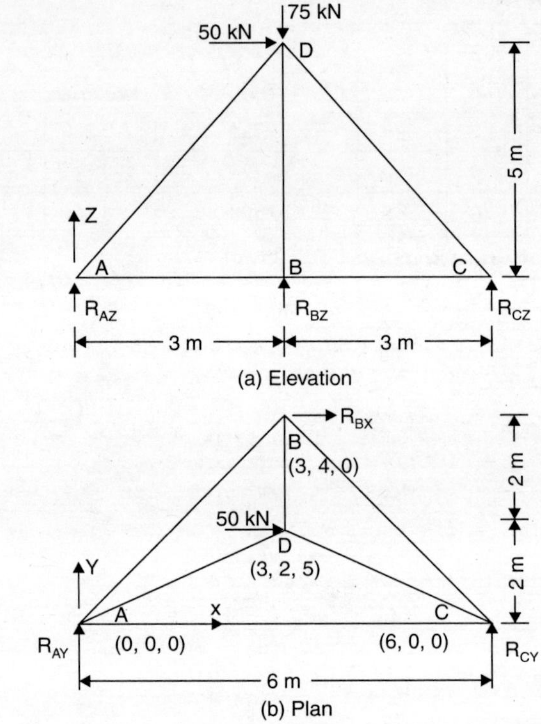

Fig. 7.10

Figure 7.10 shows the loads.

Co-ordinates of nodes and the possible 6 reactions, namely, R_{AZ}, R_{BZ}, R_{CZ}, R_{AY}, R_{CY} and R_{BX}. These six reactions can be evaluated by considering the equilibrium of the 6 member truss as a whole, using the *b* conditions of equilibrium, namely,

$$\Sigma F_X = \Sigma F_Y = \Sigma F_Z = \Sigma M_X = \Sigma M_Y = \Sigma M_Z = 0$$

$$\Sigma F_X = 0$$

$$50 + R_{BX} = 0$$

$$R_{BX} = -50 \text{ kN} \quad \ldots(1)$$
$$\Sigma F_Y = 0$$
$$R_{AY} = -R_{CY} \quad \ldots(2)$$
$$\Sigma F_Z = 0$$
$$R_{AZ} + R_{BZ} + R_{CZ} - 75 = 0 \quad \ldots(3)$$
$$\Sigma M_X = 0, \text{ about line AC.}$$
$$R_{BZ} \times 4 - 75 \times 2 = 0$$
$$R_{BZ} = 37.5 \text{ kN} \quad \ldots(4)$$
$$\Sigma M_Y = 0, \text{ about line AE (Y-axis)}$$
$$R_{CZ} \times 6 + R_{BZ} \times 3 - 75 \times 3 - 50 \times 5 = 0$$
$$R_{CZ} \times 6 = -37.5 \times 3 + 225 + 250$$
$$R_{CZ} = 60.42 \text{ kN} \quad \ldots(5)$$
$$\Sigma M_Z = 0, \text{ about Z-axis}$$
$$50 \times 2 - R_{CY} \times 6 + R_{BX} \times 4 = 0$$
$$R_{CY} = -16.67 \text{ kN} \quad \ldots(6)$$

From (3) $R_{AZ} = 75 - R_{BZ} - R_{CZ}$

$$= 75 - 37.5 - 60.42$$
$$R_{AZ} = -22.92 \text{ kN} \quad \ldots(7)$$

From (2) $R_{AY} = -(-16.67) = +16.67$ kN ...(8)

Now, we can apply the tension coefficient based equilibrium conditions to the 4 joints, A, B, C and D.

Joint D:

$$\Sigma F_X = 0$$
$$x_{DA}\, t_{DA} + x_{DB}\, t_{DB} + x_{DC}\, t_{DC} + 50 = 0$$
$$-3\, t_{DA} + 0 + 3\, t_{DC} + 50 = 0$$
$$t_{DA} = t_{DC} + 16.67 \quad \ldots(9)$$
$$\Sigma F_Y = 0$$
$$y_{DA}\, t_{DA} + y_{DB} \cdot t_{DB} + y_{DC} \cdot t_{DC} = 0$$
$$-2\, t_{DA} + 2\, t_{DB} - 2\, t_{DC} = 0$$
$$t_{DB} - 2t_{DC} = 16.67 \quad \ldots(10)$$
$$\Sigma F_Z = 0$$
$$z_{DA} \cdot t_{DA} + z_{DB} \cdot t_{DB} + z_{DC} \cdot t_{DC} - 75 = 0$$
$$-5\, t_{DA} - 5\, t_{DB} - 5\, t_{DC} - 75 = 0$$
$$t_{DA} + t_{DB} + t_{DC} = -15$$
$$t_{DB} + 2t_{DC} = -15 - 16.67$$
$$t_{DB} + 2t_{DC} = -31.67 \quad \ldots(11)$$

(10) + (11) $2t_{DB} = -15$

$$t_{DB} = -7.5 \quad \ldots(12)$$

From (11), $t_{DC} = -12.085$ ...(13)

From (9), $t_{DA} = 4.585$ ...(14)

Joint B:

$$\Sigma F_X = 0$$
$$x_{BA}\, t_{BA} + x_{BO}\, t_{BD} + x_{BC}\, t_{BC} - 50 = 0$$
$$-3\, t_{BA} + 3\, t_{BC} = 50$$
$$t_{BC} - t_{BA} = 16.67 \quad \ldots(15)$$

$$\Sigma F_Y = 0$$

$$y_{BA}\, t_{BA} + y_{BD}\, t_{BD} + y_{BC}\, t_{BC} = 0$$

$$-4\, t_{BA} - 2\, t_{BD} - 4\, t_{BC} = 0$$

$$-2\, t_{BA} - t_{BD} - 2(t_{BA} + 16.67) = 0$$

$$-4t_{BA} = +33.34 - 7.5$$

$$t_{BA} = -6.46 \qquad ...(16)$$

$$t_{BC} = 10.21 \qquad ...(17)$$

Joint A:

$$\Sigma F_X = 0$$

$$x_{AC}\, t_{AC} + x_{AB}\, t_{AB} + x_{AD}\, t_{AD} = 0$$

$$6\, t_{AC} + 3(-6.46) + 3(4.585) = 0$$

$$t_{AC} = +0.938.$$

Now that we have obtained all the 6 tension coefficients, we can tabulate them and obtain the member forces (Consequent changes in Table 7.6 for member AC).

Table 7.6 Member parameters and final forces.

Member	x_{ij} *(m)*	y_{ij} *(m)*	z_{ij} *(m)*	l_{ij}*(m)*	t_{ij} *(kN/m)*	T_{ij} *(kN)*	*Nature of force*
AB	3	4	0	5	– 6.46	– 32.300	Compressive
BC	3	– 4	0	5	10.21	51.050	Tensile
CA	– 6	0	0	6	+0.938	5.628	Tensile
AD	3	2	5	6.164	4.585	28.262	Tensile
BD	0	– 2	5	5.385	–7.500	–40.388	Compressive
CD	– 3	2	5	6.164	– 12.085	– 74.492	Compressive

Example 7.5 *Determine the forces in the members of the space truss shown in Fig. 7.11 by the tension coefficient method. The coordinates are A (2, 6, 0), B (0, 0, 0), C (6, 0, 0) and D (2, 2, 5). A, B, C are hinged supports.*

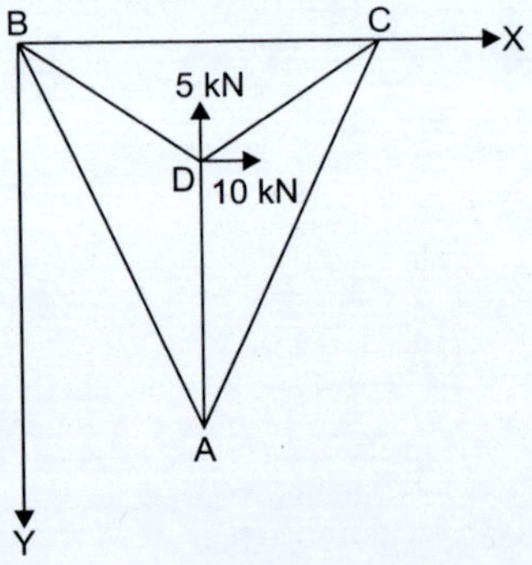

Fig. 7.11

Solution

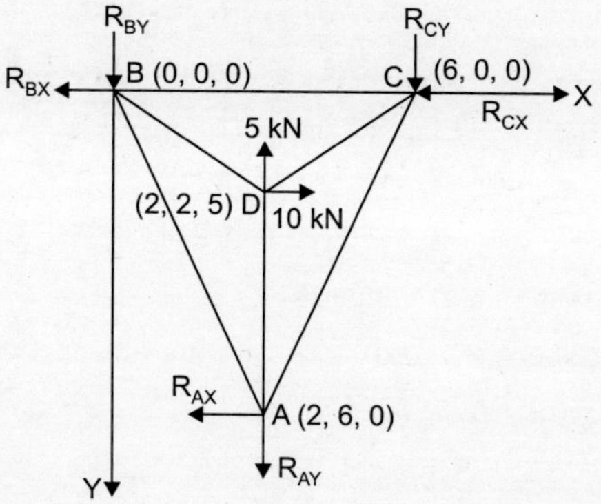

Nodal coordinates:

S. No.	*Node*	*Coordinates (m)*		
		X	***Y***	***Z***
1.	A	2	6	0
2.	B	0	0	0
3.	C	6	0	0
4.	D	2	2	5

Joint	*Member*	x_{ij} *(m)*	y_{ij} *(m)*	z_{ij} *(m)*	l_{ij} *(m)*	t_{ij} *(kN/m)*	$F_{ij} = t_{ij} \times l_{ij}$ *(kN)*	*Nature of force*
D	DA	0	4	– 5	6.403	0.833	5.334	Tensile
	DB	– 2	– 2	– 5	5.745	1.112	6.388	Tensile
	DC	4	– 2	– 5	6.708	– 1.944	– 13.04	Compressive

Consider joint D:

$$\Sigma F_X = 0,$$

$$t_{DA}(0) + t_{DB}(-2) + t_{DC}(4) + 10 = 0$$

$$-2\,t_{DB} + 4\,t_{DC} = -10$$

$$t_{DB} - 2\,t_{DC} = 5 \qquad \ldots(1)$$

$$\Sigma F_Y = 0,$$

$$t_{DA}(4) + t_{DB}(-2) + t_{DC}(-2) - 5 = 0$$

$$4\,t_{DA} - 2\,t_{DB} - 2\,t_{DC} = 5$$

$$2\,t_{DA} - t_{DB} - t_{DC} = 2.5 \qquad \ldots(2)$$

$$\Sigma F_Z = 0,$$

$$t_{DA}(-5) + t_{DB}(-5) + t_{DC}(-5) + 0 = 0$$

$$t_{DA} + t_{DB} + t_{DC} = 0 \qquad ...(3)$$

Adding (2) and (3) $\quad 3\,t_{DA} = 2.5 \quad \therefore \quad t_{DA} = 0.833$

Adding (1) and (2) $\quad 2\,t_{DA} - 3t_{DC} = 7.5$

$$2(0.833) - 3\,t_{DC} = 7.5 \quad \therefore \quad t_{DC} = -1.944$$

Substituting $\quad t_{DC} = -1.944$ in (1)

$$t_{DB} = 5 + 2\,(-1.944) \quad \therefore \quad t_{DB} = 1.112$$

(*a*) To find vertical reaction at supports

(*i*) Taking moment about BX axis (Assuming all reactions along Z-axis are upward)

$$R_{AZ}\,(6) + 5\,(5) = 0$$

$$\therefore \quad R_{AZ} = -4.167$$

i.e., $\quad R_{AZ} = 4.167\,(\downarrow)$

(*ii*) Taking moment about BY-axis

$$-R_{AZ}\,(2) + 10\,(5) - R_{CZ}\,(6) = 0$$

$$-(-4.167)\,(2) + 50 = 6\,R_{CZ}$$

$$\therefore \quad R_{CZ} = 9.722\,(\uparrow)$$

(*iii*) $\Sigma V = 0$ gives $\quad -4.167 + 9.722 + R_{BZ} = 0$

$$\therefore \quad R_{BZ} = -5.553$$

i.e., $\quad R_{BZ} = 5.553\,(\downarrow)$

(*b*) To find horizontal reactions at supports

(*i*) Taking moment about BZ-axis

$$R_{CY}\,(6) + R_{AX}\,(6) + R_{AY}\,(2) - 5(2) - 10(2) = 0$$

$$6\,R_{CY} + 6\,R_{AX} + 2\,R_{AY} = 30 \qquad ...(4)$$

(*ii*) $\Sigma F_X = 0$ gives, $-R_{BX} - R_{CX} - R_{AX} + 10 = 0$

$$R_{AX} + R_{BX} + R_{CX} = 10 \qquad ...(5)$$

(*iii*) $\Sigma F_Y = 0$ gives, $-1 + R_{AY} + R_{BY} + R_{CY} - 5 = 0$

$$R_{AY} + R_{BY} + R_{CY} = +5 \qquad ...(6)$$

Considering joint A :

$$\Sigma F_X = 0$$

$$t_{AB}\,(X_B - X_A) + t_{AC}\,(X_C - X_A) + t_{AD}\,(X_D - X_A) - R_{AX} = 0$$

$$t_{AB}\,(0 - 2) + t_{AC}\,(6 - 2) + t_{AD}\,(2 - 2) - R_{AX} = 0$$

$$-2\,t_{AB} + 4\,t_{AC} = R_{AX} \qquad ...(7)$$

$$\Sigma F_Y = 0$$

$$t_{AB}(Y_B - Y_A) + t_{AC}(Y_C - Y_A) + t_{AD}(Y_D - Y_A) + R_{AY} = 0$$

$$t_{AB}(0 - 6) + t_{AC}(0 - 6) + t_{AD}(2 - 6) + R_{AY} = 0$$

$$-6\,t_{AB} - 6\,t_{AC} - 4\,(0.833) + R_{AY} = 0$$

$$6\,t_{AB} + 6\,t_{AC} + 3.332 = R_{AY} \qquad \ldots(8)$$

Considering joint 'B' :

$$\Sigma F_X = 0$$

$$t_{BC}(X_C - X_B) + t_{BA}(X_A - X_B) + t_{BD}(X_D - X_B) - R_{BX} = 0$$

$$t_{BC}(6 - 0) + t_{BA}(2 - 0) + t_{BD}(2 - 0) - R_{BX} = 0$$

$$6\,t_{BC} + 2\,t_{BA} + 2\,t_{BD} = R_{BX}$$

$$6\,t_{BC} + 2\,(-t_{AB}) + 2\,(1.112) = R_{BX} \quad \because\ t_{AB} = -t_{BA}$$

$$6\,t_{BC} - 2\,t_{AB} + 2.224 = R_{BX} \qquad \ldots(9)$$

$$\Sigma F_Y = 0$$

$$t_{BC}(Y_C - Y_B) + t_{BA}(Y_A - Y_B) + t_{BD}(Y_D - Y_B) + R_{BY} = 0$$

$$t_{BC}(0 - 0) + t_{BA}(6 - 0) + t_{BD}(2 - 0) + R_{BY} = 0$$

$$-t_{AB}\,.\,6 + 2\,(1.112) + R_{BY} = 0 \qquad t_{BA} = -t_{AB}$$

$$6\,t_{AB} - 2.224 = R_{BY} \qquad \ldots(10)$$

Considering joint 'C' :

$$\Sigma F_X = 0$$

$$t_{CB}(X_B - X_C) + t_{CA}(X_A - X_C) + t_{CD}(X_D - X_C) - R_{CX} = 0$$

$$t_{CB}(0 - 6) + t_{CA}(2 - 6) + t_{CD}(2 - 6) - R_{CX} = 0$$

$$-t_{BC}(-6) - 4\,(-t_{AC}) - 4\,(-1.944) = R_{CX}$$

$$6\,t_{BC} + 4\,t_{AC} + 7.776 = R_{CX} \qquad \ldots(11)$$

$$\Sigma F_Y = 0$$

$$t_{CB}(Y_B - Y_C) + t_{CA}(Y_A - Y_C) + t_{CD}(Y_D - Y_C) + R_{CY} = 0$$

$$t_{CB}(0 - 0) + t_{CA}(6 - 0) + t_{CD}(2 - 0) + R_{CY} = 0$$

$$6.\,(-t_{AC}) + 2\,(-1.944) + R_{CY} = 0$$

$$6\,t_{AC} + 3.888 = R_{CY} \qquad \ldots(12)$$

Substituting the reactions values in equation (4), we get

$$6(6\,t_{AC} + 3.888) + 6(-2\,t_{AB} + 4\,t_{AC}) + 2(6\,t_{AB} + 6\,t_{AC} + 3.333) = 30$$

$$72\,t_{AC} + 30 = 30$$

$$\therefore \qquad t_{AC} = 0$$

Substituting the values of reactions in equation (5), we get

$$(-2\,t_{AB} + 4\,t_{AC}) + (6\,t_{BC} - 2\,t_{AB} + 2.224) + 6\,t_{BC} + 4\,t_{AC} + 7.776 = 10$$

$$-4\,t_{AB} + 8\,t_{AC} + 12\,t_{BC} = 0$$

$$t_{AC} = 0 \quad \therefore \qquad -4\,t_{AB} + 12\,t_{BC} = 0 \qquad \ldots(13)$$

Substituting the values in equation (6), we get

$$(6\,t_{AB} + 6\,t_{AC} + 3.33) + (6\,t_{AB} - 2.224) + (6\,t_{AC} + 3.888) = 5$$

$$12\,t_{AB} + 12\,t_{AC} = 0$$

$\therefore$ $t_{AB} = 0$

Substituting $t_{AB} = 0$ in eqn. (13) $t_{BC} = 0$

Reactions :

$$R_{AX} = 0$$
$$R_{BX} = 2.224 \text{ kN}$$
$$R_{CX} = 7.776 \text{ kN}$$
$$R_{AY} = 3.332 \text{ kN}$$
$$R_{BY} = -2.224 \text{ kN}$$
$$R_{CY} = 3.888 \text{ kN.}$$

Example 7.6 *Analyse the space truss shown in Fig. 7.12 by the method of tension coefficients and determine the member forces. Ground points A, B and C are restrained in X, Y and Z directions.*

Solution

Table 7.7 Nodal coordinates.

S. No.	*Node*	*Coordinates (m)*		
		X	***Y***	***Z***
1.	A	0	0	0
2.	B	7.2	0	0
3.	C	0	0	7.2
4.	D	3.6	4.8	0
5.	E	0	4.8	0
6.	F	0	4.8	3.6
7.	G	0	9.6	0

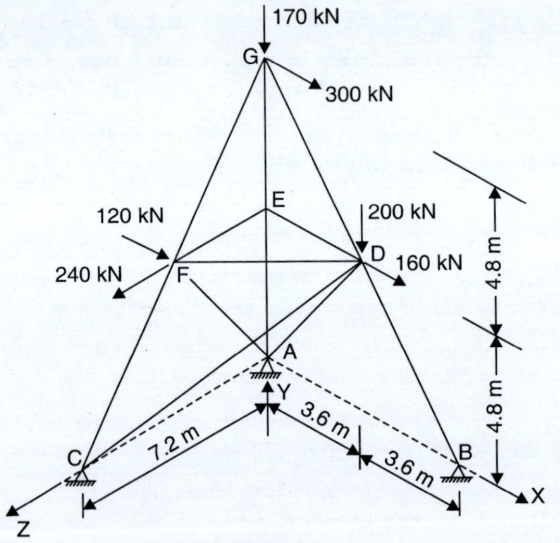

Fig. 7.12

Table 7.8 Member parameters and final forces.

Member	x_{ij} *(m)*	y_{ij} *(m)*	z_{ij} *(m)*	l_{ij}*(m)*	t_{ij} *(kN/m)*	T_{ij} *(kN)*	*Nature of force*
GF	0	– 4.8	3.6	6	0	0	—
GE	0	– 4.8	0	4.8	47.92	230.02	Tensile
GD	3.6	– 4.8	0	6	– 83.33	– 500	Compressive
EF	0	0	3.6	3.6	0	0	—
EA	0	– 4.8	0	4.8	47.92	230.02	Tensile
ED	3.6	0	0	3.6	0	0	—
FC	0	– 4.8	3.6	6	– 50.00	– 300	Compressive
FA	0	– 4.8	– 3.6	6	50.00	300	Tensile
FD	3.6	0	– 3.6	5.09	– 33.33	–169.65	Compressive
DA	– 3.6	– 4.8	0	6	1.385	8.31	Tensile
DB	3.6	– 4.8	0	6	– 143.06	– 858.36	Compressive
DC	– 3.6	– 4.8	7.2	9.37	16.67	156.20	Tensile

Determination of tension coefficients:

Joint G :

$$\Sigma F_X = 0, \text{ gives}$$

$$x_{GF}\ t_{GF} + x_{GE}\ t_{GE} + x_{GD}\ t_{GD} + 300 = 0$$

$$0\ t_{GF} + 0\ t_{GE} + 3.6\ t_{GD} + 300 = 0$$

$$t_{GD} = -\frac{300}{3.6}$$

$$t_{GD} = -\ 83.33 \text{ kN/m}$$

$$\Sigma F_Y = 0 \text{ gives,}$$

$$y_{GF}\ t_{GF} + y_{GE}\ t_{GE} + y_{GD}\ t_{GD} - 170 = 0$$

$$-\ 4.8\ t_{GF} - 4.8\ t_{GE} - 4.8\ t_{GD} - 170 = 0$$

$$-\ 4.8\ (t_{GF} + t_{GE} + t_{GD}) = 170$$

$$t_{GF} + t_{GE} - 83.33 = 170/(-\ 4.8)$$

$$t_{GF} + t_{GE} = -\ 170/4.8 + 83.33 = 47.92$$

$$t_{GF} + t_{GE} = 47.92 \qquad ...(1)$$

$$\Sigma F_Z = 0 \text{ gives,}$$

$$z_{GF}\ t_{GF} + z_{GE}\ t_{GE} + z_{GD}\ t_{GD} = 0$$

$$3.6\ t_{GF} + 0\ t_{GE} + 0\ t_{GD} = 0$$

$$3.6\ t_{GF} = 0$$

$$t_{GF} = 0$$

Therefore, from eqn. (1),

$$t_{GE} = 47.92 \text{ kN/m}$$

Joint E :

$$\Sigma F_X = 0\, t_{EG} + 0\, t_{EF} + 3.6\, t_{ED} + 0\, t_{EA} = 0$$

$$\Sigma F_Y = 4.8\, t_{EG} + 0\, t_{EF} + 0\, t_{ED} - 4.8\, t_{EA} = 0$$

$$\Sigma F_Z = 0\, t_{EG} + 3.6\, t_{EF} + 0\, t_{ED} + 0\, t_{EA} = 0$$

$$t_{ED} = 0$$

$$t_{EF} = 0$$

$$4.8 \times 47.92 - 4.8\, t_{EA} = 0$$

$$47.92 - t_{EA} = 0$$

$$t_{EA} = 47.92 \text{ kN/m}$$

Joint F :

$$\Sigma F_X = 0\, t_{FG} + 0\, t_{FE} + 3.6\, t_{FD} + 0\, t_{FA} + 0\, t_{FC} + 120 = 0$$

$$\Sigma F_Y = 4.8\, t_{FG} + 0\, t_{FE} + 0\, t_{FD} - 4.8\, t_{FA} - 4.8\, t_{FC} = 0$$

$$\Sigma F_Z = -3.6\, t_{FG} - 3.6\, t_{FE} - 3.6\, t_{FD} - 3.6\, t_{FA} + 3.6\, t_{FC} + 240 = 0$$

$$t_{FD} = -\frac{120}{3.6} = -33.33 \text{ kN/m}$$

$$4.8\,(0) - 4.8\,(t_{FA} + t_{FC}) = 0$$

$$t_{FA} + t_{FC} = 0$$

$$t_{FA} = -t_{FC} \qquad ...(2)$$

$$-3.6 \times 0 + 3.6 \times 0 - 3.6\,(-33.33) - 3.6\, t_{FA} + 3.6\, t_{FC} + 240 = 0$$

$$3.6 \times 33.33 - 3.6\,(-t_{FC}) + 3.6\, t_{FC} = -240$$

$$3.6\,(t_{FC} + t_{FC}) = -240 - 3.6 \times 33.33$$

$$3.6\,(2t_{FC}) = -240 - 3.6 \times 33.33$$

$$t_{FC} = -50 \text{ kN/m}$$

From eqn. (2) $\quad t_{FA} = 50 \text{ kN/m}$

Joint D :

$$\Sigma F_X = -3.6\,(-83.33) - 3.6 \times 0 - 3.6 \times (-33.33) - 3.6\, t_{DC} - 3.6\, t_{DA} + 3.6\, t_{DB} + 160 = 0$$

$$-3.6\, t_{DC} - 3.6\, t_{DA} + 3.6\, t_{DB} = -160 - 3.6 \times 83.33 - 3.6 \times 33.33$$

$$-t_{DC} - t_{DA} + t_{DB} = \frac{1}{3.6}[-580]$$

$$-t_{DC} - t_{DA} + t_{DB} = -161.11 \qquad ...(3)$$

$$\Sigma F_Y = 4.8 \times (-83.33) + 0 + 0 - 4.8\, t_{DC} - 4.8\, t_{DA} - 4.8\, t_{DB} - 200 = 0$$

$$-4.8\,(t_{DC} + t_{DA} + t_{DB}) = 200 + 4.8 \times 83.33$$

$$t_{DC} + t_{DA} + t_{DB} = -\frac{1}{48}(200 + 4.8 \times 83.33)$$

$$t_{DC} + t_{DA} + t_{DB} = -125 \qquad ...(4)$$

$$\Sigma F_Z = 0 + 0 + 3.6 \times (-33.33) + 7.2\, t_{DC} + 0 + 0 = 0$$

$$7.2\ t_{DC} = 3.6 \times 33.33$$
$$t_{DC} = 16.67 \text{ kN/m}$$

Substituting t_{DC},

(3) becomes,

$$-16.67 - t_{DA} + t_{DB} = -161.11$$
$$-t_{DA} + t_{DB} = -144.44 \quad ...(5)$$

(4) becomes,

$$16.67 + t_{DA} + t_{DB} = -125$$
$$t_{DA} + t_{DB} = -141.67 \quad ...(6)$$

Adding eqns. (5) and (6), $2\ t_{DB} = -286.11$

$$t_{DB} = -143.06 \text{ kN/m}$$
$$t_{DA} = -141.67 - t_{DB}$$
$$= -141.67 - (-143.06)$$
$$= 1.385 \text{ kN/m.}$$

7.8 GUIDELINES ON JOINT FORCES

It will be greatly helpful at this stage to formulate a few guide lines regarding zero forces in members at a joint. These guidelines will cut the magnitude of work involved considerably.

Take for example the joint F in Fig. 7.12.

7.8.1 Out of Plane Member

Members FG, FE, FA and FC lie in the ZY-plane. Only one member, FD is not of this plane. In this case the X-components of the external loads at F will be taken by the X-component of member FD alone. (The other member forces will have no X-component).

Take the joint E in the same figure. FE, AE and GE are in the YZ-plane. Only ED is out of YZ-plane. There are no external forces at E. Hence the force in ED will be zero.

Looking at joint E in another way, GE, DE and AE are in the X-Y plane. FE in the odd one out. It is out of the XY-plane and there is no external load. The force in FE will then be zero. From Table 7.8 we can verify that FE and ED have zero forces. We can write our guideline as under.

GL 1 : If all members in a joint but one lie in a single plane, and if there is no external force at the joint, the force in the out of plane member will be zero.

Should there be external forces at the joint, the component of the external forces normal to the plane will be taken by the normal component of the out of plane member.

7.8.2 Non Collinear Members

Suppose at joint A in Fig. 7.13, we know that BA and DA have zero forces. We are left with AE and AC. These two are not in the same line (not collinear) and in addition there are no external forces at A. In such a case forces in AE and AC are zero. This gives us our second guide line.

GL 2 : If at a joint in a space truss we know that the forces in all members except two are zero, and if these 2 members are not in the same line and if there are no external loads at the joint, then the two non collinear members will have zero forces.

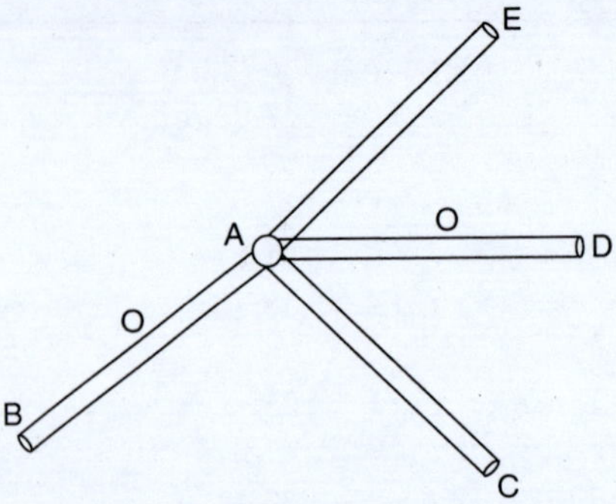

Fig. 7.13. Non collinear members.

We can apply the above two guidelines to advantage in several multimember space truss problems.

Example 7.7 *Analyse the space frame for the loading and support constraints shown in Fig. 7.14. Points B, C and D are supported on the ground. A and D are restrained along X. B is restrained along Z.*

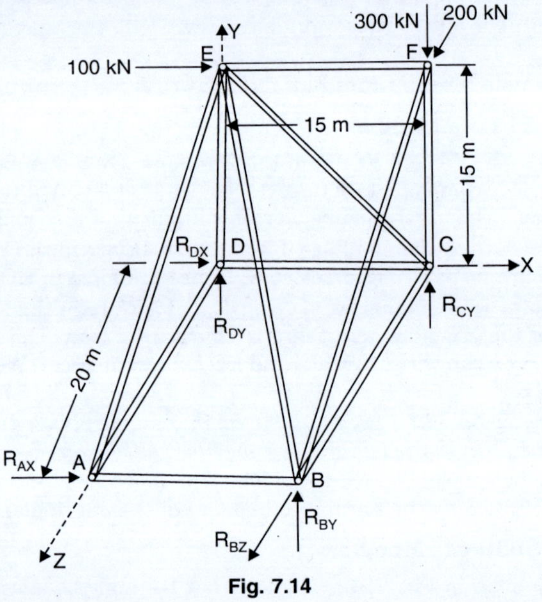

Fig. 7.14

Solution (*a*) **Indeterminacy :**

number of members m = 12

number of joints $j = 6$

$3j - 6 = 3 \times 6 - 6 = 12$ m

Since $m = 3j - 6$, the frame is stable and internally determinate.

There are 6 unknown reactions, namely, R_{AX}, R_{BZ}, R_{BY}, R_{CY}, R_{DY} and R_{DX}. These have to be evaluated from 3 moment conditions and 3 force conditions, namely,

$$\Sigma F_X = \Sigma F_Y = \Sigma F_Z = 0$$
$$\Sigma(M_x)_D = \Sigma(M_y)_D = \Sigma(M_z)_D = 0$$

Hence the frame is also externally determinate

(*b*) **Determination of support reactions :**

(*i*) Equilibrium conditions in the 3 coordinate directions :

$$R_{AX} + R_{DX} + 100 = 0 \quad ...(1)$$
$$R_{BY} + R_{CY} + R_{DY} - 300 = 0 \quad ...(2)$$
$$R_{BZ} + 200 = 0 \quad ...(3)$$
$$R_{BZ} = -200$$

(*ii*) Taking moments about the 3 coordinate axes through D :

$$M_x = 0 = -20\,R_{BY} + (200 \times 15)$$
$$R_{BY} = +150 \text{ kN}$$
$$M_y = 0 = 20\,R_{AX} - 200 \times 15 - 15\,R_{BZ} = 0$$
$$20\,R_{AX} - 200 \times 15 - 15\,(-200) = 0$$
$$R_{AX} = 0$$
$$M_z = 0 = 15\,R_{BY} + 15\,R_{CY} - (+300 \times 15) - (100 \times 15)$$
$$15\,(150) + 15\,R_{CY} - 4500 - 1500 = 0$$
$$150 + 250 + R_{DY} - 300 = 0$$
$$R_{DY} = -100 \text{ kN}$$
$$R_{AX} + R_{DX} + 100 = 0$$
$$R_{DX} = -100 - R_{AX} = -100 - 0 = -100 \text{ kN}$$
$$R_{DX} = -100 \text{ kN}$$

Determination of tension coefficients:

Joint F :

$$\Sigma F_X = -15\,t_{FE} + 0\,t_{FB} + 0\,t_{FC} = 0;\ t_{FE} = 0$$
$$\Sigma F_Y = 0\,t_{FE} - 15\,t_{FB} - 15\,t_{FC} - 300 = 0$$
$$\Sigma F_Z = 0\,t_{FE} + 20\,t_{FB} + 0\,t_{FC} + 200 = 0$$
$$t_{FB} = -\frac{200}{20} = -10 \text{ kN/m}$$
$$-15\,(-10) - 15\,t_{FC} = 300$$
$$-15\,t_{FC} = 300 - 150 = 150$$
$$t_{FC} = -10 \text{ kN/m.}$$

Table 7.9 Nodal coordinates.

S. No.	*Node*	*Coordinates (m)*		
		X	***Y***	***Z***
1.	A	0	0	20
2.	B	15	0	20
3.	C	15	0	0
4.	D	0	0	0
5.	E	0	15	0
6.	F	15	15	0

Table 7.10 Member parameters.

Member	x_{ij} *(m)*	y_{ij} *(m)*	z_{ij} *(m)*	l_{ij}*(m)*	t_{ij} *(kN/m)*	T_{ij} *(kN)*
AB	15	0	0	15	0	0
AC	15	0	– 20	25	0	0
AD	0	0	– 20	20	0	0
AE	0	15	– 20	25	0	0
BC	0	0	– 20	20	0	0
BE	– 15	15	– 20	29.15	0	0
BF	0	15	– 20	25	– 10	– 250
CD	– 15	0	0	15	6.67	100.25
CE	– 15	15	0	21.21	– 6.67	– 141.47
CF	0	15	0	15	– 10	– 150
DE	0	15	0	15	6.67	100.05
EF	15	0	0	15	0	0

Joint B :

$$\Sigma F_X = -15\, t_{BA} - 15\, t_{BE} + 0\, t_{BF} + 0\, t_{BC} = 0 \quad ...(1)$$

$$\Sigma F_Y = 0\, t_{BA} + 15\, t_{BE} + 15\, t_{BF} + 0\, t_{BC} + R_{BY} = 0 \quad ...(2)$$

$$\Sigma F_Z = 0\, t_{BA} - 20\, t_{BE} - 20\, t_{BF} - 20\, t_{BC} + R_{BZ} = 0 \quad ...(3)$$

From (2) $15\, t_{BE} + 15\,(-10) + 150 = 0$

$$t_{BE} = 0$$

From (3) $-20\,(0) - 20\,(-10) - 20\, t_{BC} + (-200) = 0$

$$-20\, t_{BC} = 200 - 200$$

$$t_{BC} = 0$$

From (1) $-15\, t_{BA} - 15(0) = 0;\ t_{BA} = 0$

Joint C :

$$\Sigma F_X = 0\, t_{CF} - 15\, t_{CE} - 15\, t_{CA} + 0\, t_{CB} - 15\, t_{CD} = 0 \quad ...(4)$$

$$\Sigma F_Y = 15\, t_{CF} + 15\, t_{CE} + 0\, t_{CA} + 0\, t_{CB} + 0\, t_{CD} + R_{CY} = 0 \quad ...(5)$$

$$\Sigma F_Z = 0\, t_{CF} + 0\, t_{CE} + 20\, t_{CA} + 20\, t_{CB} + 0\, t_{CD} = 0 \quad ...(6)$$

$$t_{CA} = 0$$

From (5) $15\,(-10) + 15\, t_{CE} + 350 = 0;\ t_{CE} = -6.67$ kN/m

From (4) $-15\,(-6.67) - 15(0) - 15\, t_{CD} = 0;\ t_{CD} = 6.67$ kN/m

Joint D :

$$\Sigma F_X = 0\, t_{DA} + 0\, t_{DE} + 15\, t_{DC} + R_{DX} = 0$$

$$t_{DC} = -\frac{R_{DX}}{15} = \frac{100}{15} = 6.67 \text{ kN/m}$$

$$\Sigma F_Y = 0\, t_{DA} + 15\, t_{DE} + 0\, t_{DC} + R_{DY} = 0$$

$$t_{DE} = -\frac{R_{Dy}}{15} = \frac{100}{15} = 6.67 \text{ kN/m}$$

$$\Sigma F_Z = 20\, t_{DA} + 0\, t_{DE} + 0\, t_{DC};\ t_{DA} = 0$$

The final forces are listed in Table 7.10.

Example 7.8 *Analyse the space truss shown in Fig. 7.15 by the method of tension coefficients and determine the forces in the members of the truss.*

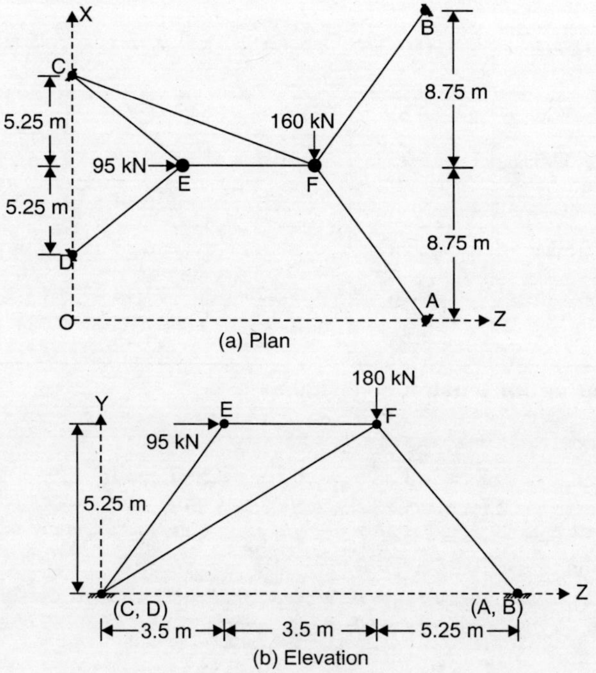

Fig. 7.15

Solution

Table 7.11 Nodal coordinates.

S. No.	*Node*	*Coordinates (m)*		
		X	***Y***	***Z***
1.	A	0	0	12.25
2.	B	17.5	0	12.25
3.	C	14	0	0
4.	D	3.5	0	0
5.	E	8.75	5.25	3.5
6.	F	8.75	5.25	7.0

Check for determinacy: Points A, B, C and D are hinged to the ground.

number of joints $j = 2$

number of members $m = 6$

m should be equal to $3j$ for space trusses emerging from a rigid frame $m = 3 \times 2 = 6$.

Hence, the frame is determinate.

Table 7.12 Member parameters.

Member	$x_{ij}(m)$	$y_{ij}(m)$	$z_{ij}(m)$	$l_{ij}(m)$	$t_{ij}(kN/m)$	$T_{ij} = t_{ij} l_{ij}(kN)$
ED	– 5.25	– 5.25	– 3.5	8.21	0	0
EC	5.25	– 5.25	– 3.5	8.21	0	0
EF	0	0	3.5	3.50	– 27.14	– 95
FC	5.25	– 5.25	– 7.0	10.20	– 6.94	– 70.79
FA	– 8.75	– 5.25	5.25	11.48	– 24.90	– 285.85
FB	8.75	– 5.25	5.25	11.48	– 2.45	– 28.126

Determination of tension co-efficients:

Joint E :

(ED, EF, EC)

$$\Sigma F_X = -5.25\, t_{ED} + 0\, t_{EF} + 5.25\, t_{EC} = 0 \qquad ...(1)$$

$$\Sigma F_Y = -5.25\, t_{ED} + 0\, t_{EF} - 5.25\, t_{EC} = 0 \qquad ...(2)$$

$$\Sigma F_Z = -3.50\, t_{ED} + 3.5\, t_{EF} - 3.50\, t_{EC} + 95 = 0 \qquad ...(3)$$

From (1) and (2) $-10.25\, t_{ED} = 0$

$\Rightarrow$ $t_{ED} = 0$

and $t_{EC} = 0$

Substituting in (3), $3.5\, t_{EF} + 95 = 0$

$$t_{EF} = -\frac{95}{3.5} = -27.14 \text{ kN/m}$$

Joint F :

(FA, FB, FC, FE)

$$\Sigma F_X = -8.75\, t_{FA} + 8.75\, t_{FB} + 5.25\, t_{FC} + 0\, t_{FE} - 160 = 0 \qquad ...(4)$$

$$\Sigma F_Y = -5.25\, t_{FA} - 5.25\, t_{FB} - 5.25\, t_{FC} + 0\, t_{FE} - 180 = 0 \qquad ...(5)$$

$$\Sigma F_Z = 5.25\, t_{FA} + 5.25\, t_{FB} - 7\, t_{FC} - 3.5\, t_{FE} = 0 \qquad ...(6)$$

Solving (5) and (6)

(5) + (6) $\quad -12.25\, t_{FC} - 3.5\, t_{FE} = 180$

$-12.25\, t_{FC} - (3.5 \times (-27.14)) = 180$

$t_{FC} = -6.94$ kN/m

Solving (4) and (5)

(4) × 0.6 $\quad -5.25\, t_{FA} + 5.25\, t_{FB} + 3.15\, t_{FC} = 96 \qquad ...(7)$

(5) × 1 $\quad -5.25\, t_{FA} - 5.25\, t_{FB} - 5.25\, t_{FC} = 180 \qquad ...(8)$

(7) + (8) $\quad -10.50\, t_{FA} - 2.10\, t_{FC} = 276$

$t_{FA} = -24.90$ kN/m

Substituting the value of t_{FA} and t_{FC} in (5)

$$-5.25(-24.90) - 5.25\, t_{FB} - 5.25(-6.94) = 180$$

$$t_{FB} = -2.446 \text{ kN/m}$$

PROBLEMS

1. Using the tension coefficient method, determine the forces in the members of the ball and socket connected truss shown in Fig. 7.16. The cross-sectional areas of members CO and DO are twice the areas of members AO and BO.

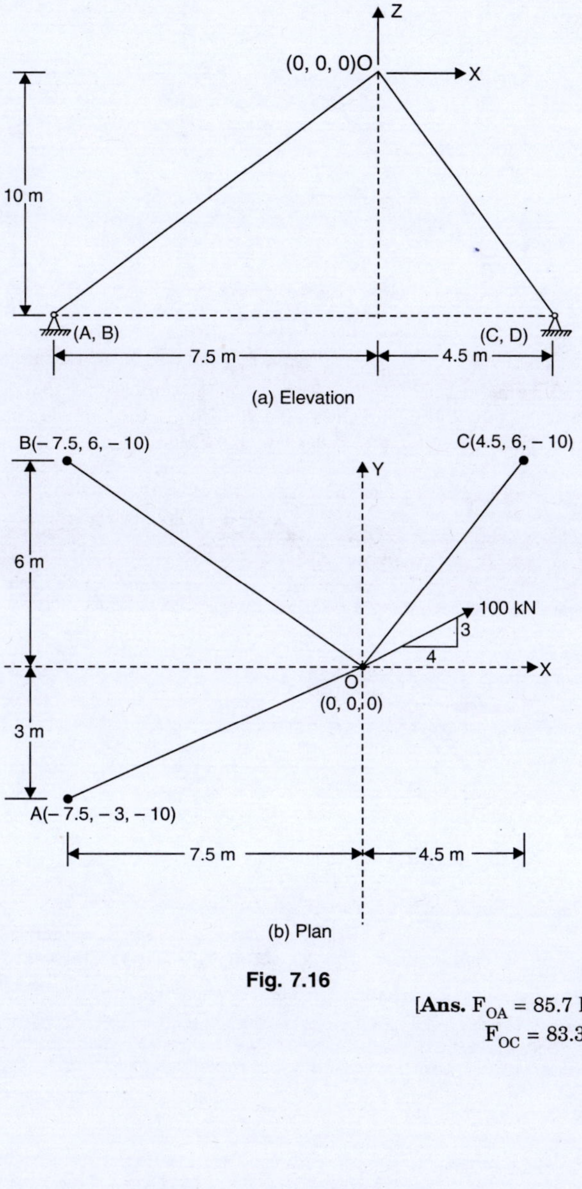

Fig. 7.16

[**Ans.** F_{OA} = 85.7 kN (Tensile); F_{OB} = 0
F_{OC} = 83.33 kN (Compressive)]

2. Figure 7.17 shows the plan of a tripod. The feet A, B and C of the tripod are in the same horizontal plane and the apex D is 3.75 m above the plane. Horizontal loads of 100 kN and 150 kN are applied at D as shown. Using the method of tension coefficients, find the forces in all the members assuming that all joints are ball and socket joints.

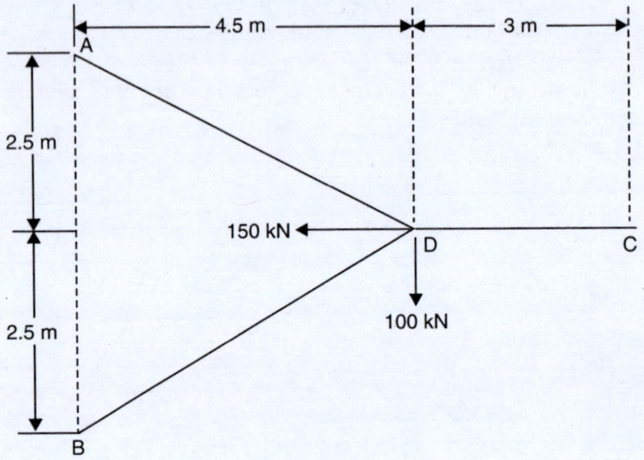

Fig. 7.17

[**Ans.** F_{DA} = 63.7 kN; F_{DC} = 96 kN and F_{DB} = – 191.1 kN]

3. The pyramidal frame in Fig. 7.18 has 7 reactions at the base as shown. Evaluate the reactions and member forces due to the load shown. In addition to the vertical supports at the four ground points, D is restrained along Y and B is restrained along X.

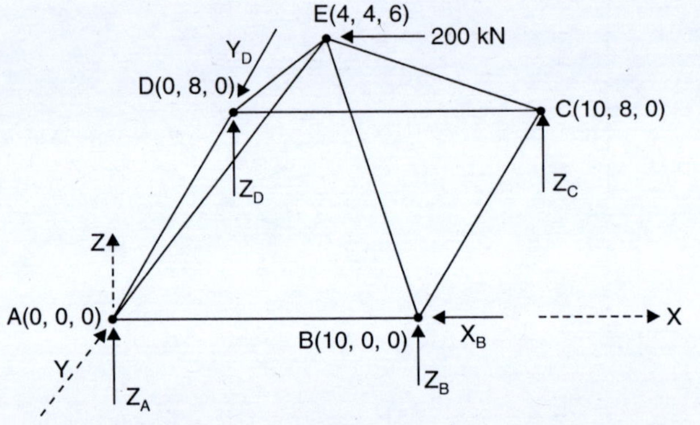

Fig. 7.18

[**Ans.** Z_A = 60; Z_B = – 60; Z_C = 60; Z_D = – 60; Y_A = 0, X_D = 100; X_B = – 100; Forces are : F_{AB} = 40; F_{AE} = – 82.4; F_{AD} = 40; F_{BE} = 94; F_{BC} = – 40; F_{CE} = 94; F_{CD} = – 60; F_{DE} = – 82.4]

4. In the frame shown in Fig. 7.19 determine the forces in members GH and AH. The frame is vertically supported at A, B, C and D (and restrained in X, Y and Z directions at all 4 supports).

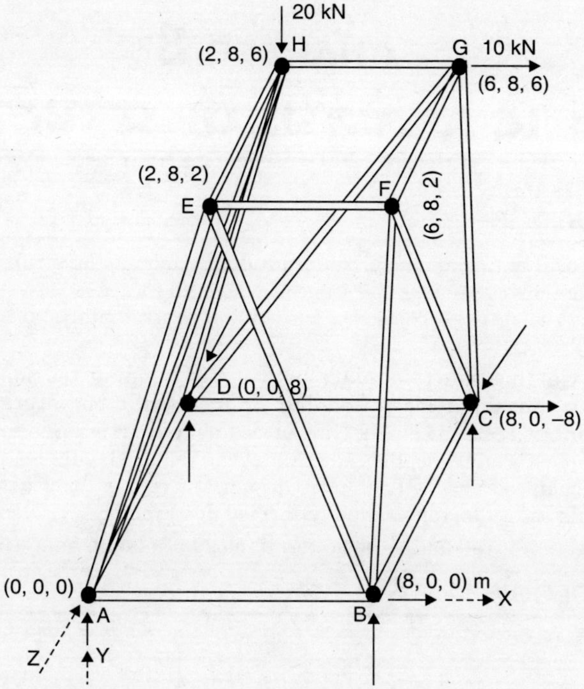

Fig. 7.19

[**Ans.** $F_{GH} = -5$ kN; $F_{AH} = -6.374$ kN]

CHAPTER 8

PLASTIC ANALYSIS OF STRUCTURES

8.1 INTRODUCTION

Most structural materials have, under gradually increasing strain, an elastic stage and a plastic stage. Plastic stage marks the stage at which an increased strain does not produce an increased stress (or resistance). The stress consequent to stretching stabilises at a value known as yield stress(σ_Y).

In beams, yielding can take place both under bending tension and under bending compression. When a beam yields in bending, an increase in curvature does not produce an increase in moment of resistance. Analysis of beams and structures made of such flexural members is called plastic analysis. Plastic analysis can tell us at what load (or load combination) a structure will collapse. This will help us design structures for a desired safety factor on limiting loads. This safety factor is usually termed 'load factor'.

The next section gives us the basic terminology of plastic analysis.

8.2 TERMINOLOGY

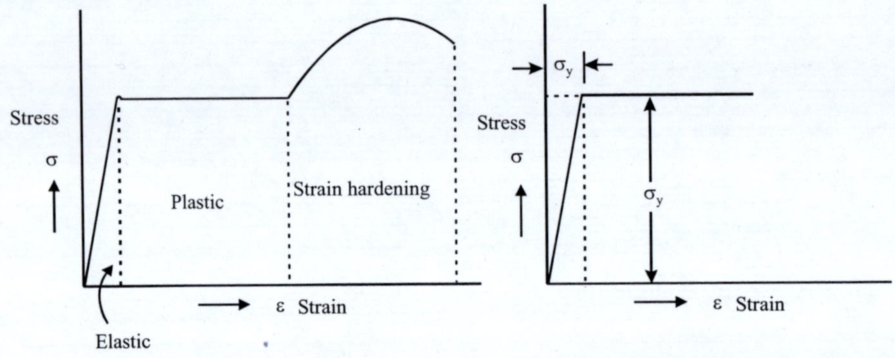

(a) Stress-strain curve for mild steel

(b) Idealised stress-strain curve for mild steel

Fig. 8.1

8.2.1 Plastic Bending of Beams

Consider a section of a beam shown in Fig. 8.2 (a), which is symmetrical about one axis(y-axis). Let the beam be subjected to increasing couples. When the moment is in the elastic range both the top and bottom fibres will be subjected to a stress less than the yield stress Fig. 8.2 (*b*). When the moment is gradually increased, at one stage the most distant fibre (say the bottom fibre) will reach the yield stress Fig. 8.2 (*c*). Now the first yield is said to occur. The moment corresponding to this stage is called the yield moment. *i.e.*, the moment at which the first yield occurs is called the yield moment M_y.

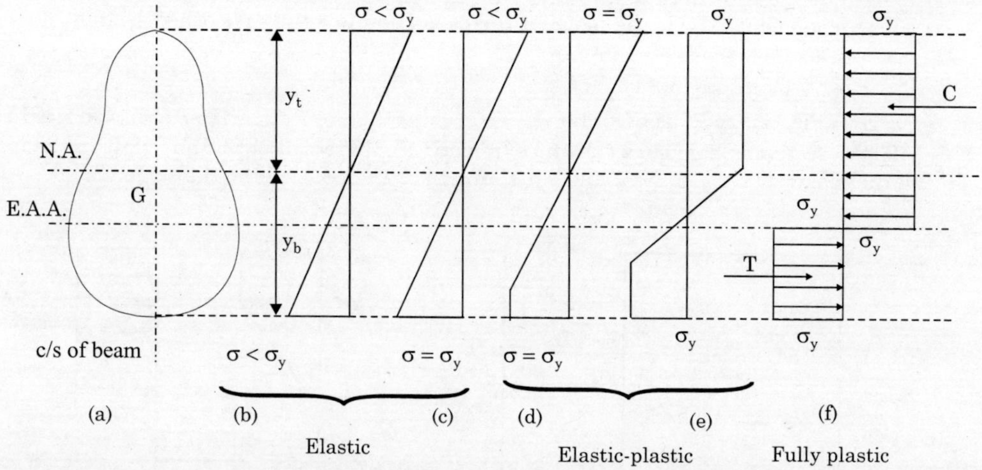

Fig. 8.2

$M_y = \sigma_y.Z$ (vide box)

where, σ_y = Yield stress

Z = Elastic section modulus.

$$\frac{M_y}{I} = \frac{\sigma_y}{y}$$

$$M_y = \sigma_y \times \frac{I}{y}$$

$$M_y = \sigma_y \times Z$$

When the moment is increased beyond the yield moment, the fibre which is already yielded will stretch further but will not take any additional stress. Hence the adjoining inner fibres will start yielding. At a certain stage the top fibre will also reach yield stress Fig. 8.2 (*d*). A shifting of the neutral axis will take place on applying moment any further. If the section is symmetrical about both axes, there will be no shifting of the neutral axis throughout the process.

When the moment is increased further, the yield from both top and bottom will spread towards the inner fibres Fig. 8.2 (*e*)

Figures 8.2 (*d*) and (*e*) correspond to elastic-plastic stage ; *i.e.*, the fibres already yielded will be plastic and the rest of the fibres will be elastic.

When the moment is further increased, there will be a stage at which all fibres from top to bottom of the section will completely yield and the section would not be able to take any further moment. The resisting moment corresponding to this fully plastic stage Fig. 8.2 (*f*) is called the plastic moment M_P.

At the fully plastic stage too, the section will have to obey the following conditions of equilibrium. $\Sigma F = 0$; ΣM = external BM.

$$\Sigma F = 0 \text{ means that } C = T,$$

$$\sigma_y \cdot A_t = \sigma_y \cdot A_C$$

Hence the areas in tension and compression shall be equal.

Plastic moment (M_P) is the moment corresponding to a fully plastified section.

$$M_P = \sigma_y \cdot Z_P$$

where, σ_y = Yield stress, and

Z_P = Plastic modulus of section.

If the section is homogeneous and symmetrical, there will not be any shift in the N.A. during the plastic state. *i.e.*, N.A. and equal area axis (E.A.A.) will be the same.

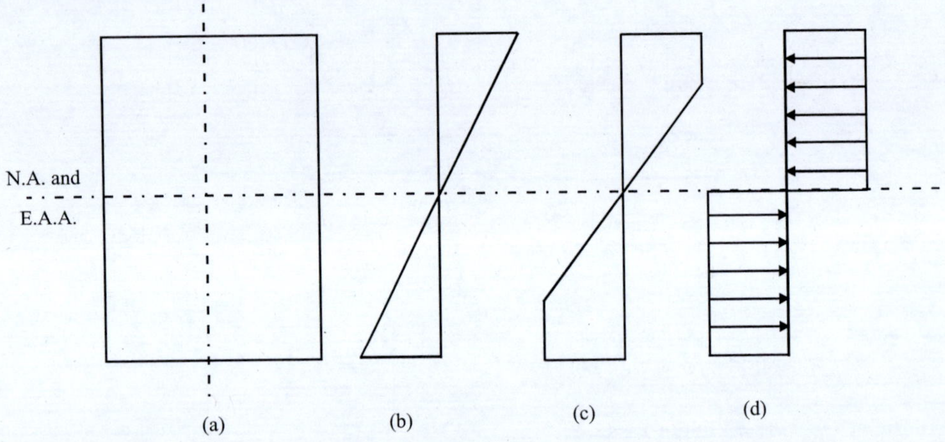

Fig. 8.3

8.2.2 Fully Plastic Moment of the Section

The fully plastic moment (M_P) of a section is defined as the maximum moment of resistance of a fully plastified or yielded cross-section in Fig. 8.2 (*f*) and Fig. 8.3 (*d*).

Assumptions made to evaluate the fully plastic moment of a section

1. Plane transverse sections remain plane and normal to the longitudinal axis after bending, the effect of shear being neglected.
2. Modulus of elasticity has the same value in tension and compression.
3. The material is homogeneous and isotropic in both the elastic and plastic state.
4. There is no resultant axial force on the beam.
   (*i.e.*, Total compression = Total tension)
5. The cross-section of the beam is symmetrical about an axis through its centroid parallel to the plane of bending.
6. Longitudinal fibres are free to expand and contract without affecting the fibres in the lateral direction.

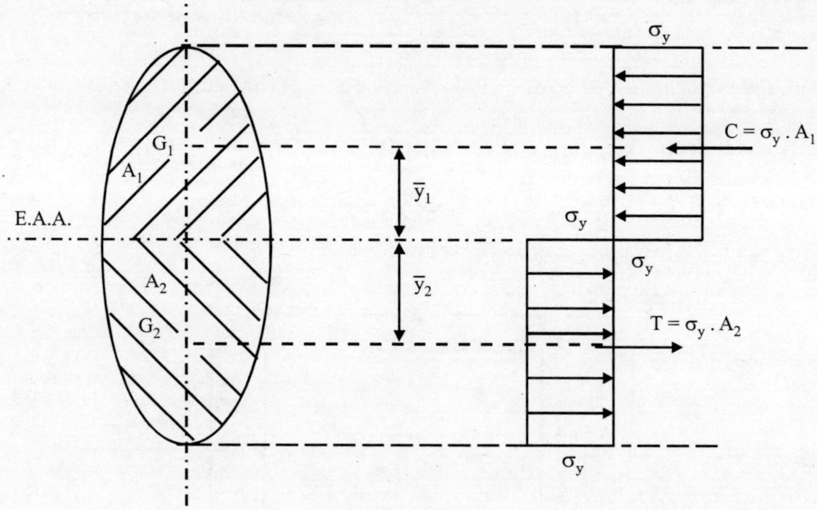

Fig. 8.4

$$C = T, \quad \sigma_y . A_1 = \sigma_y . A_2, \quad A_1 = A_2 = \frac{A}{2}$$

i.e., the area above the N.A. = The area below the N.A.

$\therefore$ The N.A. of the completely plastified section is also the 'Equal Area Axis'.

The equal area axis divides the cross-section into two equal areas.

These two equal and opposite forces, (T and C) form a couple to resist the fully plastic moment.

$$M_P = (\sigma_y.A_1).\bar{y}_1 + (\sigma_y.A_2).\bar{y}_2 = \sigma_y([A_1 . \bar{y}_1] + [A_2 . \bar{y}_2])$$

$\because$ $A_1 = A_2 = A/2$

$$M_P = \sigma_y \times \frac{A}{2}[\bar{y}_1 + \bar{y}_2]$$

$$M_P = \sigma_y . Z_P$$

where, $Z_P = \frac{A}{2}[\bar{y}_1 + \bar{y}_2]$ = Plastic modulus of the section

where, $\bar{y}_1$ = Distance of the C.G. of the area A_1 from the equal area axis.

$\bar{y}_2$ = Distance of the C.G. of the area A_2 from the equal area axis.

The plastic modulus of the section, Z_P is defined as the sum of the first moment of the areas above and below the N.A. or equal area axis. It is the resisting modulus of the completely plastified section.

8.2.3 Shape Factor (S)

The shape factor (S) is defined as the ratio of the plastic moment of a section to the yield moment of the section.

$$S = \frac{M_P}{M_y} = \frac{\sigma_y \times Z_P}{\sigma_y \times Z} = \frac{Z_P}{Z} = \frac{\text{Plastic modulus of section}}{\text{Elastic section modulus}}.$$

i.e., the shape factor is also the ratio of plastic modulus of the section to the elastic modulus of the section.

The shape factor is a geometrical property of a section and it depends only on the shape of the cross-section.

Shape factor for different sections are worked out as below.

Example 8.1 *Rectangular section.*

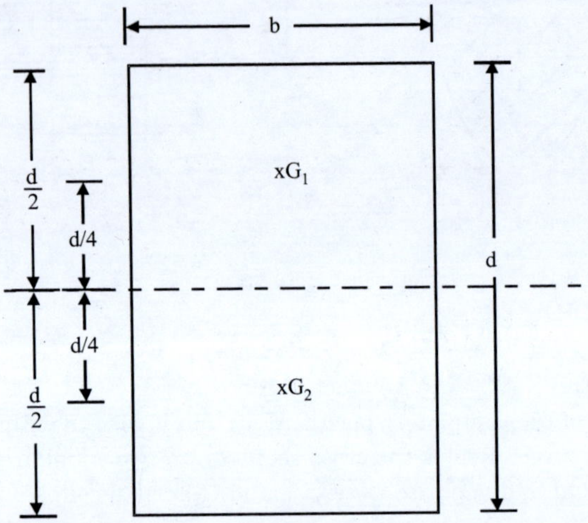

Fig. 8.5

Solution Shape factor,

$$S = \frac{Z_P}{Z}$$

$$Z_P = \text{Plastic modulus of section}$$

$$= \frac{A}{2}(\bar{y}_1 + \bar{y}_2)$$

$$= \frac{bd}{2}(d/4 + d/4) = \frac{bd}{2}(2 \times d/4) = \frac{bd^2}{4}$$

$$Z = \text{Elastic section modulus}$$

$$= \frac{bd^2}{6} \qquad \therefore \qquad S = \frac{Z_P}{Z} = \frac{\frac{bd^2}{4}}{\frac{bd^2}{6}} = 1.5$$

$\therefore$ Shape factor of rectangular section is 1.5.

Example 8.2 *Diamond section.*

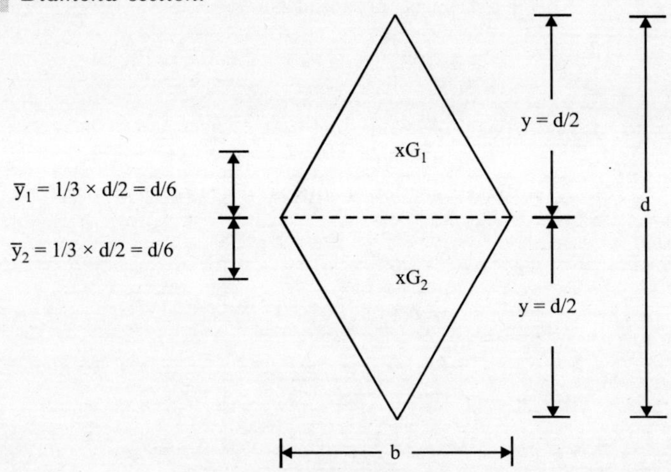

Fig. 8.6

Solution Shape factor, $S = \dfrac{Z_P}{Z}$

Z_P = Plastic modulus of section $= \dfrac{A}{2}(\bar{y}_1 + \bar{y}_2)$

$$A = \left(\frac{1}{2} \times b \times \frac{d}{2}\right) \times 2 = \frac{bd}{2}$$

$\bar{y}_1 = d/6 \qquad \bar{y}_2 = d/6$

$$\therefore \quad Z_P = \frac{(bd/2)}{2}\ (d/6 + d/6) = \frac{bd^2}{12}$$

Z = Elastic section modulus

$$= \frac{I}{y},\ I = \left[\frac{1}{12} \times b \times \left(\frac{d}{2}\right)^3\right] \times 2 = \frac{bd^3}{48},\ Z = \frac{I}{y} = \frac{\dfrac{bd^3}{48}}{\dfrac{d}{2}} = \frac{bd^2}{24}$$

$$S = \frac{Z_P}{Z} = \frac{(bd^2/12)}{(bd^2/24)} = 2$$

∴ Shape factor of diamond section is 2.

Example 8.3 *Triangle.*

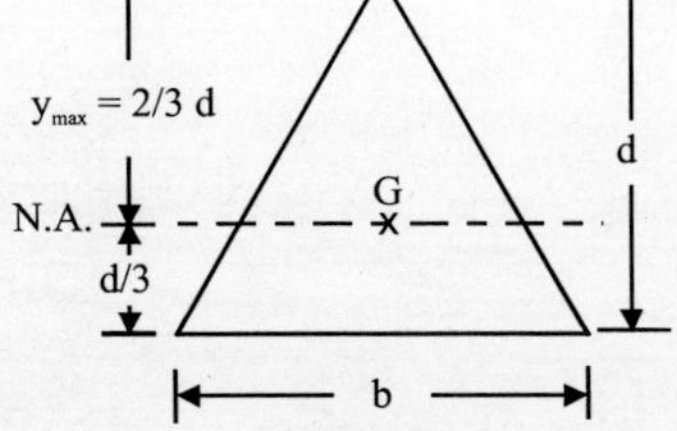

Fig. 8.7

Solution Z = elastic section modulus

$$Z = \frac{I}{y_{max}} = \frac{\left(\frac{bd^3}{36}\right)}{\left(\frac{2d}{3}\right)} = \frac{bd^2}{24}$$

$$Z_P = \text{Plastic section modulus } = \frac{A}{2}(\bar{y}_1 + \bar{y}_2)$$

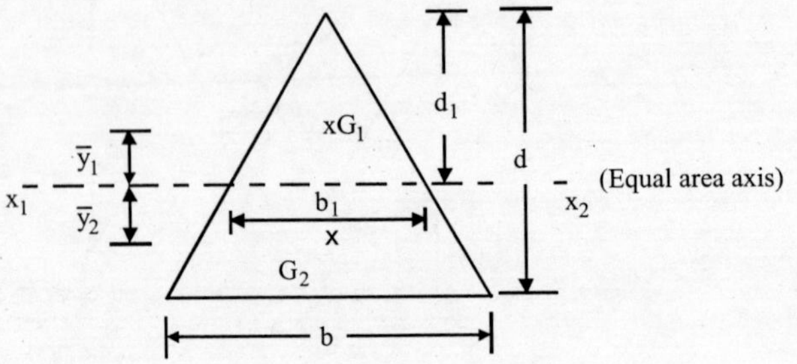

Fig. 8.8

$$\frac{b_1}{b} = \frac{d_1}{d}$$

$$A_1 = A_2 = \frac{A}{2} \qquad A_1 = A/2$$

$$\frac{1}{2}\, b_1 d_1 = \frac{1}{2} \times \frac{1}{2} bd$$

Since $\quad b_1 = \dfrac{d_1 \times b}{d}, \; \dfrac{1}{2} \times \dfrac{d_1 \times b}{d} \times d_1 = \dfrac{1}{4} bd, \quad \dfrac{d_1^{\,2} b}{2d} = \dfrac{1}{4} bd$

$$d_1^{\,2} = \frac{1}{4} bd \times \frac{2d}{b}, \; d_1^{\,2} = d^2/2$$

$\therefore \quad d_1 = \dfrac{d}{\sqrt{2}} \qquad b_1 = \dfrac{d_1 \times b}{d} = \dfrac{d/\sqrt{2} \times b}{d} \qquad b_1 = \dfrac{b}{\sqrt{2}}$

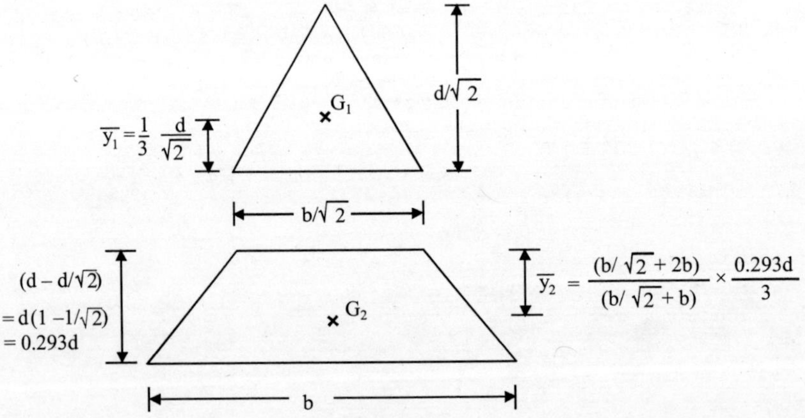

Fig. 8.9

$$\bar{y}_1 = \frac{1}{3} \times \frac{d}{\sqrt{2}} = 0.236d$$

$$\bar{y}_2 = \frac{(b/\sqrt{2} + 2b)}{(b/\sqrt{2} + b)} \times \frac{0.293d}{3} = \frac{\dfrac{b(1+2\sqrt{2})}{\sqrt{2}}}{\dfrac{b(1+\sqrt{2})}{\sqrt{2}}} \times \frac{0.293d}{3}$$

$$= \frac{(1+2\sqrt{2})}{(1+\sqrt{2})} \times \frac{0.293d}{3} = 0.155\ d \qquad Z_P = \frac{A}{2}(\bar{y}_1 + \bar{y}_2)$$

i.e.,

$$Z_P = \frac{(bd/2)}{2}(0.236d + 0.155d)$$

$$= \frac{bd}{2} \times 1/2 \times 0.391d = 0.09775\ bd^2$$

Shape factor, $S = \dfrac{Z_P}{Z} = \dfrac{0.09775\,bd^2}{bd^2/24}$

$S = 2.346.$

Example 8.4 *Triangle.*

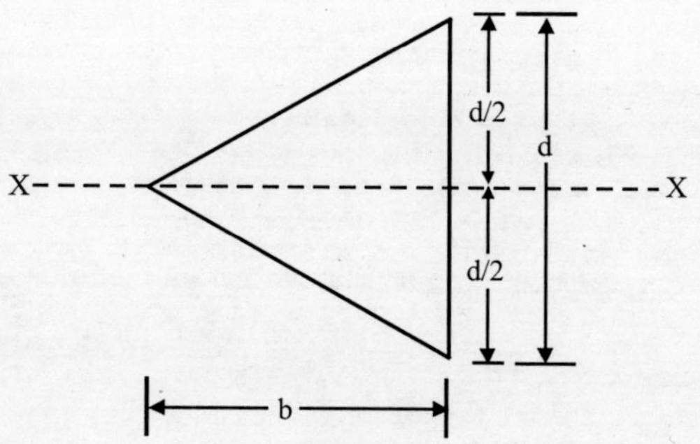

Fig. 8.10

Solution Shape factor, $S = \dfrac{Z_P}{Z}$

Z = Elastic section modulus

$$= \frac{I}{y} = \frac{\left[\dfrac{1}{12} \times b \times \left(\dfrac{d}{2}\right)^3\right] \times 2}{\left(\dfrac{d}{2}\right)} = \frac{1}{12} \times \frac{bd^3}{8} \times 2 \times 2/d = \frac{bd^2}{24}$$

Z_P = Plastic modulus of section $= \dfrac{A}{2}(\bar{y}_1 + \bar{y}_2)$

$$A = \left(\frac{1}{2} \times b \times \frac{d}{2}\right) \times 2 = \frac{bd}{2}$$

$\bar{y}_1 = \bar{y}_2 = d/6$

$$Z_P = \frac{(bd/2)}{2} \times (d/6 + d/6) = \frac{bd}{2\times 2}\frac{(d+d)}{6} = \frac{bd}{4}\left[\frac{2d}{6}\right] = \frac{bd^2}{12}$$

$$S = \frac{Z_P}{Z} = \frac{bd^2/12}{bd^2/24} = \frac{bd^2}{12} \times \frac{24}{bd^2}$$

$S = 2.$

Example 8.5 *Circular section.*

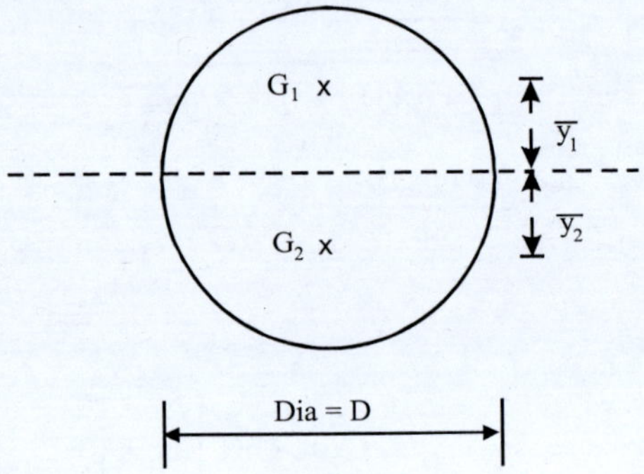

Fig. 8.11

Solution Shape factor,

$$S = \frac{Z_P}{Z}$$

Z = Elastic section modulus

$$= \frac{I}{y} = \frac{(\pi D^4/64)}{(D/2)} = \frac{\pi D^4}{64} \times \frac{2}{D} = \frac{\pi D^3}{32}$$

Z_P = Plastic section modulus

$$= \frac{A}{2}(\bar{y}_1 + \bar{y}_2),\ A = \frac{\pi D^2}{4}$$

$$\bar{y}_1 = \bar{y}_2 = \frac{4r}{3\pi} = \frac{2\times 2r}{3\pi} = \frac{2D}{3\pi}$$

$$\therefore \quad Z_P = \frac{(\pi D^2/4)}{2}\left(\frac{2D}{3\pi} + \frac{2D}{3\pi}\right) = \frac{\pi D^2}{4} \times \frac{1}{2} \times \frac{4D}{3\pi} = \frac{D^3}{6}$$

$$\therefore \quad S = \frac{Z_P}{Z} = \frac{\frac{(D^3)}{6}}{\frac{(\pi D^3)}{32}} = \frac{(D^3)}{6} \times \frac{32}{\pi D^3} = \frac{32}{6\pi}$$

$S = 1.697.$

Example 8.6 *A mild steel I-section 200 mm wide and 250 mm deep has a mean flange thickness of 20 mm and a web thickness of 10 mm. Calculate the shape factor. Find the fully plastic moment if $\sigma_y = 252\ N/mm^2$.*

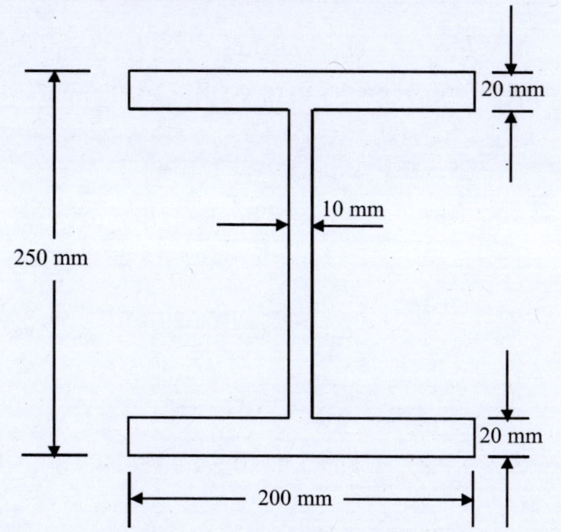

Fig. 8.12

Solution Shape factor, $S = ?$

Fully plastic moment, $M_P = ?$

$$\sigma_y = 252 \text{ N/mm}^2$$

$$S = \frac{Z_P}{Z}$$

$$Z = \text{Elastic section modulus} = \frac{I}{y}$$

$$I = \frac{200 \times 250^3}{12} - \frac{190 \times 210^3}{12} = 11378.42 \times 10^4 \text{ mm}^4$$

$$y = \frac{D}{2} = \frac{250}{2} = 125 \text{ mm}$$

$$\therefore \quad Z = \frac{11378.42 \times 10^4}{125} = 910.273 \times 10^3 \text{ mm}^3$$

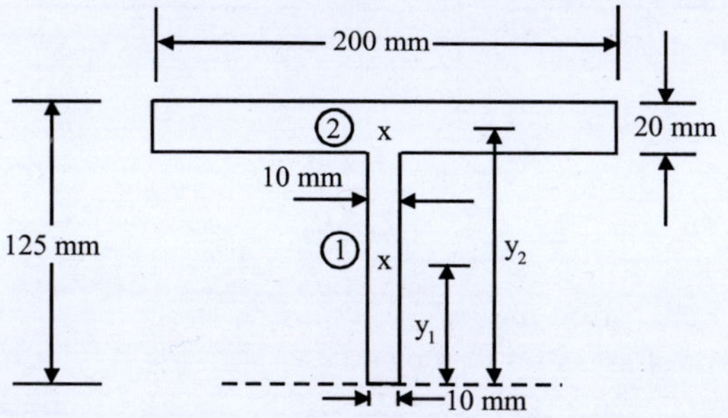

Fig. 8.13

$$Z_P = \text{Plastic section modulus} = \frac{A}{2}(\bar{y}_1 + \bar{y}_2)$$

$$A = (200 \times 20)\,2 + (210 \times 10) = 10100 \text{ mm}^2$$

$$\bar{y}_1 = \bar{y}_2 = \frac{a_1 y_1 + a_2 y_2}{a_1 + a_2}$$

$$= \frac{(105 \times 10 \times 105/2) + [200 \times 20 \times (125 - 20/2)]}{(105 \times 10) + (200 \times 20)} = 102 \text{ mm}$$

$$\therefore \quad Z_P = \frac{10100}{2}[102 + 102] = 1030.2 \times 10^3 \text{ mm}^3$$

$$\therefore \quad S = \frac{1030.2 \times 10^3}{910.273 \times 10^3} = 1.132$$

$$M_P = \sigma_y \times Z_P = 252 \times 1030.2 \times 10^3 = 259.6 \times 10^6 \text{ N mm}$$

Shape factor = 1.132

Fully plastic moment = 259.6×10^6 N mm.

Example 8.7 *Find the shape factor of the I-section given in Fig. 8.14.*

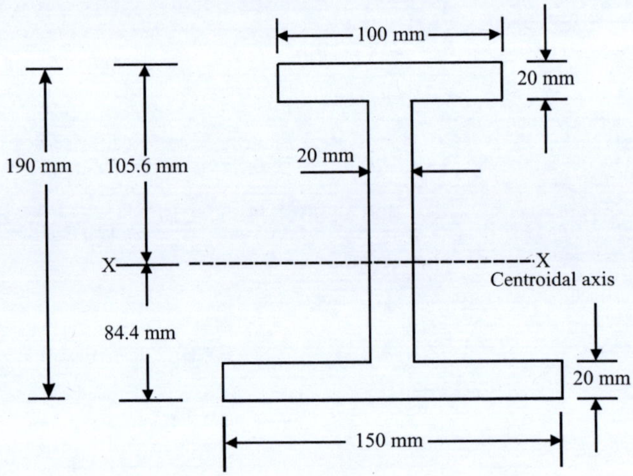

Fig. 8.14

Solution $S = \frac{Z_P}{Z}, \quad Z = \frac{I}{y_{max}}$

Location of centroidal axis :

Taking moment of areas about top,

$$y = \frac{(100 \times 20 \times 10) + [150 \times 20 \times (20 + 150/2)] + [150 \times 20 \times (20 + 150 + 20/2)]}{(100 \times 20) + (150 \times 20) + (150 \times 20)}$$

$$= 105.6 \text{ mm}$$

Moment of inertia, I

$$I = \frac{100 \times 20^3}{12} + (100 \times 20) \times (105.6 - 10)^2 + \frac{20 \times 150^3}{12} + (20 \times 150)$$

$$\times (105.6 - 95)^2 + \frac{150 \times 20^3}{12} + (150 \times 20) \times (84.4 - 10)^2$$

$$= 4101.35 \times 10^4 \text{ mm}^4$$

$$y_{max} = 105.6 \text{ mm}$$

$$\therefore \quad Z = \frac{I}{y_{max}} = \frac{4101.35 \times 10^4}{105.6} = 388.39 \times 10^3 \text{ mm}^3$$

$$Z_P = \text{Plastic section modulus} = \frac{A}{2}(\bar{y}_1 + \bar{y}_2)$$

$$A = (100 \times 20) + (150 \times 20) + (150 \times 20) = 8000 \text{ mm}^2$$

Equal area axis :

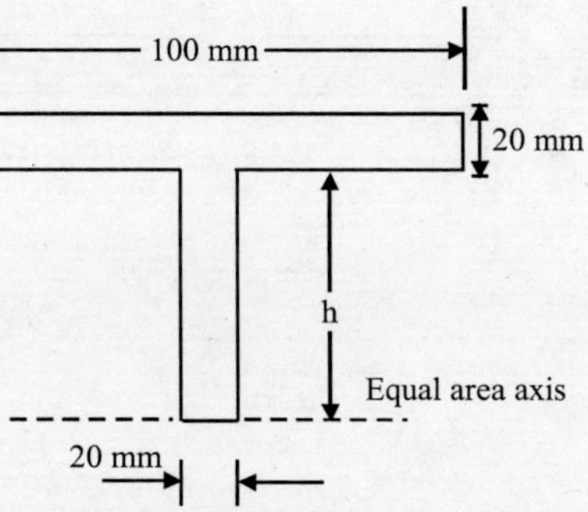

(a)

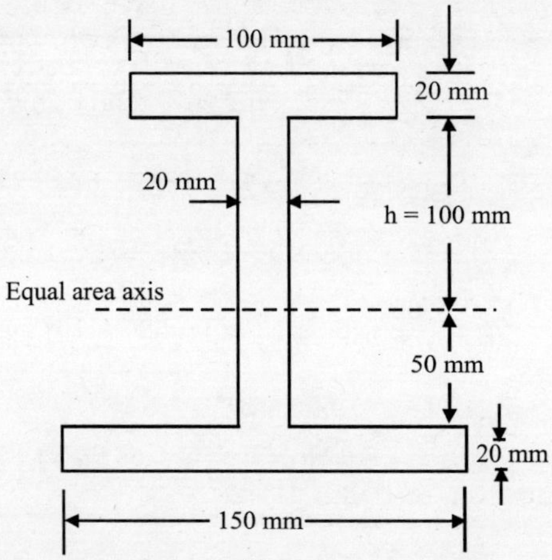

(b)

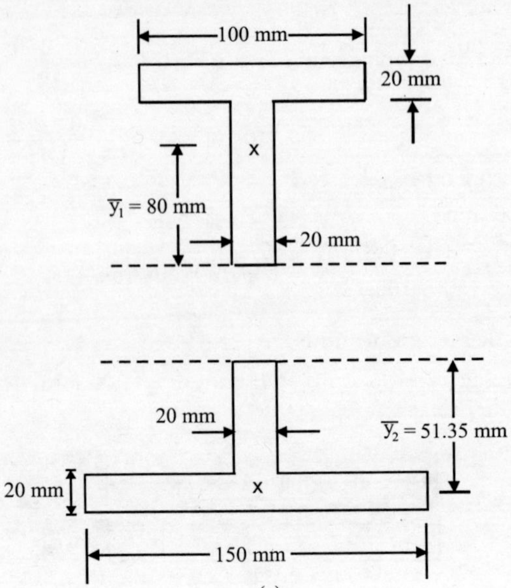

(c)

Fig. 8.15

$$\frac{8000}{2} = (100 \times 20) + 20 \times h$$

$$\therefore \quad h = \frac{\left[\dfrac{8000}{2} - (100 \times 20)\right]}{20} = 100 \text{ mm}$$

$$\overline{y}_1 = \frac{(100 \times 20) \times \left(\dfrac{100}{2}\right) + (100 \times 20) \times \left(100 + \dfrac{20}{2}\right)}{(100 \times 20) + (100 \times 20)} = 80 \text{ mm}$$

$$\overline{y}_2 = \frac{20 \times 50 \times \dfrac{50}{2} + 20 \times 150 \times \left(50 + \dfrac{20}{2}\right)}{(20 \times 50) + (20 \times 150)} = 51.25 \text{ mm}$$

$$\therefore \quad Z_P = \frac{8000}{2} [80 + 51.25] = 525 \times 10^3 \text{ mm}^3$$

$$\therefore \quad S = \frac{Z_P}{Z} = \frac{525 \times 10^3}{388.39 \times 10^3} = 1.352$$

Shape factor of the given section is 1.352.

Example 8.8 *Find the shape factor for the Tee section shown in Fig. 8.16 (a).*

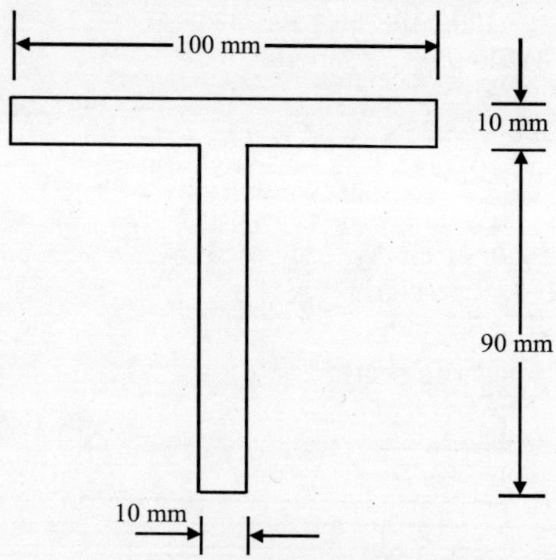

Fig. 8.16 (*a*)

Solution Shape factor, $S = \frac{Z_P}{Z} = \frac{\text{Plastic modulus of section}}{\text{Elastic modulus of section}}$

(*i*) **Elastic modulus of section :**

Location of centroid :

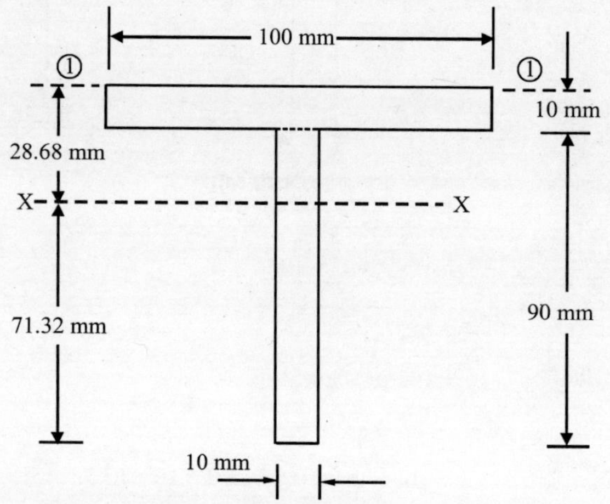

Fig. 8.16 (*b*)

Taking moment of areas about (1)–(1)

$$[100 \times 10 + 90 \times 10]\, y_t = 100 \times 10 \times 5 + 90 \times 10 \times \left(10 + \frac{90}{2}\right)$$

$$y_t = \frac{100 \times 10 \times 5 + 90 \times 10 \times 55}{100 \times 10 + 90 \times 10}$$

$$y_t = \frac{54500}{1900} = 28.68 \text{ mm}$$

$$y_b = 100 - 28.68 = 71.32 \text{ mm}$$

$$I_{XX} = \frac{100 \times 10^3}{12} + 100 \times 10 \times (23.68)^2 + \frac{10 \times 90^3}{12} + 10 \times 90 \times (71.32 - 45)^2$$

$$= 1800044 \text{ mm}^4$$

$$Z = \frac{I}{y_{max}} = \frac{1800044}{71.32} = 25239 \text{ mm}^3.$$

(*ii*) **Plastic modulus of the section :**

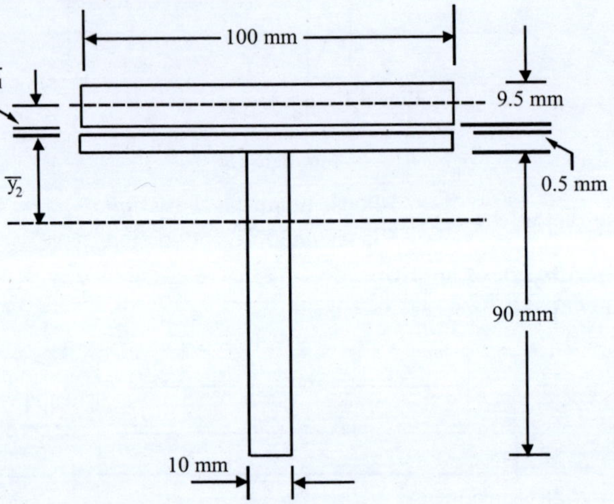

Fig. 8.16 (*c*)

Equal area axis

$$\frac{\text{Total area}}{2} = \text{Width of flange} \times h$$

$$\frac{1900}{2} = 100\ h$$

$$h = 9.5 \text{ m (from top)}$$

$$\overline{y}_1 = \frac{9.5}{2} = 4.75 \text{ mm}$$

$$\overline{y}_2 = \frac{100 \times 0.5 \times 0.25 + 90 \times 10 \times (0.5 + 45)}{100 \times 0.5 + 90 \times 10} = 43.12 \text{ mm}$$

$$Z_P = \frac{A}{2}(\overline{y}_1 + \overline{y}_2) = \frac{1900}{2}(4.75 + 43.12) = 45476.5 \text{ mm}^3$$

Shape factor $\quad S = \frac{Z_P}{Z} = \frac{45476.5}{25239} = 1.80$

$$S = 1.80.$$

Example 8.9 *Determine the shape factor of the section shown in Fig. 8.17.*

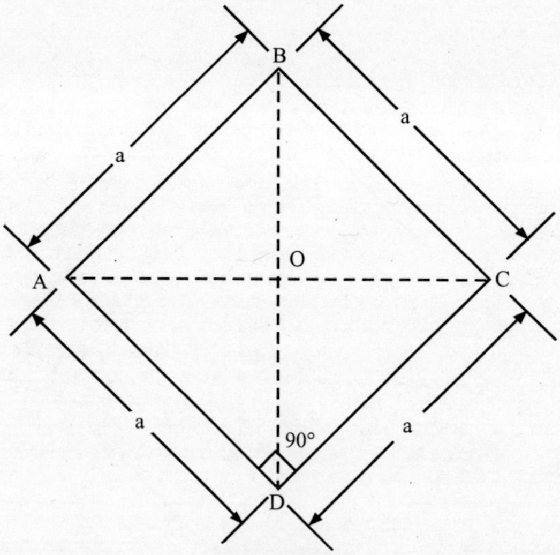

Fig. 8.17

Solution Shape factor = $\dfrac{\text{Plastic section modulus}}{\text{Elastic section modulus}}$

Plastic section modulus :

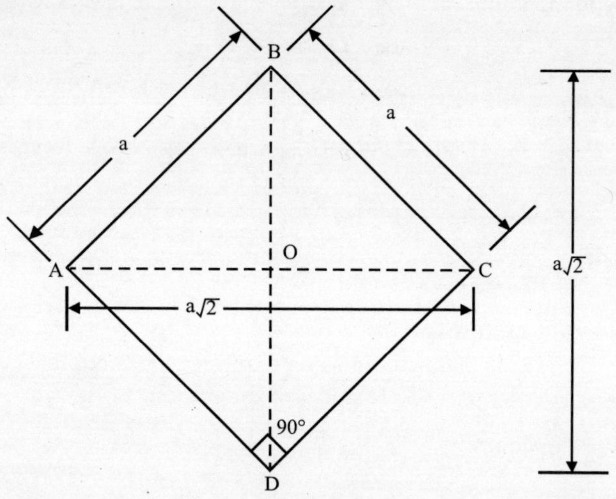

Fig. 8.17 (*a*)

Length of the diagonal AC = $a\sqrt{2}$ = BD

$$OA = OB = OC = OD = \frac{a\sqrt{2}}{2} = \frac{a}{\sqrt{2}} = h$$

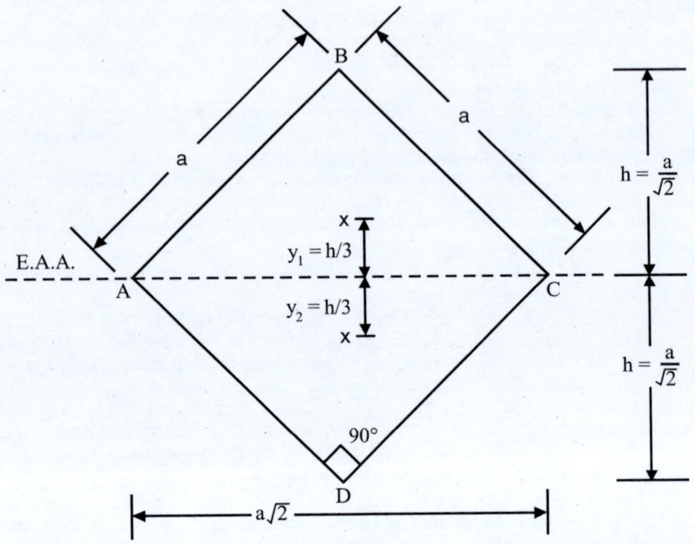

Fig. 8.17 (*b*)

$$\text{Plastic section modulus} = \frac{A}{2}(\bar{y}_1 + \bar{y}_2)$$

A = Total area of C.S. = a^2

$\bar{y}_1$ = Distance of C.G. of area above E.A.A. from E.A.A.= $\frac{h}{3}$

$\bar{y}_2$ = Distance of C.G. of area below E.A.A. from E.A.A. = $\frac{h}{3}$

$$Z_P = \frac{a^2}{2}\left[\frac{h}{3} + \frac{h}{3}\right] = \frac{a^2}{2}\left[\frac{2h}{3}\right] = \frac{a^2}{2}\left[\frac{2}{3} \times \frac{a}{\sqrt{2}}\right] \qquad \left[\because\ h = \frac{a}{\sqrt{2}}\right]$$

$$Z_P = \frac{a^3}{3\sqrt{2}}$$

Elastic section modulus :

$$Z = \frac{\text{Moment of inertia of the section about its C.G}}{\text{Distance of the extreme fibre}}$$

$$Z = \frac{I}{y}$$

$$\text{Moment of inertia of a triangle about its base} = \frac{bh^3}{12}$$

where, b = Base, h = Height.

Here $b = AC = a\sqrt{2}$

$$h = OB = \frac{a}{\sqrt{2}}$$

$$I = 2\left[\frac{a\sqrt{2}\left(\frac{a}{\sqrt{2}}\right)^3}{12}\right] = \frac{\frac{a\sqrt{2} \times a^3}{2\sqrt{2}}}{6}, \; I = \frac{a^4}{12}$$

$$y = h = \frac{a}{\sqrt{2}}$$

$$Z = \frac{I}{y} = \frac{a^4/12}{a/\sqrt{2}}, \quad Z = \frac{a^3}{6\sqrt{2}}$$

Shape factor :

$$S = \frac{Z_P}{Z} = \frac{a^3/3\sqrt{2}}{a^3/6\sqrt{2}} = 2$$

$S = 2.$

Example 8.10 *For the beam section shown in Fig. 8.18 determine the shape factor and fully plastic moment.*

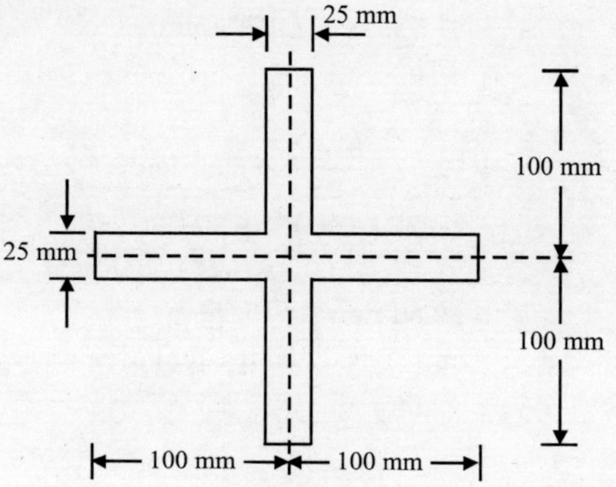

Fig. 8.18

Solution **Plastic section modulus :**

$$Z_P = \frac{A}{2}(\bar{y}_1 + \bar{y}_2)$$

A = Total area of cross-section

$= (200 \times 25) + (200 - 25) \times 25$

$A = 9375 \text{ mm}^2$

$\bar{y}_1$ = Distance of C.G. of area above E.A. axis

$\bar{y}_2$ = Distance of C.G. of area below E.A. axis

To find y_1 and y_2 :

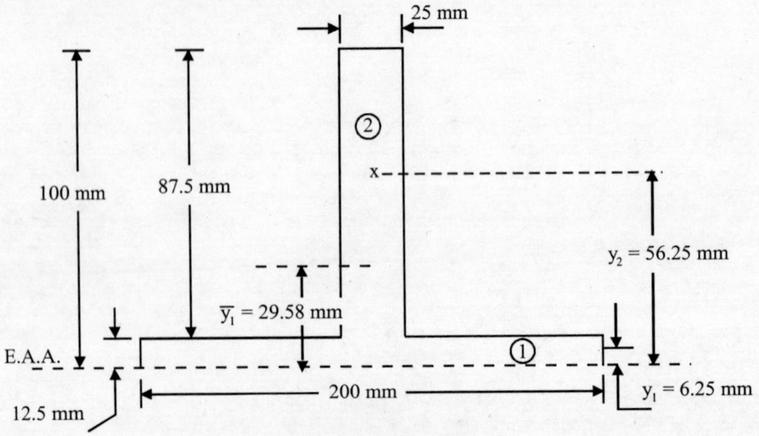

Fig. 8.18 (*a*)

$$\overline{y}_1 = \frac{A_1 y_1 + A_2 \times y_2}{A_1 + A_2}$$

$$A_1 = 200 \times 12.5 = 2500 \text{ mm}^2$$

$$A_2 = 87.5 \times 25 = 2187.5 \text{ mm}^2$$

$$y_1 = \frac{12.5}{2} = 6.25 \text{ mm}$$

$$y_2 = \left(12.5 + \frac{87.5}{2}\right) = 56.25 \text{ mm}$$

$$\overline{y}_1 = \frac{2500 \times 6.25 + 2187.5 \times 56.25}{(2500 + 2187.5)}$$

$$\overline{y}_1 = 29.58 \text{ mm}$$

Since the area of C.S. is symmetrical about equal area axis, $\overline{y}_1 = \overline{y}_2$

$$\overline{y}_2 = 29.58 \text{ mm}$$

$$Z_P = \frac{9375}{2}(29.58 + 29.58)$$

$$Z_P = 277313 \text{ mm}^3$$

Elastic section modulus :

$$Z = \frac{I}{y}$$

$$I = \left[\frac{bd^3}{12}\right] \times 2 + \frac{BD^3}{12} = \left[\frac{87.5 \times (25)^3}{12}\right] \times 2 + 25 \times \frac{(200)^3}{12}$$

$$I = 16894531.25 \text{ mm}^4$$

$$y = \frac{200}{2} = 100 \text{ mm}$$

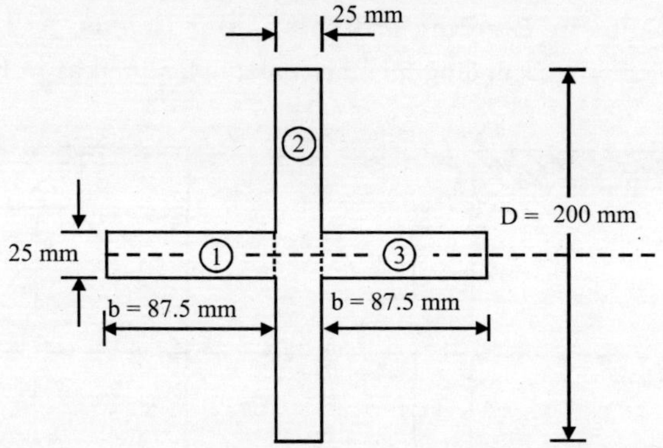

Fig. 8.18 (*b*)

$$Z = \frac{I}{y} = 168945.31 \text{ mm}^3$$

(*i*) Shape factor $= \frac{Z_P}{Z} = \frac{277313}{168945} = 1.641$

(*ii*) Plastic moment $= Z_P \times \sigma_y = 277344 \times 250 = 69.336 \times 10^6$ N mm = 69.336 kNm.

8.2.4 Plastic Hinge

Fully plastic moment is considered to have developed at any section of a structure subjected to a system of loads, when the section is completely yielded or plastified. The fibres on one side of the equal area axis of the section are in compression and tend to contract. The fibres on the other side of the axis are in tension and tend to expand. The section acts like a hinge. This hinge is known as a 'Plastic Hinge'.

The plastic hinge is defined as an yielded zone due to bending in a structural member, at which large rotations can take place at a section at a constant plastic moment, M_P.

8.2.5 Mechanism

A stable structure shall be able to resist displacement. There has to be a force corresponding to any displacement. A displacement without resistance is called a rigid body displacement. Structures have elastic displacements. Mechanisms have rigid body displacements.

When an n degree indeterminate structure develops n (plastic) hinges, it becomes determinate. Formation of an additional hinge will reduce the structure to a mechanism. The load that causes the $(n + 1)^{\text{th}}$ hinge to form is the collapse load. Because once a structure becomes a mechanism, it will collapse.

Most flexural structural frames (such as continuous beams, portals, gables or combinations thereof) can have rigid joints and mechanical hinges. Plastic hinges modify the behaviour of structures much in the same way as mechanical hinges. The only difference is that the plastic hinges permit rotation offering a constant resisting moment M_P to the rotation. So for causing a rotation θ at a plastic hinge, work has to be put in. The work W for causing a rotation θ is equal to $M_P \times \theta$.

[A mechanical hinge can therefore be treated as a plastic hinge with $M_P = 0$].

8.2.6 Elastic-Plastic Bending of Rectangular Beams

Consider a rectangular bending member partly plastified as in Fig. 8.19.

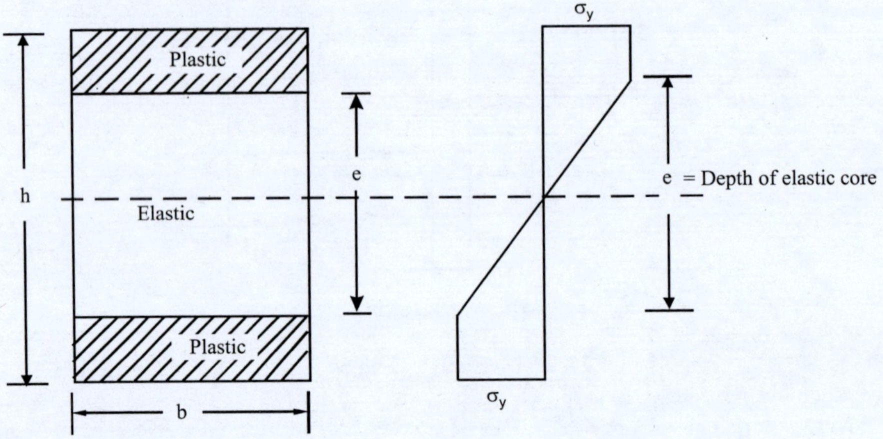

Fig. 8.19

In the elastic-plastic range, shown in Fig. 8.19 a middle layer of depth 'e' remains elastic and it is known as 'Elastic core'. The bending stress in the outer fibres is equal to the yield stress.

B.M. resisted by elastic core, (M_{core})

$$M_C = \sigma_y \times \frac{be^2}{6}$$

B.M. resisted by the fibres in the plastic region, (M_{outer})

$$M_O = \sigma_y \left[\frac{bh^2}{4} - \frac{be^2}{4} \right] \qquad \left[\because \; Z_P = \frac{bd^2}{4} \right]$$

∴ Total moment of resistance for the partially plastified section,

$$M = M_C + M_O = \sigma_y \times \frac{be^2}{6} + \sigma_y \left[\frac{bh^2}{4} - \frac{be^2}{4} \right]$$

$$= \sigma_y \left[\frac{be^2}{6} + \frac{bh^2}{4} - \frac{be^2}{4} \right] = \sigma_y \left[\frac{2be^2 + 3bh^2 - 3be^2}{12} \right]$$

$$= \sigma_y \left[\frac{3bh^2 - be^2}{12} \right] = \sigma_y \left[\frac{bh^2}{4} - \frac{be^2}{12} \right] = \sigma_y \times \frac{bh^2}{4} \left[1 - \frac{e^2}{3h^2} \right]$$

$$= \sigma_y \times Z_P \left[1 - \frac{e^2}{3h^2} \right] \Rightarrow M = M_P \left[1 - \frac{e^2}{3h^2} \right].$$

Example 8.11 *A rectangular beam section b × h is subjected to 0.82 M_P. Determine the depth of the elastic core of the section.*

Solution Given data : $M = 0.82\ M_P$.

$$M = M_P\left[1-\frac{e^2}{3h^2}\right] \qquad 0.82\ M_P = M_P\left[1-\frac{e^2}{3h^2}\right]$$

$$\frac{e^2}{3h^2} = (1-0.82) \qquad \Rightarrow \qquad e = \sqrt{0.54\ h^2}$$

Depth of elastic core, $e = 0.735\ h$.

8.3 TYPES OF MECHANISMS

A. Independent Mechanisms :

In the mechanisms sketched under, a mechanical hinge is denoted by *o* and *a* plastic hinge is denoted by •.

1. **Beam mechanisms :**

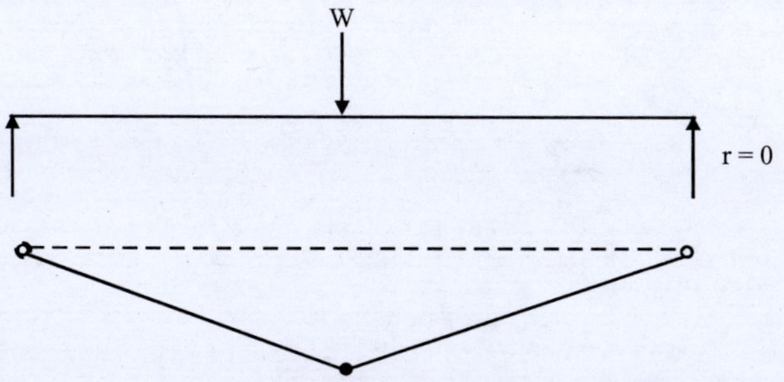

Fig. 8.20. S.S. beam.

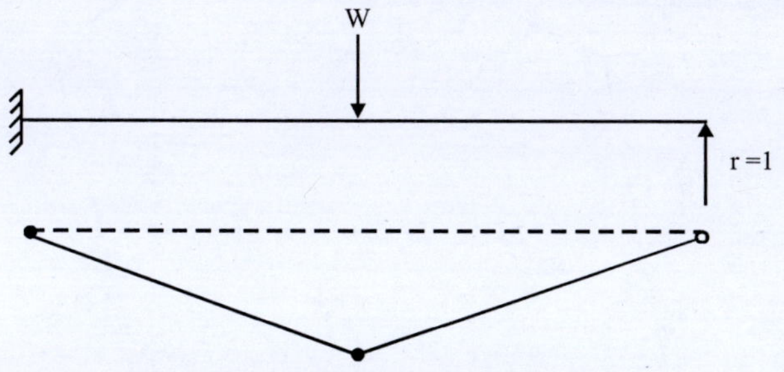

Fig. 8.21. Propped cantilever.

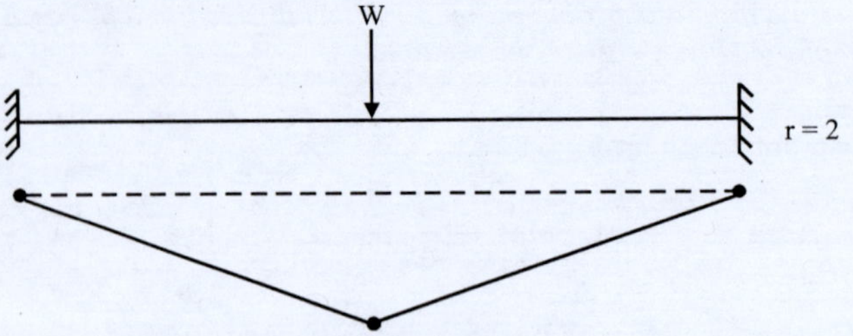

Fig. 8.22. Fixed beam.

2. **Panel mechanism or sway mechanism :**

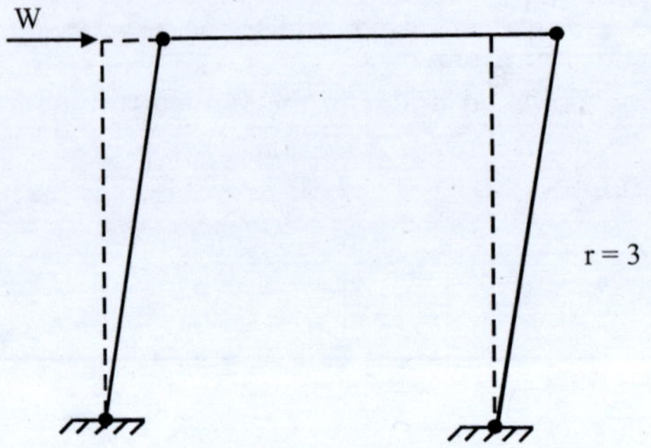

Fig. 8.23. Portal frame.

3. **Gable mechanism :**

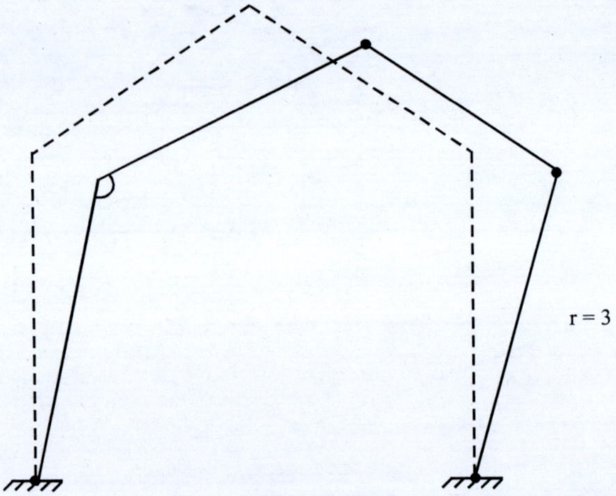

Fig. 8.24. Gable frame.

B. Composite or Combined Mechanism :

This is a combination of two or more of the above independent mechanisms.

8.4 METHODS OF PLASTIC ANALYSIS

8.4.1 Definitions

1. **Static method or Virtual work method.** This method is based on the static or lower bound theorem which states that "A load computed on the basis of an assumed equilibrium bending moment diagram in which the moments are not greater than M_P is less than or at the worst equal to the true ultimate load."

2. **Kinematic method or Mechanism method.** This method is based on the kinematic or upper bound theorem which states that "a load computed on the basis of an assumed mechanism is always be greater than or at best equal to the true ultimate load".

8.4.2 Procedure for Mechanism Method

(*a*) Locate the points of possible plastic hinges. The plastic hinges are likely to be formed at the points under concentrated loads, points of supports, points of connections or joints or points of maximum bending moment.

(*b*) Determine the number of possible independent mechanisms and the combined mechanisms.

If '*r*' is the redundancy of the structure and 'N' is the number of possible plastic hinges, then the number of independent mechanisms,

$$n = N - r$$

(*c*) Give the mechanism a consistent virtual displacement and work out the consequent displacements and rotations at strategic points. This displacement is possible because the loads move in the gravity direction and lose a certain potential energy (external work done, W_E).

The plastic hinges offer a resistance to this downward movement, since the hinges have to suffer rotations consistent with the displacement. The total work done at the plastic hinges is $\Sigma M_P \times \theta$ (internal work done W_I). If the loads are big enough so that $W_E > W_I$, collapse will occur, satisfying the need for providing the internal work W_I. Hence $W_E = W_I$ represents the collapse condition. This equation will tell us how much plastic moment is needed for the structural member to resist collapse. Conversely it will also tell us what is the collapse load for a given M_P in a structure.

(*d*) This procedure has a draw back. If there are '*n*' possible mechanisms, we end up with '*n*' values of ultimate loads. The least ultimate load is the correct one. The corresponding mechanism is the likely one. If we overlook a possible mechanism, we would be overestimating the collapse load.

We can only say that the actual ultimate load cannot be more than the least value obtained by this method. Hence this is called an *upper bound method.* This will give a designer a nagging fear that the structure may collapse at a value lower than one computed by him/her.

8.5 LOAD FACTOR

8.5.1 Introduction

For a designer of steel or RC structural frames, plastic analysis provides a new philosophy of safety. In the elastic design methods, the concern is about keeping the stresses at working loads to a value about half the yield stress. Then we say that the factor of safety is 2.

$$\text{Factor of safety} = \frac{\text{Yield stress}}{\text{Working stress}}$$

The concept of factor of safety is questioned by some experts. (*a*) If the factor of safety is two, does it mean that the structure would be able to take twice the working load with out failing, or reaching yield point stress at some locations ? (*b*) How much additional load can it sustain before it collapses ?

For a steel structure the answer to question (*a*) is 'yes'. For an RC structure (which is not linear elastic, the answer can only be 'may be'.

To find the answer to question (*b*) we have to take recourse to plastic analysis. Here we recognize the fact that even after yielding the structure has a considerable load carrying capacity. For a given load pattern (which increases monotonically) we can evaluate the collapse load W_u. When we divide W_u by a pre decided load factor, we get the working load W. Thus

Load factor $\lambda = W_u / W$.

8.5.2 Why Load Factor

The advantage of using the concept of load factor is mostly psychological. A layman who might be the owner of a structure understands "load" rather than "stress" much more readily. If we have carried out collapse load analysis of a structure, we can tell the client, "The structure is designed to take a total load of 850 kN safely. It will fail only when the load reaches 2550 kN". Try telling this to a layman in terms of working stress and yield stress !

Reaching yield stress is not failure, or collapse. But reaching W_u is collapse, disaster. So the value of load factor has to be much more than factor of safety. For steel structures factor of safety is usually 2, but load factor can be fixed as any thing from 3 to 6, depending upon the importance of the structure, consequences of collapse, expectation of life span etc.

8.5.3 Limitations of Load Factor Concept

The concept of load factor has its limitations.

(*i*) The analysis procedure does not give us any clue if at a load $W_u \div$ load factor, the structure behaves well, meaning, whether the stresses are within limit. So we have to check the stresses at crucial points by conventional elastic method.

(*ii*) It was mentioned in section 8.5.1 that the loads are expected to increase monotonically. This is a peculiar and unrealistic assumption. Suppose there are 3 loads on a structure at working stress level. If the loads are 120, 180 and 300 kN in value, and if the load factor is 4, we expect a load combination of 480 kN, 720 kN and 1200 kN to cause collapse. This implies that all the loads increase at a constant multiple of loads at safe level. It is hard to think why the 300 kN load cannot reach 1400 kg while 120 becomes 480.

The assumption of monotonic increase in loading is a simplistic, naive assumption. But it is convenient and so we stick to it.

Example 8.12 *A beam of span 6 m is to be designed for an ultimate u.d.l. of 25 kN/m. The beam is simply supported at the ends. Design a suitable I section using plastic theory, assuming* $\sigma_y = 250\ N/mm^2$.

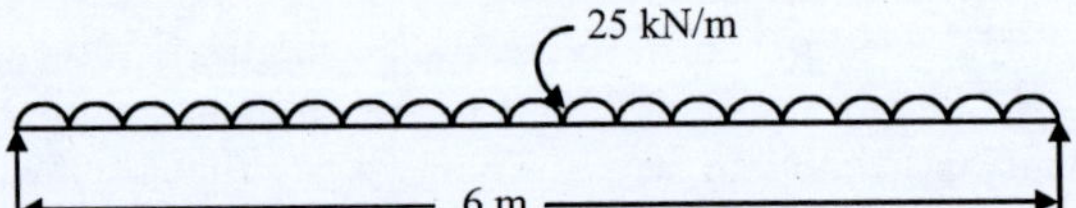

(a)

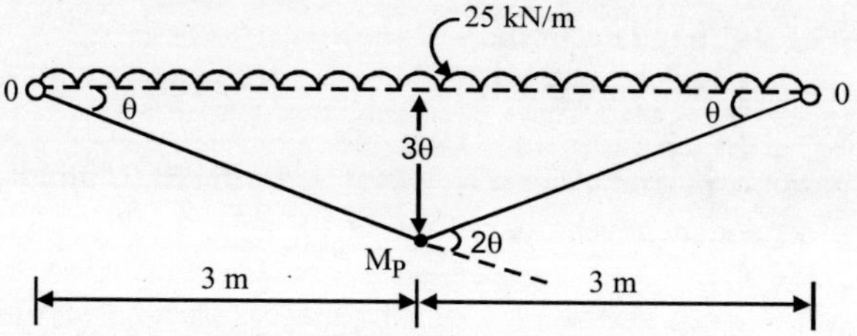

(b)

Fig. 8.25

Solution The given beam is a statically determinate structure. For the formation of a mechanism, only one plastic hinge is needed. The possible location is the mid-span. The mechanism is as shown in Fig. 8.25 (*a*). Let us assume M_P is the required plastic moment.

To check if the given load is capable of causing collapse of the beam by developing a hinge at midspan, let us try giving a trial displacement to the beam Fig. 8.25 (*b*).

Internal work done $= 0 + M_P \times 2\theta + 0 = 2M_P \times \theta$

External work done = Load intensity × Area of triangle under the load

$= 25 \times (\frac{1}{2} \times 6 \times 3\theta) = 225\theta$

Equating I.W.D. to E.W.D., $2M_P\theta = 225\theta$

∴ The required value of $M_P = \dfrac{225\theta}{2\theta} = 112.5$ kNm

We know that, $M_P = \sigma_y \times Z_P$

∴ $$Z_P = \frac{M_P}{\sigma_y} = \frac{1125 \times 10^6 \text{ N.mm}}{250 \text{ N/mm}^2} = 4.5 \times 10^5 \text{ mm}^3$$

Assuming the shape factor for I-section as 1.15

$$S = \frac{Z_P}{Z}$$

∴ $$Z = \frac{Z_P}{S} = \frac{4.5 \times 10^5}{1.15} = 391.304 \times 10^3 \text{ mm}^3$$

∴ Adopt ISLB 275 @ 330 N/m (from steel tables).

Example 8.13 *A continuous beam ABC is loaded as shown in Fig. 8.26. Determine the required M_P if the load factor is 3.2.*

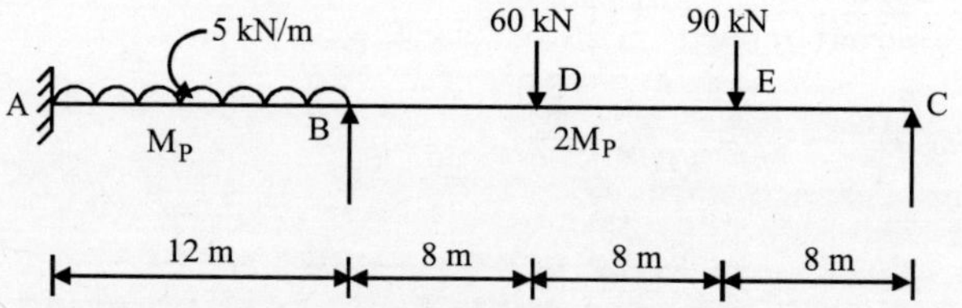

Fig. 8.26

Solution Degree of static indeterminacy of the structure,

r = Vertical reactions + end moments – 2 [For beams with vertical loading only]

= 3 + 1 – 2 = 2

Number of possible plastic hinges, N = 5 (at A, B, in AB, D and E)

Number of independent mechanisms, n = N – r = 5 – 2 = 3

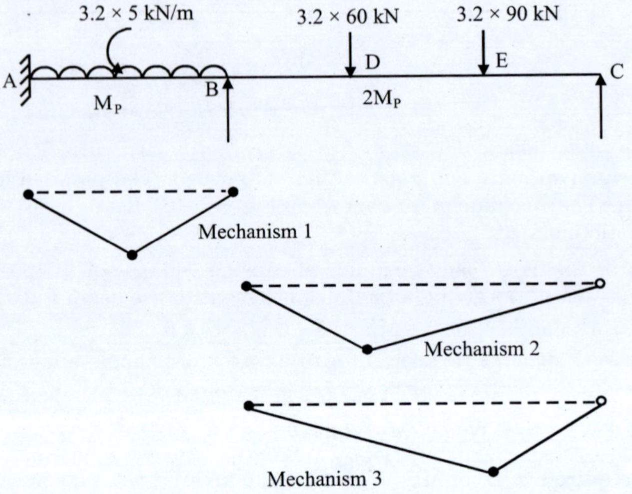

Mechanism (1) :

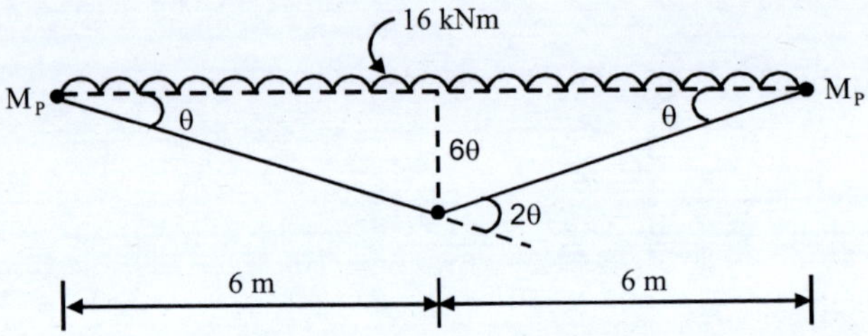

External work done, $W_E = 16 \times 1/2 \times 12 \times 6\theta = 576\ \theta$

Internal work done, $W_I = M_P.\theta + M_P.2\theta + M_P.\theta = 4M_P\theta$

Equating internal and external work done,

$$4M_P\theta = 576\ \theta$$

$$M_P = \frac{576}{4} = 144\ \text{kNm}$$

$$M_P = 144\ \text{kNm} \qquad ...(1)$$

Mechanism (2) :

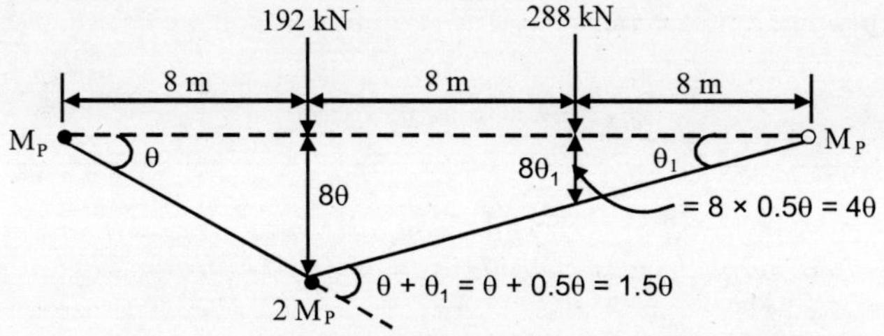

$$8\theta = 16\theta_1$$

$\therefore$ $$\theta_1 = \frac{8\theta}{16} = 0.5\theta$$

External work done, $W_E = 192 \times 8\theta + 288 \times 4\theta = 2688\ \theta$

Internal work done, $W_I = M_P \times \theta + 2M_P \times 1.5\theta + 0 = 4M_P\theta$

Equating internal and external work done,

$$4M_P.\theta = 2688\ \theta$$

$\therefore$ $$M_P = \frac{2688}{4} = 672\ \text{kNm}$$

$$M_P = 672\ \text{kNm} \quad \text{...(2)}$$

Mechanism (3) :

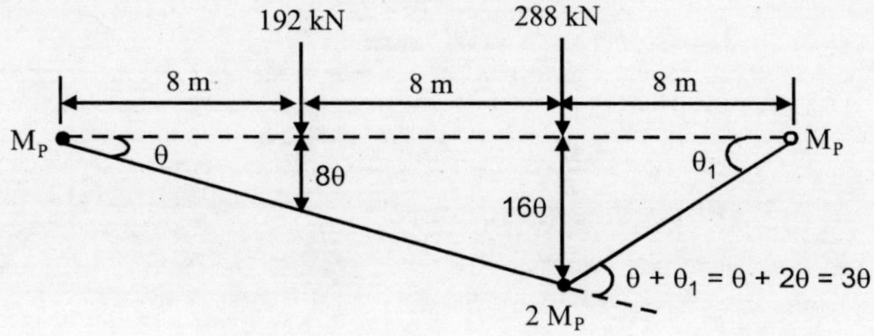

$$16\theta = 8\theta_1 \qquad \therefore \qquad \theta_1 = \frac{16\theta}{8} = 2\theta$$

External work done, $W_E = 192 \times 8\theta + 288 \times 16\theta = 6144\ \theta$

Internal work done, $W_I = M_P \times \theta + 2M_P \times 3\theta + 0 = 7M_P\theta$

Equating external and internal work done,

$$6144\theta = 7M_P\theta$$

$\therefore$ $$M_P = \frac{6144}{7} = 877.71\ \text{kNm}$$

$$M_P = 877.71\ \text{kNm} \quad \text{...(3)}$$

From eqns. (1), (2) and (3)

$\therefore$ The required plastic moment of the beam section shall be, $M_P = 877.71$ kNm (Highest value from eqns. (1), (2) and (3))

Note :

For collapse load. Least value (*i.e.*) the minimum load at which collapse will occur is to be taken.

For fully plastic moment. Highest value *i.e.,* the maximum moment for which the structure is to be designed is to be taken.

Example 8.14 *A beam fixed at both ends is subjected to a udl 'W' on the right half portion. Determine the value of collapse load W_C. The beam is of uniform plastic moment M_P.*

Solution

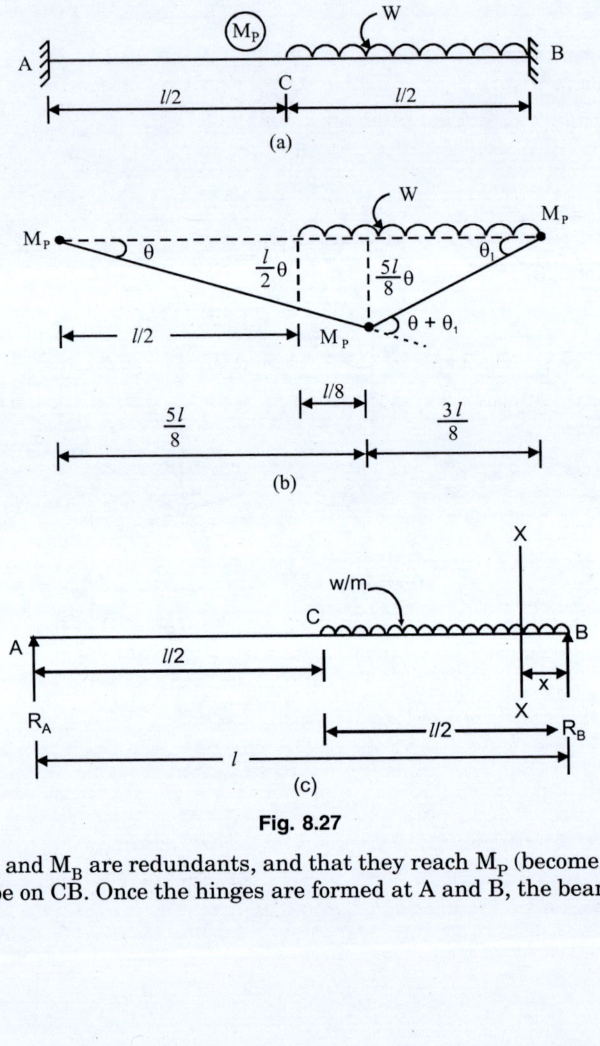

Fig. 8.27

Assume M_A and M_B are redundants, and that they reach M_P (become plastic). The third hinge is likely to be on CB. Once the hinges are formed at A and B, the beam would behave as a S.S. beam.

Taking moments about B, $R_A \times l - W_C \times \frac{l}{4} = 0$

$$R_A \times l = W_C \times \frac{l}{4}$$

$$R_A = \frac{W_C}{4} \quad R_B = \frac{3}{4}W_C$$

Considering a section XX at a distance 'x' from B,

B.M. at XX, $M_x = R_B \times x - \frac{2W_C}{l} \times x \times \frac{x}{2} = \frac{3}{4}W_C \times x - W_C \times \frac{x^2}{l}$

M_x will be maximum when $\frac{dM_x}{dx} = 0$

$$\frac{d}{dx}\left[\frac{3}{4}W_C \times x - W_C \times \frac{x^2}{l}\right] = 0 \qquad \frac{3}{4}W_C - \frac{W_C}{l} \times 2x = 0$$

$$x = \frac{3l}{8}$$

Hence the 3rd hinge will form at a section $\frac{3l}{8}$ from B.

$$\frac{5l}{8} \times \theta = \frac{3l}{8} \times \theta_1 \qquad \frac{5}{8} \times l \times \theta \times \frac{8}{3l} = \theta_1,$$

$$\theta_1 = \frac{5\theta}{3} \qquad \theta + \theta_1 = \theta + \frac{5\theta}{3} = \frac{3\theta + 5\theta}{3} = \frac{8\theta}{3}$$

External work done = Intensity of udl × Area swept by the load.

$$= \left[\frac{2W_C}{l}\right] \times \left[\frac{1}{2} \times \frac{3l}{8} \times \frac{5l\theta}{8} + 1/2\left(\frac{l\theta}{2} + \frac{5l\theta}{8}\right) \times \frac{l}{8}\right]$$

$$= \frac{2W_C}{l} \times \left[\frac{15l^2\theta}{128} + \frac{9l^2\theta}{128}\right] = \frac{2W_C}{l} \times \frac{24l^2\theta}{128} = \frac{3}{8}W_C l\theta$$

Internal work done = $M_P \times \theta + M_P \times \frac{8\theta}{3} + M_P \times \frac{5\theta}{3} = \frac{16M_P \times \theta}{3}$

Equating internal and external work done,

$$\frac{16M_P\theta}{3} = \frac{3W_C\, l\theta}{8}$$

$$M_P = \frac{3W_C l\theta}{8} \times \frac{3}{16\theta} \qquad M_P = \frac{9}{128}W_C \times l$$

The collapse load is, $W_C = \frac{128}{9} \times \frac{M_P}{l}$.

Example 8.15 *A beam fixed at both ends is subjected to two concentrated loads, each at 1/3rd points of the span. Determine the collapse load for the beam in terms of its* M_P.

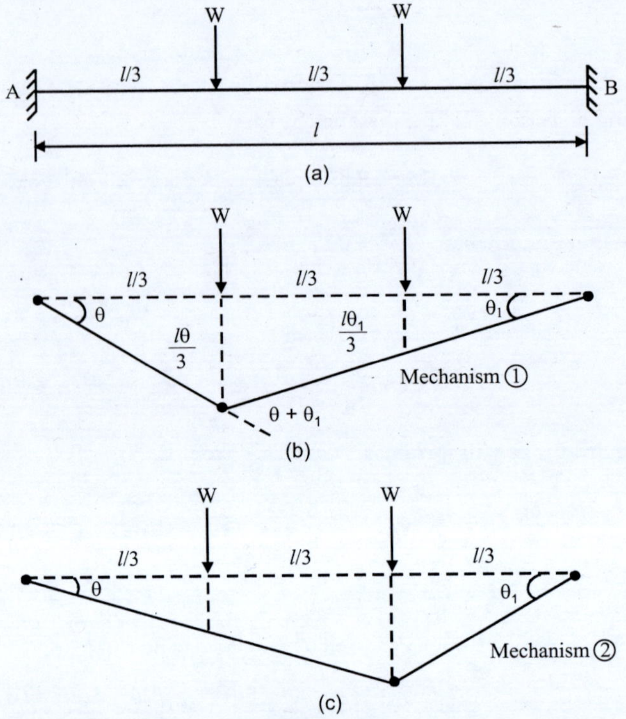

Fig. 8.28

Solution Degree of static indeterminacy = r = 2

Number of possible plastic hinges = N = 4

∴ Number of independent mechanisms, $n = N - r = 4 - 2 = 2$

Mechanism (1) :

$$\frac{l\theta}{3} = \frac{2l\theta_1}{3} \qquad \therefore \quad \theta_1 = \frac{\theta}{2} = 0.5\theta$$

$$\therefore \quad \theta + \theta_1 = \theta + 0.5\theta = 1.5\theta$$

External work done $= W \times \frac{l\theta}{3} + W \times \frac{l}{3} \times 0.5\theta = \frac{1.5\, Wl\theta}{3}$

Internal work done $= M_P \times \theta + M_P \times 1.5\theta + M_P \times 0.5\theta = 3\, M_P \times \theta$

Equating external work done and internal work done,

$$\frac{1.5\, Wl\theta}{3} = 3\, M_P \times \theta$$

$$\therefore \quad W_C = \frac{3M_P \times \theta \times 3}{1.5\, l\theta} = \frac{6M_P}{l}$$

Due to symmetry, in mechanism (2) also, $W_C = \dfrac{6M_P}{l}$

$\therefore$ Collapse load $= \dfrac{6M_P}{l}$.

Example 8.16 *Analyse the propped cantilever loaded as shown and determine the collapse load.*

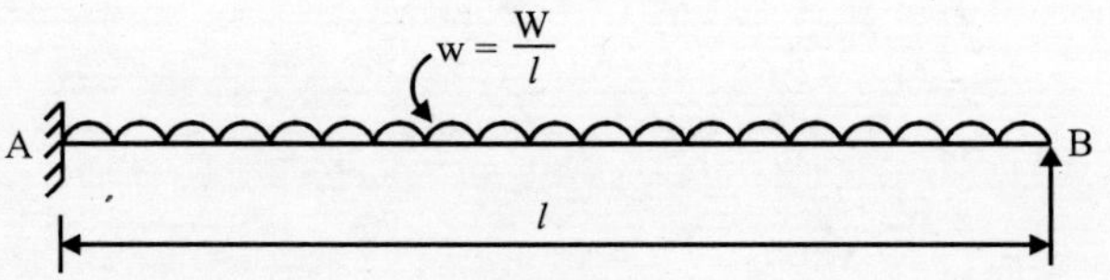

Fig. 8.29

Solution

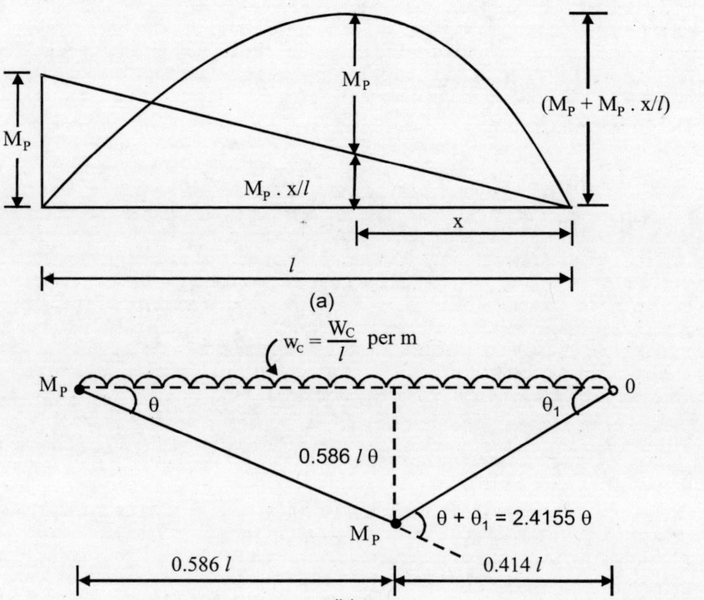

(a)

(b)

Consider the moment at A as redundant and that it reaches M_P. The second hinge will form where the net positive b.m. is maximum.

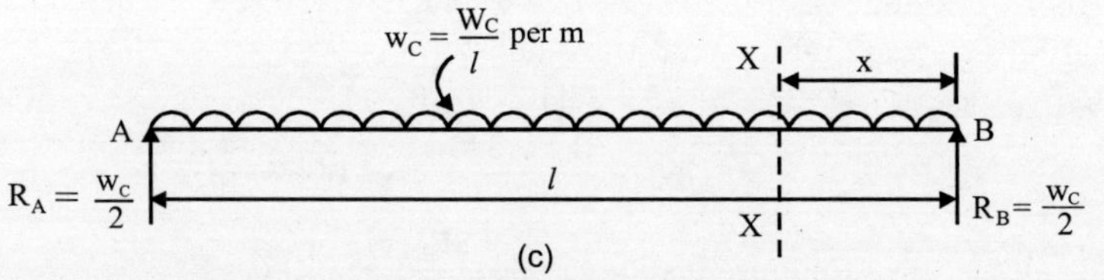

(c)

Fig. 8.30

$$M_x = \frac{W_C \times x}{2} - \frac{W_C \times x^2}{2l}$$

From Fig. 8.30 (*a*) and (*b*),

$$M_P + \frac{M_P \times x}{l} = \frac{W_C \times x}{2} - \frac{W_C \times x^2}{2l}$$

$$M_P \left[1 + \frac{x}{l}\right] = \frac{W_C}{2} \times x - \frac{W_C \times x^2}{2l}$$

$$M_P \left[\frac{l + x}{l}\right] = \frac{W_C}{2} \times x \times \frac{(l - x)}{l}$$

$$M_P = \frac{W_C}{2} \times x \times \frac{(l - x)}{l} \times \frac{l}{(l + x)}$$

$$M_P = \frac{W_C}{2} \times x \frac{(l - x)}{(l + x)}$$

i.e.,
$$M_P = \frac{W_C}{2} \frac{(lx - x^2)}{(l + x)}$$

For 'M_P' to be maximum, $\frac{dM_P}{dx} = 0$

$$\frac{dM_P}{dx} = \frac{W_C}{2}\left[\frac{(l + x)(l - 2x) - (lx - x^2)(1)}{(l + x)^2}\right] = 0$$

i.e.,
$$(l + x)(l - 2x) - (lx - x^2) = 0$$

$$l^2 - 2lx + lx - 2x^2 - lx + x^2 = 0$$

$$l^2 - 2lx - x^2 = 0$$

$$x^2 + 2lx - l^2 = 0$$

$$x = \frac{-2l \pm \sqrt{8l^2}}{2}$$

$$x = 0.414\, l$$

Mechanism : (Fig. 8.30 (*b*))

$$0.586\, l\theta = 0.414\, l\theta_1$$

$\therefore$
$$\theta_1 = \frac{0.586\theta}{0.414} = 1.4155\theta$$

$\therefore$
$$\theta + \theta_1 = \theta + 1.4155\,\theta = 2.4155\,\theta$$

External work done $= \frac{W_C}{l} \times \frac{1}{2} \times l \times 0.586\, l\theta = 0.293\, W_C\, l\theta$

Internal work done $= M_P \times \theta + M_P \times 2.4155\theta + 0 = 3.4155\, M_P\, \theta$

Equating external and internal work done,

$$0.293\, W_C\, l\theta = 3.4155\, M_P\, \theta$$

$\therefore$
$$W_C = \frac{3.4155\, M_P}{0.293\, l} = 11.656\, \frac{M_P}{l}$$

$$\text{Collapse load} = 11.656\, \frac{M_P}{l}.$$

Example 8.17 *A two span continuous beam ABC has span lengths AB = 6 m and BC = 6 m and carries a udl of 30 kN/m completely covering the spans AB and BC. A and C are simple supports. If the load factor is 1.80 and the shape factor is 1.15 for the 'I' section, find the section modulus needed. Assume yield stress for the material as 250 N/mm².*

Solution

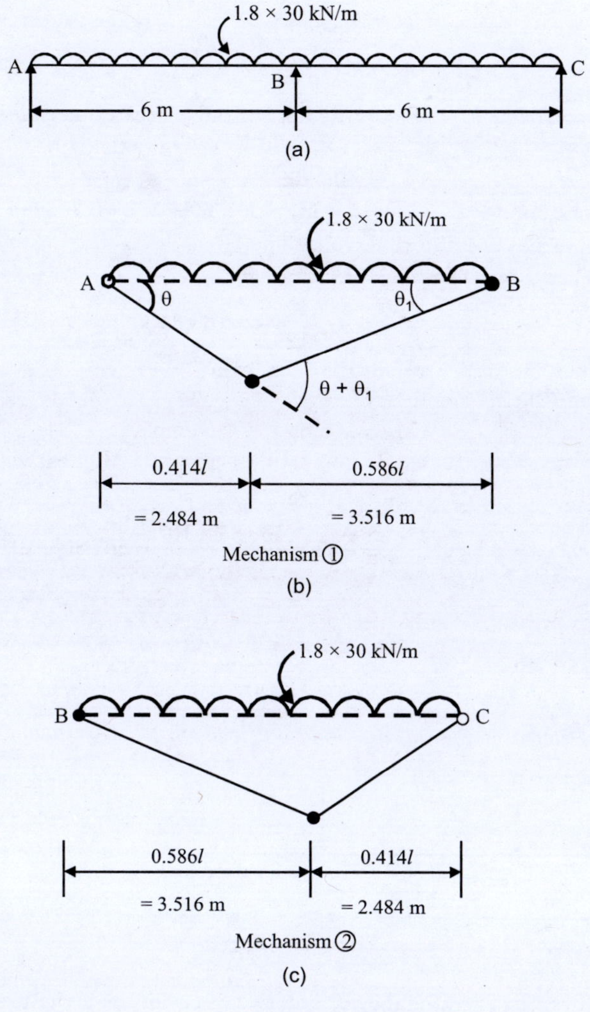

Fig. 8.31

Shape factor, $S = 1.15$

Yield stress, $\sigma_y = 250 \text{ N/mm}^2$

Degree of static indeterminacy, $r = 1$

Number of possible plastic hinges, $N = 3$

∴ Number of independent mechanisms, $n = N - r = 3 - 1 = 2$

Mechanism (1) :

$$2.484\theta = 3.516\theta_1$$

$$\theta_1 = \frac{2.484\theta}{3.516} = 0.706\theta$$

∴ $\theta + \theta_1 = \theta + 0.706\theta = 1.706\theta$

External work done = intensity of udl × area of mechanism below udl

$= (1.8 \times 30) \times (1/2 \times 6 \times 2.484\theta)$

$= 402.408\theta$

Internal work done $= 0 + M_P \times 1.706\theta + M_P \times 0.706\theta = 2.412\ M_P\theta$

Equating external and internal work done,

$$402.408\theta = 2.412\ M_P.\theta$$

∴ $$M_P = \frac{402.408\theta}{2.412\theta} = 166.84 \text{ kNm}$$

i.e., $M_P = 166.48 \times 10^6$ N.mm ...(1)

Mechanism (2) : By symmetry,

$M_P = 166.84 \times 10^6$ N.mm ...(2)

∴ $M_P = 166.84 \times 10^6$ N.mm (Highest value among 1 and 2)

We know that, $M_P = \sigma_y \times Z_P$

∴ $$Z_P = \frac{M_P}{\sigma_y} = \frac{166.84 \times 10^6}{250} = 667.36 \times 10^3 \text{ mm}^3$$

Also, $$S = \frac{Z_P}{Z}$$

∴ $$Z = \frac{Z_P}{S} = \frac{667.36}{1.15} \times 10^3 \text{ mm}^3 = 580 \times 10^3 \text{ mm}^3.$$

Example 8.18 *Determine the collapse load of the beam load as shown in Fig. 8.32.*

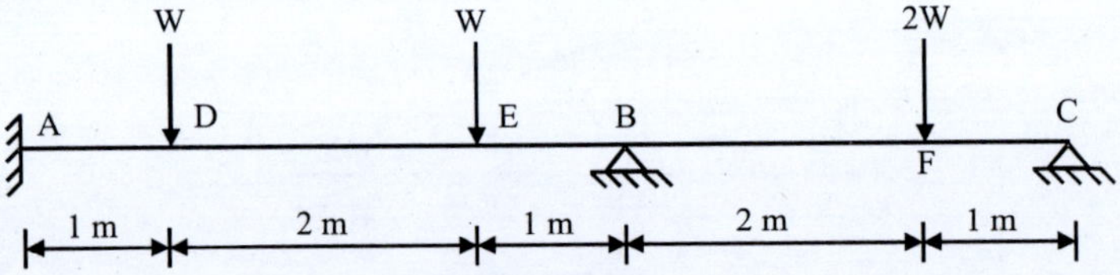

Fig. 8.32

Solution The degree of redundancy or indeterminacy of the structure = r

r = Vertical reactions + end moments–No. of useful equilibrium equations

$= 3 + 1 - 2 = 2$; No. of possible hinges, N = 5 (At A, D, E, B and F)

No. of independent mechanisms = $N - r = 5 - 2 = 3$

Mechanism (1) :

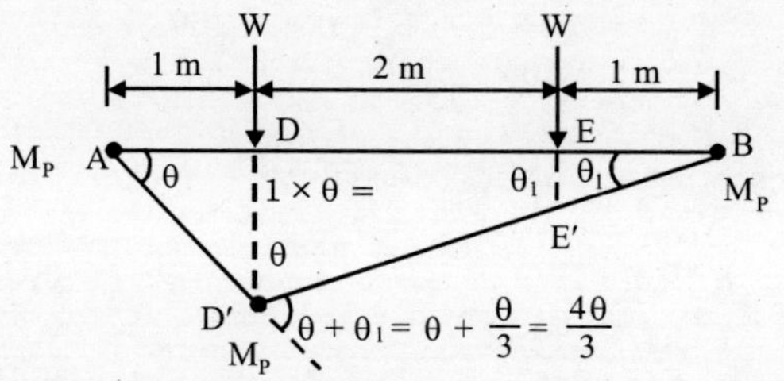

Let the beam collapse by the formation of plastic hinges at A, B and D.

$$1.\theta = 3\theta_1 = DD'$$

$$\theta_1 = \frac{\theta}{3}$$

External work done = Σ Forces × Displacements

$$= W \times (DD') + W \times (EE') = W\theta + W\theta_1$$

$$= W\theta + \frac{W\theta}{3} = \frac{4W\theta}{3} \quad ...(1)$$

Internal work done = [Plastic moment × Angle of rotation] at A, B and D.

$$= M_P\theta + M_P(\theta + \theta_1) + M_P(\theta_1)$$

$$= M_P\theta + M_P\left(\theta + \frac{\theta}{3}\right) + M_P\left(\frac{\theta}{3}\right)$$

$$\text{I.W.D.} = \frac{8}{3} M_P\theta \quad ...(2)$$

Equating (1) and (2), $\frac{4}{3} W\theta = \frac{8}{3} M_P.\theta$

$$W = 2M_P \quad ...(I)$$

Mechanism (2) :

Let the beam collapse by the formation of plastic hinges at A, B and E

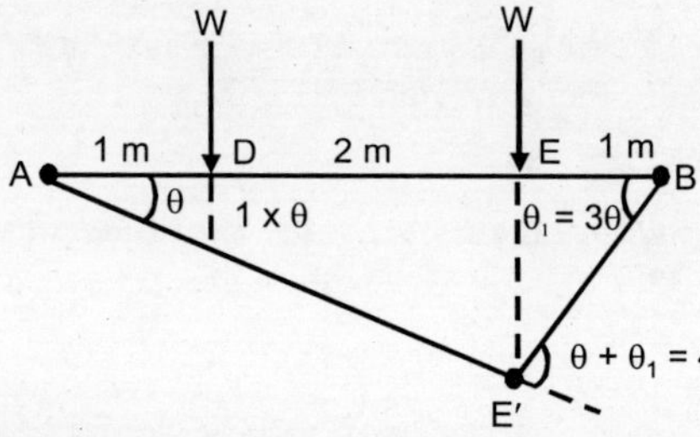

$$\theta_1 = 3\theta$$

External work done = (Load × displacement) at D and E

$$= W \times DD' + W(EE') = W\theta + W\theta_1 = W\theta + W(3\theta)$$

$$\text{E.W.D.} = 4W\theta \quad ...(3)$$

Internal work done = [Plastic moment × Angle of rotation] at A, E and B

$= M_P \times \theta + M_P(4\theta) + M_P(3\theta)$

$= M_P\theta + 4M_P\theta + 3M_P\theta$

I.W.D. $= 8M_P\theta$ ...(4)

Equating internal work done and external work done,

I.W.D. = E.W.D.

$8\ M_P\theta = 4W\theta$

$$W = \frac{8M_P\theta}{4\theta}$$

$W = 2M_P$ ...(II)

Mechanism (3) :

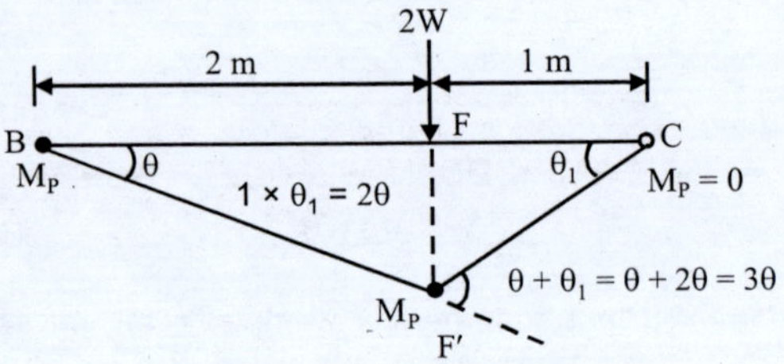

Let the beam collapse by the formation of plastic hinges at B and F (with structural hinge at C.)

External work done = (Load × displacement) of F

E.W.D. $= 2W \times 2\theta = 4W\theta$ ...(5)

Internal work done = (Plastic moment × rotation) at B and F

$= M_P\theta + M_P\ (3\theta)$

I.W.D. $= 4M_P\theta$ ...(6)

E.W.D. = I.W.D.

(4) = (5)

$4W\theta = 4\ M_P\theta$

$W = M_P$ ...(III)

The least value of load among I, II and III is M_P

Hence, collapse load, $W_C = M_P$.

Example 8.19 *Find the collapse load W_C for the continuous beam shown in Fig. 8.33. The beam has uniform plastic moment M_P.*

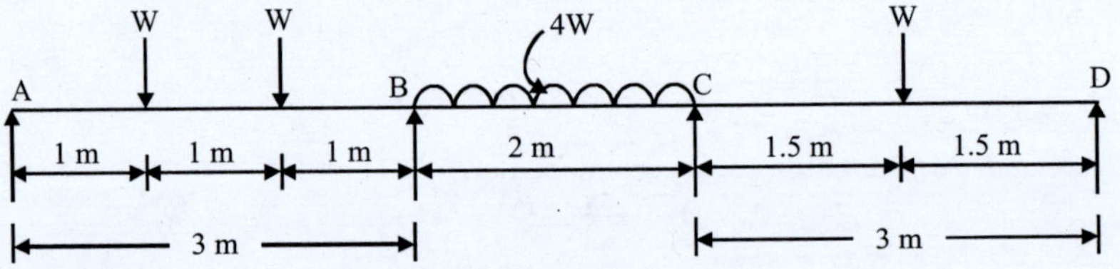

Fig. 8.33

Solution

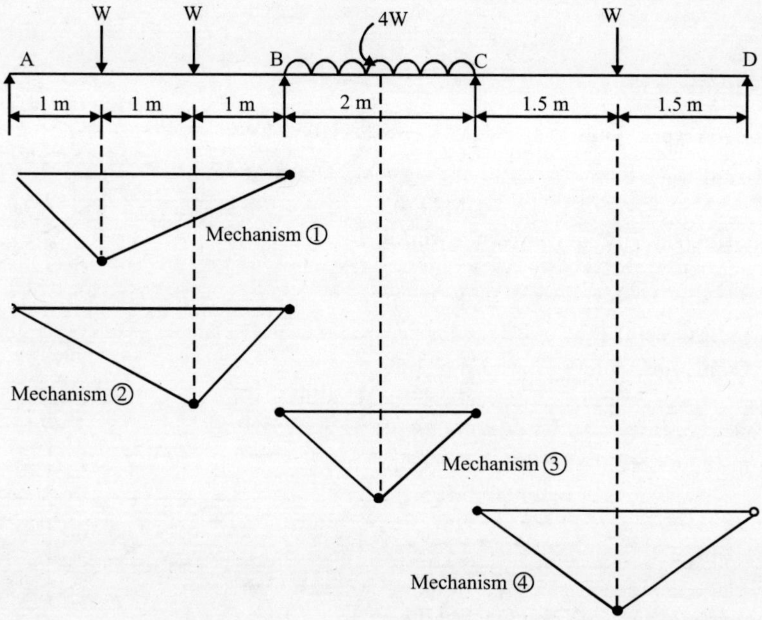

Fig. 8.34

Degree of indeterminacy, I = V.R. + E.M. – 2 = 4 + 0 – 2 = 2

No. of possible plastic hinges = 6

No. of independent mechanisms = 6 – 2 = 4

Mechanism (1) :

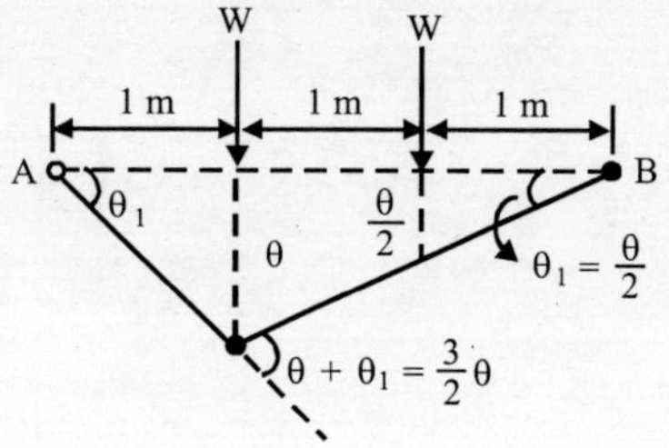

$$1 \times \theta = 2\theta_1,\ \theta_1 = \frac{\theta}{2}$$

$$\theta + \theta_1 = \theta + \frac{\theta}{2} = \frac{3\theta}{2}$$

External work done $= W \times \theta + W \times \dfrac{\theta}{2}$

$$= W\left(\theta + \frac{\theta}{2}\right) = \frac{3}{2}W\theta$$

Internal work done $= M_P \times \dfrac{3\theta}{2} + M_P \times \dfrac{\theta}{2} = M_P\left(\dfrac{3\theta}{2} + \dfrac{\theta}{2}\right) = M_P \times 2\theta$

Equating external and internal works

$$\frac{3}{2}W\theta = M_P \times 2\theta,\ W = M_P \times 2\theta \times \frac{2}{3\theta},\ W_C = \frac{4M_P}{3} \quad \text{...(1)}$$

Mechanism (2) :

Since this case is very similar to mechanism 1,

$$W_C = \frac{4M_P}{3} \quad ...(2)$$

Mechanism (3) :

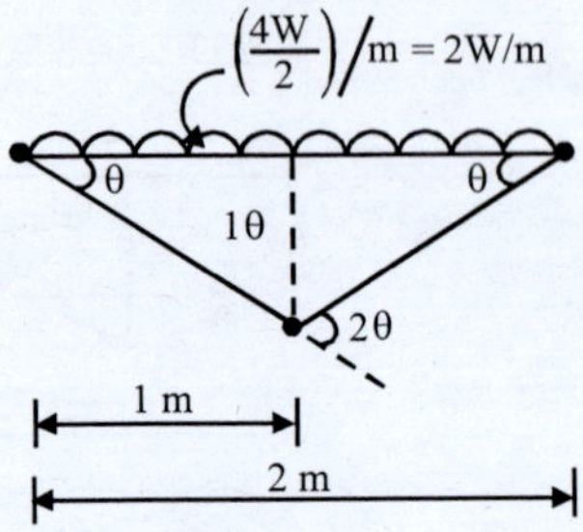

External work done $= 2W \times \frac{1}{2} \times 2 \times 1\theta = 2W\theta$

Internal work done $= M_P \times \theta + M_P \times 2\theta + M_P \times \theta$

$= 4M_P \times \theta$

Equating external and internal works

$$2W\theta = 4M_P\theta$$

$$W_C = 2M_P \quad ...(3)$$

Mechanism (4) :

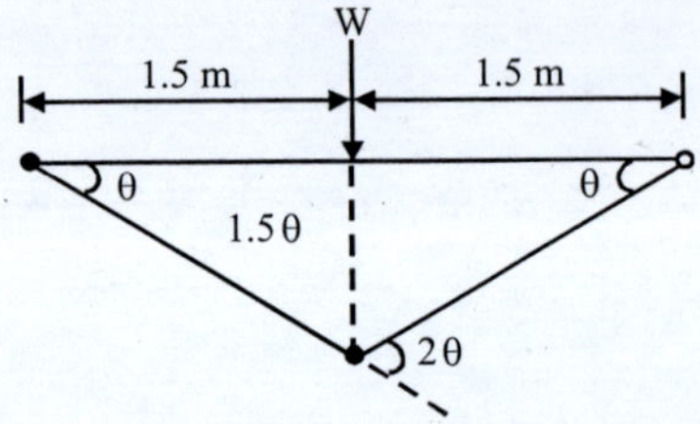

External work done $= W \times 1.5\theta$

Internal work done $= M_P \times \theta + M_P \times 2\theta$

$= 3M_P \times \theta$

Equating external and internal work done

$$W \times 1.5\theta = 3M_P \times \theta$$

$$W_C = \frac{2M_P}{1.5} = 2M_P$$

$$W_C = 2M_P \quad ...(4)$$

Comparing (1) to (4), the least value of

$$W_C = \frac{4}{3} M_P.$$

Example 8.20 *Find the collapse load for the frame, shown in Fig. 8.35.*

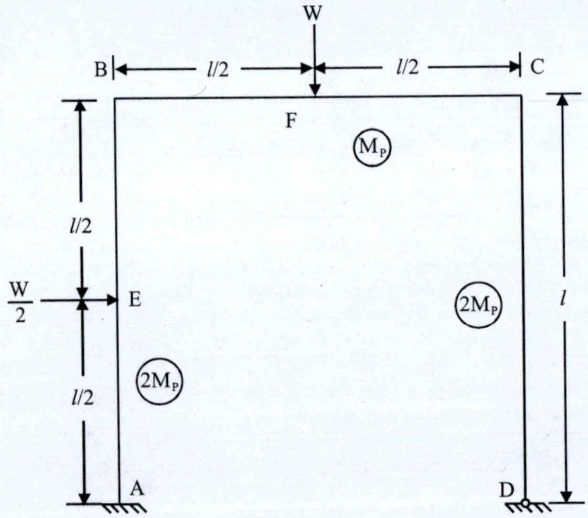

Fig. 8.35

Solution 1. Number of independent mechanisms :

Number of independent mechanisms

= No. of possible plastic hinges – Degree of redundancy

= (N – r)

Degree of redundancy, r = 3 (No. of closed loops) – No. of releases

= 3(1) – 1 = 2

(Since the end 'D' is not restricted against rotation, no. of releases = 1)

No. of possible plastic hinges, N = 5 (at A, B, C, E, F)

No. of independent mechanisms r = N – r = 5 – 2 = 3

Three numbers of independent mechanisms are possible. They are

(*i*) Beam mechanism.

(*ii*) Column mechanism.

(*iii*) Panel mechanism or sway mechanism.

In addition to the above 3, there will be the combined mechanisms.

(*i*) **Beam mechanism :**

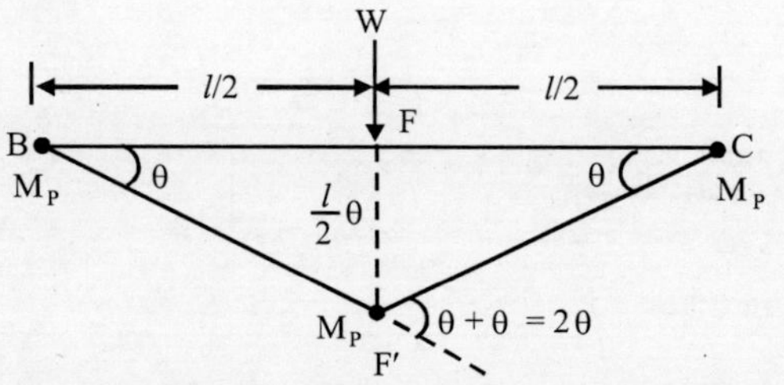

External work done = (Load × Displacement) at F = W × FF′ = W × $\frac{\theta l}{2}$

$$\text{E.W.D.} = \frac{Wl\theta}{2} \quad ...(1)$$

Internal work done = (Plastic moment × Rotation) at B, C, F

$$= M_P\theta + M_P(\theta + \theta) + M_P\theta$$

$$\text{I.W.D.} = 4M_P\theta \quad ...(2)$$

$$\text{E.W.D.} = \text{I.W.D.}$$

$$\frac{Wl\theta}{2} = 4M_P$$

$$W = \frac{8M_P}{l} \quad ...(I)$$

(*ii*) **Column mechanism :**

Plastic hinges are formed at A, B and E

External work done = [Load × Displacement] at E = $\frac{W}{2} \times \frac{\theta l}{2}$

$$\text{E.W.D.} = \frac{Wl\theta}{4} \quad ...(3)$$

Internal work done = (Plastic moment × Rotation) at A, B, E

$$= 2\,M_P\theta + M_P\theta + 2M_P(\theta + \theta)$$

$$\text{I.W.D.} = 7M_P\theta \quad ...(4)$$

$$\text{E.W.D.} = \text{I.W.D.}$$

$$\frac{Wl\theta}{4} = 7M_P\theta$$

$$W = \frac{28M_P}{l} \quad ...(II)$$

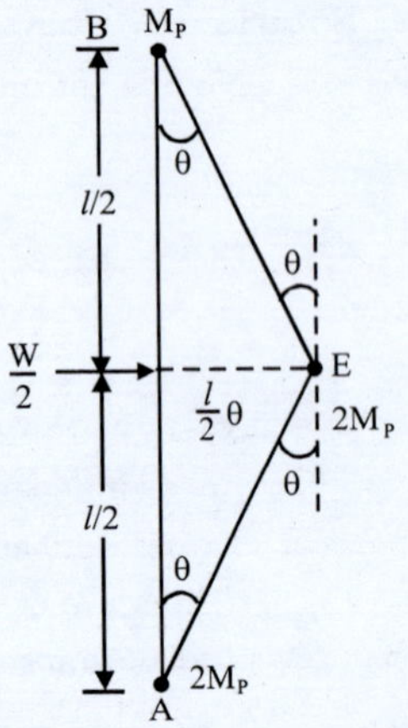

(*iii*) **Panel mechanism :**

Plastic hinges are formed at A, B and C and structural hinge at D.

External work done = (Load × Displacement) at E

$$= \frac{W}{2} \times \frac{\theta l}{2}$$

$$\text{E.W.D.} = \frac{Wl\theta}{4} \quad ...(5)$$

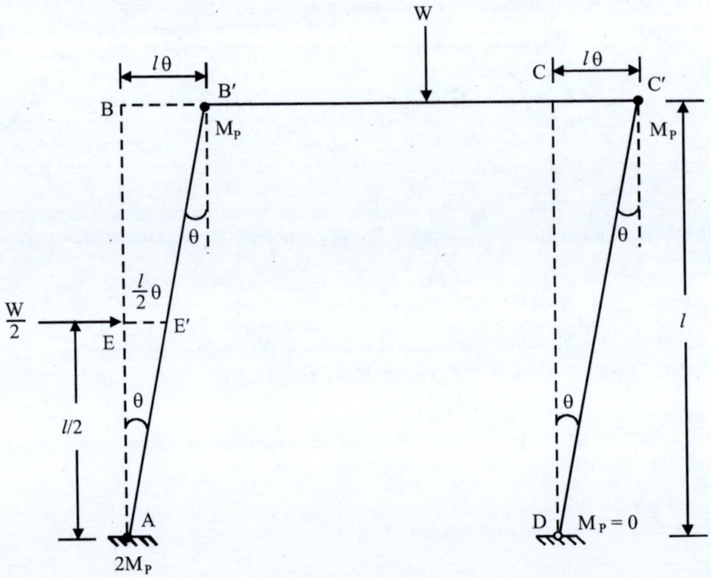

Internal work done = (Plastic moment × Rotation) at A, B and C

$$= 2M_P \times \theta + M_P \times \theta + M_P \times \theta$$

$$\text{I.W.D.} = 4M_P.\theta \quad ...(6)$$

$$\text{E.W.D.} = \text{I.W.D.}$$

$$\frac{Wl.\theta}{4} = 4M_P.\theta, \quad W = \frac{16M_P}{l} \quad \text{...(III)}$$

2. **Combined mechanism :**

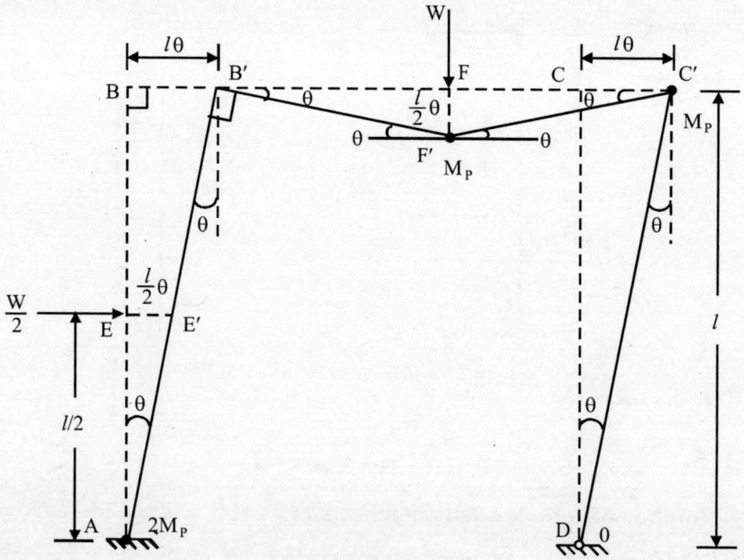

Plastic hinges are formed at A, F and C

External work done = (Load × Displacement) at E and F

$$= \left[\frac{W}{2} \times \frac{\theta l}{2} + W \times \frac{\theta l}{2}\right] = \frac{3Wl\theta}{4} \quad \text{...(7)}$$

Internal work done = (Plastic moment × Rotation) at A, F and C

$$= 2M_P\theta + M_P(\theta + \theta) + M_P(\theta + \theta)$$

$$\text{I.W.D.} = 6\ M_P\theta \quad \text{...(8)}$$

$$\text{E.W.D.} = \text{I.W.D.}$$

$$\frac{3Wl\theta}{4} = 6\ M_P.\theta$$

$$W = \frac{8M_P}{l} \quad \text{...(IV)}$$

From (I), (II), (III) and (IV), the least value of $W = \frac{8M_P}{l}$

Hence collapse load, $W_C = \frac{8M_P}{l}$.

Example 8.21 *Determine the collapse load for the frame shown in Fig. 8.36. M_P is the same for all members.*

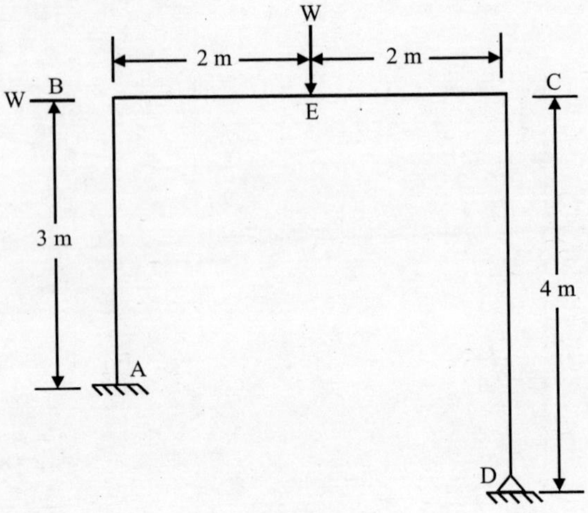

Fig. 8.36

Solution **Number of independent mechanisms :**

Degree of redundancy of the frame = No. of closed loops × 3 – No. of releases.

$r = 1 \times 3 - 1$ (1 release (rotation of joint D))

$r = 2$

No. of plastic hinges = N = 4 (at A, B, C, E)

No. of independent mechanisms = $N - r = 4 - 2 = 2$

1. **Beam mechanism :**

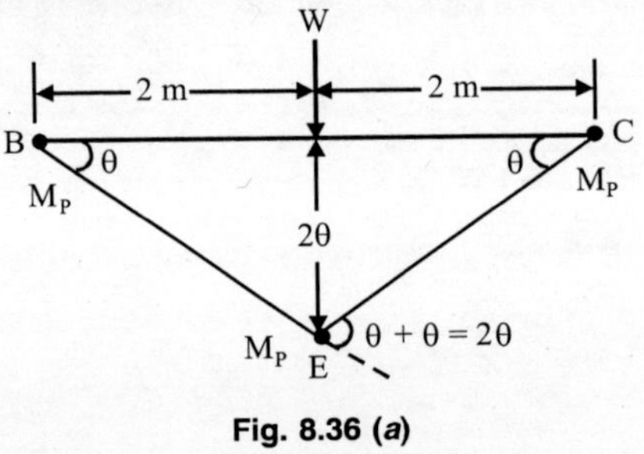

Fig. 8.36 (a)

Let the frame collapse due to the formation of plastic hinges at B, C and E

External work done = Load × Displacement = W × 2θ

$$\text{E.W.D.} = 2W\theta \quad \text{...(1)}$$

Internal work done = [Plastic moment × Angle of rotation] at B, E, C

$$= M_P\theta + M_P(2\theta) + M_P\theta = 4\,M_P\theta$$

$$\text{I.W.D.} = 4\,M_P\theta \quad \text{...(2)}$$

$$\text{E.W.D.} = \text{I.W.D.}$$
$$2W\theta = 4\,M_P\theta$$
$$W = 2M_P \qquad \text{...(I)}$$

2. **Panel mechanism or Sway mechanism :**

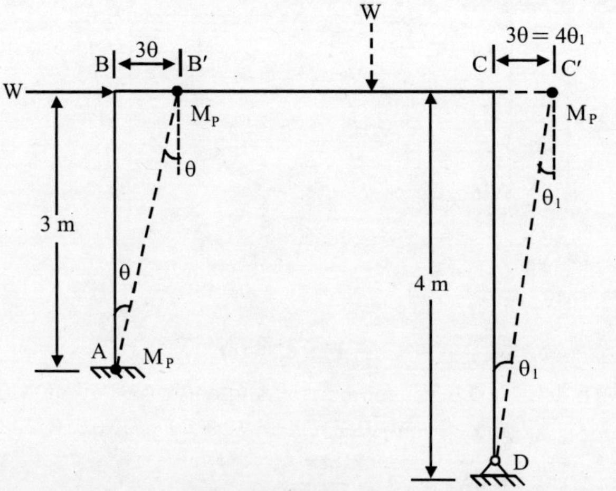

Fig. 8.36 (*b*)

Plastic hinges are formed at A, B and C.

External work done = (Force × Displacement) at B = $W \times 3\theta$

$$\text{E.W.D.} = 3W\theta \qquad \text{...(3)}$$

Internal work done = (Plastic moment × Angle of rotation) at A, B and C

$$= M_P\theta + M_P\theta + M_P\theta_1 = 2M_P\theta + M_P\left(\frac{3\theta}{4}\right) \qquad [3\theta = 4\theta_1]$$

$$\text{I.W.D.} = \frac{11}{4}\,M_P\theta \qquad \text{...(4)}$$

$$\text{I.W.D.} = \text{E.W.D}$$

$$3W\theta = \frac{11}{4}M_P\theta$$

$$W = \frac{11}{12}\,M_P \qquad \text{...(II)}$$

3. **Combined mechanism :**

Plastic hinges are formed at A, E and C.

External work done = (Force × Displacement) at B and E

$$= W \times 3\theta + W \times 2\theta$$

$$\text{E.W.D.} = 5\,W\theta \qquad \text{...(5)}$$

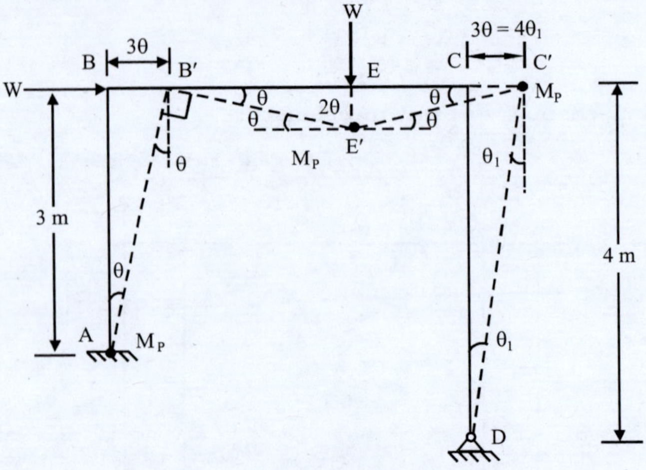

Fig. 8.36 (c)

Internal work done = (Plastic moment × Angle of rotation) at A, E and C

$= M_P\theta + M_P(\theta + \theta) + M_P(\theta + \theta_1) = 4M_P\theta + M_P\,(\theta_1)$

$= 4M_P\theta + M_P\left(\dfrac{3\theta}{4}\right)$ $\left[\because\ \theta_1 = \dfrac{3\theta}{4}\right]$

I.W.D. = 4.75 $M_P\theta$ ...(6)

E.W.D. = I.W.D. $5W\theta = 4.75M_P\theta$, $W = 0.95\ M_P$ ...(III)

From (I), (II) and (III) the least value of W is $\dfrac{11}{12}$ M_P, *i.e.*, 0.92 M_P, $W_C = 0.92\ M_P$.

And the structure will fail by panel mechanism.

Example 8.22 *Determine the collapse load for the frame shown in Fig. 8.37.*

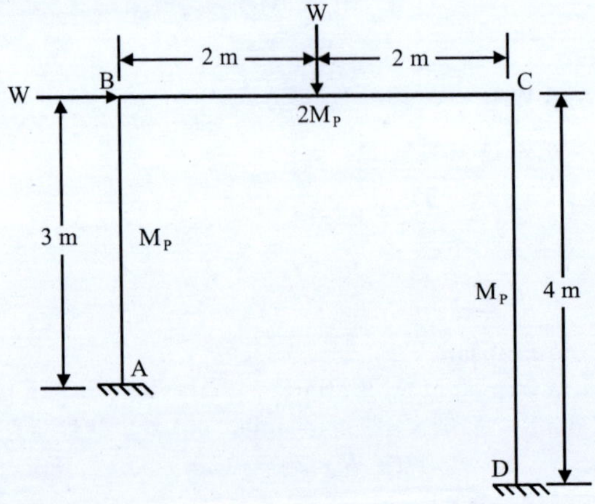

Fig. 8.37

Solution Degree of static indeterminacy = No. of closed loops × 3 – No. of releases

$= 1 \times 3 - 0 = 3$

No. of possible plastic hinges $= 5$

No. of independent mechanisms $= 5 - 3 = 2$

Mechanism (1) : Beam mechanism :

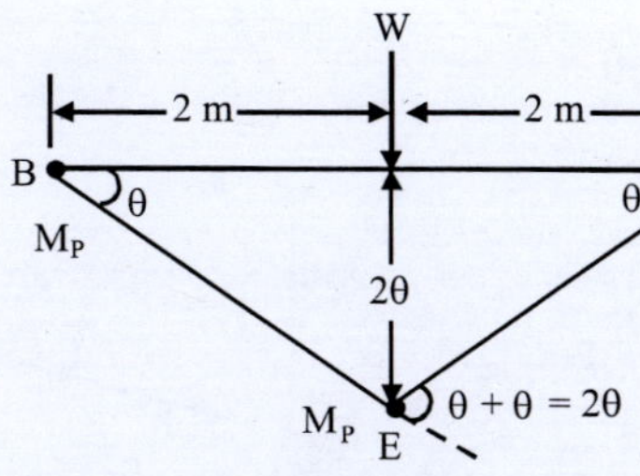

Fig. 8.37 (*a*)

External work done $= W \times 2\theta$

Internal work done $= M_P \times \theta + 2M_P \times 2\theta + M_P\theta = 6M_P \times \theta$

Equating external and internal works

$$W \times 2\theta = 6M_P \times \theta$$

$$W = \frac{6M_P}{2} = 3M_P$$

$$W_C = 3M_P \quad \text{...(1)}$$

Mechanism (2) : Panel mechanism :

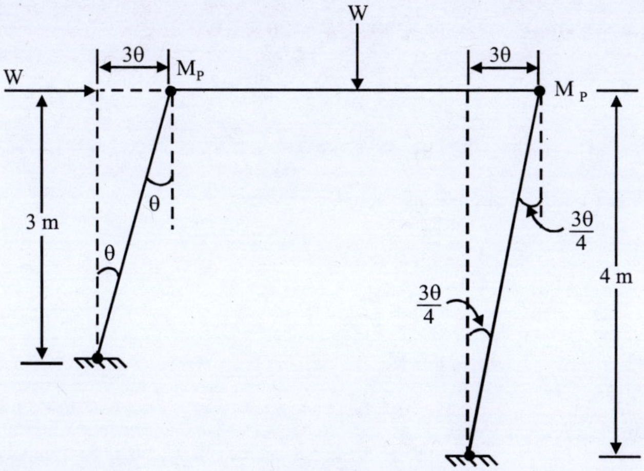

Fig. 8.37 (*b*)

External work done $= W \times 3\theta$

Internal work done $= M_P \times \theta + M_P \times \theta + M_P \times \frac{3\theta}{4} + M_P \times \frac{3\theta}{4} = \frac{7}{2} M_P \times \theta$

Equating external and internal works

$$W \times 3\theta = \frac{7}{2} M_P \times \theta, \quad W = \frac{7}{2 \times 3} M_P$$

$$W_c = \frac{7}{6} M_P \qquad \text{...(2)}$$

Mechanism (3) : Composite mechanism :

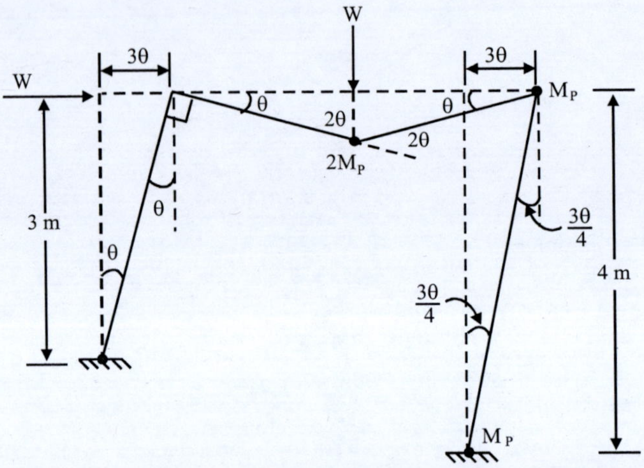

Fig. 8.37 (*c*)

External work done $= W \times 3\theta + W \times 2\theta = 5W\theta$

Internal work done $= M_P \times \theta + 2M_P \times 2\theta + M_P \left(\theta + \frac{3\theta}{4}\right) + M_P \times \frac{3\theta}{4}$

$$= \left[M_P\, \theta \left[1 + 4 + \frac{7}{4} + \frac{3}{4}\right]\right] = \frac{30}{4} M_P \times \theta = \frac{15}{2} M_P\, \theta.$$

Equating external and internal works

$$5W\theta = \frac{15}{2} M_P \times \theta$$

$$W_C = \frac{15}{2 \times 5} M_P$$

$$W_C = 1.5\, M_P \qquad \text{...(3)}$$

Comparing (1), (2) and (3),

the collapse load

$W_C = \frac{7}{6} M_P$ and the frame will fail by sway mechanism.

Example 8.23 *Find the collapse load for the portal frame loaded as shown in Fig. 8.38.*

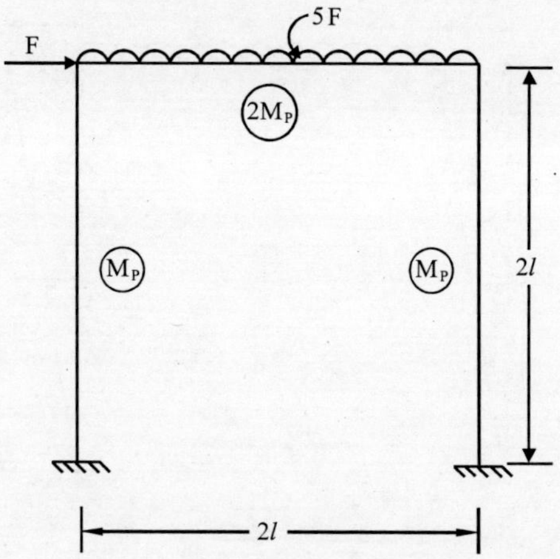

Fig. 8.38

Solution Degree of redundancy, $r = 3 \times$ No. of loops – No. of releases $= 3 \times 1 - 0 = 3$

No. of possible plastic hinges = 5

∴ No. of independent mechanisms = 5 – 3 = 2

1. **Beam mechanism :**

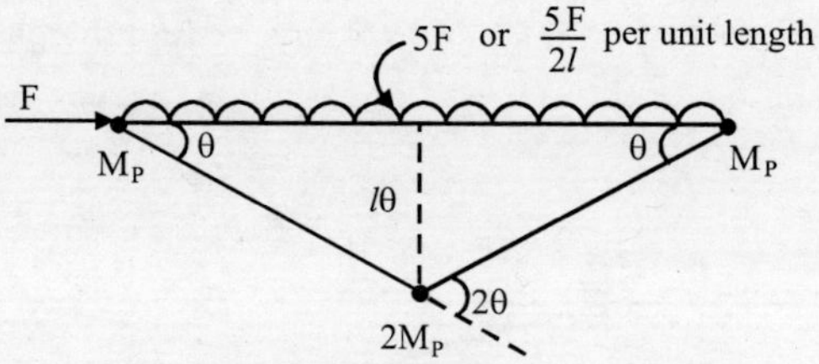

Fig. 8.38 (*a*)

External work done = Intensity of udl × Area of mechanism below udl

$$= \frac{5F}{2l}\left(\frac{1}{2} \times 2l \times l\theta\right) = \frac{5}{2}Fl\theta$$

Internal work done $= M_P \times \theta + 2M_P \times 2\theta + M_P\theta = 6M_P \times \theta$

Equating the works done

$$\frac{5}{2}Fl\theta = 6M_P\theta$$

$$F = \frac{2.4M_P}{l} \qquad ...(1)$$

2. **Sway mechanism :**

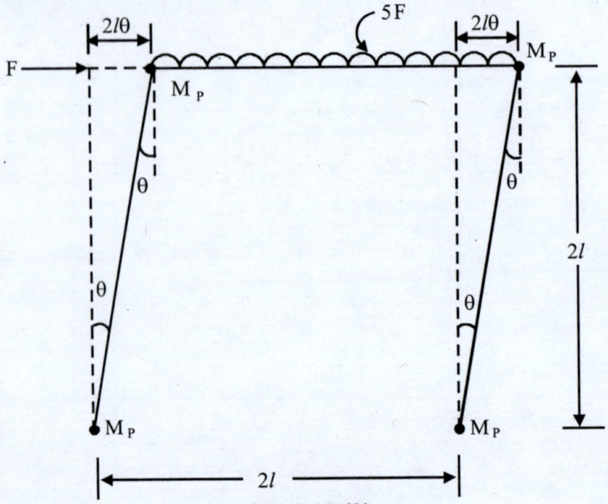

Fig. 8.38 (*b*)

External work done $= F \times 2l\theta = 2Fl\theta$

Internal work done $= M_P \times \theta + M_P \times \theta + M_P \times \theta + M_P \times \theta = 4M_P \times \theta$

Equating the works, $2Fl\theta = 4M_P\theta$

$$F = \frac{2M_P}{l} \quad ...(2)$$

3. **Combined mechanism :**

This is an unsymmetrical mechanism. We have to find the position of hinge in the beam by maximising M_P.

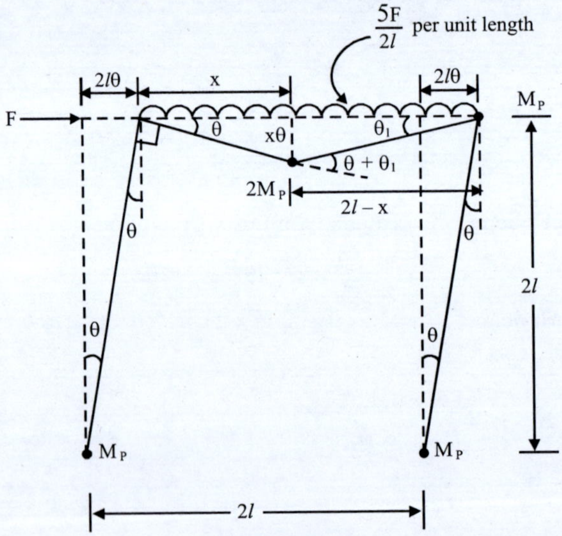

Fig. 8.38 (*c*)

$$x\theta = (2\,l - x)\,\theta_1 \qquad \Rightarrow \qquad \theta_1 = \left[\frac{x}{2l - x}\right]\theta$$

$$\theta + \theta_1 = \theta + \frac{x}{2l - x}\theta = \left[\frac{2l - x + x}{2l - x}\theta\right]$$

$$\theta + \theta_1 = \frac{2l\theta}{2l - x}$$

External work done $= F \times 2l\theta + \dfrac{5F}{2l}\left[\dfrac{1}{2} \times 2l \times x\theta\right]$

$$= 2Fl\theta + \frac{5}{2}F \times \theta = \left[2Fl + \frac{5}{2}Fx\right]\theta$$

Internal work done $= M_P\theta + 2M_P(\theta + \theta_1) + M_P(\theta + \theta_1) + M_P\,\theta$

$$= M_P\theta + 2M_P\left[\frac{2l\theta}{2l - x}\right] + M_P\left[\frac{2l\theta}{2l - x}\right] + M_P\theta$$

$$= 2M_P\theta + 3M_P\left[\frac{2l\theta}{2l - x}\right] = M_P\theta\left[\frac{2(2l - x) + 6l}{2l - x}\right]$$

$$= M_P\theta\left[\frac{4l - 2x + 6l}{2l - x}\right] = M_P\theta\left[\frac{10l - 2x}{2l - x}\right]$$

Equating the works,

$$\left[2Fl + \frac{5}{2}Fx\right]\theta = M_P\theta\left[\frac{10l - 2x}{2l - x}\right]$$

$$M_P = \frac{F(2l + 2.5x)(2l - x)}{(10l - 2x)} = \frac{F(4l^2 - 2lx + 5lx - 2.5x^2)}{(10l - 2x)}$$

$$= \frac{(4Fl^2 + 3Flx - 2.5\,Fx^2)}{(10l - 2x)} \qquad \text{...(A)}$$

For this moment to be maximum, $\dfrac{dM_P}{dx} = 0$

$$\frac{dM_P}{dx} = \frac{(10l - 2x)(3Fl - 5Fx) - (4Fl^2 + 3Flx - 2.5Fx^2)(-2)}{(10l - 2x)^2} = 0$$

$\therefore \quad 30Fl^2 - 50Flx - 6Flx + 10Fx^2 + 8Fl^2 + 6Flx - 5Fx^2 = 0.$

$$5Fx^2 - 50Flx + 38Fl^2 = 0$$

$$x^2 - 10lx + 7.6l^2 = 0$$

$$x = \frac{10l \pm \sqrt{(-10l)^2 - 4 \times 1 \times 7.6}}{2} = \frac{10l \pm 8.343l}{2}$$

$$x = 0.829l$$

Substituting, $x = 0.829\,l$ in (A)

$$M_P = \frac{4Fl^2 + 3Fl(0.829l) - 2.5F(0.829l)^2}{10l - 2(0.829l)}$$

$$M_P = 0.572\,\frac{Fl^2}{l} = 0.572\,Fl.$$

$$F = \frac{M_P}{0.572l}$$

$$F = 1.75\,\frac{M_P}{l} \qquad \text{...(3)}$$

From equations (1), (2), (3)

Collapse load $F_C = 1.75\,\dfrac{M_P}{l}$.

Example 8.24 *Find the collapse load for the loaded portal frame shown in Fig. 8.39.*

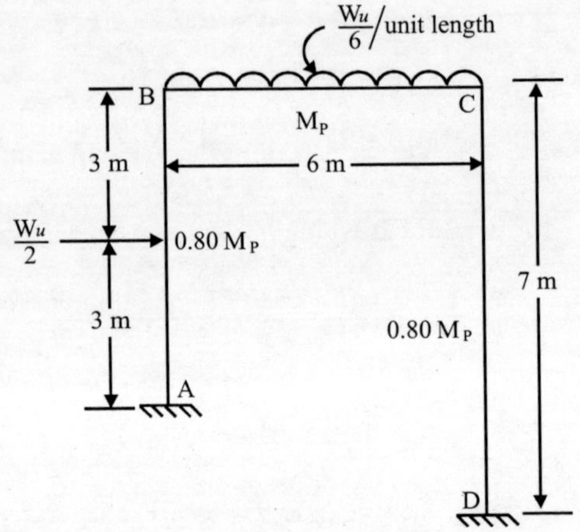

Fig. 8.39

$M_P = 1 \times 10^5$ Nm for beams

$M_P' = 0.80\ M_P = 0.80 \times 10^5$ Nm for columns.

Solution Degree of redundancy, $r = 1 \times 3 - 0 = 3$

No. of possible plastic hinges $= 6$

No. of independent mechanism $= 6 - 3 = 3$

1. **Beam mechanism :**

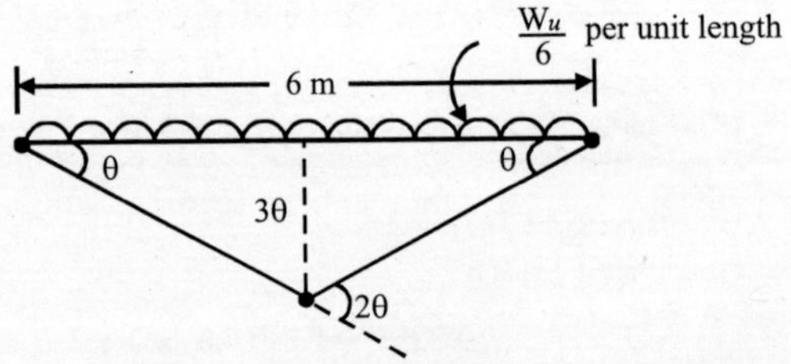

Fig. 8.39 (*a*)

External work done = $\frac{W_u}{6} \times \left(\frac{1}{2} \times 6 \times 3\theta\right) = \frac{3}{2} W_u \theta$

Internal work done = $(0.8 \times 10^5 \times \theta) + (1 \times 10^5 \times 2\theta) + (0.8 \times 10^5\, \theta)$

$= 3.6 \times 10^5\, \theta.$

Equating the works, $\frac{3}{2} W_u \theta = 3.6 \times 10^5 \theta$

$W_u = 2.4 \times 10^5$ N ...(1)

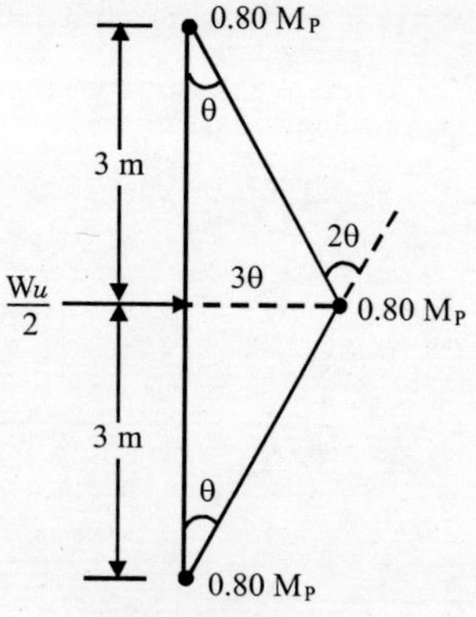

Fig. 8.39 (*b*)

2. **Column mechanism :**

External work done, $W_E = \frac{W_u}{2}(3\theta) = \frac{3}{2} W_u \theta$

Internal work done, $W_I = (0.8 \times 10^5 \times \theta)$

$+ (0.8 \times 10^5 \times 2\theta) + (0.8 \times 10^5\, \theta)$

$= 3.2 \times 10^5\, \theta.$

Equating the works, $W_E = W_I$

$\frac{3}{2} W_u \theta = 3.2 \times 10^5\, \theta$

$W_u = 2.13 \times 10^5$ N ...(2)

3. **Sway mechanism :**

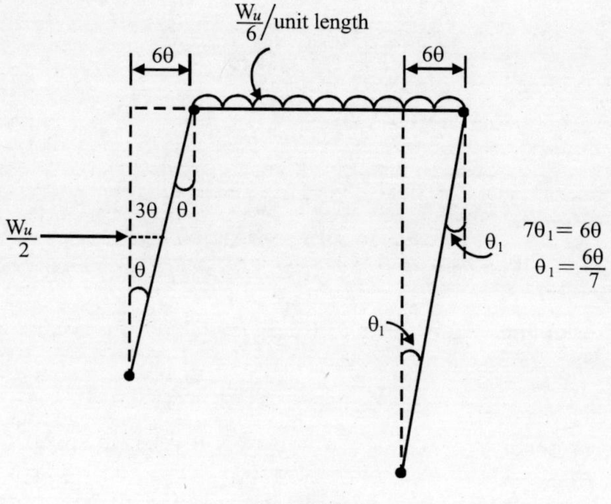

Fig. 8.39 (*c*)

External work done, $W_E = \frac{W_u}{2}(3\theta) = \frac{3}{2} W_u \theta$

Internal work done, $W_I = (0.8 \times 10^5 \times \theta) + (0.8 \times 10^5 \times \theta) + (0.8 \times 10^5\, \theta_1) + (0.8 \times 10^5\, \theta_1)$

$= 2 \times (0.8 \times 10^5\, \theta) + 2 \times 0.8 \times 10^5 \left[\frac{6}{7}\theta\right] = 2.97 \times 10^5\, \theta.$

Equating the works, $W_E = W_I$

$$\frac{3}{2} W_u\theta = 2.97 \times 10^5 \theta$$

$$W_u = 1.98 \times 10^5 \text{ N} \qquad ...(3)$$

4. **Combined mechanism :**

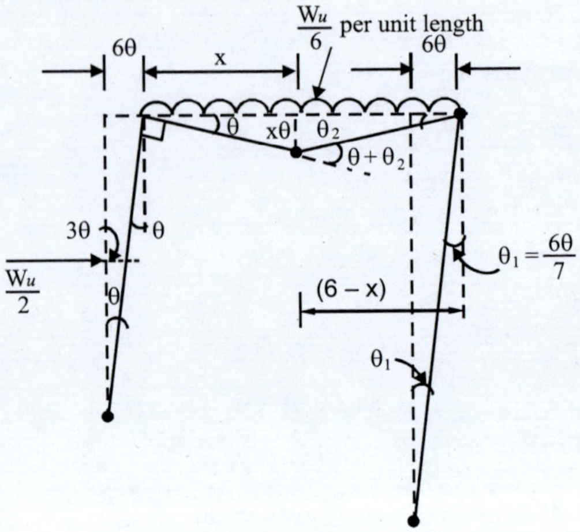

Fig. 8.39 (d)

$$x\theta = (6 - x)\theta_2 \qquad \Rightarrow \qquad \theta_2 = \frac{x\theta}{6-x}$$

$$\theta + \theta_2 = \theta + \frac{x\theta}{6-x} = \left[\frac{(6-x+x)\theta}{(6-x)}\right] = \frac{6\theta}{6-x}$$

$$\theta_1 + \theta_2 = \frac{6\theta}{7} + \frac{x\theta}{6-x} = \frac{(36-6x+7x)\theta}{7(6-x)} = \frac{(36+x)\theta}{7(6-x)}$$

External work done, $W_E = \frac{W_u}{2} \times 3\theta + \frac{W_u}{6}\left[\frac{1}{2} \times 6 \times x\theta\right] = \frac{3}{2} W_u\theta + \frac{1}{2} W_u \times x\theta$

$$= \left[\frac{3}{2} W_u + \frac{1}{2} W_u\, x\right]\theta = \frac{W_u\theta}{2}[3+x]$$

Internal work done, $W_I = (0.8 \times 10^5 \theta) + 1 \times 10^5 (\theta + \theta_2) + 0.8 \times 10^5 (\theta_1 + \theta_2) + (0.8 \times 10^5 \theta_1)$

$$= (0.8 \times 10^5 \theta) + 1 \times 10^5 \left[\frac{6\theta}{6-x}\right] + 0.8 \times 10^5 \left[\frac{(36+x)\theta}{7(6-x)}\right] + \left[0.8 \times 10^5 \left(\frac{6\theta}{7}\right)\right]$$

$$= 1 \times 10^5 \theta \left[0.8 + \frac{6}{(6-x)} + \frac{0.8(36+x)}{7(6-x)} + 0.8 \times \frac{6}{7}\right]$$

$$= 1 \times 10^5 \theta \left[1.486 + \frac{6}{6-x} + \frac{0.8(36+x)}{7(6-x)}\right]$$

$$= 1 \times 10^5 \theta \left[\frac{62.412 - 10.402x + 42 + 28.8 + 0.8x}{7(6-x)}\right]$$

$$= 1 \times 10^5 \theta \left[\frac{133.212 - 9.602x}{7(6-x)}\right] = \left[\frac{19.03 - 1.37x}{(6-x)}\right] \times 10^5 \theta$$

Equating the works,

$$\frac{W_u \theta}{2}(3+x) = \left[\frac{19.03 - 1.37x}{(6-x)}\right] \times 10^5 \theta, \quad W_u = \frac{2(19.03 - 1.37x)\, 10^5}{(6-x)(3+x)}$$

$$W_u = \frac{(38.06 - 2.74x)\, 10^5}{(18 + 3x - x^2)} \qquad \text{...(A)}$$

For this W_u to be maximum,

$$\frac{dW_u}{dx} = 0, \quad \frac{dW_u}{dx} = \frac{(18 + 3x - x^2)(-2.74 \times 10^5) - (38.06 - 2.74x)10^5(3 - 2x)}{(18 + 3x - x^2)} = 0$$

$\therefore$ $-(49.32 + 8.22x - 2.74x^2)10^5 - (114.18 - 76.12x - 8.22x + 5.48x^2)10^5 = 0$

$-49.32 - 8.22x + 2.74x^2 - 114.18 + 84.34x - 5.48x^2 = 0$

$-2.74x^2 + 76.12x - 163.5 = 0$

dividing by -2.74, $x^2 - 27.78x + 59.67 = 0$

$$x = \frac{27.78 \pm \sqrt{(-27.78)^2 - (4 \times 1 \times 59.67)}}{2} = \frac{27.78 \pm 23.09}{2} = 2.346 \text{ m.}$$

Substituting, $x = 2.346$ m in (A)

$$W_u = \frac{[38.06 - 2.74(2.346)]10^5}{[18 + 3(2.346) - (2.346)^2]}, \quad W_u = 1.62 \times 10^5 \text{ N} \qquad \text{...(4)}$$

From equations (1), (2), (3) and (4)

Collapse load, $W_u = 1.62 \times 10^5$ N.

PROBLEMS

1. Calculate the shape factor of the hollow rectangular section shown in Fig. 8.40.

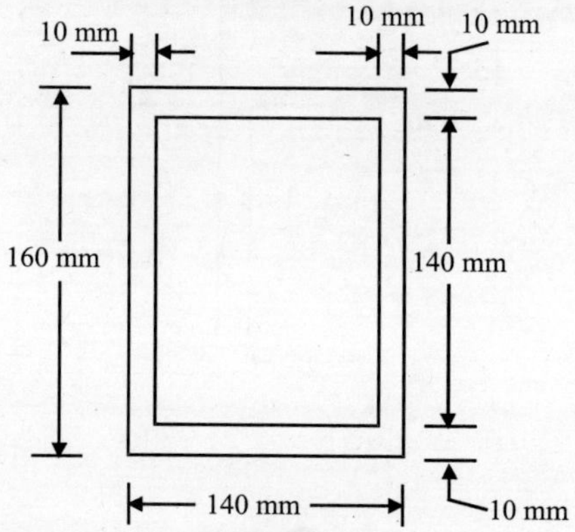

Fig. 8.40

[**Ans.** S = 1.211]

2. Find the shape factor of the I-section shown in Fig. 8.41.

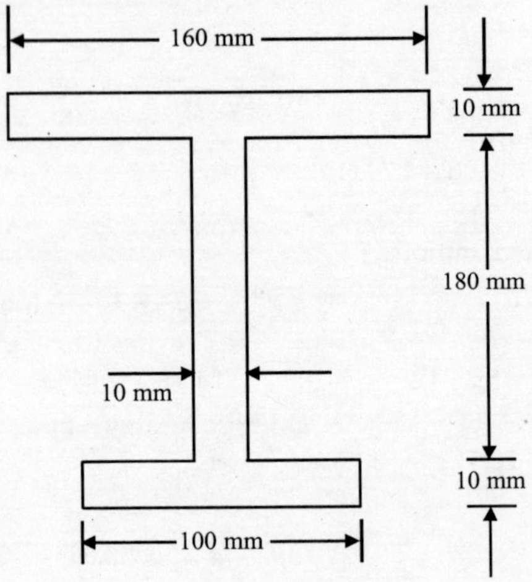

Fig. 8.41

[**Ans.** S = 1.3031]

3. Find the shape factor and plastic moment of the steel section (I-section) the details of which are, shown in Fig. 8.42.

Assume yield stress of steel = 250 N/mm^2

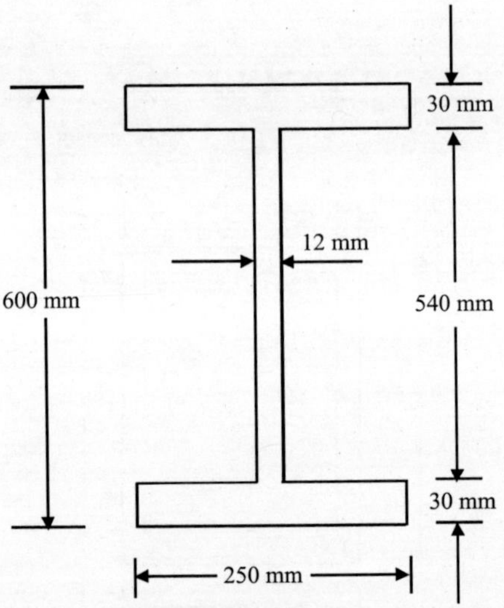

Fig. 8.42

[**Ans.** Shape factor, S = 1.12, Plastic moment, M_P = 1287.45 kNm]

4. Determine the fully plastic moment capacity of the T-section shown in Fig. 8.43 Yield stress, $\sigma_y = 250$ N/mm^2.

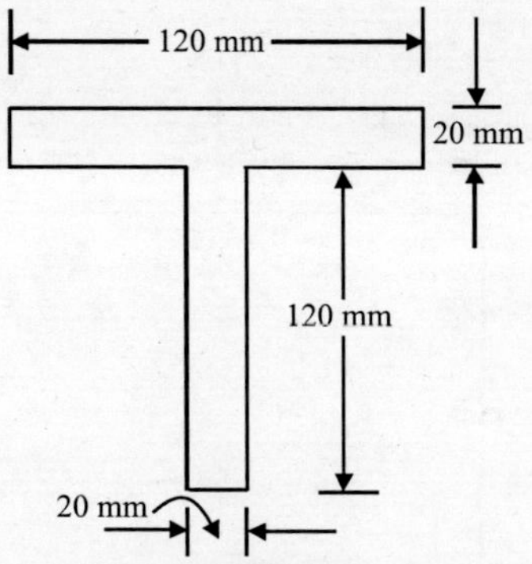

Fig. 8.43

[**Ans.** $M_P = 42$ kNm]

5. A propped cantilever of 4 metres span carries a uniformly distributed load of 20 kN/m. A 225 mm × 150 mm joist, (thickness of flange 9.9 mm, thickness of web 6.4 mm) is used for the beam. Calculate the load factor, taking the yield stress as 260 N/mm^2. State the location of the plastic hinge in the span at collapse loading. [**Ans.** 3.664 ; at 2.344 m from fixed end]

6. An encastre steel beam of 8 m span, carries a uniform load of 40 kN/m over the right hand half of the span. Taking the load factor as 1.75, select a suitable section. The shape factor for the section may be assumed to be 1.15 and the yield stress for steel at 260 N/mm^2.

[**Ans.** $Z = 526.7 \times 10^3$ mm^3]

7. Find out the collapse load for the continuous beam of uniform section and loaded as shown in Fig. 8.44, in terms of its plastic moment M_P.

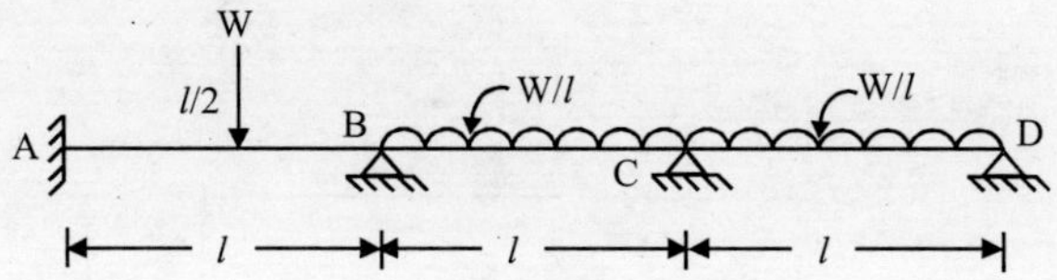

Fig. 8.44

$$\left[\textbf{Ans. } W_C = \frac{8M_P}{l}\right]$$

8. Find the collapse load factor for the continuous beam shown in Fig. 8.45.

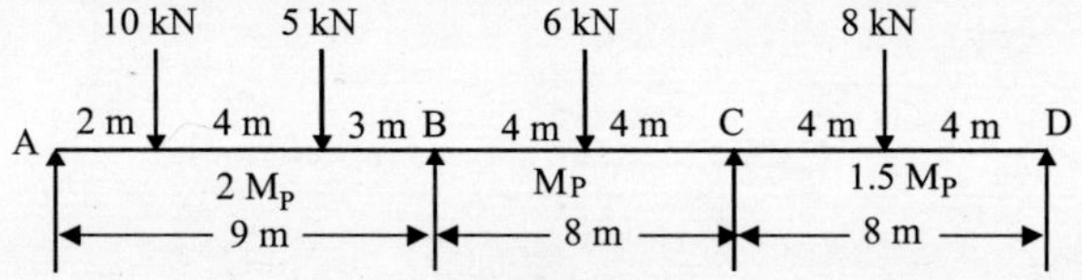

Fig. 8.45

$$\left[\textbf{Ans. } \lambda = \frac{2M_P}{l}\right]$$

9. Find out the plastic moment for the given frame of uniform cross-section under the applied factored loads as shown in Fig. 8.46.

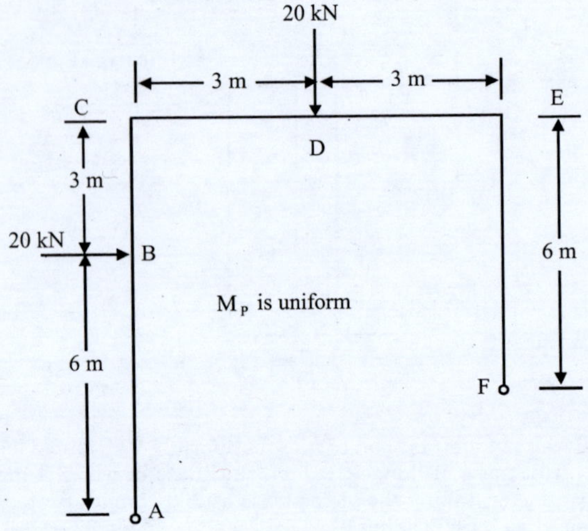

Fig. 8.46

[**Ans.** M_P = 80 kNm]

10. Determine the required plastic moment capacity of the section to be adopted for the frame shown in Fig. 8.47. The loads shown are the working loads. Load factor λ is to be taken as 1.75. Plastic moment capacity for all the members is the same.

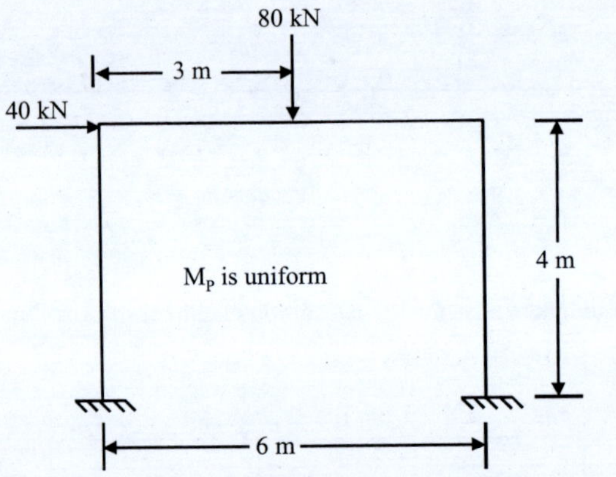

Fig. 8.47

[**Ans.** M_P = 116.67 kNm ; by combined mechanism]

11. Determine the fully plastic moment capacity required for the portal frame shown in Fig. 8.48.

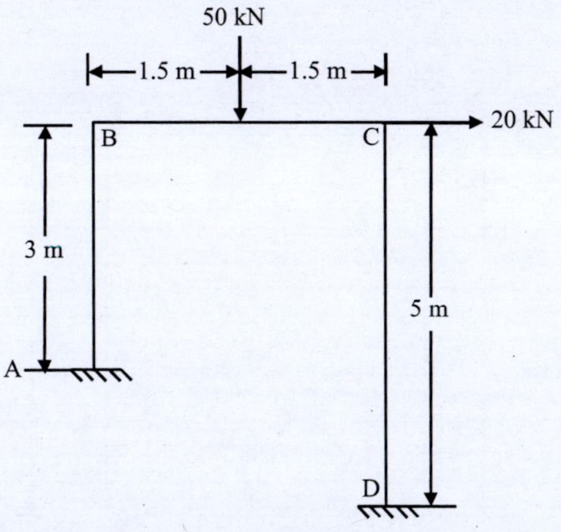

Fig. 8.48

[**Ans.** M_P = 25.96 kNm]

12. Find the value of W at collapse for the portal frame loaded as shown in Fig. 8.49.

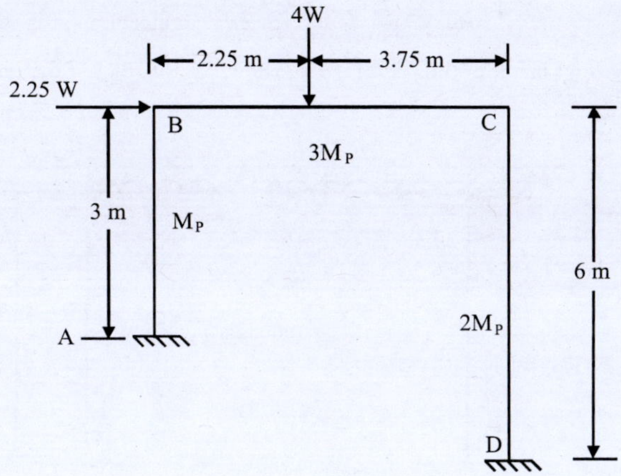

Fig. 8.49

[**Ans.** 0.565 M_P in combined mechanism]

13. Determine the load factor λ for the frame shown in Fig. 8.50.

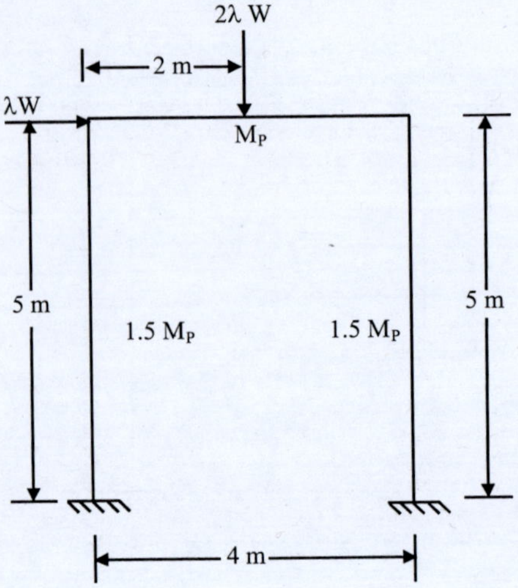

Fig. 8.50

$$\left[\textbf{Ans. } \lambda = \frac{7M_P}{9W} \text{ and the critical mechanism is the combined mechanism}\right]$$

14. Determine the load factor of the portal frame shown in Fig. 8.51 if the plastic moment capacity of all members is 100 kNm.

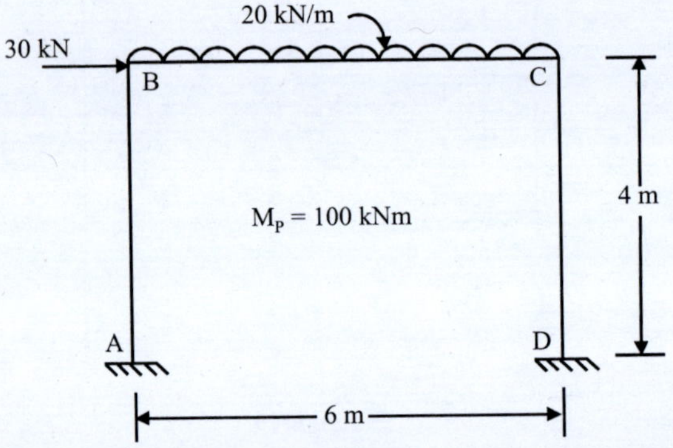

Fig. 8.51

[**Ans.** Load factor, λ = 1.996]

CHAPTER 9
REDUNDANT FRAMES

9.1 INTRODUCTION

This chapter deals with pin jointed redundant frames. Pin jointed frames, usually made of steel (rarely, made up of RC, prestressed concrete, aluminium, FRP or wood) are employed for covering large spans. Roof trusses of spans in excess of 10 m, bridge trusses, components of cranes and hoisting devices, and towers are examples of pin jointed frames, determinate or redundant.

Determinate frames have only one advantage. They are easy to analyse. But the fact is that we use "pin joints" only for analysis. In practice, we find that realising pin joints is quite expensive. Hence, invariably all 'pin jointed' trusses are jointed by rivets or bolts or more commonly by welding, using a gusset plate at the junction of members to facilitate connection. (The analyst is normally not told about it lest he should feel hurt!). The altered scheme of connections is going to change the forces in the members, but this is conveniently ignored. In important constructions however, an analysis is made for 'secondary stresses' arising out of the departure from the 'pin joint' assumption.

A determinate, pin jointed structure has a major drawback, from the safety point of view. Reliability engineers would tell us that if there are 12 members in a truss, and if each member has a reliability of 0.99, the net reliability is $(0.99)^{12} = 0.886$. This is because, even if any one of the members of the truss fails, the whole truss would fail. There is no second line of defense.

In an indeterminate truss, we add more structural members than the minimum required for stability. (This can also be done by adding additional external constraints, in which case the truss would be *externally* indeterminate).

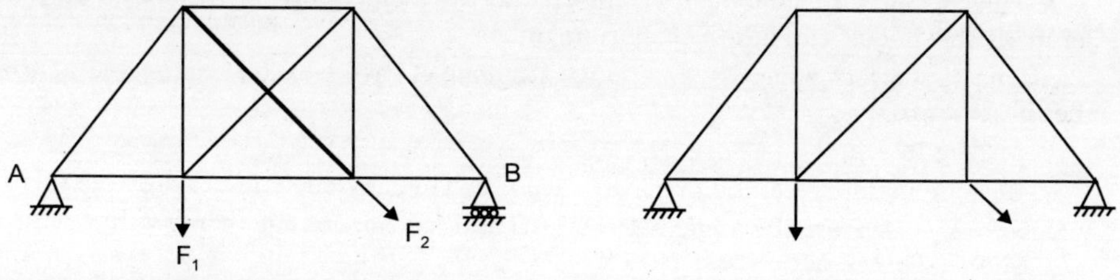

(a) Internally indeterminate externally determinate

(b) Externally indeterminate internally determinate

Fig. 9.1

9.2 REDUNDANCY

9.2.1 External Redundancy

Basically, for plane frames, there are 3 conditions of overall equilibrium. They are,

(*i*) $\Sigma H = 0$ (algebraic sum of all horizontal forces must be zero)

(*ii*) $\Sigma V = 0$ (algebraic sum of all vertical forces must be zero) and

(*iii*) $\Sigma M = 0$ (algebraic sum of all moments about any point in the plane of the truss must be zero)

Since we consider only structures in equilibrium, these 3 conditions must always be satisfied. So we always get 3 equations of equilibrium involving all forces on the truss. The forces would include external active forces like live loads, dead loads, wind loads, etc. and all passive reactions from the supports.

Referring to Fig. 9.1 (*a*), there can only be 3 possible reactions, namely, a horizontal force and a vertical force at A, and a vertical force alone at B (V_A, H_A and V_B). These, and the known active forces F_1, F_2, etc., shall obey the 3 conditions of equilibrium. So we can find V_A, H_A and V_B. The truss in Fig. 9.1 (*a*) is therefore *externally determinate.*

The truss in Fig. 9.1 (*b*) however will have 4 external reactions, V_A, V_B, H_A and H_B. Again there are only 3 conditions of equilibrium involving these 4 unknowns. We cannot solve for the unknowns from the conditions of equilibrium alone. Thus, the frame in Fig. 9.1(*b*) is *externally statically indeterminate*. Since the number of unknown reactions is *one* more than the number of equations available, the degree of external indeterminacy is *one.*

9.2.2 Internal Redundancy

Redundant means excess. If a frame requires only 9 members to ensure stability, and there are actually 10 members, we say one member is redundant. (If a team requires only 11 players for completeness and there are actually 12 players, we say one player is redundant).

But a redundant member is not a waste. It can serve a number of useful purposes. It can strengthen the truss. It can reduce the forces induced in the other members due to the external loading. It can continue to keep the truss stable when one of the other members fail (just as the 12th man can take the field in case one of the other eleven players get hurt).

Now, how do we decide how many members are just adequate?

A truss is made up of members connected at the 2 ends to other members. The so called pin joints at the 'nodes' can join 2 or more members.

A truss is perfect when it has enough number of members (m) connected at the right number of joints (j).

Referring to Fig. 9.2,

(*a*) has one member and no joint. $m = 1, j = 0$. This obviously is not a frame / truss.

(*b*) $m = 2, j = 1$. Here the angle between AC and AB can change to any value. ABC has no shape. This is not a truss either.

(*c*) $m = 3, j = 3$. This is the most elementary truss. ABC has a shape and can retain the shape except for elastic deformations.

So, a *truss* must have at least 3 members and 3 joints. If this truss has to become a *structure* it must have adequate number of supports (and support reactions like V_A, V_B and H_A as shown in Fig. 9.2(*h*)). The truss in Fig. 9.2(*h*) is internally and externally determinate and stable.

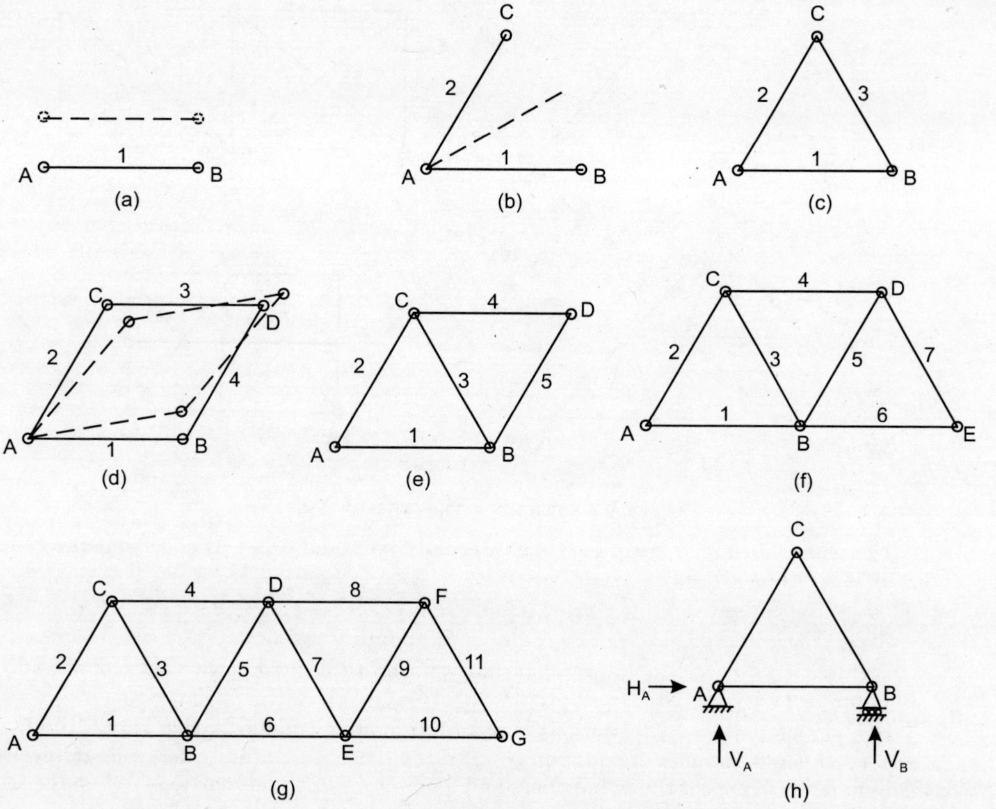

Fig. 9.2

In Fig. 9.2(*d*), $m = 4$ and $j = 4$. This quadrilateral cannot retain its shape and is free to change its geometry as shown in dotted lines. To make this perfect, we have to introduce one more truss member, say along BC (see Fig. 9.2(*e*))

In Fig. 9.2(*e*) $m = 5;\ j = 4$

In Fig. 9.2(*f*) $m = 7;\ j = 5$

In Fig. 9.2(*g*) $m = 11;\ j = 7$

We can see that starting from the triangle in Fig. 9.2(*c*) we have to have 2 more members to create each additional joint like, D, E, F and G.

So the relation between m and j for a perfect determinate truss is

$$m = 2j - 3 \qquad \text{...(9.1)}$$

If m is greater than $2j - 3$, by a number i, then i is the degree of redundancy or indeterminacy. We can easily verify in the cases Fig. 9.2 (*c*), (*e*), (*f*) and (*g*) that $m = 2j - 3$. They are all internally determinate. The frame in Fig. 9.2 (*d*) wants one more member. It is one degree *deficient*.

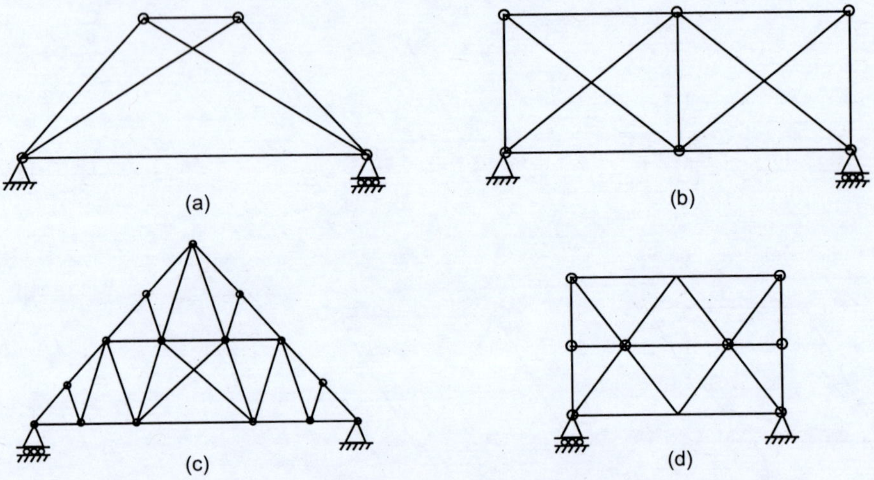

Fig. 9.3. Internally indeterminate trusses.

The trusses in Fig. 9.3(*a*) to (*d*) are internally redundant. Let us check out the redundancy by using the formula

$$ii = m - (2j - 3) \qquad ...(9.2)$$

Please notice that all 4 are externally determinate.

Let us not go under the impression that we are fitting the known values of m and j to get equation (9.1).

In a pin jointed truss, 2 equations of equilibrium are available at each joint ($\Sigma H = 0$ and $\Sigma V = 0$). For j joints, number of equations $= 2j$. If the truss is externally determinate, we will have m unknown forces for the m members and 3 unknown external reactions. When the truss is determinate,

Number of unknown forces = number of reactions

$$m + 3 = 2j \; ; \; m = 2j - 3$$

Table 9.1. Finding degrees of internal redundancy (ii).

Case (Fig. 9.3)	*m*	*j*	*2j – 3*	*ii*
(*a*)	6	4	5	1
(*b*)	11	6	9	2
(*c*)	30	15	27	3
(*d*)	19	10	17	2

9.2.3 Combination of Internal and External Redundancies

A frame can be *both internally and externally* redundant. There are some grey areas in the combination. The truss in Fig. 9.4(*a*) is attached to a wall (or column).

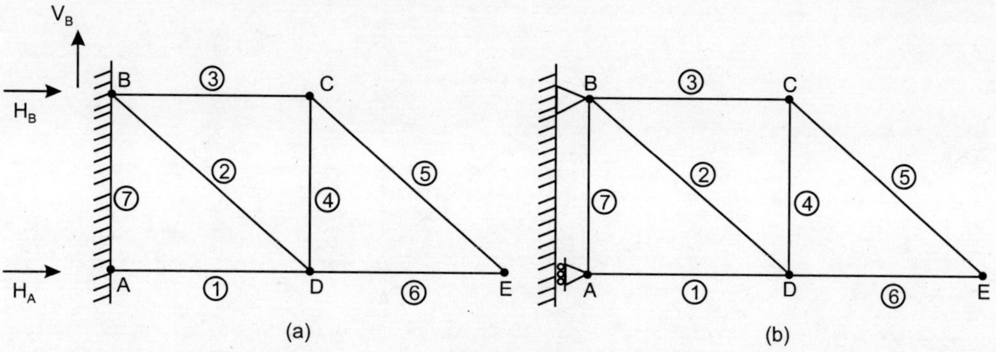

Fig. 9.4

Externally it has only 3 reactions, H_A, H_B and V_B. V_A, the vertical force at A has to be absent since all members in a pin jointed frame will take only axial forces (tension or compression). So, the reaction at A is necessarily horizontal. Thus $V_A = 0$. We conclude that the truss is externally determinate.

We also find that $m = 6$ and $j = 5$, $2j - 3 = 7$ which lets us conclude that the truss is deficient. This is not the case. The part of the wall AB between hinges A and B must also be treated as a member. So $ii = 7 - (2 \times 5 - 3) = 0$; determinate and stable.

This truss can be visualised in another way. A and B, let us say are ground points. AB has infinite rigidity and does not deform. If we connect (1) and (2) and hinge them at D, we get a stable triangle. Then (3) and (4) anchored at B and D are connected at C. (5) and (6) are anchored at C and D and connected at E. Thus we go on making stable additions, adding 2 members and one joint in every step. Thus for a perfect frame emerging from 2 ground (or wall) stations,

$$m = 2j$$

By our new reasoning, $m = 6$ and $j = 3$. Then $i = m - 2j$, where j does not include ground joints. The truss in Fig. 9.4(b) falls under the routine case.

Number of external forces = 3; externally determinate.

Indeterminacy, $\quad ii = m - (2j - 3)$

$$ii = 7 - (2 \times 5 - 3) = 0.$$

9.2.4 Reaction Components Possible

Unlike in beams and flexure frames, plane pin jointed frames have only 2 kinds of support conditions, namely hinge and rollers.

In Fig. 9.5, we have hinges at A and B. Each hinge can develop 2 reactions, horizontal and vertical. C and D can have only one reaction each.

Also, A and B have zero degree of freedom and C and D have one degree of freedom each. It must be noted that each of these reactions V_A to H_D, can be either positive or negative. Table 9.2 lists a large number of trusses and gives information about their degrees of redundancy.

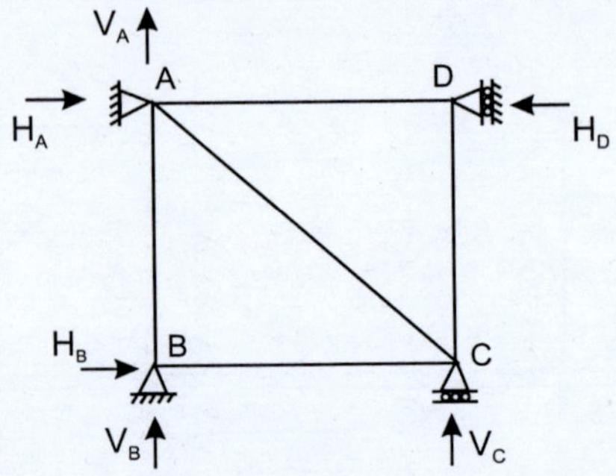

Fig. 9.5

Table 9.2. Determination of degrees of redundancy.

r = Possible reaction components ; R = Number of condition equations available

S. No.	*Truss/Frame*	*Internal Redundancy (ii)*	*External Redundancy (ie)*	*Total degree of static indeterminacy (I = ii + ie)*
1.	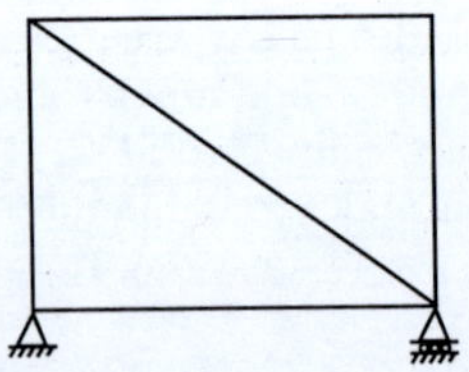	$m = 5; j = 4$ $ii = m - (2j - 3)$ $= 5 - (2 \times 4 - 3)$ $= 0$	$r = 3$; R = 3 $ie = r - \text{R}$ $= 3 - 3$ $= 0$	0
2.	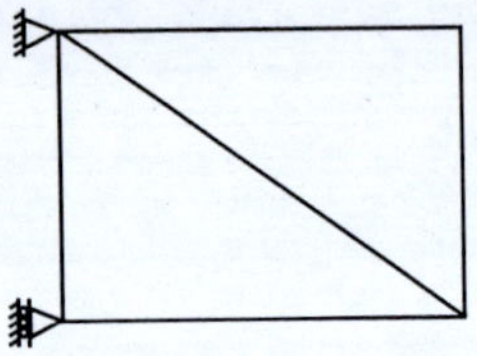	$m = 5; j = 4$ $ii = 0$	$r = 3$; R = 3 $\therefore \quad ie = 0$	0
3.	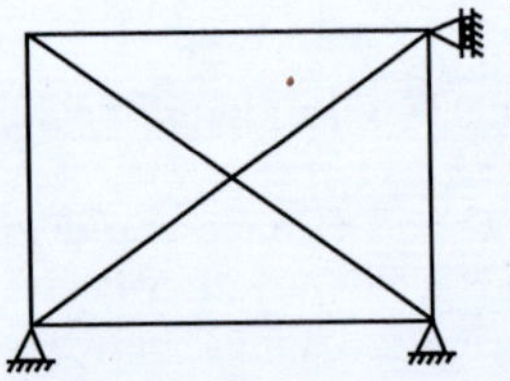	$m = 6; j = 4$ $ii = 6 - (2 \times 4 - 3)$ $= 1$	$r = 5$; R = 3 $ie = r - R$ $= 5 - 3$ $= 2$	$1 + 2 = 3$

4.	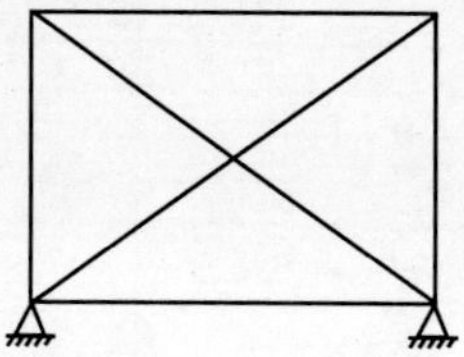	$m = 6; j = 4$ $ii = 1$	$r = 4$; R = 3 $ie = 4 - 3$ $= 1$	$1 + 1 = 2$
5.	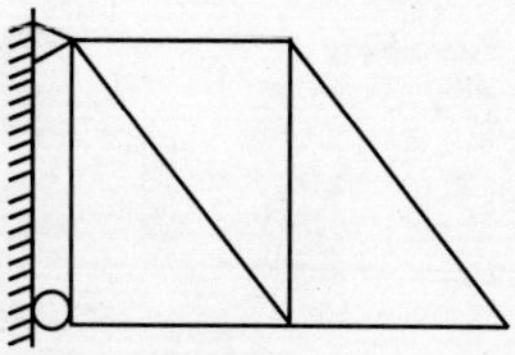	$m = 7; j = 5$ $ii = m - (2j - 3)$ $= 7 - (2 \times 5 - 3)$ $= 0$	$ie = 0$	$0 + 0 = 0$
6.	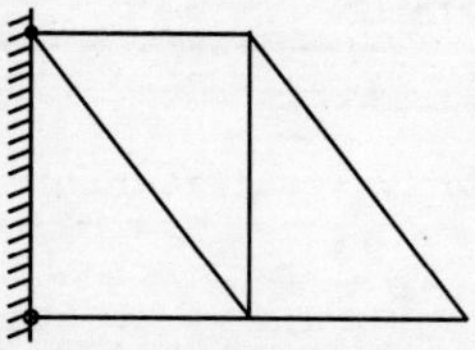	$m = 6; j = 3$ $ii = m - 2j$ $= 6 - (2 \times 3)$ $= 0$	$ie = 0$	0
7.	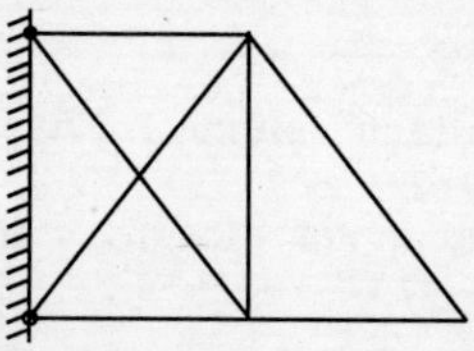	$m = 7; j = 3$ $ii = m - 2j$ $= 7 - (2 \times 3)$ $= 1$	$ie = 0$	$1 + 0 = 1$
8.	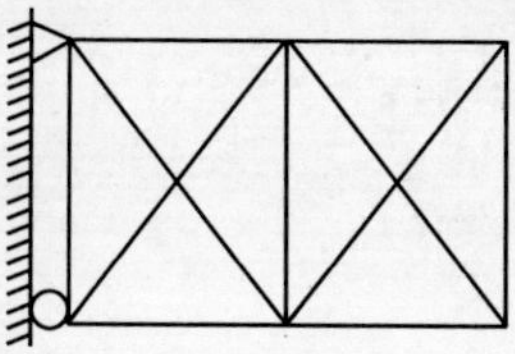	$m = 11; j = 6$ $ii = m - (2j - 3)$ $= 11 - (2 \times 6 - 3)$ $= 2$	$ie = 0$	$2 + 0 = 2$
9.	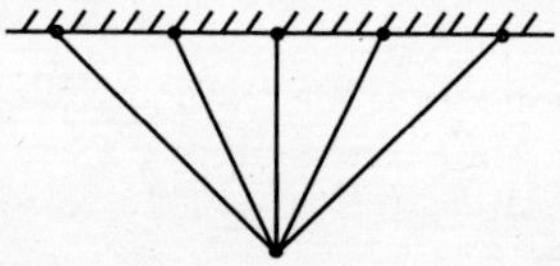	$m = 5; j = 1$ $ii = m - 2j$ $= 5 - (2 \times 1)$ $= 3$	$ie = 0$	$3 + 0 = 3$

10.	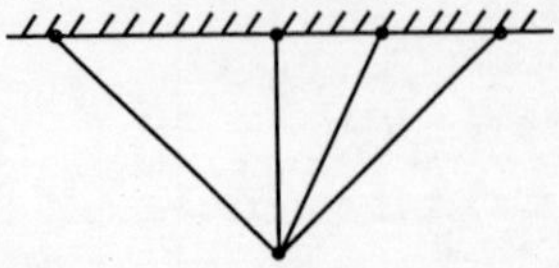	$m = 4; j = 1$ $ii = m - 2j$ $= 4 - (2 \times 1)$ $= 2$	$ie = 0$	$2 + 0 = 2$
11.	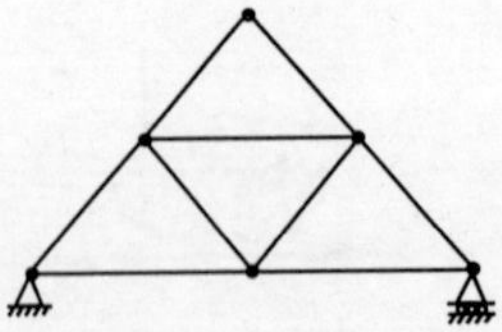	$m = 9; j = 6$ $ii = m - (2j - 3)$ $= 9 - (2 \times 6 - 3)$ $= 0$	$ie = 3 - 3$ $= 0$	$0 + 0 = 0$

9.3 METHODS OF SOLVING REDUNDANT FRAMES

9.3.1 General

The following are some of the methods of solving for forces in redundant pin jointed plane frames.

1. Matrix flexibility method
2. Matrix stiffness method
3. Maxwell's unit load method
4. Castigliano's strain energy method

Among these, the first two methods have been covered in Volume 1, Chapters 4 and 5. The other two would be dealt with here.

Actually, both method 3 and method 4 make use of strain energy concept to solve for the redundant member forces or redundant reactions.

9.3.2 Strain Energy Method

A stressed truss member carries a strain energy of $\frac{P^2 l}{2AE}$, where P is the tension or compression (Bending moments, shears and torsion are ruled out in truss members) in the member. Thus the total strain energy stored in a pin jointed frame, be it determinate or indeterminate is given by

$$U = \sum \frac{P^2 l}{2AE} \quad \text{...(9.3)}$$

the summing being carried out for all members.

But P values will not be readily available for all members if the frame is indeterminate. In this case, if there are i indeterminate members, let us remove them all and assume that the indeterminate forces are $T_1, T_2, ..., T_n,.....$ We can get P values in terms of T_n. In other words, the P values are functions of T_n $(n = 1, i)$.

Castigliano's theorem states that in any indeterminate structure, the indeterminate forces take such values as to make the total strain energy a minimum. Hence,

$$\frac{\partial U}{\partial T_1} = \frac{\partial U}{\partial T_2} = ... = \frac{\partial U}{\partial T_n} = 0, \text{ for all } n \text{ ......}$$

Thus if there are i redundants, we get i equations like $\frac{\partial U}{\partial T_n} = 0$, from which we can compute all T_n values, namely all indeterminate forces. This method of solving redundant frames is called strain energy method. This involves differentiation of the strain energy with respect to each T_n. In the strain energy equation (9.3), only P is dependant on T_n. Hence,

$$\frac{\partial U}{\partial T_n} = \sum 2P \frac{\partial P}{\partial T_n} \frac{l}{2AE} = 0$$

$$\sum \frac{P \frac{\partial P}{\partial T_n} l}{AE} = 0 \qquad \text{...(9.4)}$$

P is the force in a truss member due to the given loading and due to T_n, the redundant forces. In a linear structure like a pin jointed frame internal forces will be proportional to the applied force(s). In other words, force P will carry a first power term of T_n. In this case differentiation with T_n will just amount to dropping T_n or replacing T_n with unity. For example, if

$$P = 17.2 + 1.71\, T_1,$$

$$\frac{\partial U}{\partial T_1} = 1.71$$

In other words, $\frac{\partial U}{\partial T_1}$ is the value of force P if the load T_1 were equal to unity.

9.3.3 Maxwell's Unit Load Method

The last concept we arrived at in strain energy method is made use of in the so called unit load method attributed to Clark Maxwell.

Supposing we have to solve for the forces in a one degree redundant truss, having 6 members (including the redundant). Then,

1. Remove the redundant member and compute the internal forces F in the other 5 members due to the applied loads.
2. Apply a unit tensile force corresponding to what the redundant member would apply at its two nodal joints. Compute the 5 internal forces k. Let T be the force in the redundant member.

Then,

$$\sum_{n=1}^{6} \frac{Pkl}{AE} = 0 \qquad \text{...(9.5)}$$

since k is in fact $\frac{\partial P}{\partial T}$. This comes with a little hitch. P values are in fact dependent on k values.

For member 1, if F_1 is the force when T is not acting and k_1, the force due to (T = 1), then

$$P_1 = F_1 + k_1 T$$

$$P_n = F_n + k_n T \qquad n = 1, 5$$

For the sixth member, $F_n = 0$, $k_n = 1$ and $P_n = T$

Equation (9.5) then becomes

$$\sum_1^6 \frac{(F + kT)\, kl}{AE} = 0; \; \sum_1^6 \frac{Fkl}{AE} + \sum_1^6 \frac{k^2 l}{AE} \times T = 0$$

Hence, $$T = -\frac{\sum_{1}^{6} \frac{Fkl}{AE}}{\sum_{1}^{6} \frac{k^2 l}{AE}} \qquad ...(9.6)$$

The denominator should contain 6 terms. For the 6th member, the redundant, $k = 1$ and $F = 0$. To be more explicit,

$$\sum_{1}^{6} \frac{Fkl}{AE} = \sum_{1}^{5} \frac{Fkl}{AE} \qquad \text{since } F_6 = 0$$

$$\sum_{1}^{6} \frac{k^2 l}{AE} = \sum_{1}^{5} \frac{k^2 l}{AE} + \frac{l_6}{A_6 E_6} \qquad \text{since } k_6 = 1$$

$$T = \frac{-\sum_{1}^{5} \frac{Fkl}{AE}}{\sum_{1}^{5} \frac{k^2 l}{AE} + \frac{l_0}{A_0 E_0}} \qquad ...(9.7)$$

This equation is valid for a 6 member truss with one internal redundant member. $\frac{l_0}{A_0 E_0}$ relates to the redundant member.

For a m member redundant truss with one member (internally) redundant,

$$T = \frac{-\sum_{1}^{m-1} \frac{Fkl}{AE}}{\sum_{1}^{m-1} \frac{k^2 l}{AE} + \frac{l_0}{A_0 E_0}} \qquad ...(9.8)$$

and $$P_n = F_n + k_n T \qquad ...(9.9)$$

Example 9.1 applies this procedure to solve a 6 member truss with $i = 1$.

Examples 9.2 and 9.3 show how an externally indeterminate truss is solved.

Example 9.1 *Find the stresses in all the members of the frame given in Fig. 9.6 in which the cross-sectional areas of vertical members are 3000 mm² each and those of all other members are 2200 mm². E = 2 × 10⁵ N/mm².*

Please note that in solving problems of redundant trusses some books make use of equation (9.6). Here particulars of *all the members* including redundants, (the forces in primary structure, forces due to unit loads, A, E, l, etc.) are tabulated and processed. Some books make use of equations (9.8) and (9.9). Here redundants are not tabulated, but dealt with as additional terms in the denominator. Both methods are equally valid. In this book equation (9.6) would be used in most of the examples.

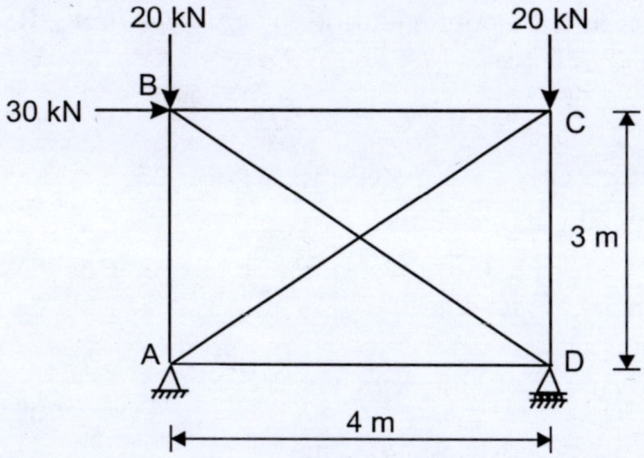

Fig. 9.6

Solution Degree of indeterminacy, $m = 6 \,;\, j = 4$

Internal indeterminacy, $ii = m - (2j - 3)$

$= 6 - (2 \times 4 - 3)$

$= 1$

External indeterminacy, *ie* $= r - R$

$= 3 - 3$

$= 0$

Hence the frame / truss is internally indeterminate to the first degree and externally determinate. The truss will be first made determinate by removing member AC. The primary structure is shown in Fig. 9.7.

Reactions :

$\Sigma H = 0$ gives $H_A = 30$ kN ($\leftarrow$)

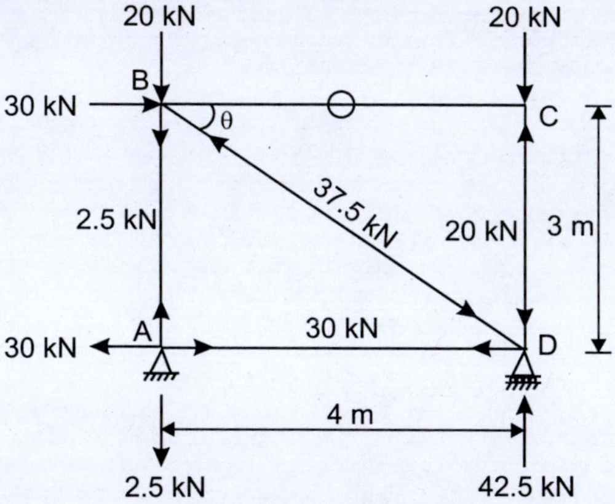

Fig. 9.7 (*a*) Primary structure with F values obtained.

Taking moments about D and equating to zero, (assuming R_A upwards)

$$R_A \times 4 + 30 \times 3 - 20 \times 4 = 0$$

$$R_A = \frac{-90 + 80}{4} = -2.5 \text{ kN}$$

$\therefore$ $R_A = 2.5$ kN (↓) ; $\Sigma V = 0$ gives, $R_D - 20 - 20 - 2.5 = 0$

$$R_D = 20 + 20 + 2.5 = 42.5 \text{ kN } (\uparrow)$$

Determination of F values : (Initially assume the force all the members to be tensile, to start with)

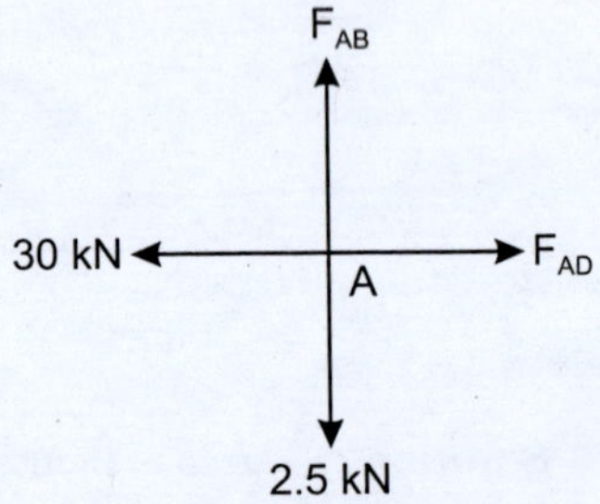

Fig. 9.7 (*b*)

Joint A :

$\Sigma V = 0$ gives $F_{AB} = 2.5$ kN (tensile)

$\Sigma H = 0$ gives $F_{AD} = 30$ kN (tensile)

Joint C :

$F_{CB} = 0$ since there is no other horizontal force.

$\Sigma V = 0$ gives

$$F_{CD} + 20 = 0$$

$$F_{CD} = -20$$

$$F_{CD} = 20 \text{ kN (Comp.)}$$

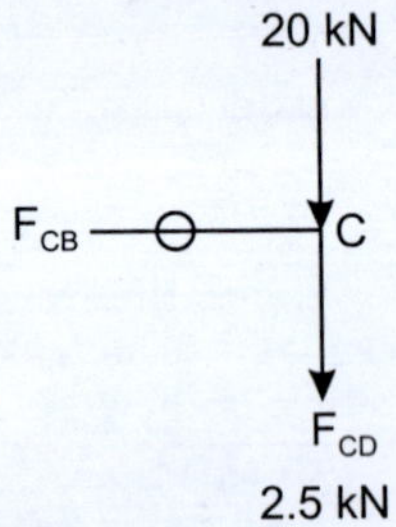

Fig. 9.7 (*c*)

Joint B :

$$\tan\theta = \frac{3}{4}$$

$$\theta = 36° \, 52'$$

$$\cos\theta = 0.8$$

$$90° - \theta = 53° \, 08'$$

$$\cos(90° - \theta) = 0.6$$

$\Sigma V = 0$ gives

$$F_{BD} \cos(90° - \theta) + 20 + 2.5 = 0$$

(↓) (↓) (↓)

$$F_{BD} = -\frac{22.5}{\cos(90° - \theta)} = -37.5 \text{ kN}$$

$$F_{BD} = 37.5 \text{ kN (Comp.)}$$

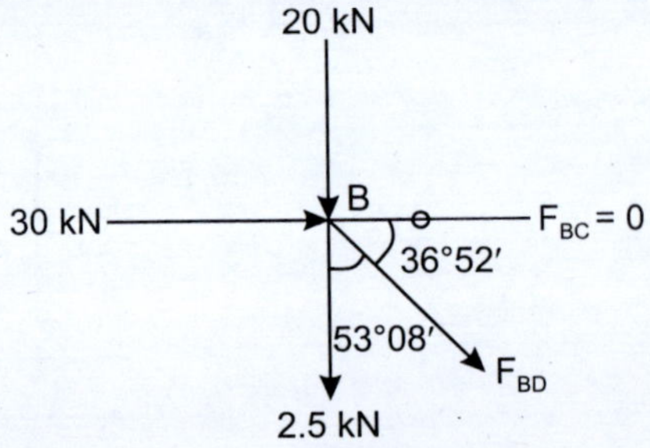

Fig. 9.7 (*d*)

Determination of k values:

Removing the external forces and applying unit tensile force along the direction of the redundant AC, *k* values will be obtained.

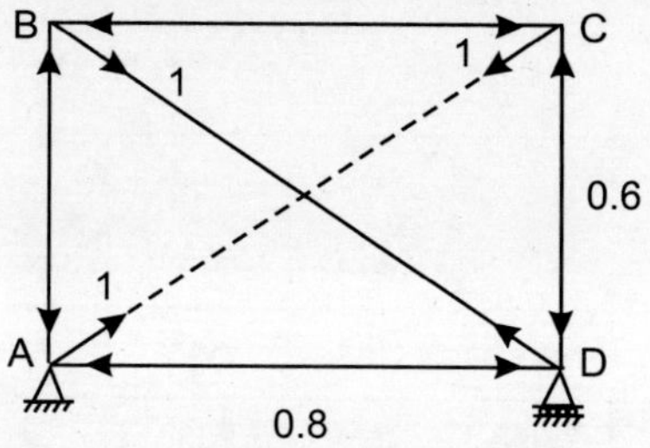

Fig. 9.8. *k* values.

Joint A :

$\Sigma V = 0$ gives

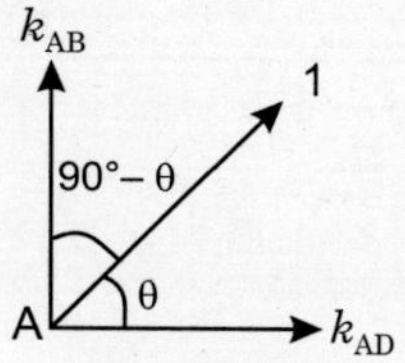

Fig. 9.8 (*a*)

$k_{AB} + 1 \times \cos(90° - \theta) = 0$

$k_{AB} = -0.6$

$k_{AB} = 0.6$ (Comp.)

$\Sigma H = 0$ gives

$k_{AD} + 1 \times \cos\theta = 0$

$k_{AD} = -0.8$

$k_{AD} = 0.8$ (Comp.)

Joint B :

$\Sigma V = 0$ gives

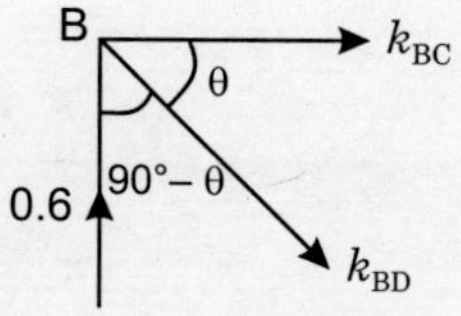

Fig. 9.8 (*b*)

$k_{BD} \times \cos(90° - \theta) - 0.6 = 0$

$$k_{BD} = \frac{0.6}{\cos(90° - \theta)} = 1$$

$k_{BD} = 1.00$ (Tensile)

$\Sigma H = 0$ gives

$k_{BC} + k_{BD}\cos\theta = 0$

$k_{BC} + 1 \times \cos\theta = 0$

$k_{BC} = -0.8$

$k_{BC} = 0.8$ (Comp.)

Joint C :

$\Sigma V = 0$ gives

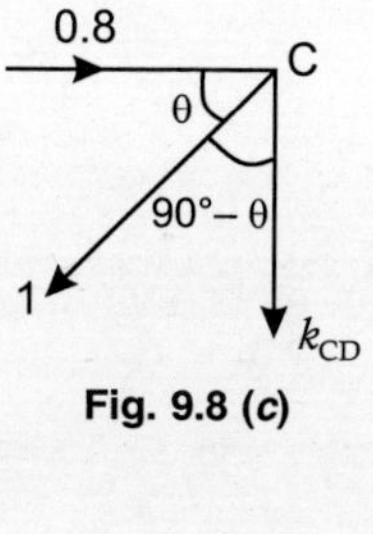

Fig. 9.8 (*c*)

$k_{CD} + 1 \times \cos(90° - \theta) = 0$

$k_{CD} = -0.6$

$k_{CD} = 0.6$ (Comp.)

Table 9.3

$E = 2 \times 10^5$ N/mm^2 = 200 kN/mm^2

S.No.	Member	F (kN)	k	l (mm)	A (mm^2)	$\frac{Fkl}{A}$	$\frac{k^2 l}{A}$	Final force $P = F + kT$ (kN)
1.	AB	2.5	– 0.6	3000	3000	–1.5	0.36	–6.866
2.	BC	0	– 0.8	4000	2200	0	1.16	–12.488
3.	CD	– 20	– 0.6	3000	3000	12.0	0.36	–29.366
4.	DA	30	– 0.8	4000	2200	–43.63	1.16	17.512
5.	BD	– 37.5	1.0	5000	2200	–85.23	2.27	–21.89
6.	AC	0	1.0	5000	2200	0	2.27	+15.61
					Σ	–118.36	7.58	

Force in the redundant member AC, $T = -\dfrac{\sum \dfrac{Fkl}{AE}}{\sum \dfrac{k^2 l}{AE}}$

$\because$ E is uniform, $\frac{1}{E}$ in numerator and denominator gets cancelled

$$T = \frac{-(-118.36)}{7.58} = \frac{+118.36}{7.58} = +15.61$$

P_{AC} = 15.61 kN (Tensile)

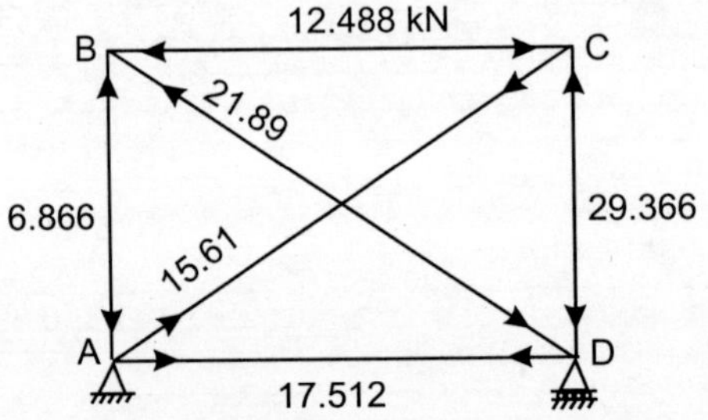

Fig. 9.9. Final forces.

Forces in other members are calculated using P = F + kT and presented in the last column of Table 9.3.

Example 9.2 *Determine the forces in the members of the truss shown to Fig. 9.10. AE is constant for all the members.*

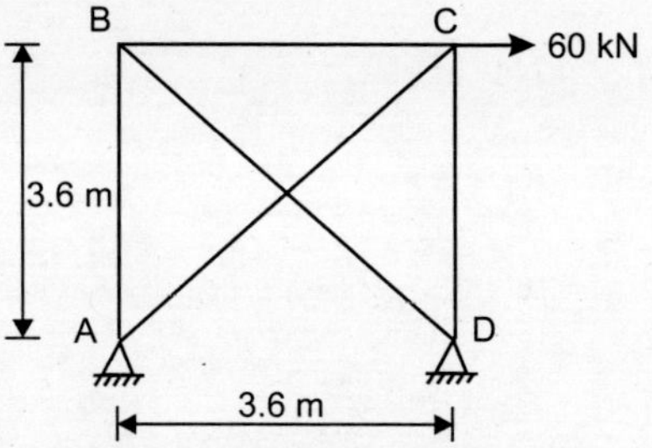

Fig. 9.10

Solution Degree of internal indeterminacy, $ii = m - (2j - 3)$

$$= 5 - (2 \times 4 - 3)$$
$$= 0$$

Degree of external indeterminacy, $ie = r - R$

$$= 4 - 3$$
$$= 1$$

Hence the structure is externally indeterminate to the first degree. The problem will be solved by using Maxwell's unit load method. First, the structure will be made determinate by replacing the hinged support by a roller support at D (*ie*) the horizontal reaction at D will be treated as the redundant. Hence the primary structure will be as shown in Fig. 9.11.

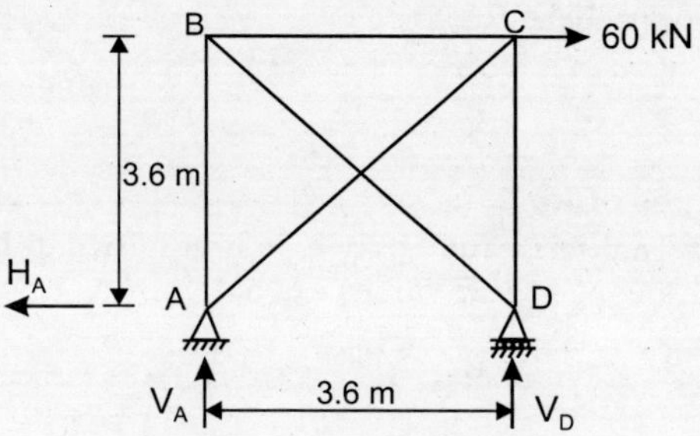

Fig. 9.11. Primary structure.

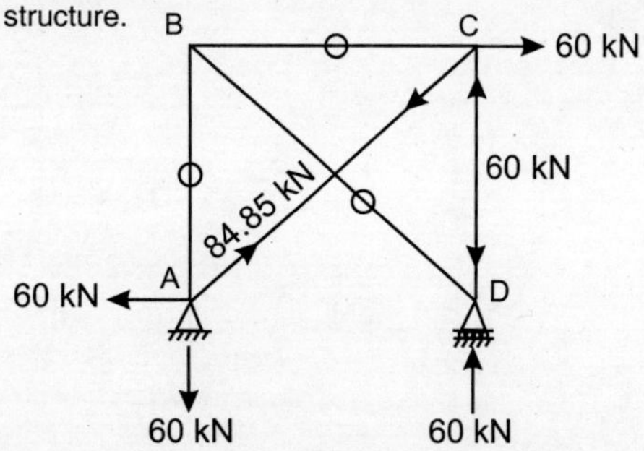

Fig. 9.12. Reactions and F values.

Determination of values of F :

Reactions :

$\Sigma H = 0$ yields $H_A = 60$ kN ($\leftarrow$)

Equating moments about D to zero,

$V_A \times 3.6 + 60 \times 3.6 = 0$

$V_A = -60$ kN

(*ie*) $V_A = 60$ kN ($\downarrow$)

$\Sigma V = 0$ yields

$V_D = 60$ kN ($\uparrow$)

Joint A :

$\Sigma H = 0$ yields

$F_{AC} \cos 45° - 60 = 0$

$F_{AC} = 84.85$ kN (Tensile)

$\Sigma V = 0$ yields

$F_{AB} + F_{AC} \cos 45° - 60 = 0$

$F_{AB} + 84.85 \cos 45° - 60 = 0$

$F_{AB} = 0$

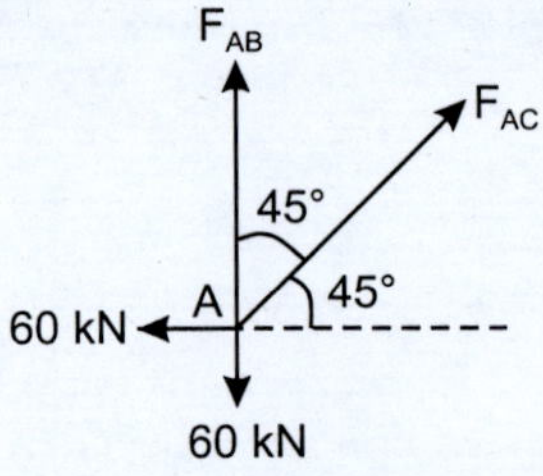

Fig. 9.12 (*a*)

Joint D :

$\Sigma H = 0$ yields, $F_{DB} \cos 45° = 0$

$F_{DB} = 0$

$\Sigma V = 0$ yields

$F_{DC} + 60 = 0$

$\therefore \quad F_{DC} = -60$ kN

(*ie*) $F_{DC} = 60$ kN (Comp.)

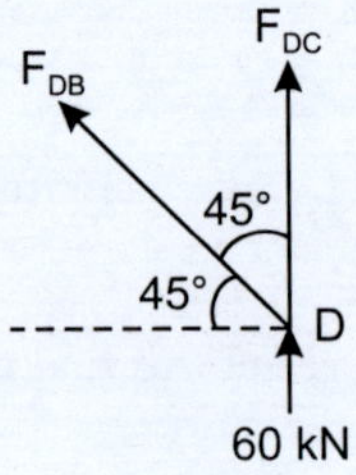

Fig. 9.12 (*b*)

Joint C :

$\Sigma H = 0$ yields

$-F_{CB} + 60 - 84.85 \cos 45° = 0$

$F_{CB} = 0$

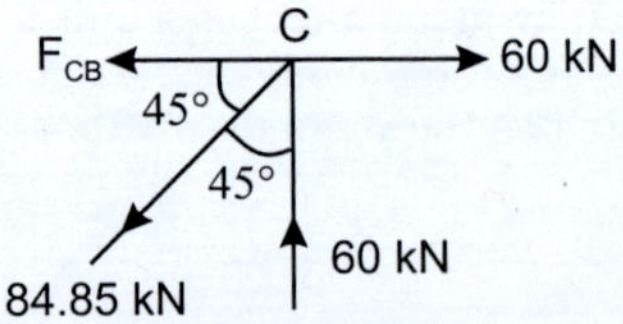

Fig. 9.12 (*c*)

Determination of values of k :

Removing the external force and applying a unit horizontal force at D (*i.e.* in the direction of the assumed redundant), the *k* values are obtained as below :

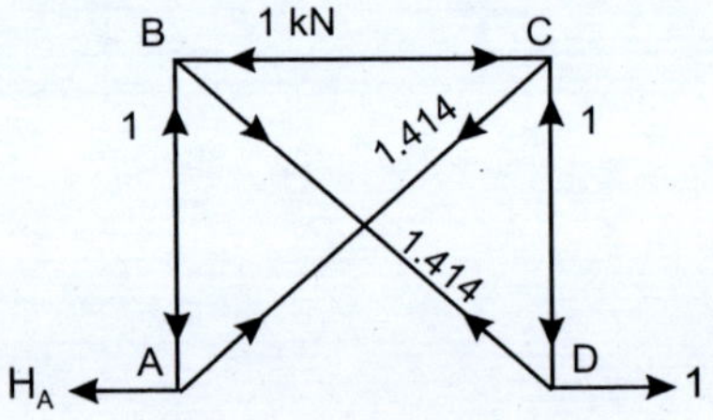

Fig. 9.13. *k* values.

$\Sigma H = 0$ yields

$H_A = 1$

Joint A :

$\Sigma H = 0$ yields

$k_{AC} \cos 45° - 1 = 0$

$k_{AC} = 1/\cos 45° = 1.414$ kN

$k_{AC} = 1.414$ kN (Tensile)

$\Sigma V = 0$ yields

$k_{AB} + k_{AC} \cos 45° = 0$

$k_{AB} = -1.414 \cos 45° = -1$

$\therefore \quad k_{AB} = 1$ kN (Comp.)

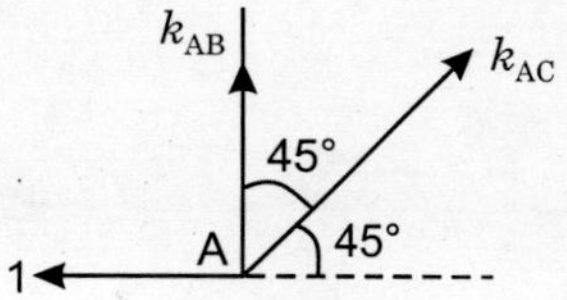

Fig. 9.13 (*a*)

Joint D :

$\Sigma H = 0$ gives

$-k_{DB} \cos 45° + 1 = 0$

$k_{DB} = 1/\cos 45° = 1.414$ kN (Tensile)

$\Sigma V = 0$ yields

$k_{DC} + k_{DB} \cos 45° = 0$

$k_{DC} = -1.414 \cos 45° = -1$

$k_{DC} = 1$ kN (Comp.)

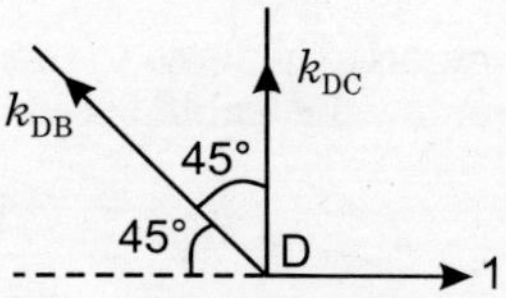

Fig. 9.13 (*b*)

Joint C :

$\Sigma H = 0$ gives

$-k_{CB} - 1.414 \cos 45° = 0$

$k_{CB} = -1.414 \cos 45° = -1$

$k_{CB} = 1$ kN (Comp.)

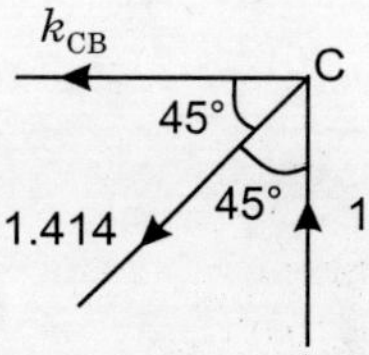

Fig. 9.13 (*c*)

Table 9.4

S.No.	*Member*	*F (kN)*	*k*	*Length, l(m)*	*Fkl*	k^2l	*Final force P = F + kT (kN)*
1.	AB	0	–1	3.6	0	3.6	26.53
2.	BC	0	–1	3.6	0	3.6	26.53
3.	CD	–60	–1	3.6	216	3.6	–33.47
4.	AC	84.85	1.414	5.09	610.69	10.18	47.34
5.	BD	0	1.414	5.09	0	10.18	–37.51
				Total	826.69	31.16	

Horizontal reaction at D, $T = -\dfrac{\sum \dfrac{Fkl}{AE}}{\sum \dfrac{k^2 l}{AE}}$

$$= -\frac{\sum Fkl}{\sum k^2 l} \quad \text{(AE is same for all members)}$$

$$= -\frac{826.69}{31.16}$$

$$T = -26.53$$

∴ Horizontal reaction at D = 26.53 kN (←)

Force in other members :

Force in AB = F + kT

$$P_{AB} = 0 + (-1)(-26.53) = 26.53 \text{ kN (Tensile)}$$

Similarly the force in members BC, CD, AC and BD are calculated and presented in the last column of the Table 9.4.

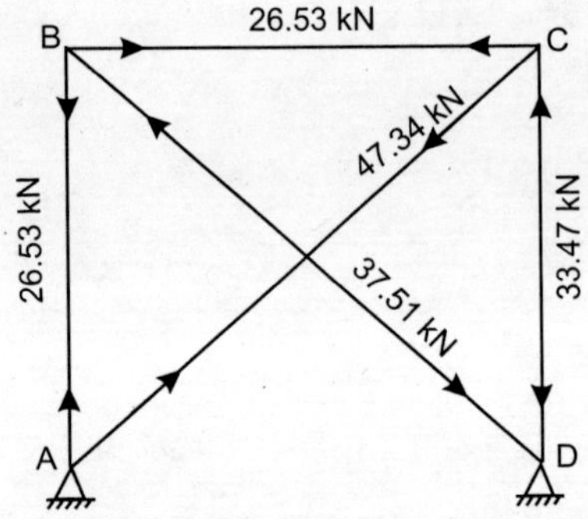

Fig. 9.14. Final forces.

Example 9.3 *A two span continuous truss is loaded as shown in Fig. 9.15. All the members are of the same material and have the same cross-sectional area. Find the reaction at the central support C. Also find the forces in all the members of the truss.*

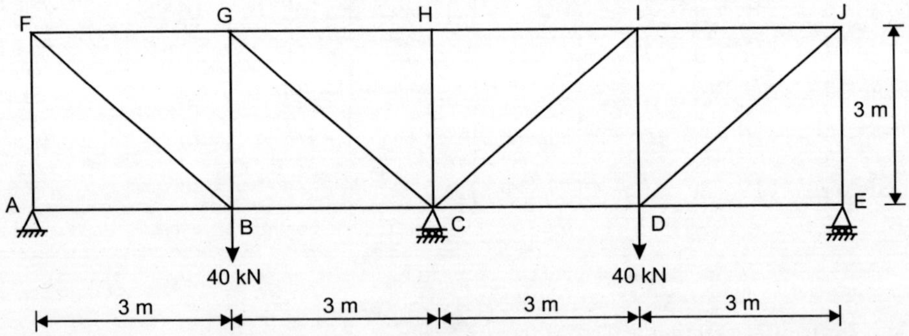

Fig. 9.15

Solution Degree of static indeterminacy :

$$m = 17; j = 10$$
$$ii = m - (2j - 3)$$
$$= 17 - (2 \times 10 - 3)$$
$$= 0$$
$$ie = r - R$$
$$= 4 - 3$$
$$= 1$$

Hence the truss is internally determinate and externally indeterminate to the first degree. The truss will be made determinate by removing the support at C.

Determinate of F values :

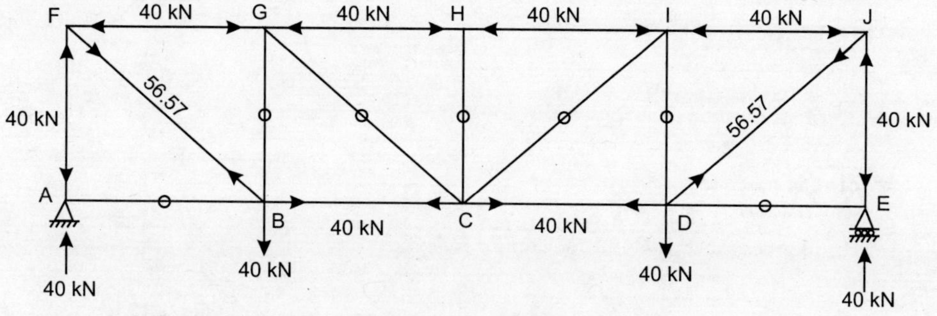

Fig. 9.16. F values.

By symmetry, $R_A = R_E = 40$ kN

Joint A :

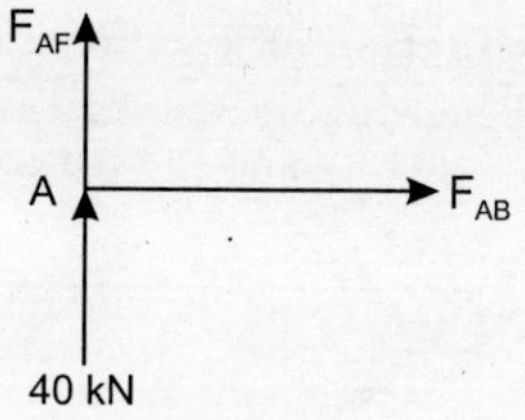

Fig. 9.16 (*a*)

$$\Sigma H = 0 \text{ gives}$$
$$F_{AB} = 0$$
$$\Sigma V = 0 \text{ gives}$$
$$F_{AF} + 40 = 0$$
$$F_{AF} = -40 \text{ kN}$$
$$\therefore \quad F_{AF} = 40 \text{ kN (Comp.)}$$

Joint F :

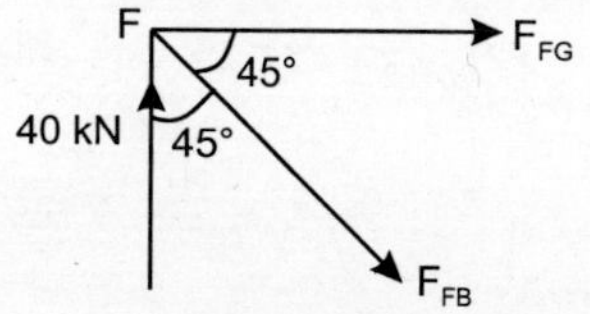

Fig. 9.16 (*b*)

$$\Sigma V = 0 \text{ gives}$$
$$-F_{FB} \cos 45° + 40 = 0$$
$$F_{FB} = 40/\cos 45° = 56.57 \text{ kN (Tensile)}$$
$$\Sigma H = 0 \text{ gives}$$
$$F_{FG} + F_{FB} \cos 45° = 0$$
$$F_{FG} + 56.57 \cos 45° = 0$$
$$F_{FG} = -40 \text{ kN}$$
$$F_{FG} = 40 \text{ kN (Comp.)}$$

Joint B :

$\Sigma V = 0$ gives

$56.57 \cos 45° + F_{BG} - 40 = 0$

$F_{BG} = 40 - 40 = 0$

$\Sigma H = 0$ gives

$F_{BC} - 56.57 \cos 45° = 0$

$F_{BC} = 40$ kN (Tensile)

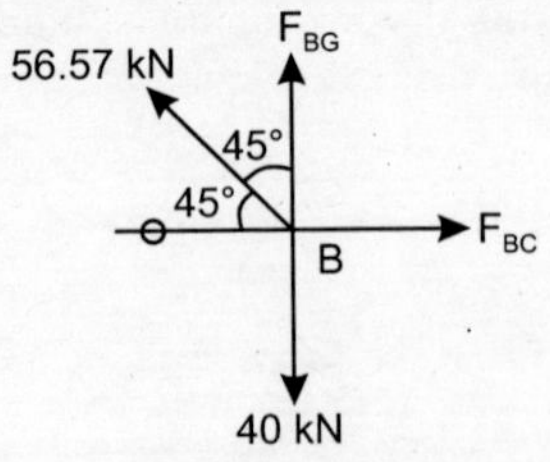

Fig. 9.16 (*c*)

Joint G :

$\Sigma V = 0$ gives

$F_{GC} = 0$

$\Sigma H = 0$ gives

$40 + F_{GH} + F_{GC} \cos 45° = 0$

$F_{GH} = -40$ kN $\qquad (\because F_{GC} = 0)$

$F_{GH} = 40$ kN (Comp.)

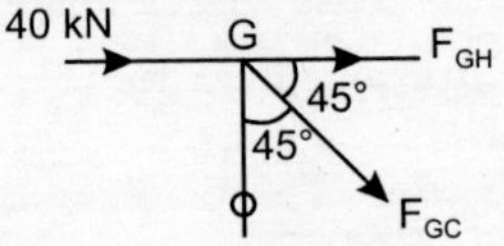

Fig. 9.16 (*d*)

Joint H :

$\Sigma V = 0$ gives $\quad F_{HC} = 0$

$\Sigma H = 0$ gives $\quad F_{HI} = 40$ kN (Comp.)

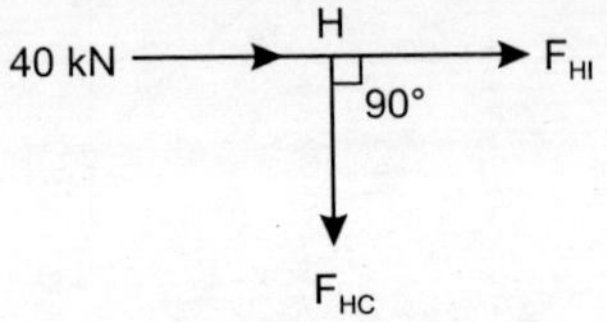

Fig. 9.16 (*e*)

By symmetry, the values of F to the right of HC will be a mirror reflection of the values to the left of HC.

Determination of k values :

Removing the external loads, a unit load will be applied along the direction of redundant reaction R_C and the truss will be analysed for *k* values.

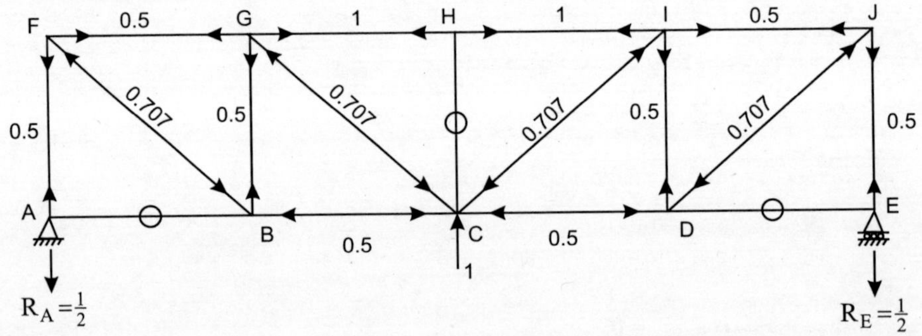

Fig. 9.17. *k* values.

By symmetry, $R_A = R_E = \dfrac{1}{2}$

Joint A :

$\Sigma H = 0$ gives

$k_{AB} = 0$

$\Sigma V = 0$ gives

$k_{AF} - 1/2 = 0$

$k_{AF} = 0.5$ (Tensile)

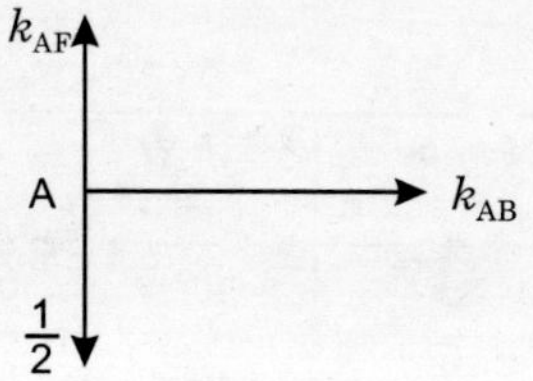

Fig. 9.17 (*a*)

Joint F :

$\Sigma V = 0$ gives

$k_{FB} \cos 45° + 0.5 = 0$

$k_{FB} = -0.5/\cos 45° = -0.707$

$k_{FB} = 0.5$ (Comp.)

$\Sigma H = 0$ gives

$k_{FG} + k_{FB} \cos 45° = 0$

$k_{FG} + (-0.707) \cos 45° = 0$

$k_{FG} = 0.5$ (Tensile)

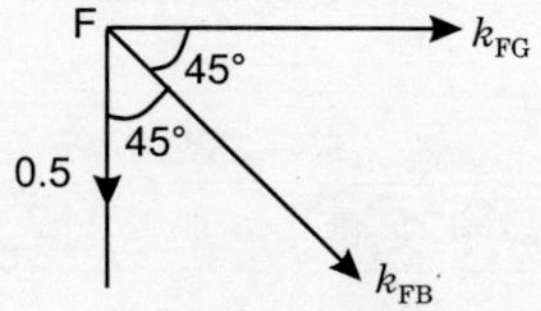

Fig. 9.17 (*b*)

Joint B :

$\Sigma V = 0$ gives

$k_{BG} - 0.707 \cos 45° = 0$

$k_{BG} = 0.5$ (Tensile)

$\Sigma H = 0$ gives

$k_{BC} + 0.707 \cos 45° = 0$

$k_{BC} = -0.5$

$k_{BC} = 0.5$ (Comp.)

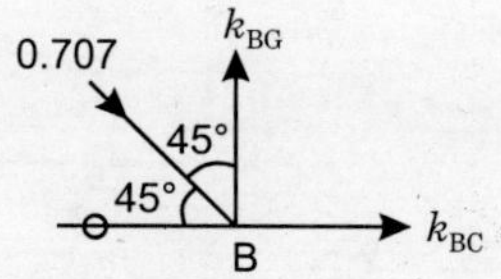

Fig. 9.17 (*c*)

Joint G :

$\Sigma V = 0$ gives

$k_{GC} \cos 45° + 0.5 = 0$

$k_{GC} = -0.5/\cos 45° = -0.707$

$k_{GC} = 0.707$ (Comp.)

$\Sigma H = 0$ gives

$k_{GH} + k_{GC} \cos 45° - 0.50 = 0$

$k_{GH} + (-0.707) \cos 45° - 0.50 = 0$

$k_{GH} = 1$ kN (Tensile)

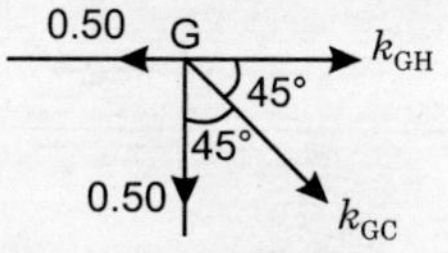

Fig. 9.17 (*d*)

Joint H :

$\Sigma V = 0$ gives

$k_{HC} = 0$

$\Sigma H = 0$ gives

$k_{HI} = 1$ (Tensile)

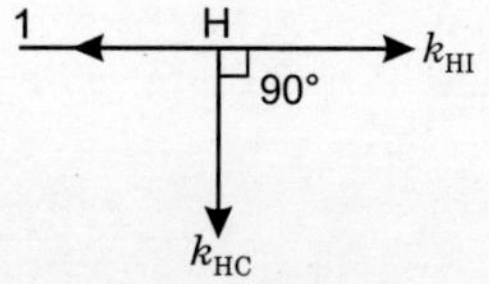

Fig. 9.17 (*e*)

By symmetry, the values of k to the right of HC will be a mirror reflection of the values to the left of HC.

Table 9.5

S.No.	*Member*	*F (kN)*	*k*	*l(mm)*	*Fkl*	k^2l	*Final force* $P = F + kT$ *(kN)*
1.	AF	– 40	0.5	3	– 60	0.75	– 17.07
2.	FG	– 40	0.5	3	– 60	0.75	– 17.07
3.	GH	– 40	+ 1.0	3	– 120	3.00	5.87
4.	HI	– 40	+ 1.0	3	– 120	3.00	5.87
5.	IJ	– 40	0.5	3	– 60	0.75	– 17.07
6.	JE	– 40	0.5	3	– 60	0.75	– 17.07
7.	ED	0	0	3	0	0	0
8.	DC	40	– 0.5	3	– 60	0.75	17.07
9.	CB	40	– 0.5	3	– 60	0.75	17.07
10.	BA	0	0	3	0	0	0
11.	FB	56.57	– 0.707	4.243	– 169.70	2.12	24.14
12.	BG	0	0.5	3	0	0.75	22.94
13.	GC	0	– 0.707	4.243	0	2.12	– 32.43
14.	CH	0	0	3	0	0	0
15.	CI	0	– 0.707	4.243	0	2.12	– 32.43
16.	ID	0	0.5	3	0	0.75	22.94
17.	DJ	56.57	– 0.707	4.243	– 169.70	2.12	24.14
		Total			– 939.4	20.48	

Redundant reaction at C, $T = -\dfrac{\sum \dfrac{Fkl}{AE}}{\sum \dfrac{k^2l}{AE}}$

Since A and E are same for all the members, $T = -\dfrac{\sum Fkl}{\sum k^2l}$

$$= -\frac{(-939.4)}{20.48} = 45.87 \text{ kN}$$

$$T = 45.87 \text{ kN}$$

$$\therefore \quad R_C = 45.87 \text{ kN } (\uparrow)$$

Force in member AF = F + kT

= – 40 + 0.5 × 45.87

= – 17.07 kN

All the final forces are listed in the last column of Table 9.5.

9.4 LACK OF FIT

In a determinate frame, a slight deviation in the length of a fabricated member would not even be observed. Of course it will produce a change in shape, but it will not create stresses.

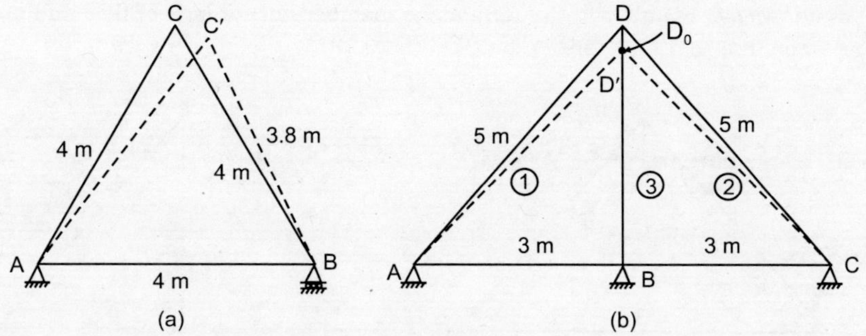

Fig. 9.18

For example, in Fig. 9.18(a), AB, BC and AC are to be theoretically 4 m long. Suppose, during fabrication BC has been made 3.8 m (*i.e.,* 0.2 m shorter than specification), C will not be at the desired location, but shift to C′, so that AC′ = AC = 4 m and BC′ = 3.8 m. In other words, after assemblage, the members will keep their original length. There would be no change in length, no strain, no stress and no internal force.

This is not the case with indeterminate frames. In Fig. 9.18(b) we have a 3 member indeterminate frame. The theoretical dimensions are, AB = BC = 3 m, AD = CD = 5 m and BD = 4 m.

Suppose BD has been made 0.2 m too short. Let us assemble the correct members first. AD and CD will joint at the right place, D. BD which is by mistake too short can not connect B and D. We will only have a member BD′ = (3.8 m), leaving a gap of 0.2 m.

Now we will have to force-fit this member. This will involve pressing D down (creating a compression in DA and DC) and pulling D′ up (creating tension on BD′) till D and D′ meet a common point D_0. This implies that even in the absence of an external load, members AD, BD and CD would get stressed.

Actually these stresses can be termed as prestress and can be advantageously adopted to reduce the danger of buckling in long compression members. In fact the spokes of bicycle wheels are pretentioned. Even at no load conditions all the spokes are in tension. So when a rider mounts the bike the compression induced in the vertical spokes only reduce the pretension and there is no net compressive force.

The forces induced in truss members and/or supports, due to lack of fit can be worked out as under.

Equation (9.8) reads as

$$T = \frac{\sum_{1}^{m-1} \frac{Fkl}{AE}}{\sum_{1}^{m-1} \frac{k^2 l}{AE} + \frac{l_0}{A_0 E_0}} = \frac{\sum_{1}^{m} \frac{Fkl}{AE}}{\sum_{1}^{m} \frac{k^2 l}{AE}}$$

Let us modify this to suit the lack-of-fit case in Fig. 9.18(*b*).

Let us remove member 3, the redundant member with a lack of fit δ and apply a unit force corresponding to a unit tension on BD (Fig. 9.19).

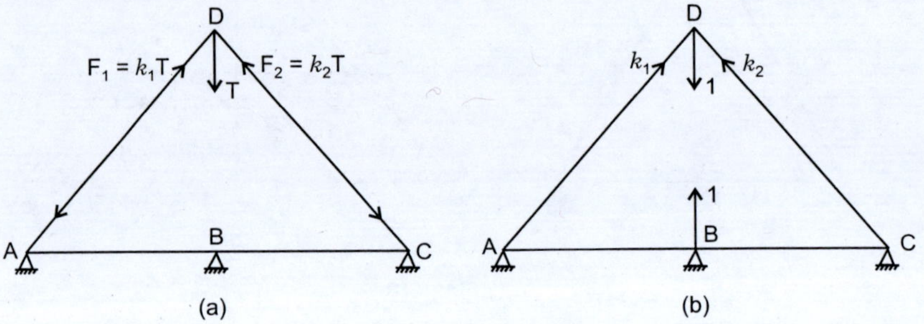

Fig. 9.19

If T is going to be the tension in BD, then the downward displacement of D is $\sum \frac{Fkl}{AE}$.

Obviously $F = T \times k$. Hence $\sum \frac{Fkl}{AE} = T \times \frac{k^2 l}{AE}$. The upward displacement of D′ is the elongation of 3 $= T \times \frac{l_0}{A_0 E_0}$.

Then,

$$\delta = T \times \sum \frac{k^2 l}{AE} + T \times \frac{l_0}{A_0 E_0}$$

$$T = \frac{\delta}{\sum_{1}^{m-1} \frac{k^2 l}{AE} + \frac{l_0}{A_0 E_0}} \qquad \text{...(9.10)}$$

Example 9.4 makes use of this equation.

Example 9.4 *While fabricating the pin jointed frame shown in Fig. 9.20, the member AC was the last member to be fitted and was found to be 1.5 mm short of the required length. Find the forces induced in all the members of the frame when the member AC is forced into position. The diagonal members are each 1000 mm² in area, while the remaining members are 2000 mm² in area. Take* $E = 2 \times 10^5$ *N/mm²*.

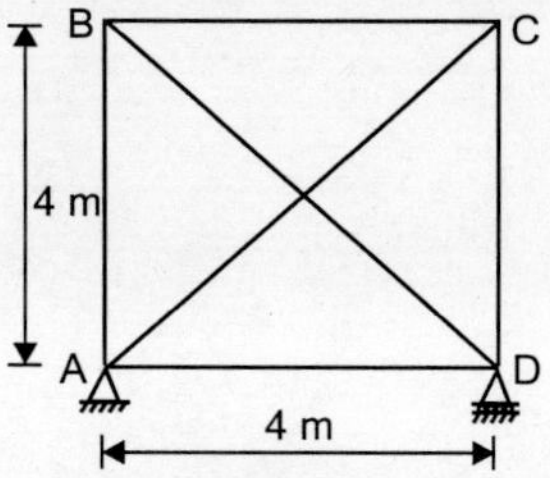

Fig. 9.20

Solution $m = 6; j = 4$

Degree of internal indeterminacy, $I = m - (2j - 3)$

$= 6 - (2 \times 4 - 3)$

$= 1$

Degree of external indeterminacy, $E = r - R$

$= 3 - 3$

$= 0$

Hence the structure is internally indeterminate to the first degree. The member AC will be treated as the redundant and the structure will be made determinate by removing it. A unit force will be applied along the member AC and the truss will be analysed for k values.

Determination of values of k :

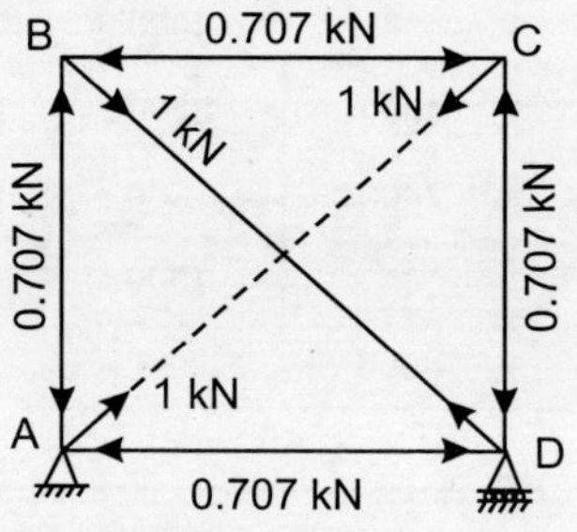

Fig. 9.21. *k* values.

Joint A :

$\Sigma H = 0$ gives

$k_{AD} + 1 \times \cos 45° = 0$

$k_{AD} = -0.707$ kN

$k_{AD} = 0.707$ kN (Comp.)

$\Sigma V = 0$ gives

$k_{AB} + 1 \times \cos 45° = 0$

$k_{AB} = -0.707$ kN

$k_{AB} = 0.707$ kN (Comp.)

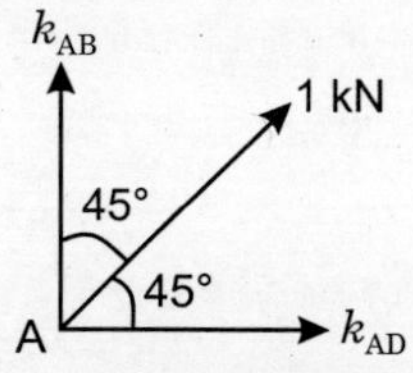

Fig. 9.21 (*a*)

Joint B :

$\Sigma V = 0$ gives

$0.707 - k_{BD} \cos 45° = 0$

$k_{BD} = 0.707 / \cos 45° = 1$ (Tensile)

$\Sigma H = 0$ gives

$k_{BC} + k_{BD} \cos 45° = 0$

$k_{BC} = -1 \times \cos 45° = -0.707$

$k_{BC} = 0.707$ kN (Comp.)

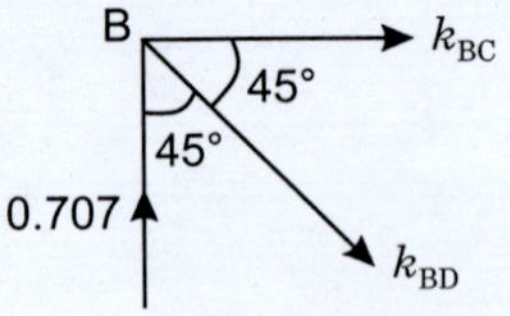

Fig. 9.21 (*b*)

Joint C :

$\Sigma V = 0$

$-k_{CD} - 1 \times \cos 45° = 0$

$k_{CD} = -0.707$

$k_{CD} = 0.707$ kN (Comp.)

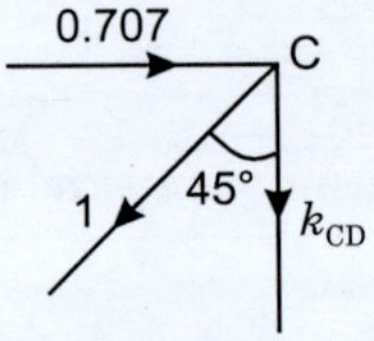

Fig. 9.21 (*c*)

Table 9.6

$E = 2 \times 10^5$ N/mm² = 200 kN/mm²

S.No.	*Member*	*k (kN)*	*Length, l (mm)*	*Area, A (mm²)*	k^2l	$\frac{k^2l}{AE}$	*Final forces P = k × T (kN)*
1.	AB	– 0.707	4000	2000	2000	0.005	– 13.902
2.	AD	– 0.707	4000	2000	2000	0.005	– 13.902
3.	BC	– 0.707	4000	2000	2000	0.005	– 13.902
4.	BD	1	5657	1000	5657	0.028	19.663
5.	CD	– 0.707	4000	2000	2000	0.005	– 13.902
					Σ	0.048	

Force in any redundant member, $T = \pm \dfrac{\delta}{\sum \dfrac{k^2 l}{AE} + \dfrac{l_0}{A_0E_0}}$

$\delta = +1.5$ mm, since AC is shorter than its required length

$$T = \frac{1.5}{0.048 + \dfrac{5657}{1000 \times 200}} = 19.663 \text{ kN}$$

∴ Force in member AC = 19.663 kN (Tensile)

Force in AB = $kT = -0.707 \times 19.663$

$P_{AB} = -13.902$ kN

$P_{AB} = 13.902$ kN (Comp.)

Similarly, forces in members AD, BC, BD and CD are calculated and presented in the last column of Table 9.6.

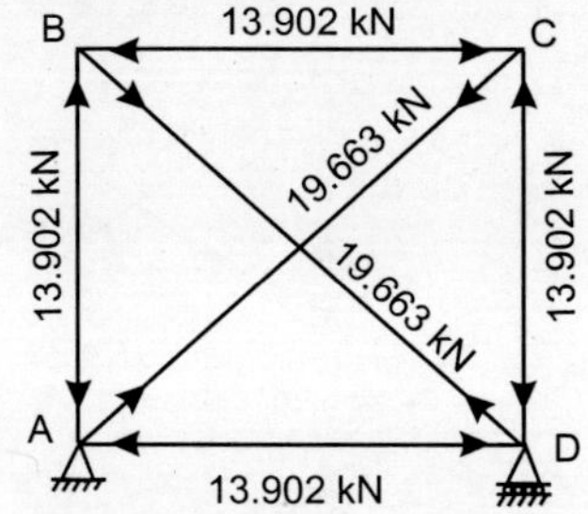

Fig. 9.22. Final forces.

9.5 TEMPERATURE STRESSES

Like in the case of lack of fit, changes in member lengths due to change of temperature would produce no internal forces in determinate frames. But redundant frames will develop forces (and stresses) due to temperature change.

If only one member develops change of length due to temperature change, the problem becomes very similar to the lack of fit problem. Equation (9.8) would be used, taking $\delta = -\,l_0\,\alpha T$, if T is the increase in temperature.

Now when will one member alone suffer a change of temperature? This can happen when direct sunlight heats one or more members, or when some welding is done on a truss member, or when an AC duct is in contact with a member.

When all members of an internally redundant truss suffer change of length due to the same temperature rise or drop, the truss would enlarge or shrink as a whole and not develop internal stresses. But when it comes to external indeterminacy, the situation is different. The distance between hinges anchored to ground may not change at the same rate as the truss members and can cause temperature stresses.

Examples 9.5 and 9.6 tackle problems of temperature changes in trusses. They employ the modified version of equation (9.8) as under.

$$T = \pm \frac{l_0 \alpha t}{\sum_{1}^{n-1} \frac{k^2 l}{AE} + \frac{l_0}{A_0 E_0}} \qquad \text{...(9.11)}$$

Here l_0, A_0 and E_0 refer to the (redundant) member which suffers change of temperature. Of the + or – sign prefixing R.H.S, we use – sign when temperature increases and + sign when temperature drops.

Example 9.5 *The frame shown in Fig. 9.23 is supported from a rigid wall at D and E. All members have the same cross-sectional area of 1060 mm². If there is a rise of temperature of member BC by 32°C, determine the resulting forces in the members. Coefficient of expansion α = 0.000012 per degree centigrade.*

$$E = 2.08 \times 10^5 \text{ N/mm}^2.$$

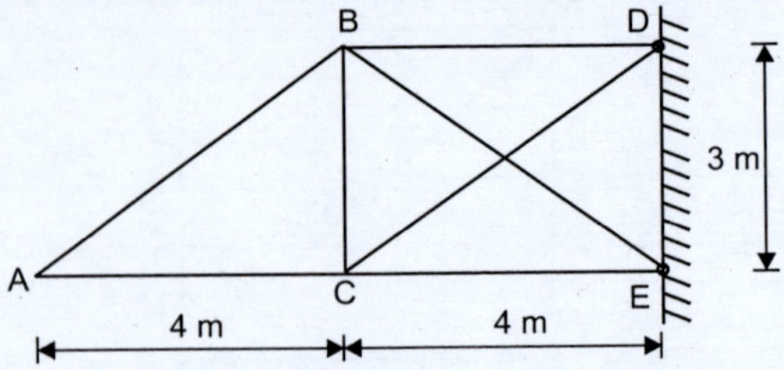

Fig. 9.23

Solution Degree of internal indeterminacy : $m = 7$; $j = 3$

$$ii = m - 2j \qquad \text{(frame emerging from ground)}$$
$$= 7 - 2 \times 3 = 1$$

Hence, the frame is internally indeterminate to the first degree. The member BC will be treated as redundant and the structure will be made determinate by removing it. A unit tensile force will be applied along member BC and the frame is analysed for k values.

Determination of values of k :

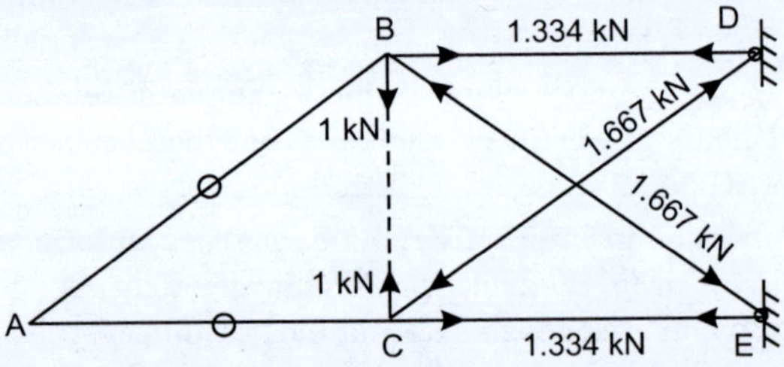

Fig. 9.24. *k* values.

$$\tan \theta = \frac{3}{4}$$
$$\theta = 36°52'$$
$$90° - \theta = 53°08'$$

Joint A :

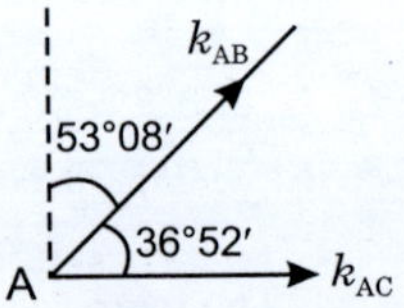

Fig. 9.24 (*a*)

$$\Sigma V = 0 \text{ gives}$$
$$k_{AB} \cos 53°08' = 0$$
$$k_{AB} = 0$$
$$\Sigma H = 0 \text{ gives}$$
$$k_{AC} + k_{AB} \cos 36°52' = 0$$
$$k_{AC} = 0 \qquad (\because k_{AB} = 0)$$

Joint C :

$$\Sigma V = 0 \text{ gives}$$
$$k_{CD} \cos 53°08' + 1 = 0$$

$k_{CD} = -1/\cos 53°08' = -1.667$ kN

$k_{CD} = 1.667$ kN (Comp.)

$\Sigma H = 0$ gives

$k_{CE} + k_{CD} \cos 36°52' = 0$

$k_{CE} + (-1.667) \cos 36°52' = 0$

$k_{CE} = 1.334$ kN (Tensile)

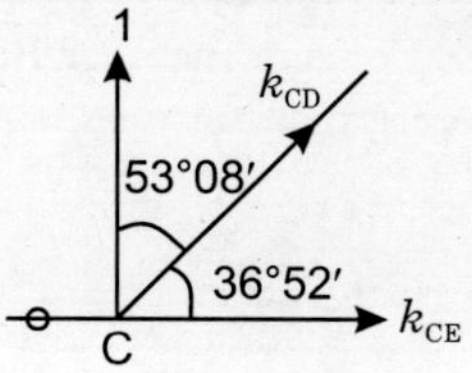

Fig. 9.24 (*b*)

Joint B :

$\Sigma V = 0$ gives

$-k_{BE} \cos 53°08' - 1 = 0$

$k_{BE} = -1/\cos 53°08' = -1.667$

$k_{BE} = 1.667$ kN (Comp.)

$\Sigma H = 0$ gives

$k_{BD} + k_{BE} \cos 36°52' = 0$

$k_{BD} + (-1.667) \cos 36°52' = 0$

$k_{BD} = 1.334$ kN (Tensile)

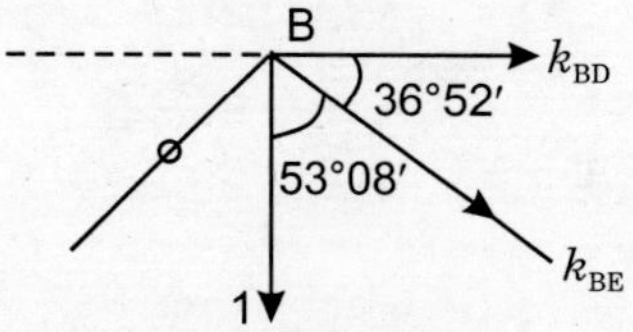

Fig. 9.24 (*c*)

Table 9.7

S.No.	*Member*	*k (kN)*	*Length, l (m)*	k^2l	*Final forces P = kT (kN)*
1.	AB	0	5	0	0
2.	BD	1.334	4	7.118	– 7.525
3.	EC	1.334	4	7.118	– 7.525
4.	CA	0	4	0	0
5.	BE	– 1.667	5	13.894	9.404
6.	CD	– 1.667	5	13.894	9.404
Total				42.024	

Force in the redundant member, $T = \pm \dfrac{l_0 \alpha t}{\sum_1^6 \dfrac{k^2 l}{AE} + \dfrac{l_0}{A_0E_0}}$

– sign will be used since the member is subjected to rise in temperature

$$E = 2.08 \times 10^5 \text{ N/mm}^2 = 208 \text{ kN/mm}^2$$

$$l_0 = 3 \text{ m} = 3000 \text{ mm}$$

$$T = \frac{-3000 \times 0.000012 \times 32}{\left(\dfrac{42.024 \times 1000 + 3000}{1060 \times 208}\right)}$$

$$= \frac{-3000 \times 0.000012 \times 32 \times 1060 \times 208}{45024}$$

$$= -5.641 \text{ kN}$$

∴ Force in member BC, P_{BC} = 5.641 kN (Comp.)

Forces in other members are calculated using P = kT and presented in the last column of Table 9.7.

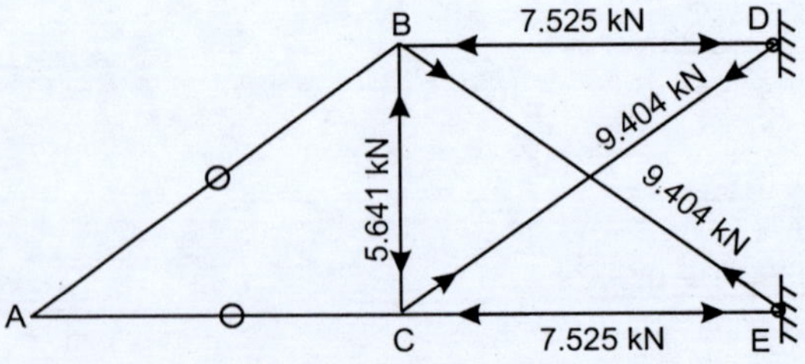

Fig. 9.25. Final forces.

Example 9.6 *Determine the forces in all the members of the frame shown in Fig. 9.26 due to a fall in temperature of member BC by 28°C. All the members have a cross-sectional area of 1300 mm². Coefficient of expansion, α = 0.000012 per degree centigrade. E = 2 × 10⁵ N/mm².*

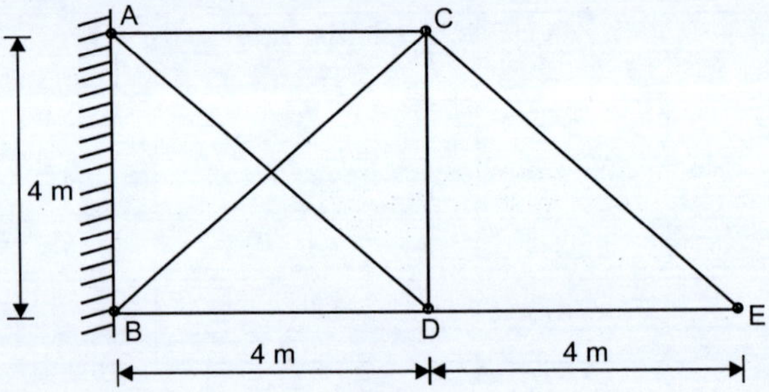

Fig. 9.26

Solution Degree of internal indeterminacy : $m = 7$; $j = 3$

$$ii = m - 2j$$
$$= 7 - 2 \times 3 = 1$$

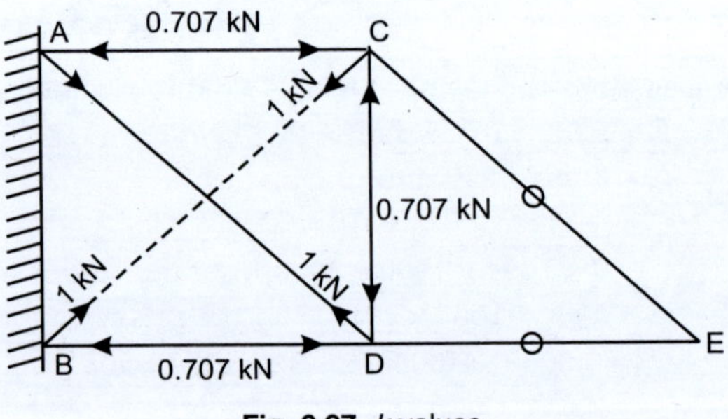

Fig. 9.27. k values.

Hence the frame is internally indeterminate to the first degree. The member BC will be treated as redundant and the structure will be made determinate by removing it. A unit force will be applied along the member BC and the frame will be analysed for k values (Fig. 9.27).

Determination of values of k :

Joint E :

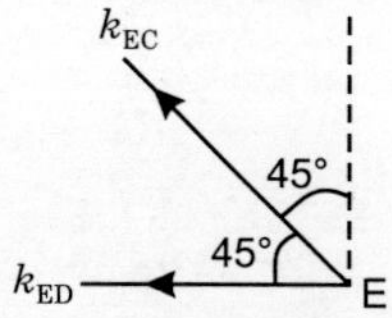

Fig. 9.27 (*a*)

$$\Sigma V = 0 \text{ gives}$$
$$k_{EC} \cos 45° = 0$$
$$k_{EC} = 0$$
$$\Sigma H = 0 \text{ gives}$$
$$k_{ED} + k_{EC} \cos 45° = 0$$
$$k_{ED} = 0 \quad \text{(Because of } k_{EC} = 0\text{)}$$

Joint C :

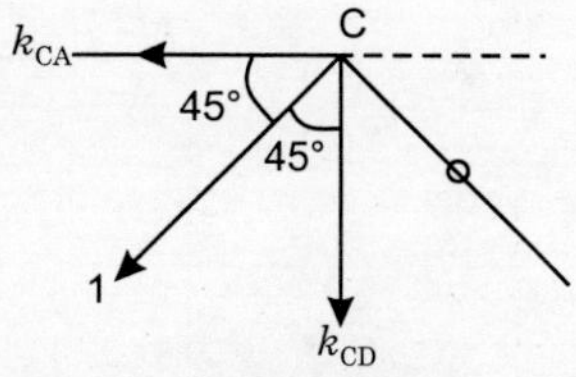

Fig. 9.27 (*b*)

$$\Sigma H = 0 \text{ gives}$$
$$- k_{CA} - 1 \times \cos 45° = 0$$
$$k_{CA} = - 0.707 \text{ kN}$$
$$k_{CA} = 0.707 \text{ (Comp.)}$$
$$\Sigma V = 0 \text{ gives}$$
$$- k_{CD} - 1 \times \cos 45° = 0$$
$$k_{CD} = - 0.707 \text{ kN}$$
$$k_{CD} = 0.707 \text{ (Comp.)}$$

Joint D :

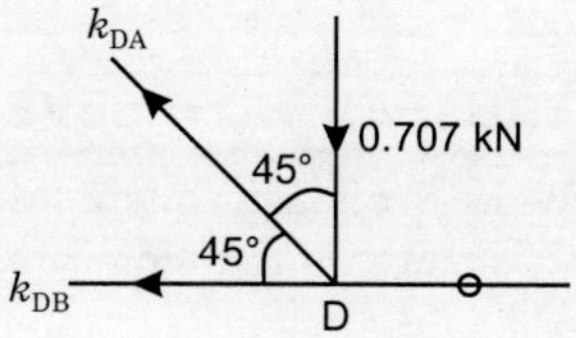

Fig. 9.27 (*c*)

$$\Sigma V = 0 \text{ gives}$$
$$- 0.707 + k_{DA} \cos 45° = 0$$
$$k_{DA} = 0.707/\cos 45° = 1 \text{ kN}$$
$$k_{DA} = 1 \text{ kN (Tensile)}$$
$$\Sigma H = 0 \text{ gives}$$
$$- k_{DB} - k_{DA} \cos 45° = 0$$
$$k_{DB} = - 1 \times \cos 45° = - 0.707$$
$$k_{DB} = 0.707 \text{ kN (Comp.)}$$

Table 9.8

S.No.	*Member*	*k (kN)*	*Length, l (m)*	k^2l	*Final forces P = kT (kN)*
1.	AC	– 0.707	4	2	– 20.18
2.	CE	0	5.657	0	0
3.	ED	0	4	0	0
4.	DB	– 0.707	4	2	– 20.18
5.	AD	1	5.657	5.657	28.54
6.	CD	– 0.707	4	2	– 20.18
		Total		11.657	

Force in the redundant member, $T = \pm \dfrac{l_0 \alpha t}{\sum \dfrac{k^2 l}{AE} + \dfrac{l_0}{A_0 E_0}}$

+ sign will be used, since the member is subjected to *fall* in temperature.

$t = 28°$; $l = 5657$ mm; $E = 200$ kN/m²

$$T = + \frac{5.657 \times 10^3 \times 0.000012 \times 28}{\left(\dfrac{11.657 \times 1000 + 5.657 \times 1000}{1300 \times 200} \right)}$$

$= 28.54$ kN

Force in the member BC = 28.54 kN (Tensile)

Force in other members are calculated, using $P = kT$ and are presented in the last column of Table 9.8.

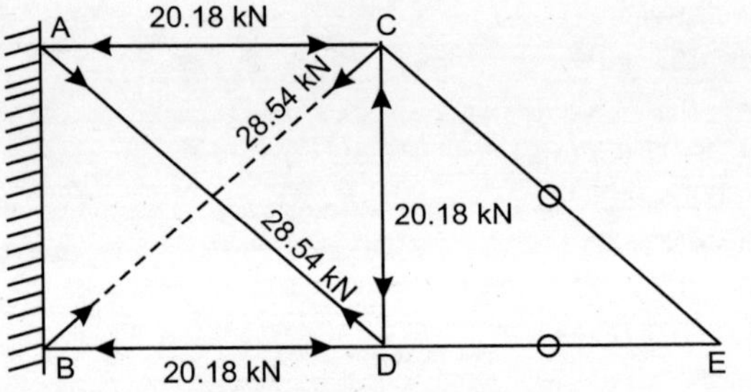

Fig. 9.27 (*d*). Final forces.

9.6 TRUSSES WITH SEVERAL REDUNDANCIES

While dealing with trusses with more than one redundancy, we have to

(*i*) remove all redundant members to create the primary structure

(*ii*) analyse the primary structure to get internal forces F_i

(*iii*) apply unit loads along each redundant member, one at a time and analyse the primary structure to get sets of values of k_1, k_2, etc.

Now we will go back to the basics to decide how to deal with multiple redundants.

From equation (9.4), if T_1, T_2, etc. are the redundants

$$\sum P \frac{\partial P}{\partial T_1} \frac{l}{AE} = 0$$

$$\sum P \frac{\partial P}{\partial T_2} \frac{l}{AE} = 0, \text{ etc.}$$

Let us confine our further discussions for the case of 2 redundants (Their final forces being T_1 and T_2).

Since $P = F + k_1 T_1 + k_2 T_2$ for every member,

$$\frac{\partial P}{\partial T_1} = k_1 \quad \text{and} \quad \frac{\partial P}{\partial T_2} = k_2$$

Hence, $$\sum (F + k_1T_1 + k_2T_2)k_1 \frac{l}{AE} = 0 \text{ and}$$

$$\sum (F + k_1T_1 + k_2T_2)k_2 \frac{l}{AE} = 0$$

These can be rewritten as

$$\sum Fk_1 \frac{l}{AE} + T_1 \sum {k_1}^2 \frac{l}{AE} + T_2 \sum k_1k_2 \frac{l}{AE} = 0 \qquad ...(9.12)$$

and $$\sum Fk_2 \frac{l}{AE} + T_1 \sum k_1k_2 \frac{l}{AE} + T_2 \sum {k_2}^2 \frac{l}{AE} = 0 \qquad ...(9.13)$$

Here, once the analyses of the truss has been made on the primary structure for

(*i*) external loading,

(*ii*) unit load along T_1 and ,

(*iii*) unit load along T_2.

the only unknowns would be T_1 and T_2. Solving equations (9.12) and (9.13), we would get T_1 and T_2 the redundant forces. Hence all member forces can be obtained.

Example 9.7 shows how this is implemented.

Example 9.7 *Find the forces in all the members of the pin jointed frame work shown in Fig. 9.28. All the members have the same area of cross-section and Young's modulus.*

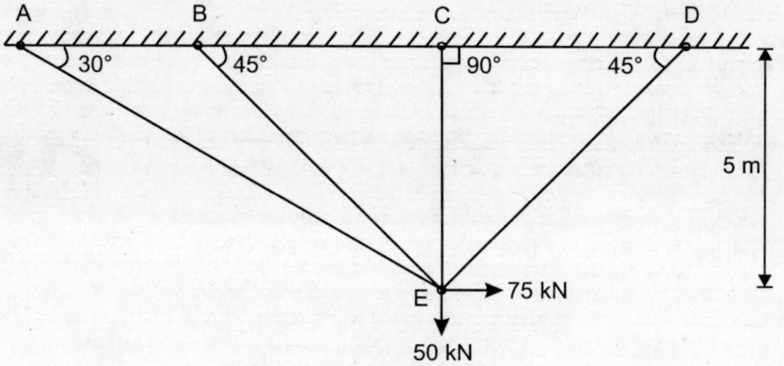

Fig. 9.28

Solution Degree of internal indeterminacy : $m = 4$; $j = 1$

$$ii = m - 2j$$
$$= 4 - 2 \times 1 = 2$$

Therefore, the structure is statically (internally) indeterminate to the second degree. The structure will be made determinate by removing any two members, say BE and CE as shown in Fig. 9.29.

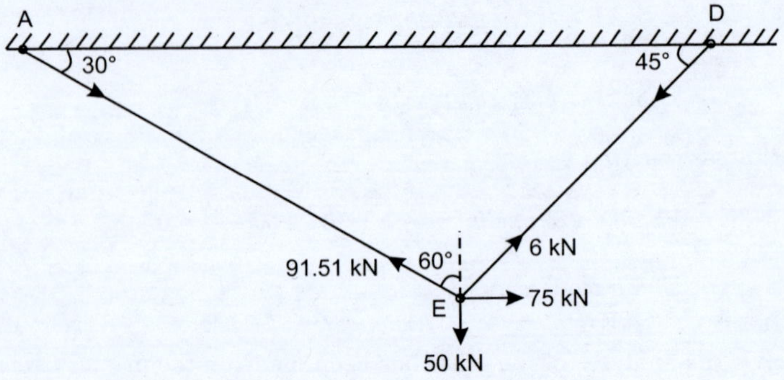

Fig. 9.29. Primary structure.

Determination of F values :

Joint E :

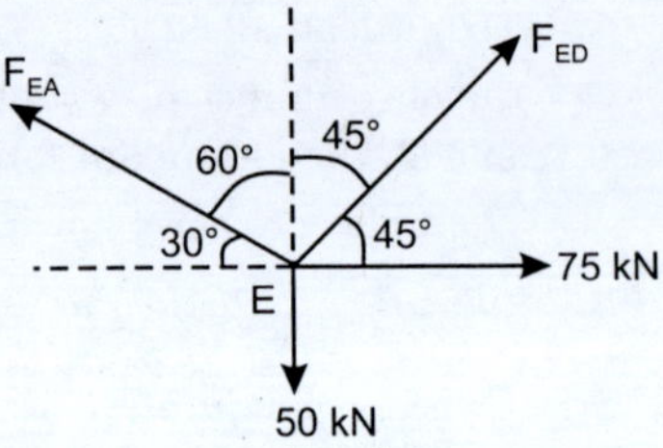

Fig. 9.29 (*a*)

$\Sigma V = 0$ gives

$F_{EA} \cos 60° + F_{ED} \cos 45° - 50 = 0$

(↑) (↑) (↓)

$0.5\ F_{EA} + 0.707\ F_{ED} = 50$ ...(1)

$\Sigma H = 0$ gives

$F_{ED} \cos 45° + 75 - F_{EA} \cos 30° = 0$

(→) (→) (←)

$0.707\ F_{ED} - 0.866\ F_{EA} = -75$ ...(2)

$0.707\ F_{ED} + 0.5\ F_{EA} = 50$ ...(1)

(2) – (1) $-1.366\ F_{EA} = -125$

$F_{EA} = 91.51$ kN (Tensile)

Substituting the value of F_{EA} in (1)

$0.5 \times 91.51 + 0.707\ F_{ED} = 50$

$0.707\ F_{ED} = 50 - 0.5 \times 91.51$

$F_{ED} = 6$ kN (Tensile)

Determination of k_1 values :

Removing the external forces and applying unit tensile (or comp.) force along the direction of BE, k_1 values will be obtained.

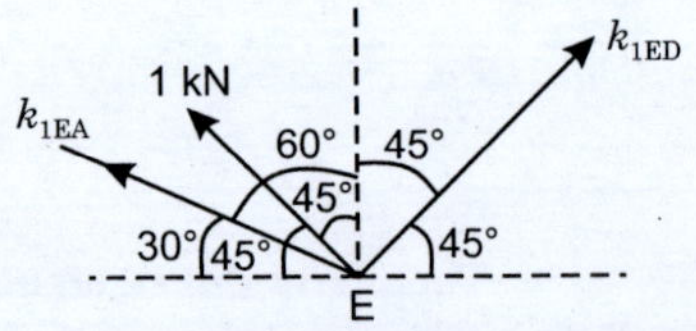

Fig. 9.29 (*b*)

$\Sigma V = 0$ gives

$k_{1EA} \cos 60° + k_{1ED} \cos 45° + 1 \times \cos 45° = 0$

$0.5\ k_{1EA} + 0.707\ k_{1ED} = -0.707$ ...(3)

$\Sigma H = 0$ gives

$-k_{1EA} \cos 30° + k_{1ED} \cos 45° - 1 \times \cos 45° = 0$

$$-0.866\, k_{1EA} + 0.707\, k_{1ED} = 0.707 \quad ...(4)$$

$$0.5\, k_{1EA} + 0.707\, k_{1ED} = -0.707 \quad ...(3)$$

(4) – (3) $\Rightarrow$ $-1.366\, k_{1EA} = 1.414$

$k_{1EA} = -1.035$

$k_{1EA} = 1.035$ (Comp.)

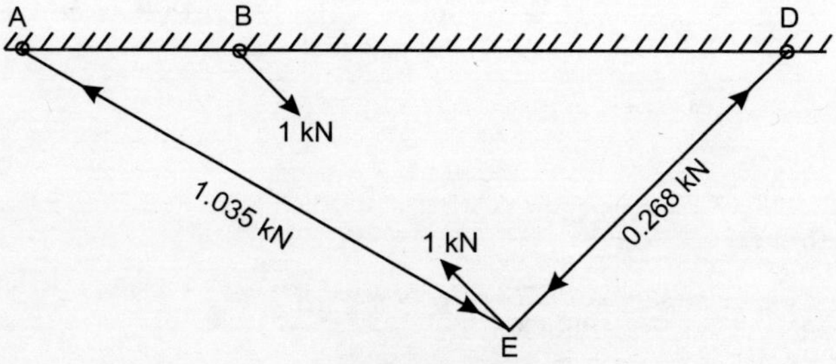

Fig. 9.30. k_1 values.

Substituting in (3)

$0.5 \times (-1.035) + 0.707\, k_{1ED} = -0.707$

$0.707\, k_{1ED} = -0.707 + 0.5 \times 1.035$

$k_{1ED} = -0.268$

$k_{1ED} = 0.268$ (Comp.)

Determination of k_2 values :

Removing the external forces and applying unit tensile (or comp.) force along the direction of EC, k_2 values will be obtained.

$\Sigma H = 0$ gives

$k_{2ED} \cos 45° - k_{2EA} \cos 30° = 0$

$0.707\, k_{2ED} - 0.866\, k_{2EA} = 0 \quad ...(5)$

$\Sigma V = 0$ gives

$k_{2ED} \cos 45° + k_{2EA} \cos 60° + 1 = 0$

$0.707\, k_{2ED} + 0.5\, k_{2EA} = -1 \quad ...(6)$

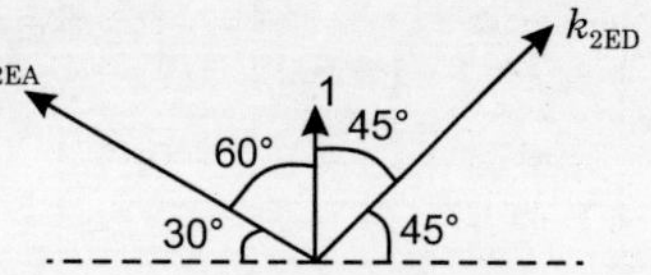

Fig. 9.30 (*a*)

$0.707\, k_{2ED} - 0.866\, k_{2EA} = 0 \quad ...(5)$

$0.707\, k_{2ED} + 0.5\, k_{2EA} = -1 \quad ...(6)$

(5) – (6) $\Rightarrow$ $-1.366\, k_{2EA} = 1$

$k_{2EA} = -0.732$

$k_{2EA} = 0.732$ kN (Comp.)

Substituting in (5)

$0.707\, k_{2ED} - 0.866\,(-0.732) = 0$

$$k_{2ED} = \frac{-0.866 \times 0.732}{0.707} = -0.897$$

$k_{2ED} = 0.897$ kN (Comp.)

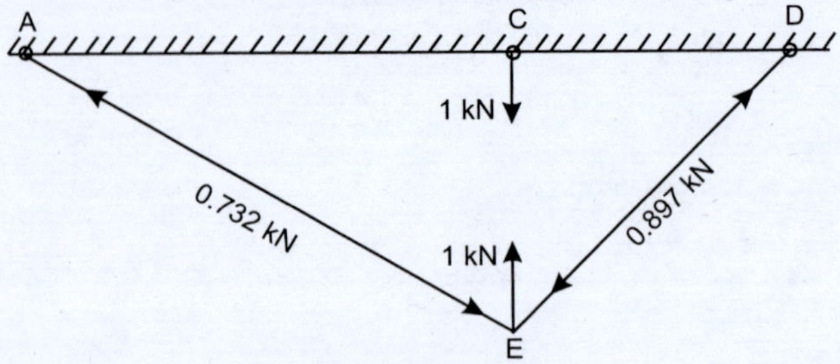

Fig. 9.31. k_2 values.

Length of members:

$$\frac{5}{AE} = \sin 30° \; ; \; \frac{5}{\sin 30°} = AE; \; AE = 10 \text{ m}$$

$$\frac{5}{BE} = \sin 45° \; ; \; \frac{5}{\sin 45°} = BE; \; BE = 7.071 \text{ m} = ED$$

CE = 5 m

Table 9.9

Members	*F*	k_1	k_2	*l (m)*	Fk_1l	Fk_2l	k_1^2l	k_2^2l	k_1k_2l	*Final forces* $P = F + T_1k_1 + T_2k_2$
AE	91.51	– 1.035	– 0.732	10	– 947.13	– 669.85	10.710	5.36	7.58	33.23
DE	6.0	– 0.268	– 0.897	7.071	– 11.37	– 38.06	0.508	5.69	1.70	– 22.91
BE	0	1	0	7.071	0	0	7.071	0	0	42.49
CE	0	0	1	5.0	0	0	0	5.0	0	+19.54
Total					– 958.5	– 707.91	18.289	16.05	9.28	

If T_1 and T_2 are the forces in redundants BE and CE respectively, equations (9.12) and (9.13) take the form

$$-958.5 + 18.289\, T_1 + 9.28\, T_2 = 0 \quad \text{...(1)}$$

$$-707.91 + 9.28\, T_1 + 16.05\, T_2 = 0 \quad \text{...(2)}$$

(2) × (9.28 / 16.05) ⇒ $-409.38 + 5.367\, T_1 + 9.28\, T_2 = 0$ ...(3)

(1) – (3) ⇒ $-549.1 + 12.922\, T_1 = 0$

$$T_1 = 42.49$$

Apply T_1 in (2)

$$16.05\, T_2 = 707.91 - 9.28 \times 42.49 = +313.57$$

$$T_2 = +19.54$$

$$P_1 = 91.51 - 1.035\,(42.49) - 0.732\,(+19.54) = 33.23$$

$$P_2 = 6.0 - 0.268\ (42.49) - 0.897\ (+19.54) = -\ 22.91$$

The final forces are listed in Fig. 9.32.

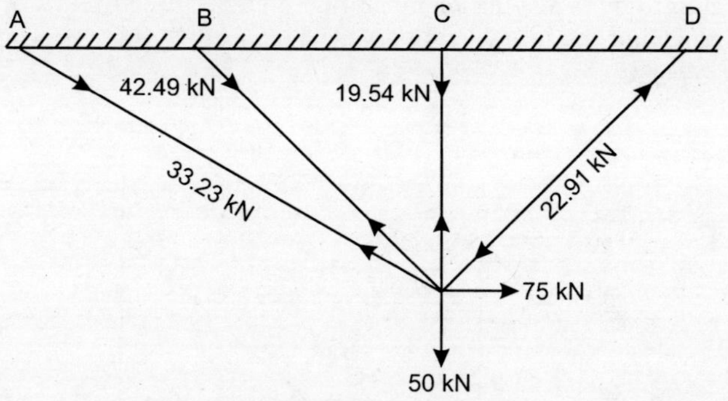

Fig. 9.32. Final forces.

Example 9.8 *Determine the forces in the members of the truss shown in Fig. 9.33, if the member BC is short by length 2 mm. E = 2 × 10^5 N/mm^2. Area of the members in mm^2 are given in brackets.*

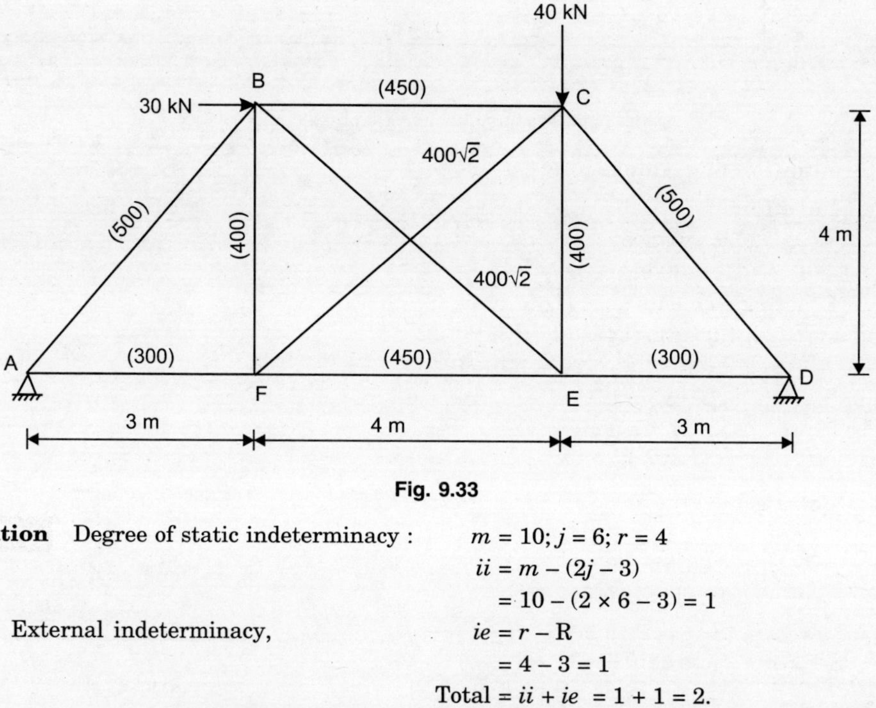

Fig. 9.33

Solution Degree of static indeterminacy : $m = 10; j = 6; r = 4$

$$ii = m - (2j - 3)$$
$$= 10 - (2 \times 6 - 3) = 1$$

External indeterminacy, $ie = r - \text{R}$

$$= 4 - 3 = 1$$

$$\text{Total} = ii + ie\ = 1 + 1 = 2.$$

Hence, the structure is internally indeterminate to the first degree and externally indeterminate to the first degree, the total indeterminacy being two. The frame will be made internally determinate by removing member BC (member having lack of fit) and externally determinate by replacing the hinged support at D by a roller support. The primary structure is shown in Fig. 9.34.

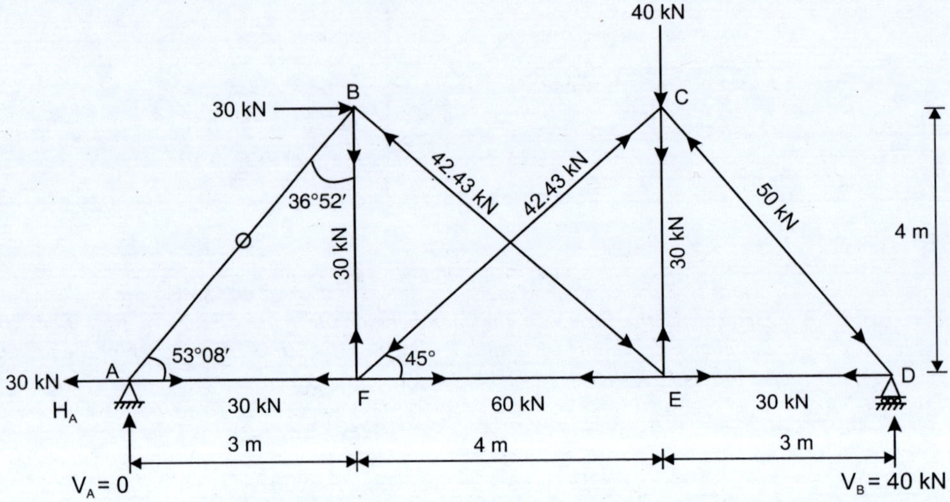

Fig. 9.34. Primary structure with F values obtained.

Determination of F values :

Reactions :

$\Sigma H = 0$ gives

$H_A = 30$ kN (←)

Equating moments about D to zero,

$(V_A \times 10) + (30 \times 4) - (40 \times 3) = 0$

$10V_A = -120 + 120 = 0$

$\therefore \quad V_A = 0$

$\Sigma V = 0$ gives

$V_B = 40$ kN (↑)

Joint A :

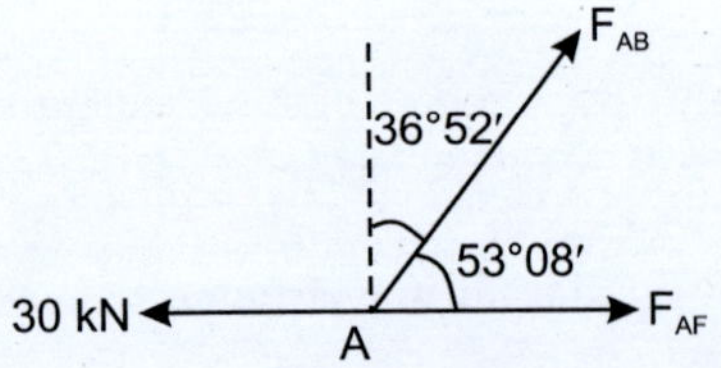

Fig. 9.34 (*a*)

$\Sigma V = 0$ gives

$F_{AB} \cos 36°52' = 0$

$F_{AB} = 0$

$\Sigma H = 0$ gives

$F_{AF} + F_{AB} \cos 53°08' - 30 = 0$

$F_{AF} = 30$ kN (Tensile)

Joint B :

$\Sigma H = 0$ gives

$F_{BE} \cos 45° + 30 = 0$

$F_{BE} = -30/\cos 45° = -42.43$ kN

$F_{BE} = 42.43$ kN (Comp.)

$\Sigma V = 0$ gives

$F_{BF} + F_{BE} \cos 45° = 0$

$F_{BF} + (-42.43) \cos 45° = 0$

$F_{BF} = 30$ kN (Tensile)

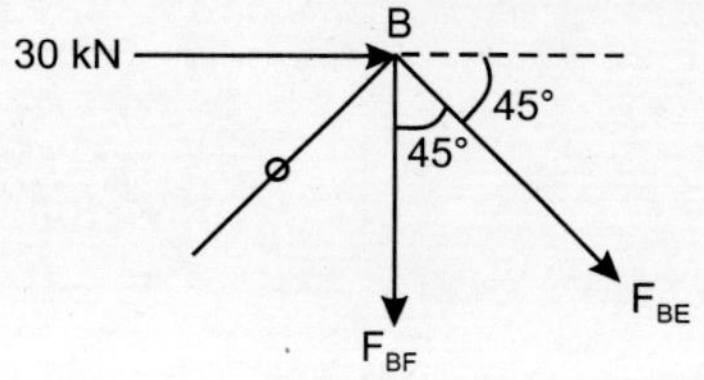

Fig. 9.34 (*b*)

Joint F :

$\Sigma V = 0$ gives

$F_{FC} \cos 45° + 30 = 0$

$F_{FC} = -30/\cos 45° = -42.43$ kN

$F_{FC} = 42.43$ kN (Comp.)

$\Sigma H = 0$ gives

$F_{FE} + F_{FC} \cos 45° - 30 = 0$

$F_{FE} - 42.43 \cos 45° - 30 = 0$

$F_{FE} = 30 + 30 = 60$ kN (Tensile)

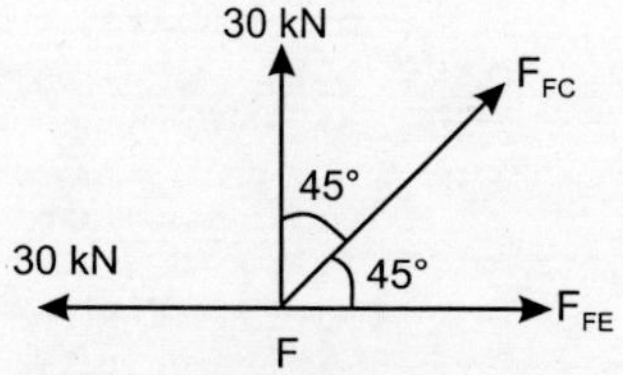

Fig. 9.34 (*c*)

Joint E :

$\Sigma V = 0$ gives

$F_{EC} - 42.43 \cos 45° = 0$

$F_{EC} = 30$ kN (Tensile)

$\Sigma H = 0$ gives

$F_{ED} + 42.43 \cos 45° - 60 = 0$

$F_{ED} + 30 - 60 = 0$

$\therefore \quad F_{ED} = 30$ kN (Tensile)

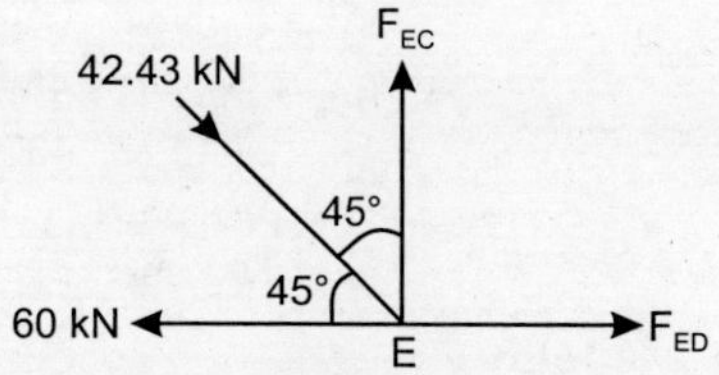

Fig. 9.34 (*d*)

Joint D :

$\Sigma H = 0$ gives

$F_{DC} \cos 53°08' + 30 = 0$

$F_{DC} = -30 / \cos 53°08' = -50$ kN

$\therefore \quad F_{DC} = 50$ kN (Comp.)

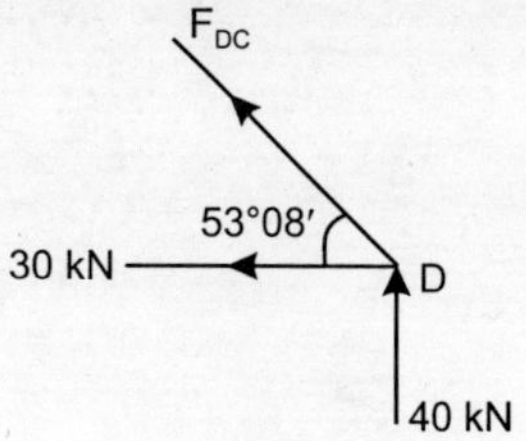

Fig. 9.34 (*e*)

Determination of k_1 values :

Removing the external loads, unit tensile (or compressive) load will be applied along the member BC and the frame will be analysed for k_1 values.

At A, there is no load or reaction(s)

$\therefore \quad \Sigma V = 0$ gives $k_{1AB} = 0$

$\Sigma H = 0$ gives $k_{1AF} = 0$

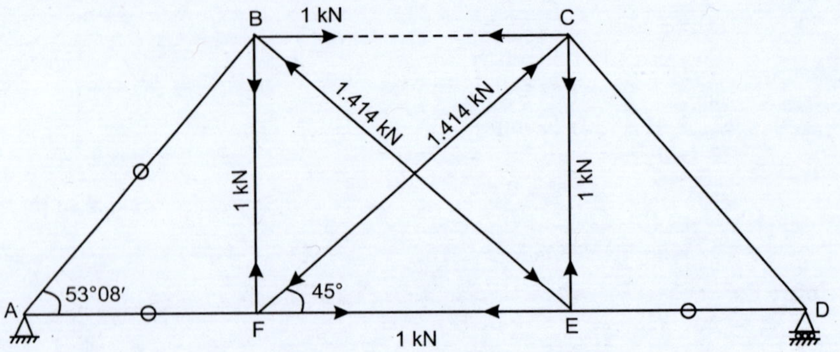

Fig. 9.35. k_1 values.

Joint B :

ΣH = 0 gives

$k_{1BE} \cos 45° + 1 = 0$

$k_{1BE} = -1/\cos 45° = -1.414$

$k_{1BE} = 1.414$ kN (Comp.)

ΣV = 0 gives

$k_{1BF} + k_{1BE} \cos 45° = 0$

$k_{1BF} - 1.414 \cos 45° = 0$

$k_{1BF} = 1$

$k_{1BF} = 1$ (Tensile)

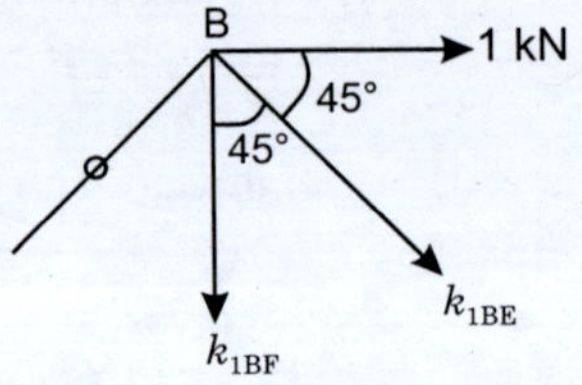

Fig. 9.35 (*a*)

Joint F :

ΣV = 0 gives

$k_{1FC} \cos 45° + 1 = 0$

$k_{1FC} = -1/\cos 45° = -1.414$

$k_{1FC} = 1.414$ kN (Comp.)

ΣH = 0 gives

$k_{1FE} + k_{1FC} \cos 45° = 0$

$k_{1FE} - 1.414 \cos 45° = 0$

$k_{1FE} = 1$ (Tensile)

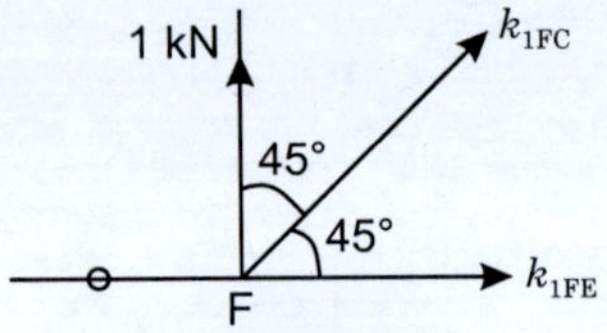

Fig. 9.35 (*b*)

Joint C :

ΣH = 0 gives

$k_{1CD} \cos 53°08' + 1.414 \cos 45° - 1 = 0$

$k_{1CD} \cos 53°08' = -1 + 1 = 0$

$\therefore \quad k_{1CD} = 0$

ΣV = 0 gives

$k_{1CE} - 1.414 \cos 45° + k_{1CD} \cos 36°52' = 0$

$k_{1CE} - 1 + 0 = 0$

$k_{1CE} = 1$ (Tensile)

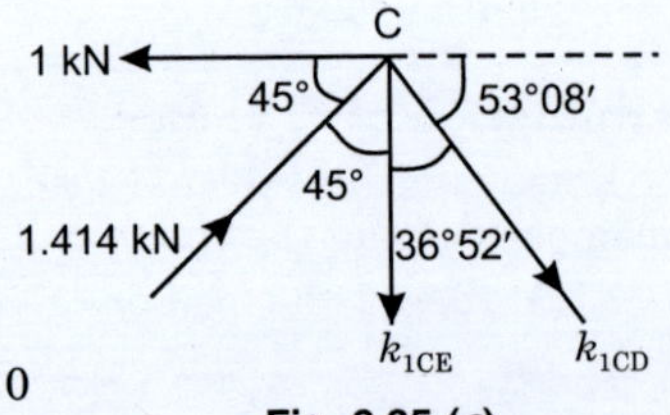

Fig. 9.35 (*c*)

Joint E :

$\Sigma H = 0$ gives

$k_{1ED} + 1.414 \cos 45° - 1 = 0$

$k_{1ED} + 1 - 1 = 0$

$k_{1ED} = 0$

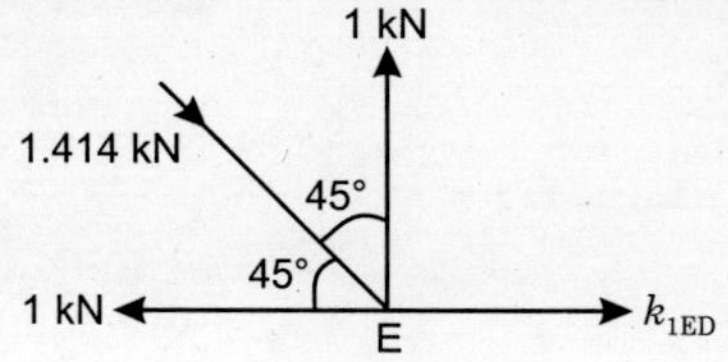

Fig. 9.35 (*d*)

Determination of k_2 values :

Removing the external loads and applying unit horizontal force at D, k_2 values will be obtained.

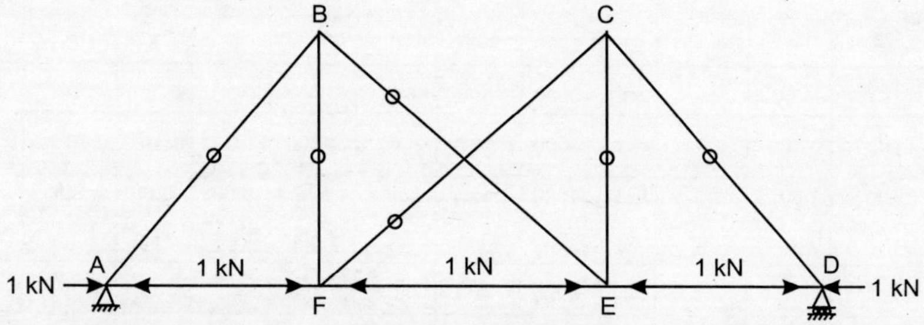

Fig. 9.36. k_2 values.

$\Sigma H = 0$ gives

$H_A = 1\ (\rightarrow)$

Joint A :

$\Sigma V = 0$ gives

$k_{2AB} = 0$

$\Sigma H = 0$ gives

$k_{2AF} = 1$ (Comp.)

Joint B :

$\Sigma H = 0$ gives

$k_{2BE} = 0$

$\Sigma V = 0$ gives

$k_{2BF} = 0$

Joint F :

$\Sigma V = 0$ gives

$k_{2FC} = 0$

$\Sigma H = 0$ gives

$k_{2FE} = 1$ (Comp.)

Joint C :

$\Sigma H = 0$ gives

$k_{2CD} = 0$

$\Sigma V = 0$ gives

$k_{2CE} = 0$

Joint D :

$\Sigma H = 0$ gives

$k_{2DE} = 1$ (Comp.)

Table 9.10

S.No.	Member	F(kN)	k_1	k_2	l (mm)	A (mm²)	$\frac{Fk_1l}{A}$	$\frac{Fk_2l}{A}$	$\frac{k_1^2l}{A}$	$\frac{k_2^2l}{A}$	$\frac{k_1k_2l}{A}$	Final forces $P = F + k_1T_1 + k_2T_2$
1.	AB	0	0	0	5000	500	0	0	0	0	0	0
2.	CD	– 50	0	0	5000	500	0	0	0	0	0	– 50
3.	DE	30	0	– 1	3000	300	0	– 300	0	10	0	– 17.37
4.	EF	60	1	– 1	4000	450	533.3	– 533.3	8.89	8.89	– 8.89	– 13.82
5.	FA	30	0	– 1	3000	300	0	– 300	0	10	0	– 17.37
6.	BF	30	1	0	4000	400	300	0	10	0	0	3.55
7.	BE	– 42.43	– 1.414	0	5657	566	599.44	0	19.98	0	0	– 5.03
8.	CF	– 42.43	– 1.414	0	5657	566	599.44	0	19.98	0	0	– 5.03
9.	CE	30	1	0	4000	400	300	0	10	0	0	3.55
10.	BC	0	1	0	4000	450	0	0	8.89	0	0	– 26.45 (T_1)
						Total	2332.61	–1133.3	77.84	28.89	– 8.89	$T_2 = 47.37$

$$\sum \frac{Fk_1l}{A} + T_1 \sum \frac{k_1^2l}{A} + T_2 \sum \frac{k_1k_2l}{A} = 0$$

$$\sum \frac{Fk_2l}{A} + T_1 \sum \frac{k_1k_2l}{A} + T_2 \sum \frac{k_2^2l}{A} = 0$$

$$2332.61 + T_1 \times 77.74 - 8.89\,T_2 = 0 \qquad \text{...(1)}$$

$$-1133.33 + T_1 \times (-8.89) + 28.89\,T_2 = 0 \qquad \text{...(2)}$$

(1) × 28.89/8.89 $\quad 7580.28 + T_1 \times 252.63 - 28.89\,T_2 = 0 \qquad$ ...(3)

(2) + (3) $\quad 6447 + 243.74\,T_1 = 0$

$$T_1 = -26.45$$

From (2) $T_2 = (1133.33 + 235.14) / 28.89 = 47.37$

The final forces are computed and entered in the last column of Table 9.10 and shown in Fig. 9.37.

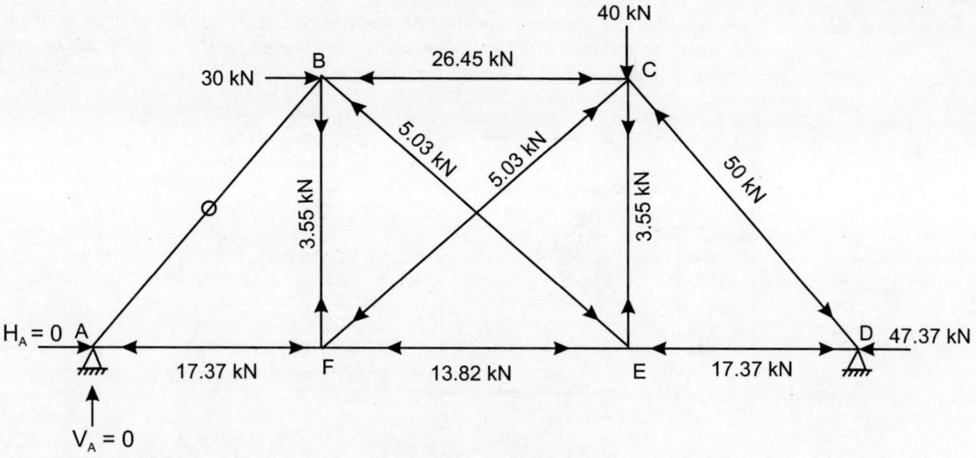

Fig. 9.37. Final forces.

9.7 STRAIN ENERGY METHOD

So far we were dealing with unit load method which in fact is a disguised version of strain energy minimization method. This became obvious when we saw that k_1, k_2, etc. due to unit loads were only the values of $\delta P / \delta T_1$ and $\delta P / \delta T_2$ (see the derivations of equations (9.12) and (9.13). Here we remove the disguise and call a derivative.

In the strain energy method of solving redundant frames, we remove the redundant member, apply on the primary structure a force T (the final force in the redundant member) and compute the member forces P due to external loading plus T. These P values will be in terms of T.

Next we differentiate P with respect to T and write down $\delta P / \delta T$ values for each member (This obviously corresponds to k values in unit load method).

To minimize the strain energy $U = \sum \frac{P^2 l}{2AE}$, we differentiate with respect to T. Thus

$$\frac{\partial U}{\partial T} = 0 = \sum_{1}^{n} P \frac{\partial P}{\partial T} \frac{l}{AE}$$

The summation would be done for all members including the redundant.

9.7.1 Several Redundants

If there are more redundants say T_1, T_2, ..., etc. then every P_i is a function of T_1, T_2,....

$$U = \sum \frac{P^2 l}{2AE}$$

$\delta U / \delta T_1$, $\delta U / \delta T_2$, ..., etc. must each be equal to zero. So there would be as many energy minimization equations as there are unknowns. So we can solve for each and every T_i.

Now let us go to Example 9.9.

Example 9.9 *A braced cantilever truss is loaded as shown in Fig. 9.38. All the members are of the same material and have the same cross-sectional area. Find the axial force in the member AD.*

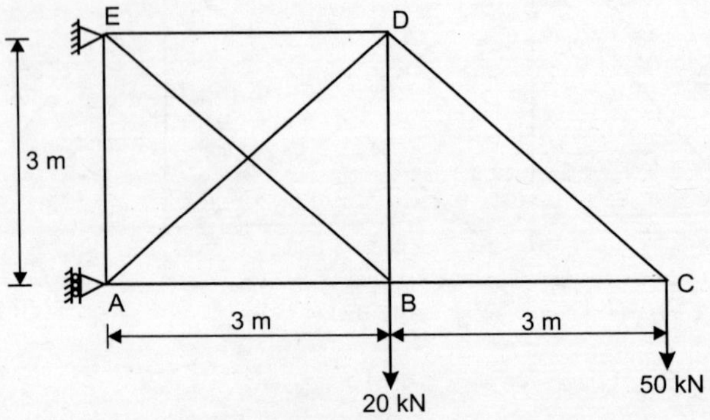

Fig. 9.38

Solution Degree of static indeterminacy : $m = 8; j = 5$

$$I = m - (2j - 3)$$
$$= 8 - (2 \times 5 - 3) = 8 - 7 = 1$$

The truss is externally determinate and internally indeterminate to the first degree. Treating the member AD as redundant, and removing member AD, the primary structure is as shown in Fig. 9.39. Applying a tensile force T at A and D along the member AD, the truss is analysed as shown below :

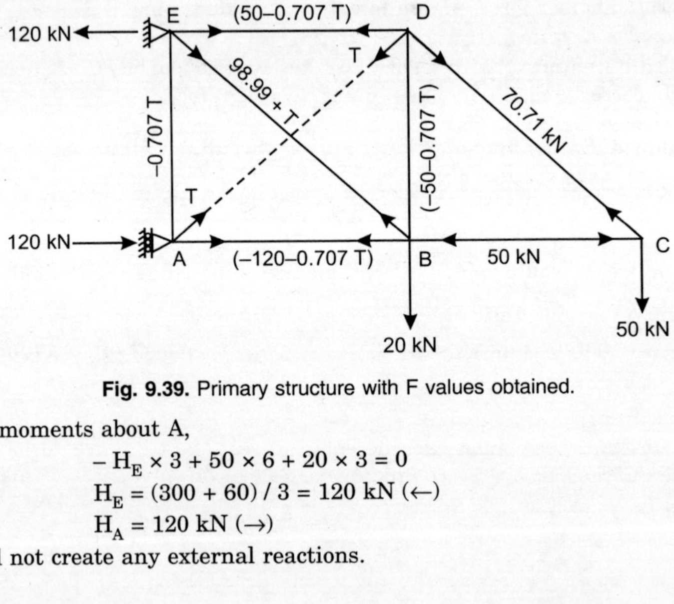

Fig. 9.39. Primary structure with F values obtained.

Taking moments about A,

$$-H_E \times 3 + 50 \times 6 + 20 \times 3 = 0$$
$$H_E = (300 + 60) / 3 = 120 \text{ kN } (\leftarrow)$$

So, $H_A = 120 \text{ kN } (\rightarrow)$

T-T will not create any external reactions.

Joint C :

$\Sigma V = 0$ gives

$F_{CD} \cos 45° - 50 = 0$

$F_{CD} = 50/\cos 45° = 70.71$

$F_{CD} = 70.71$ kN (Tensile)

$\Sigma H = 0$ gives

$F_{CB} + F_{CD} \cos 45° = 0$

$F_{CB} + 70.71 \cos 45° = 0$

$F_{CB} = -50$ kN

$F_{CB} = 50$ kN (Comp.)

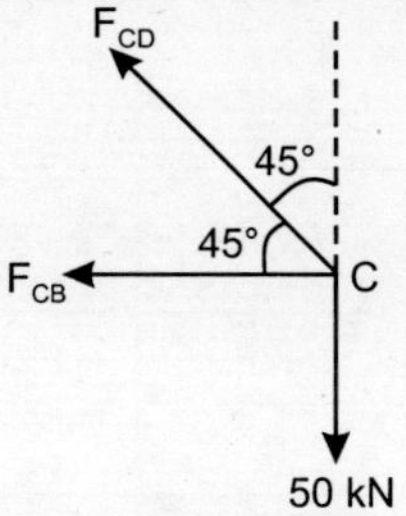

Fig. 9.39 (*a*)

Joint D :

$\Sigma H = 0$ gives

$-F_{DE} - T \cos 45° + 70.71 \cos 45° = 0$

$F_{DE} + 0.707\,T = 50$

$F_{DE} = 50 - 0.707\,T$ ...(1)

$\Sigma V = 0$ gives

$F_{DB} + T \cos 45° + 70.71 \cos 45° = 0$

$F_{DB} + 0.707\,T = -50$

$F_{DB} = -50 - 0.707\,T$ ...(2)

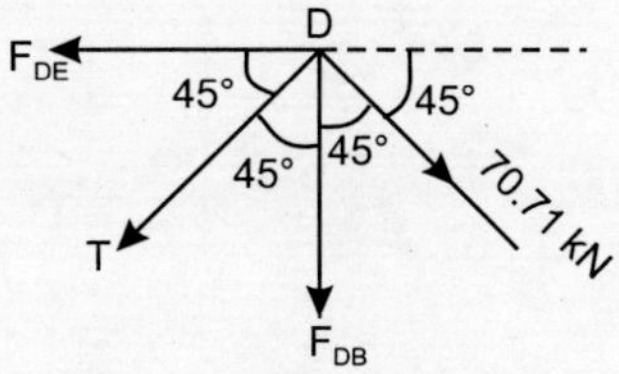

Fig. 9.39 (*b*)

Joint B :

$\Sigma V = 0$ gives

$F_{BE} \cos 45° - 50 - 0.707\,T - 20 = 0$

$F_{BE} \cos 45° = 70 + 0.707\,T$

$F_{BE} = 98.99 + T$ ...(3)

$\Sigma H = 0$ gives

$F_{BA} + F_{BE} \cos 45° + 50 = 0$

$F_{BA} + (98.99 + T) \cos 45° + 50 = 0$

$F_{BA} + 70 + 0.707\,T + 50 = 0$

$F_{BA} = -120 - 0.707\,T$ ...(4)

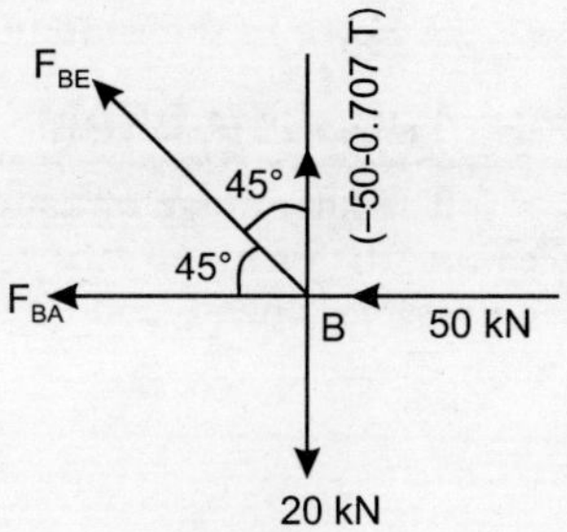

Fig. 9.39 (*c*)

Joint A :

$\Sigma V = 0$ gives

$F_{AE} + T \cos 45° = 0$

$F_{AE} = -T \cos 45°$

$F_{AE} = -0.707\,T$ ...(5)

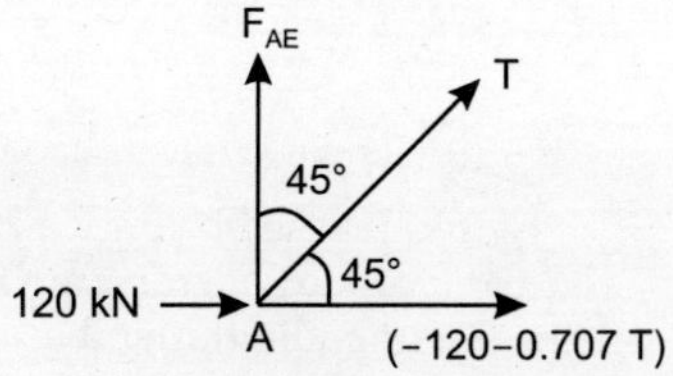

Fig. 9.39 (*d*)

All the members are of the same material and have the same cross-sectional area *i.e.*, AE is same for all the members.

Table 9.11

Redundant R=T

S.No.	Member	F (kN)	$\frac{\partial F}{\partial R}$	l (m)	$F\frac{\partial F}{\partial R}l$
1.	AB	– 120 – 0.707 T	– 0.707	3	(120 + 0.707 T) × 0.707 × 3
2.	BC	– 50	0	3	0
3.	CD	70.71	0	4.243	0
4.	DE	50 – 0.707 T	– 0.707	3	(50–0.707 T) × (– 0.707) × 3
5.	EA	– 0.707 T	– 0.707	3	0.707 T × 0.707 × 3
6.	EB	98.99 + T	1	4.243	(98.99 + T) × 4.243
7.	BD	– 50 – 0.707 T	– 0.707	3	(50 + 0.707 T) × 0.707 × 3
8.	AD	T	1	4.243	4.243 T

$$\frac{\partial U}{\partial R} = 0 = \sum F \frac{\partial F}{\partial R} \frac{l}{AE}$$

$$= \frac{1}{AE}\,[254.52 + 1.5T - 106.05 + 1.5T + 1.5T + 420.01 + 4.243T + 106.05 + 1.5T] = 0$$

$$674.53 + 14.486\,T = 0$$

$$T = -\,46.56 \text{ kN}$$

∴ Force in the redundant member AD = 46.56 kN (Comp.)

9.8 TRUSSED BEAMS

The advantage of using strain energy method is that we can simultaneously take into account all forms of strain energy, meaning that energy stored in members in tension, compression, shear, bending and torsion. This is called upon when we deal with a trussed beam.

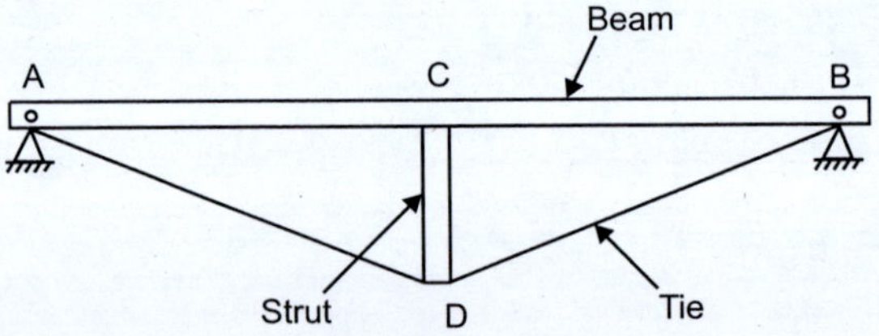

Fig. 9.40. Trussed beam.

In a trussed beam we combine a beam (bending member) with a truss (made of tension and compression members). Figure 9.40 shows a trussed beam. Here AB is a beam. It is a flexure (or bending) member and would store energy in flexure given by

$$U_b = \int \frac{m^2 dx}{2EI}, \text{ where } m \text{ is the bending moment at any position given by } x.$$

AD, BD and CD are direct stress members whose strain energy $U_d = \frac{P^2 l}{2AE}$.

Actually AB can also be in direct compression. We can either take or leave that part of U since the AE of the beam would be large and $\frac{P^2 l}{2AE}$ is bound to be small.

9.8.1 Indeterminacy

The beam part of the structure is determinate. We are adding 3 members to it hinged at D.

So, $j = 1$; $m = 3$; $i = m - 2j = 1$.

It is customary and convenient to take strut CD as the redundant. (If there are 2 struts, each can be taken as a redundant). All we have to do is to minimize the total strain energy. Hence, $\delta U/\delta R = 0$, where R is the force in the strut. Example 9.10 shows how this is done.

Trussed beams do not exactly add to the beauty of a structure. Trussing is a method of strengthening an existing beam.

Example 9.10 *A trussed beam consists of the following:*

(i) a timber beam 150 mm wide, 200 mm deep, simply supported over a span of 5 m.

(ii) a C.I. strut 1.25 m long with a cross-sectional area of 800 mm².

(iii) two tie rods each with a cross-sectional area of 600 mm² connecting the bottom of the strut and the ends of the beam.

The beam carries a uniformly distributed load of 20 kN/m over its entire span. Find out the thrust in the strut. Take E for timber, C.I. and steel as 1×10^4 N/mm², 1.5×10^5 N/mm² and 2×10^5 N/mm² respectively.

Solution The beam AB will be subjected to bending moment and axial force. The strut CD and ties AD and DB will be subjected to axial forces only.

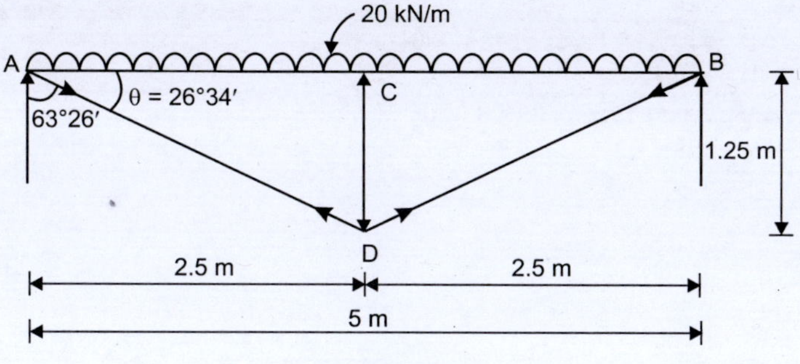

Fig. 9.41

$$\theta = \tan^{-1}\frac{1.25}{2.5} = 26°34'$$

Timber beam :

Size of the timber beam 150 mm × 200 mm

$$\text{M.I.} = \frac{150 \times 200^3}{12} = 1 \times 10^8 \text{ mm}^4$$

Modulus of elasticity of the timber beam, $E = 1 \times 10^4$ N/mm^2.

$$EI = 1 \times 10^8 \times 1 \times 10^4 \text{ Nmm}^2$$

$$= \frac{1\times 10^8 \times 1\times 10^4}{1000\times 1000^2} = 1000 \text{ kNm}^2$$

$$AE = 150 \times 200 \times 1 \times 10^4 \text{ N}$$

$$= \frac{150\times 200\times 10^4}{1000} = 3 \times 10^5 \text{ kN}$$

Tie rods :

$$AE = 600 \times 2 \times 10^5 \text{ N}$$

$$= \frac{1200\times 10^5}{1000} = 120000 \text{ kN}$$

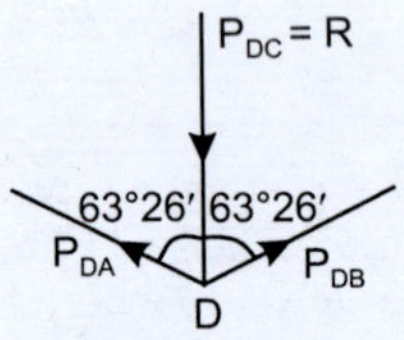

Fig. 9.41 (*a*)

Strut :

$$AE = 800 \times 1.5 \times 10^5 \text{ N} = 120000 \text{ kN}$$

Let the force R in the strut be treated as the redundant. We can now find all forces and bending moments and use the condition $\frac{\partial F}{\partial R} = 0$ to determine R.

Joint D :

$$P_{DA} \cos 63°26' + P_{DB} \cos 63°26' - R = 0$$

$$P_{DA} = P_{DB}$$

$$2 P_{DA} \cos 63°26' = R$$

$$P_{DA} = \frac{R}{2\times \cos 63°26'} = 1.118 \text{ R}$$

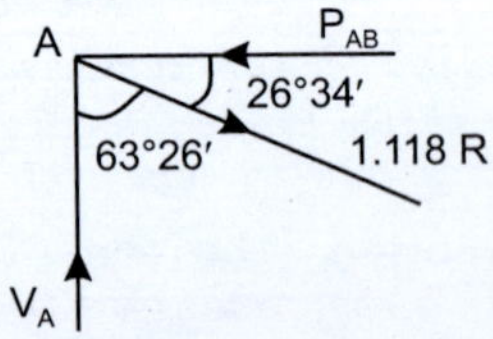

Fig. 9.41 (*b*)

Joint A :

$$P_{AB} - 1.118 \text{ R} \cos 26°34' = 0$$

$$P_{AB} = R$$

$$V_A - 1.118 \text{ R} \cos 63°26' = 0$$

$$V_A = 0.5 \text{ R}$$

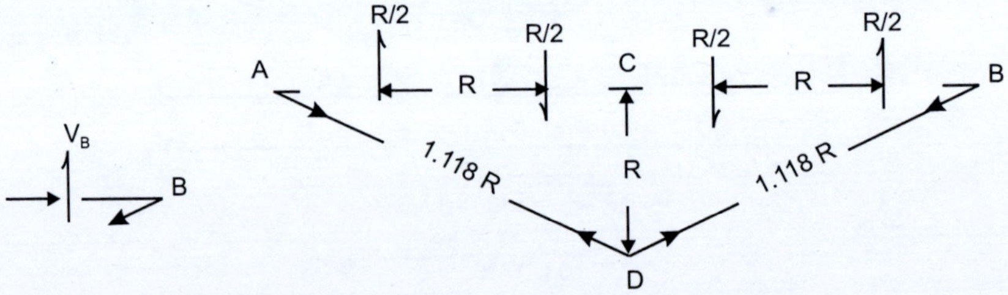

Fig. 9.42 (*a*). Forces in the members due to R only.

$$V_A = V_B = (50 - 0.5R)$$

$$M_x = R_A x - \frac{wx^2}{2} = (50 - 0.5R)x - 20 \times \frac{x^2}{2}$$

$$M_x = 50x - 0.5Rx - 10x^2$$

$$\frac{\partial M_x}{\partial R} = -0.5x = \frac{-x}{2}$$

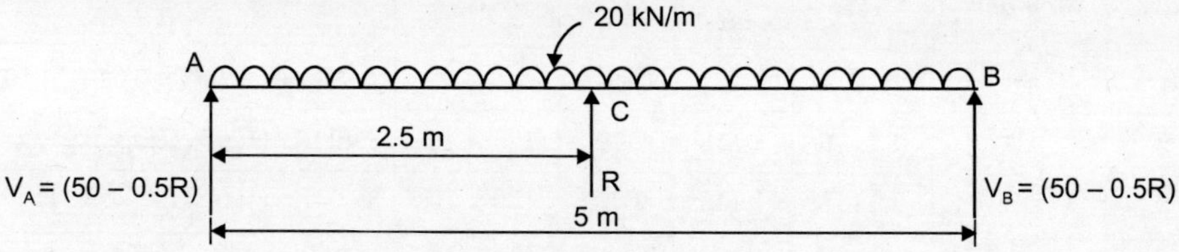

Fig. 9.42 (*b*). Beam action.

Bending :

Strain energy due to bending, $U_{AB} = 2\int_0^{2.5} \frac{Mx^2\,dx}{2EI}$

$$\frac{\partial U_{AB}}{\partial R} = 2\int_0^{2.5} 2M_x \frac{\partial M_x}{\partial R}\cdot\frac{1}{2EI}\cdot dx$$

$$= \frac{2}{EI}\int_0^{2.5} (50x - 0.5Rx - 10x^2)\left(\frac{-x}{2}\right)\cdot dx$$

$$= \frac{1}{1000}\int_0^{2.5} (10x^3 + 0.5Rx^2 - 50x^2)\,dx$$

$$= \frac{1}{1000}\left[10\left(\frac{x^4}{4}\right)_0^{2.5} + 0.5\,R\left(\frac{x^3}{3}\right)_0^{2.5} - 50\left(\frac{x^3}{3}\right)_0^{2.5}\right]$$

$$= \frac{1}{1000}\left[\frac{10}{4}(2.5^4) + \frac{0.5}{3}R\,(2.5^3) - \frac{50}{3}(2.5^3)\right]$$

$$= \frac{1}{1000}[97.66 + 2.604R - 260.42]$$

$$\frac{\partial U_{AB}}{\partial R} = \frac{1}{1000}[2.604R - 162.76]$$

i.e.,
$$\frac{\partial U_b}{\partial R} = \frac{1}{1000}[2.604R - 162.76]$$

Axial forces :

Strain energy due to axial forces, $U = \sum \frac{P^2 l}{2AE}$

$$\therefore \quad \frac{\partial U_a}{\partial R} = \sum P\cdot\frac{\partial P}{\partial R}\cdot\frac{l}{AE} \qquad \because\ P, \frac{\partial P}{\partial R} \text{ and AE are constants.}$$

Table 9.12

S.No.	Member	P	$\frac{\partial P}{\partial R}$	l (m)	AE	$P\frac{\partial P}{\partial R}\cdot\frac{l}{AE}$
1.	AB	–R	–1	5	3×10^5	$5R / (3 \times 10^5) = 0.0167 \times 10^{-3}$ R
2.	AD	1.118R	1.118	2.795	120×10^3	$3.494R / (120 \times 10^3) = 2.911 \times 10^{-5}$ R
3.	DB	1.118R	1.118	2.795	120×10^3	$3.494R / (120 \times 10^3) = 2.911 \times 10^{-5}$ R
4.	CD	–R	–1	1.25	120×10^3	$1.25R / 120 \times 10^3 = 0.1042 \times 10^{-4}$ R

$$\sum P \cdot \frac{\partial P}{\partial R}\frac{l}{AE} = 8.544 \times 10^{-5} R$$

$$\therefore \quad \frac{\partial U_a}{\partial R} = 8.544 \times 10^{-5} R$$

$$\frac{\partial U_b}{\partial R} + \frac{\partial U_a}{\partial R} = 0$$

$$\Rightarrow \quad \frac{1}{1000}[2.604R - 162.76] + \frac{8.544R}{10^5} = 0$$

$$\frac{2.604R}{1000} + \frac{0.08544R}{1000} = \frac{162.76}{1000}$$

$$\therefore \quad R(2.604 + 0.0854) = 162.76$$

Solving,
$$R = \frac{162.76}{2.689}$$

$$R = \boxed{60.52 \text{ kN}}$$

PROBLEMS

1. Find the forces in all the members of a truss loaded as shown in Fig. 9.43. All the members are of the same material having same cross-sectional area.

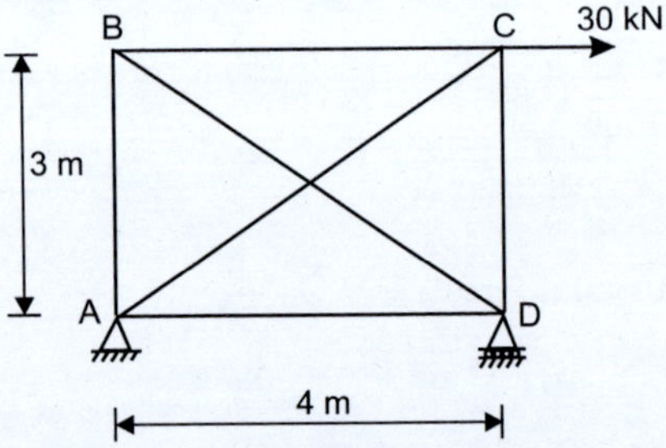

Fig. 9.43

[**Ans.** P_{AB} = 5.38 kN, P_{BC} = 7.17 kN, P_{CD} = – 17.12 kN, P_{DA} = 7.17 kN, P_{AC} = 28.53 kN, P_{BD} = – 8.96 kN]

2. Determine the forces in all the members of the structure loaded as shown in Fig. 9.44. AE is same for all the members.

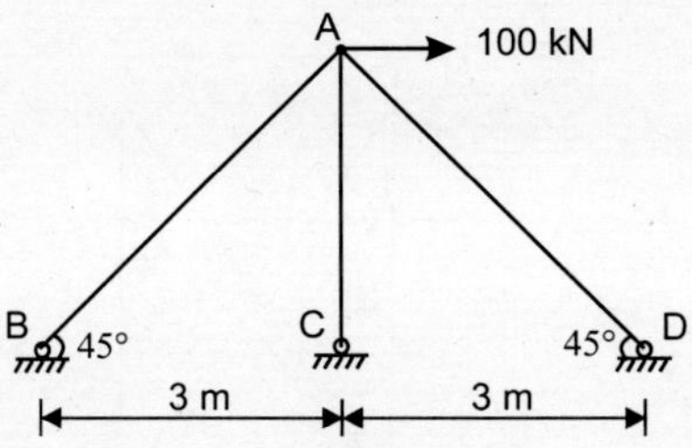

Fig. 9.44

[**Ans.** P_{AB} = 70.71 kN, P_{AC} = 0, P_{AD} = – 70.71 kN]

3. In the truss shown in Fig. 9.45, the member BE was the last member to be fixed and it was found that BE was too long by 1.2 mm. Find the force in BE. A = 1000 mm^2 and E = 2 × 10^5 N/mm^2 for all the members.

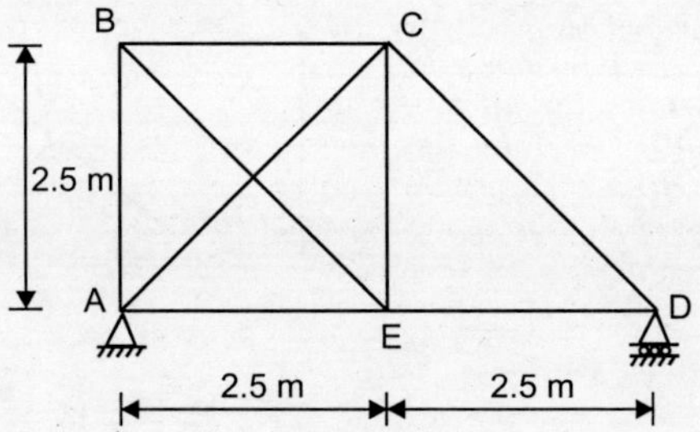

Fig. 9.45

[**Ans.** P_{BE} = – 20 kN]

4. In the truss shown in Fig. 9.46, the member BD is initially shorter by 10 mm than its required length. Area = 1000 mm^2 and E = 2 × 10^5 N/mm^2 for all the members. Find the forces in all the members.

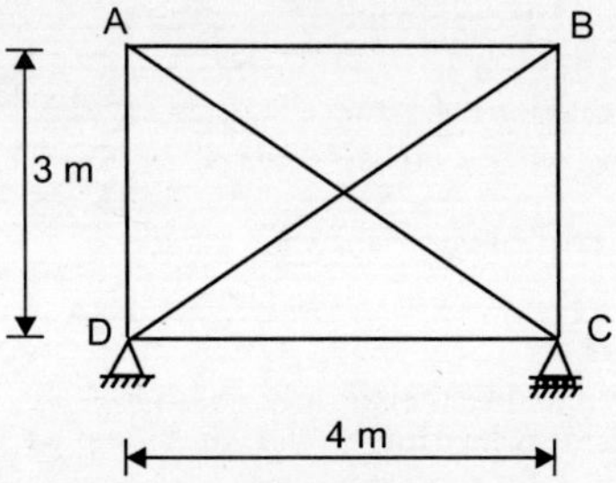

Fig. 9.46

[**Ans.** P_{AB} = – 92.59 kN, P_{BC} = – 69.44 kN, P_{CD} = – 92.59 kN, P_{DA} = – 69.44 kN, P_{AC} = 115.74 kN, P_{BD} = 115.74 kN]

5. Find the force in the member BD of the frame shown in Fig. 9.47, if the member BD is subjected to a rise in temperature of 16.67°C. A = 1000 mm^2 and E = 2 × 10^5 N/mm^2 for all the members. α = 12 × 10^{-6}/°C.

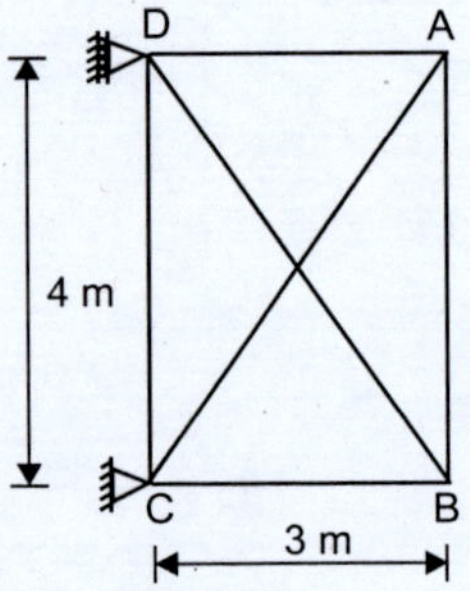

Fig. 9.47

[**Ans.** P_{BD} = – 11.574 kN]

6. In the frame shown in Fig. 9.48, the member BD is subjected to a fall in temperature of 30°C. A = 1000 mm^2 and E = 2.047 × 10^5 N/mm^2 for all the members. α = 12 × 10^{-6}/°C. Find the force in the member BD.

[**Ans.** P_{BD} = 22.74 kN (Comp.)]

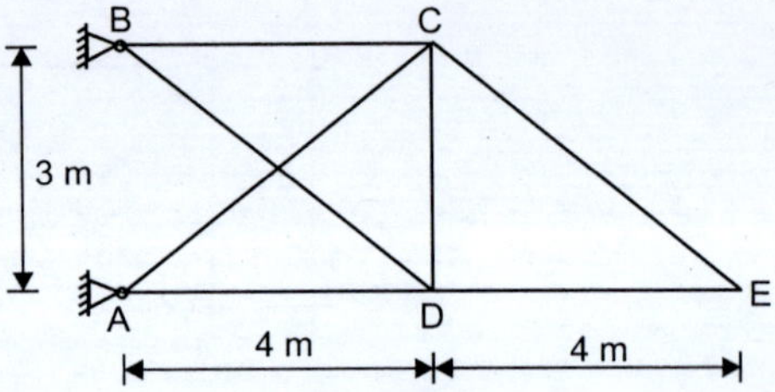

Fig. 9.48

7. Find the horizontal reaction at D of the pin jointed structure shown in Fig. 9.49. The cross-sectional areas of vertical members are 3000 mm^2 each and those of all other members are 2200 mm^2. Both supports are hinged.

[**Ans.** H_D = 20.458 kN (←)]

8. A trussed beam consists of the following :

(*i*) A timber beam 120 mm wide, 240 mm deep, simply supported over a span of 5 m.

(*ii*) A CI strut 1.25 m long with a cross-sectional area of 900 mm^2 and

(*iii*) Two tie rods each with a cross-sectional area of 500 mm^2 connecting the bottom of the strut and the ends of the beam.

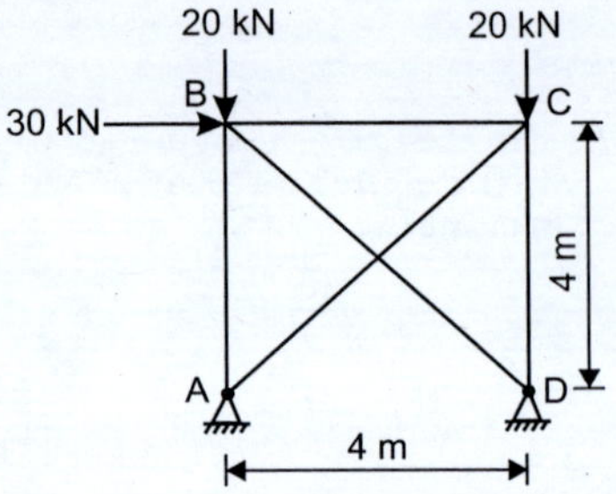

Fig. 9.49

The beam carries a uniformly distributed load of 30 kN/m over its entire span. Find out the thrust in the strut. Take E for timber, C.I. and steel as 1 × 10^4 N/mm^2, 1.5 × 10^5 N/mm^2 and 2 × 10^5 N/mm^2 respectively. [**Ans.** R = 87.95 kN]

CHAPTER 10

RAPID AND APPROXIMATE ANALYSIS OF BUILDING FRAMES

10.1 INTRODUCTION

These days, practically all major buildings are framed structures. For buildings exceeding 2 storeys, we seldom adopt load bearing wall construction. For buildings exceeding 5 storeys, it is better to resort to computer based analysis procedures. Approximate analysis procedures are useful for analysing building frames upto 5 storeys.

The approximate procedures outlined in this chapter can always be adopted for a rapid check on the extensive results spewed by computers using analysis softwares. Softwares are of course wonderful things. But the weak link is the user — software interface. The user is not sure if he/she fed the data the way the programmer intended. Hence several counter checks about the plausibility of the results given by the computer are called for otherwise we might end up with absurd results. Rapid methods would serve as a countercheck admirably.

In rapid methods it is comfortable to treat vertical loads and horizontal loads (on frames) separately. Several acceptable approximations are made in order to lend ease and quickness to the analytical process.

10.2 VERTICAL LOADS

Vertical loads primarily affect beam moments. The column bending moments induced are of secondary order. Column thrusts however are directly linked to vertical loads.

Whatever be the number of storeys, it is customary (and permissible) to analyse only a part of the frame, termed ***substitute frame***. The moment distribution method which is widely employed in rapid methods is also approximated. We employ what is known as ***Two cycle moment distribution***, which terminates the iterative process after 2 carry overs and 2 balancings. (It implies that we are satisfied with whatever accuracy we get after 2 iterations).

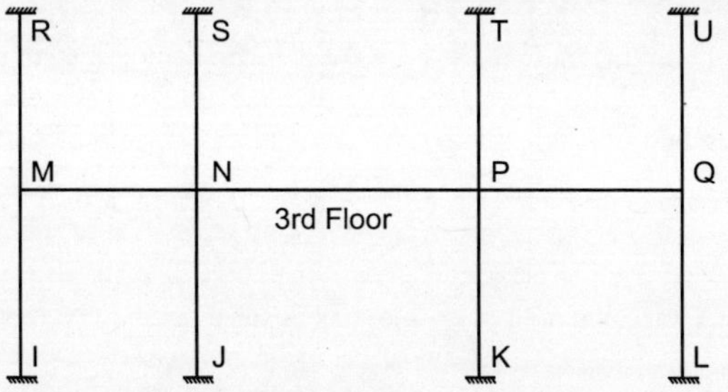

Fig. 10.1. Substitute frame for 3rd floor beams.

Suppose we are analysing a 5-storey 3 bay building frame. If we want to compute the bending moments on the beams at the 3rd floor level, we can substitute the 5 × 3 panel frame by a simplified frame in Fig. 10.1.

Is this correct ? No. But it is reasonably OK. We treat the columns above and below the floor as fixed at the far ends.

Beams however are elastically connected to columns and are treated as such. But if the beams join shear walls, it is alright to treat them as fixed (Fig. 10.2).

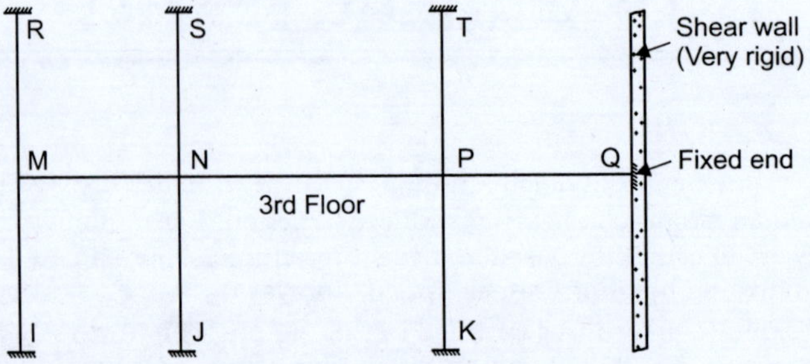

Fig. 10.2. Substitute frame, beam 'fixed' to shear wall.

10.3. LOADING CASES

When we design beams, the main consideration is the bending moment. The span bending moments are usually sagging (tension at bottom) and the support bending moments are invariably hogging (tension at top).

10.3.1 Span Moments

For arriving at critical mid-span bending moment in any given span, we have to choose the position of live loads to be at the most adverse location. The concept of influence lines (described in Chapter 3) will come to our aid. Figure 10.3 shows the ILD and the critical loading position for live loads. Dead loads, however, would always be present in all the spans.

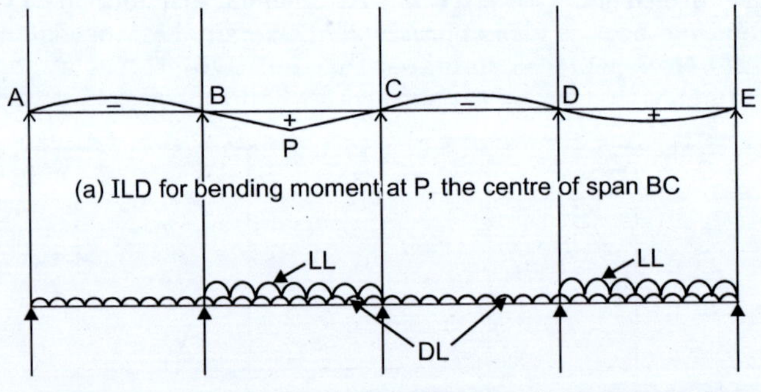

Fig. 10.3

10.3.2 Support Moment

If we want to design a typical support section of a beam, the loading shall be as shown in Fig. 10.4.

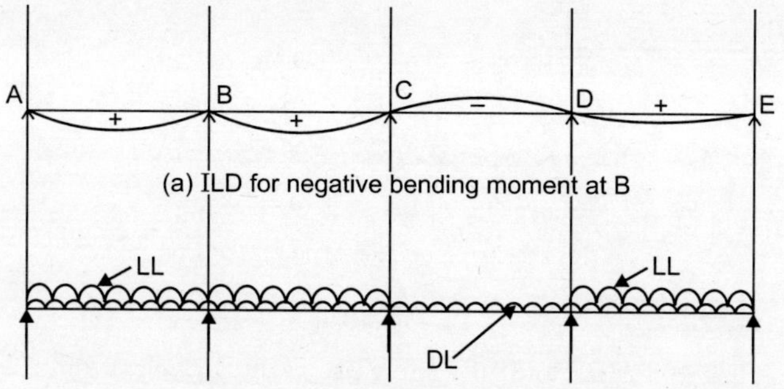

(a) ILD for negative bending moment at B

(b) Positions of LL to create max bending moment at B(in beam).

Fig. 10.4

Therefore for maximum span bending moment, the span in question and all alternate spans shall be laden with LL. For maximum support bending moment, the 2 spans adjoining the support and every alternate span must have LL. This will be adopted in analysing the substitute building frames under 2 cycle moment distribution.

10.3.3 Column Bending Moments

Columns are primarily compression members. But the presence of bending moments cannot be ignored. This will have to be necessarily taken into account while designing columns since bending moment and thrust interact to produce higher stresses in columns. We can arrive at strategic live load positions by extending the concepts of influence lines further. Figure 10.5 shows the IL for 2 kinds of bending moment in 2 typical columns. We have to remember we are now discussing only vertical loads and these come primarily on beams only. So from the ILD we will choose beams showing downward ordinates for carrying live loads transferring critical support moments to columns.

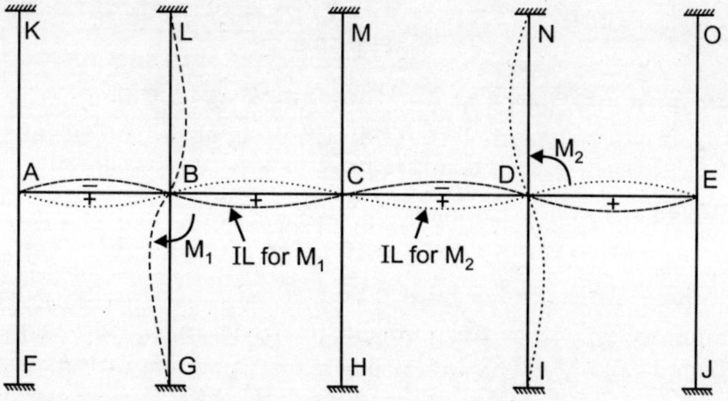

(a) - - - IL for M_1 · · · · IL for M_2

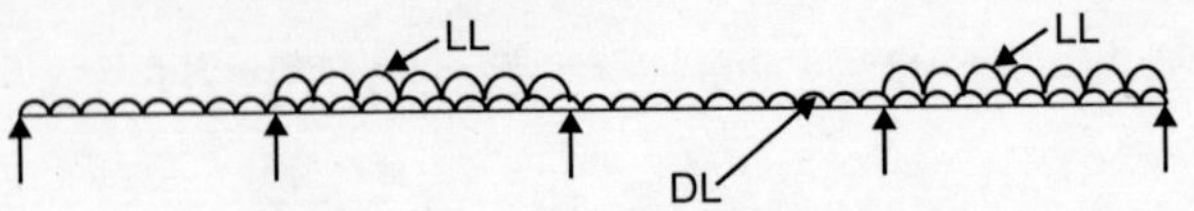

(b) Loading for maximizing M_1 (on BG)

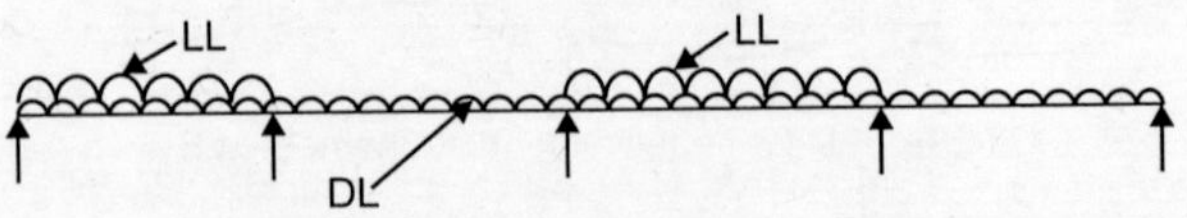

(c) Loading for maximizing M_2 (on DN)

Fig. 10.5

Examples 10.1, 10.2, 10.3 and 10.4 make use of the above concepts to determine critical bending moments is beams and columns.

Example 10.1 *In a multistoreyed building, the frame shown in Fig. 10.6 are spaced at 4 m intervals. Dead load from the slab is 3 kN/m² and the live load is 5 kN/m². Analyse the beam BC for midspan positive bending moment. Self weight of the beams may be ignored. Use two-cycle moment distribution method.*

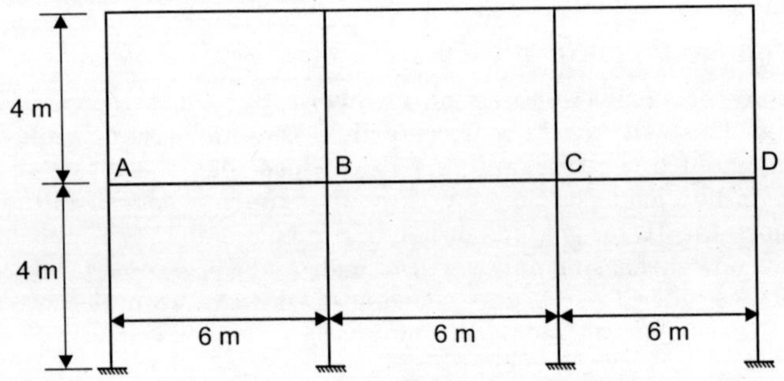

Fig. 10.6

Solution The frames are spaced at 4 m intervals. Consider one metre length of the beam ABCD. It can be assumed that half the span of slab on either side will transfer the load to this beam. Self weight of the beams is ignored.

Dead load from slab = Intensity of DL × 1 × 4

= 3 × 1 × 4 = 12 kN/m

Live load from slab = 5 × 1 × 4 = 20 kN/m

For maximum midspan positive moment in BC, span BC will carry DL + LL and spans AB and CD will carry DL only. The substitute frame is shown in Fig. 10.7.

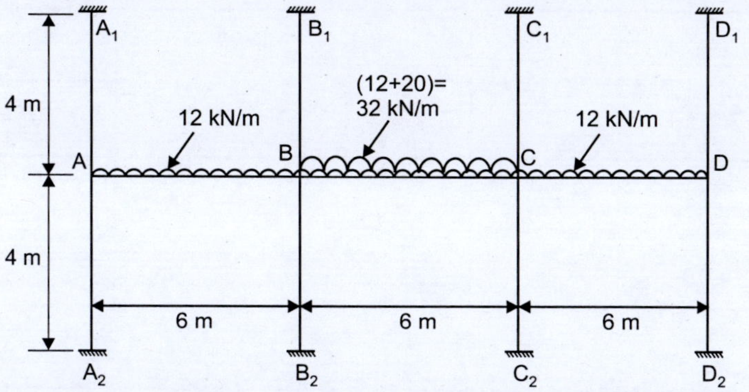

Fig. 10.7. Substitute frame with loading for span moment on BC.

Fixed end moments :

Span AB :

$$M_{FAB} = -\frac{wl^2}{12} = -\frac{12 \times 6^2}{12} = -36 \text{ kNm}$$

$$M_{FBA} = \frac{wl^2}{12} = \frac{12 \times 6^2}{12} = 36 \text{ kNm}$$

Span BC :

$$M_{FBC} = -\frac{32 \times 6^2}{12} = -96 \text{ kNm}$$

$$M_{FCB} = \frac{32 \times 6^2}{12} = 96 \text{ kNm}$$

Span CD :

$$M_{FCD} = -\frac{12 \times 6^2}{12} = -36 \text{ kNm}$$

$$M_{FDC} = \frac{12 \times 6^2}{12} = 36 \text{ kNm}$$

Distribution factors :

Uniform moment of inertia is assumed for all the members. Hence I can be dropped. Using symmetry, the moment distribution is done for half the beam ABCD as below :

Joint	*Member*	*Relative stiffness*	*Sum*	*Distribution factors*
A	AA_1	$\frac{1}{4}$	$\frac{1}{4}+\frac{1}{6}+\frac{1}{4}=\frac{2}{3}$	$\frac{1}{4} \div \frac{2}{3} = \frac{3}{8} = 0.375$
	AB	$\frac{1}{6}$		$\frac{1}{6} \times \frac{3}{2} = \frac{1}{4} = 0.25$
	AA_2	$\frac{1}{4}$		$\frac{1}{4} \times \frac{3}{2} = \frac{3}{8} = 0.375$

B	BA	$\frac{1}{6}$		$\frac{1}{6} \div \frac{3}{4} = \frac{2}{9} = 0.222$
	BB_1	$\frac{1}{4}$		$\frac{1}{4} \div \frac{3}{4} = \frac{1}{3} = 0.333$
	BC	$\frac{1}{2}\,\frac{1}{6}$	$\frac{1}{6} + \frac{1}{4} + \frac{1}{12} + \frac{1}{4} = \frac{3}{4}$	$\frac{1}{12} \div \frac{3}{4} = \frac{1}{9} = 0.111$
	BB_2	$\frac{1}{4}$		$\frac{1}{4} \div \frac{3}{4} = \frac{1}{3} = 0.333$

Joint	***A***	***B***	
Member	AB	BA	BC
Distribution factors	0.25	0.222	0.111
Fixed end moments Balance	– 36 9.00	36 13.32	– 96.00 12.00
Carry over Balance		4.50 – 1.0	– 6.66 0.33
Final moments			– 89.84

Note. The sum of the moments at joint B *i.e.,* BA and BC or at C *i.e.,* CB and CD will not be zero in the table because the moment distribution to columns are not shown in the table.

Free body diagram of span BC :

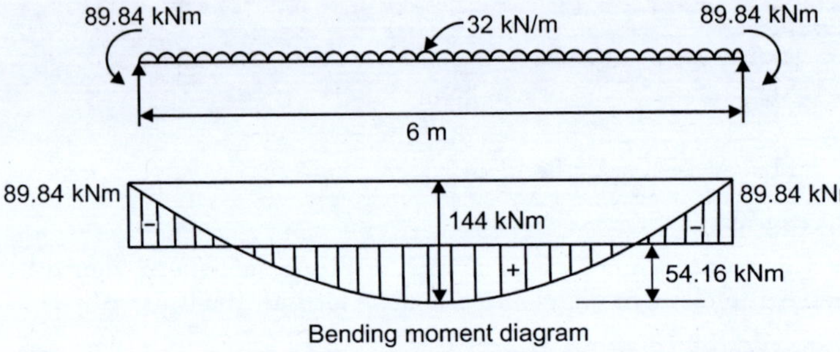

Fig. 10.8

Free moment at the centre of BC = $\frac{wl^2}{8} = \frac{32 \times 6^2}{8}$ = 144 kNm

Therefore design moment at midspan of BC = 144 – 89.84 = 54.16

Midspan positive moment in BC = 54.16 kNm.

Example 10.2 *In a multistoreyed building, the frames shown in Fig. 10.9 are spaced at 3.5 m intervals. Dead load from the slab is 3 kN/m² and live load is 5 kN/m². Analyse the beam BC for mid span negative bending moment. Self weight of beams of 4 m span may be taken as 4 kN/m and that of 6 m span beams may be taken as 5 kN/m. Use two cycle moment distribution method. The stiffnesses of the members are indicated against each member.*

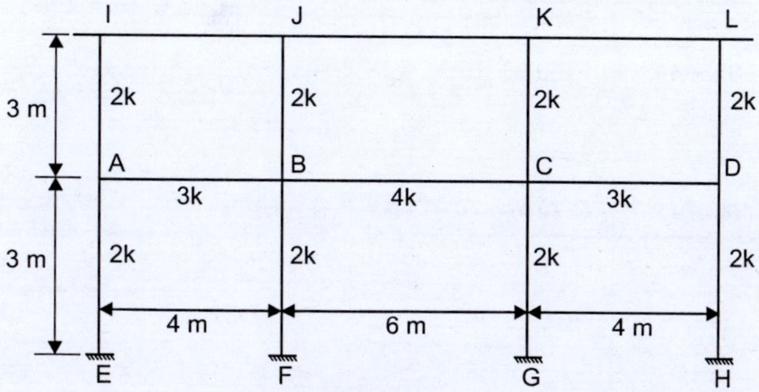

Fig. 10.9

Solution The frames are spaced at 3.5 m intervals. Consider 1 m length of the beam ABCD.

The beam will carry the load transmitted by $\left(\frac{3.5}{2} = \right)$ 1.75 m length of the slab on either side.

Self weight of the beams AB and CD = 4 kN/m

Self weight of the beam BC = 5 kN/m

Dead load from the slab = 3 kN/m²

Live load from the slab = 5 kN/m²

Dead load (from slab + beam) in AB and CD = 1 × 3.5 × 3 + 4 = 14.5 kN/m

Dead load (from slab + beam) in BC = 1 × 3.5 × 3 + 5 = 15.5 kN/m

Live load from the slab per m run of beam = 1 × 3.5 × 5 = 17.5 kN/m

For maximum midspan negative bending moment in BC, the span BC will be loaded with dead load only and the spans AB and CD will be loaded with dead load + live load Fig. 10.10.

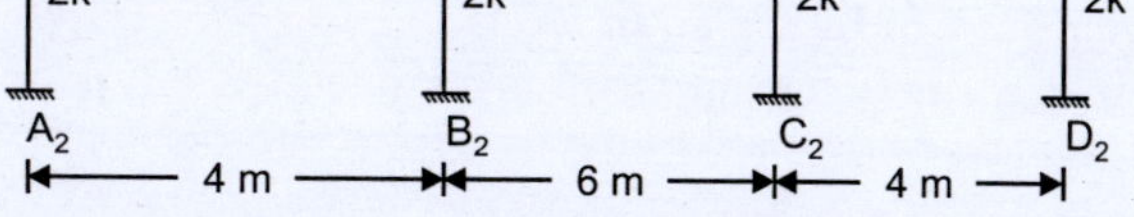

Fig. 10.10

Substitute frame with loading for max negative moment at midspan of BC.

Fixed end moments :

$$M_{FAB} = -\frac{wl^2}{12} = -\frac{32 \times 4^2}{12} = -42.67 \text{ kNm} \qquad M_{FBA} = 42.67 \text{ kNm}$$

$$M_{FBC} = \frac{-15.5 \times 6^2}{12} = -46.5 \text{ kNm} \qquad M_{FCB} = 46.5 \text{ kNm}$$

$$M_{FCD} = -\frac{32 \times 4^2}{12} = -42.67 \text{ kNm} \qquad M_{FDC} = 42.67 \text{ kNm}$$

Distribution factors :

Joint	*Member*	*Relative Stiffness*	*Sum*	*Distribution factors*
A	AA_1	2k	7k	2k/7k = 0.286
	AB	3k		3k/7k = 0.428
	AA_2	2k		2k/7k = 0.286
B	BA	3k	11k	3k/11k = 0.272
	BB_1	2k		2k/11k = 0.182
	BC	4k		4k/11k = 0.364
	BB_2	2k		2k/11k = 0.182

By symmetry, the distribution factors for the joints C and D will be mirror reflections of those of B and A.

Joint	*A*	*B*		*C*		*D*
Member	AB	BA	BC	CB	CD	DC
Distribution factors	0.428	0.272	0.364	0.364	0.272	0.428
Fixed and moments Balance	– 42.67 18.26	42.67 1.04	– 46.5 1.39	46.5 – 1.39	– 42.67 – 1.04	42.67 – 18.26
Carry over Balance	0.52 – 0.22	9.13 – 2.29	– 0.70 – 3.07	0.70 3.07	– 9.13 2.29	– 0.52 0.22
Final moments			– 48.88	48.88		

$$\text{Simply supported beam moment in span BC} = \frac{wl^2}{8} = \frac{15.5 \times 6^2}{8} = 69.75 \text{ kNm}$$

Free body diagram and BMD for span BC :

It is found that the net moment in BC is not negative, but positive.

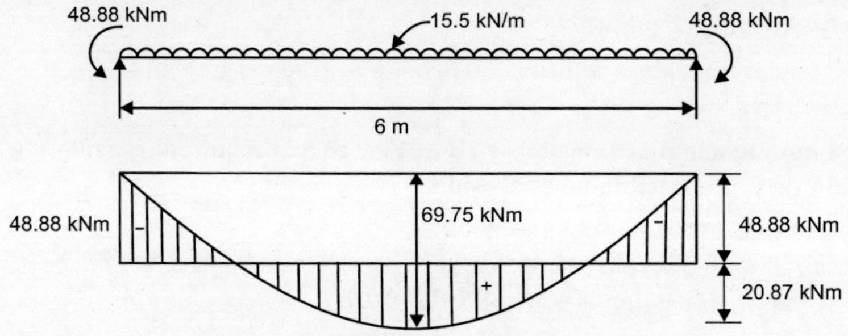

Fig. 10.11. Bending moment diagram.

Example 10.3 *A multistoreyed building consists of 4 storeyed 3 bay frames spaced at 3 m c/c. Live load on floor slab is 3 kN/sq m and dead load is 3.5 kN/sq m. The spans of the beams from left to right are 6 m, 4 m and 4 m respectively. Storey height is 3.5 m. Moment of inertia of the beams is 1.5 times that of columns. Self weight of the beams is 3.5 kN/m. Determine the maximum moment in the beam at the junction of first span and second span of an intermediate floor.*

Solution The frames are spaced at 3.5 m intervals. Consider one metre length of the beam ABCD. Half span load on slabs on either side will be transferred to the beam.

Self weight of beams AB, BC, CD = 3.5 kN/m

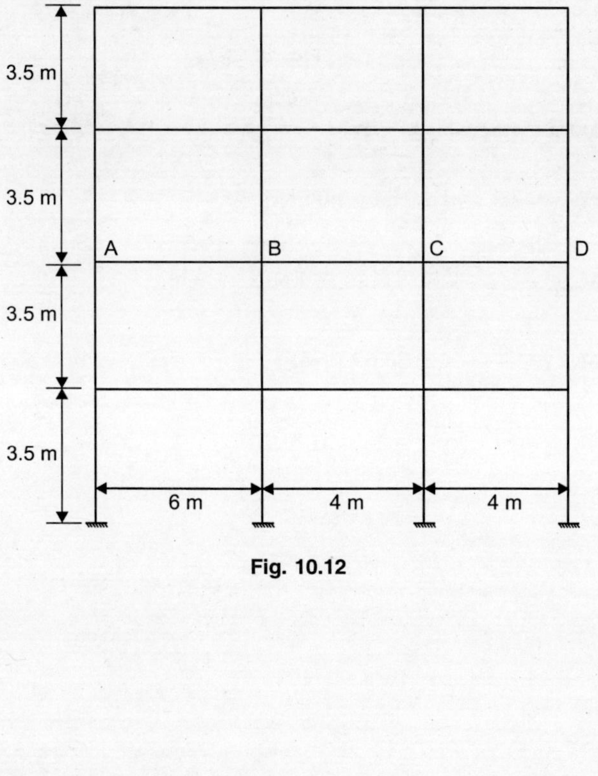

Fig. 10.12

Load per m run of the beam:

Dead load from slabs and beam = 1 × 3.5 × 3 + 3.5 = 14 kN/m

Live load from slabs = 1 × 3 × 3 = 9 kN/m

Maximum moment at the support B in Fig. 10.8 is required (considering ABCD as the intermediate floor proposed to be analysed).

For maximum moment at B :

Spans AB and BC with DL + LL and span CD with DL only will be considered. The substitute frame with loading is shown in Fig. 10.7.

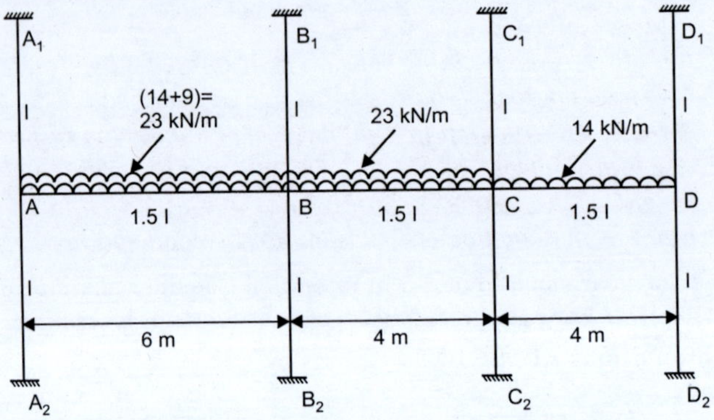

Fig. 10.13

$$M_{FAB} = -\frac{wl^2}{12} = -\frac{23 \times 6^2}{12} = -69 \text{ kNm}$$

$$M_{FBA} = \frac{wl^2}{12} = \frac{23 \times 6^2}{12} = 69 \text{ kNm}$$

$$M_{FBC} = -\frac{23 \times 4^2}{12} = -30.67 \text{ kNm}$$

$$M_{FCB} = \frac{23 \times 4^2}{12} = 30.67 \text{ kNm}$$

$$M_{FCD} = -\frac{14 \times 4^2}{12} = -18.67 \text{ kNm}$$

$$M_{FDC} = \frac{14 \times 4^2}{12} = 18.67 \text{ kNm}$$

Distribution factors :

Joint	*Member*	*Relative stiffness*	*Sum*	*Distribution factors*
A	AA_1	$\frac{I}{3.5}$	$\frac{I}{3.5}+\frac{1.5I}{6}+\frac{I}{3.5}$	$\frac{I}{3.5}\times\frac{42}{34.5I}=0.348$
	AB	$\frac{1.5I}{6}$	$=\frac{12I+10.5I+12I}{42}$	$\frac{1.5I}{6}\times\frac{42}{34.5I}=0.304$
	AA_2	$\frac{I}{3.5}$	$=\frac{34.5I}{42}$	$\frac{I}{3.5}\times\frac{42}{34.5I}=0.348$
B	BA	$\frac{1.5I}{6}$		$\frac{1.5I}{6}\times\frac{84}{100.5I}=0.209$
	BB_1	$\frac{I}{3.5}$	$\frac{21I+24I+31.5I+24I}{84}$	$\frac{I}{3.5}\times\frac{84}{100.5I}=0.239$
	BC	$\frac{1.5I}{4}$	$=\frac{100.5I}{84}$	$\frac{1.5I}{4}\times\frac{84}{100.5I}=0.313$
	BB_2	$\frac{I}{3.5}$		$\frac{I}{3.5}\times\frac{84}{100.5I}=0.239$
C	CB	$\frac{1.5I}{4}$		0.284
	CC_1	$\frac{I}{3.5}$	$\frac{(31.5+24+31.5+24)I}{84}$	0.216
	CD	$\frac{1.5I}{4}$	$=\frac{111I}{84}$	0.284
	CC_2	$\frac{I}{3.5}$		0.216
D	DC	$\frac{1.5I}{4}$	$\frac{(31.5+24+24)2I}{84}$	0.396
	DD_1	$\frac{I}{3.5}$	$=\frac{79.5I}{84}$	0.302
	DD_2	$\frac{I}{3.5}$		0.302

Two cycle moment distribution:

Joint	*A*	*B*		*C*		*D*
Member	AB	BA	BC	CB	CD	DC
Distribution factors	0.304	0.209	0.313	0.284	0.284	0.396
Fixed end moments	– 69.00	69.00	– 30.67	30.67	– 18.67	18.67
Balance	20.98	– 8.01	– 12.00	– 3.41	– 3.41	– 7.39
Carry over	– 4.01	10.49	– 1.71	– 6.00	– 3.70	– 1.71
Balance	1.22	– 1.84	– 2.75	2.75	2.75	0.68
Final moments		69.64	– 47.13			

Maximum moment at joint (in beams) B = 69.64 kN/m.

Example 10.4 *Analyse the frame in example 10.2 for moments in columns. Use two cycle moment distribution method.*

Solution **(Fig. 10.9—Example 10.2).** For maximum moment in columns, alternate spans should be loaded with live load and dead load. Here, spans AB and CD will be loaded with live and dead load and span BC with only dead load and will be analysed *first*.

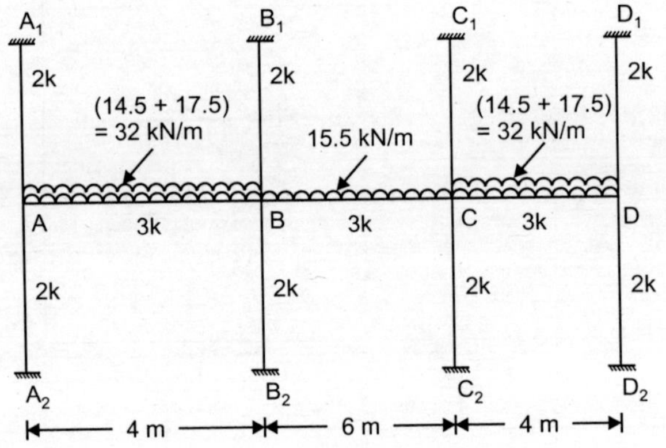

Fig. 10.14

Fixed end moments :

M_{FAB} = – 42.67 kNm $\quad$ M_{FBA} = 42.67 kNm

M_{FBC} = – 46.5 kNm $\quad$ M_{FCB} = 46.5 kNm

M_{FCD} = – 42.67 kNm $\quad$ M_{FDC} = 42.67 kNm

Distribution factors :

Calculated and available in example 10.2

Moment distribution :

Joint	*A*			*B*				*C*				*D*		
Member	AE	AI	AB	BA	BF	BJ	BC	CB	CG	CK	CD	DC	DL	DH
DF	.286	.286	.428	.272	.182	.182	.364	.364	.182	.182	.272	.428	.286	.286
FEM Balance CO	 11.44	 11.44	– 42.67 18.26	42.67 1.05	 0.70	 0.70	– 46.5 1.4	46.5 – 1.4	 – 0.7	 – 0.7	– 42.67 – 1.05	42.67 – 18.26	 – 11.44	 – 11.44
(*) CO Balance	– 0.15	– 0.15	*	*	– 1.53	– 1.53	*	*	1.53	1.53	*	*	0.15	0.15
Final moments (column only)	11.29	11.29			– 0.83	– 0.83			0.83	0.83			– 11.29	– 11.29

* Since we are interested only in the column moments, the CO moments to AB and DC, and the distribution moments against AB and DC are not shown.

Moment distribution :

Joint	***A***			***B***				***C***				***D***		
Member	AE	AI	AB	BA	BF	BJ	BC	CB	CG	CK	CD	DC	DL	DH
DF	.286	.286	.428	.272	.182	.182	.364	.364	.182	.182	.272	.428	.286	.286
FEM			– 19.33	19.33			– 99.00	99.00			– 19.33	19.33		
Dist.	5.52	5.52	8.27	21.67	14.50	14.50	29.0	– 29.0	– 14.5	– 14.5	– 21.67	– 8.27	– 5.52	– 5.52
(*) CO Balance	– 3.09	–3.09	*	*	1.89	1.89	*	*	– 1.89	– 1.89	*	*	3.09	3.09
Final moments	2.43	2.43			16.38	16.38			– 16.38	16.38			– 2.43	–2.43

* Since we need only column moments, entries for beams are not shown in the last step.

Second Analysis will be done with spans AB and CD with dead load only and the span BC with live load and dead load.

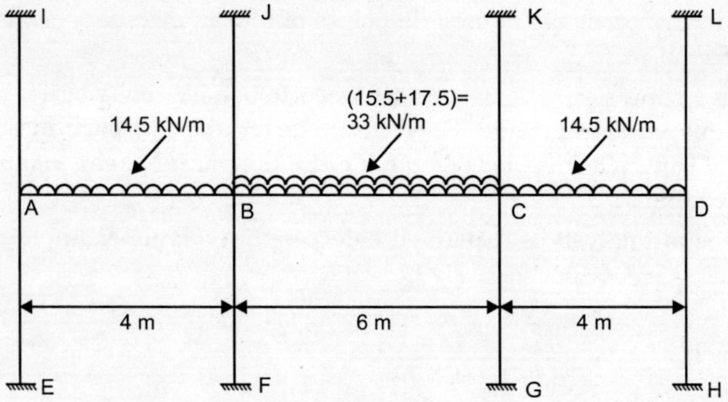

Fig. 10.15

Fixed end moments :

$$M_{FAB} = -\frac{wl^2}{12} = -\frac{14.5 \times 4^2}{12} = -19.33 \text{ kNm} = M_{FCD}$$

$$M_{FBA} = \frac{wl^2}{12} = \frac{14.5 \times 4^2}{12} = 19.33 \text{ kNm} = M_{FDC}$$

$$M_{FBC} = -\frac{wl^2}{12} = -\frac{33 \times 6^2}{12} = -99 \text{ kNm}; \; M_{FDC} = 99 \text{ kNm}$$

10.4. LATERAL LOADS

Building frames are subjected to horizontal forces due to wind pressure and seismic effects. These horizontal forces introduce direct forces on columns (tensile forces on windward columns and compressive forces on leeward columns and zero forces on axial column, if any, lying on the axis of symmetry besides bending moments in all the members of the frame.

The effects of horizontal forces are comparatively small in most of the cases. Hence an elaborate and rigorous method of analysis is not required. And rigorous analysis is also laborious.

In the analysis for horizontal loads, it is assumed that horizontal forces are transferred to joints, where the beams and exterior columns on the windward side meet. The bending moments and shear forces due to the horizontal forces can be determined by approximate methods. The methods commonly adopted for the analysis of frames subjected to horizontal loads are:

1. The portal method,
2. The cantilever method, and
3. The factor method.

10.4.1 The Portal Method

This approximate method of analysis is more suited for low rise building frames. In this method, the following assumptions are made. The frame is made less indeterminate (because of these assumptions) and the moments, shear forces and axial forces are easily obtained.

Assumptions made :

(*i*) Under the action of horizontal forces, the frames will deflect. The points of contraflexure occur at the middle points of all the members *i.e.,* beams and columns of the frame.

(*ii*) Frames having several bays are assumed to behave like a series of single bay portal frames placed side by side. Horizontal shear taken by each interior column, at the points of contraflexure, is thus double the horizontal shear taken by each of the external columns.

The method of analysis is explained below with a simple example of a two storey two bay frame.

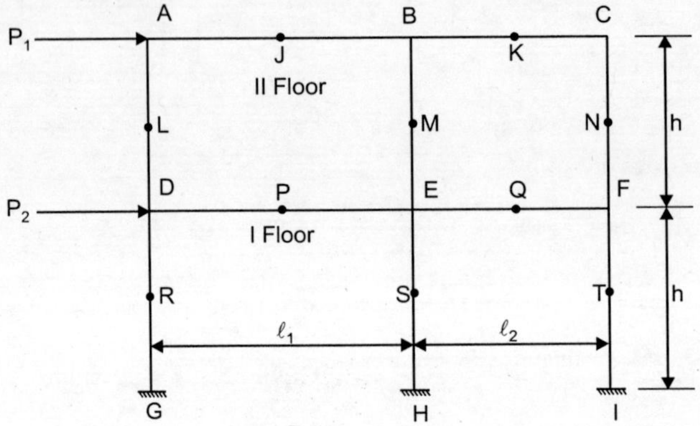

Fig. 10.16 (*a*)

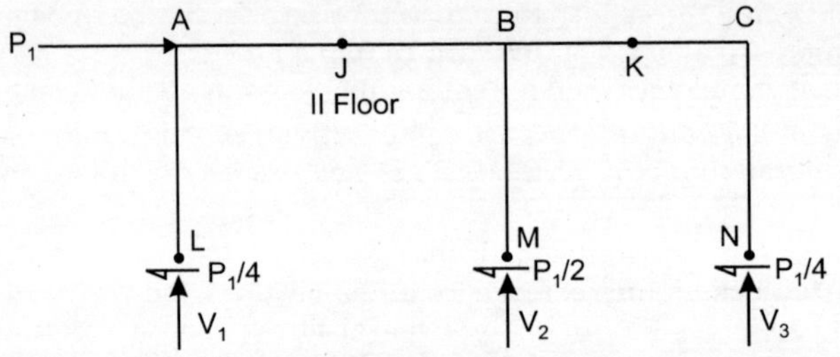

Fig. 10.16 (*b*)

Figure 10.16 (*a*) shows the frame. The black dots are (the assumed) points of contraflexure. This assumption is equivalent to assuming hinges at those points.

Without those hinges the structure is 12 degree indeterminate. Assumption 1 introduces 10 hinges and removes 10 degrees of redundancy. Because of assumption 2 we get the shears at the 2 levels namely at the middle of first storey columns and at the middle of second storey columns. Because of these assumptions we have 2 systems (Fig. 10.16 (*b*)) and (Fig. 10.16(*c*)) that have only 3 unknowns each (the 3 vertical reactions) that can be determined from the conditions of equilibrium.

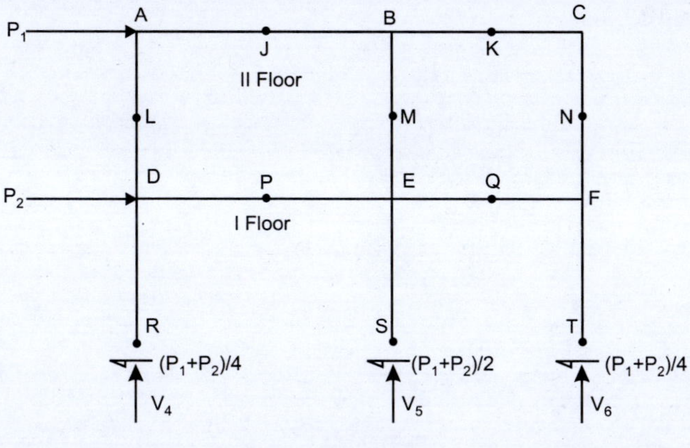

Fig. 10.16 (*c*)

Once we know the 6 reactions at the base hinges, we can find the shear forces and thrusts in each of the hinges by considering the equilibrium of the free bodies of each ***'bent'*** (see Fig. 10.17). That will help us evaluate the bending moments at the joints.

But we must keep in mind that the method is approximate, abetted by a number of assumptions.

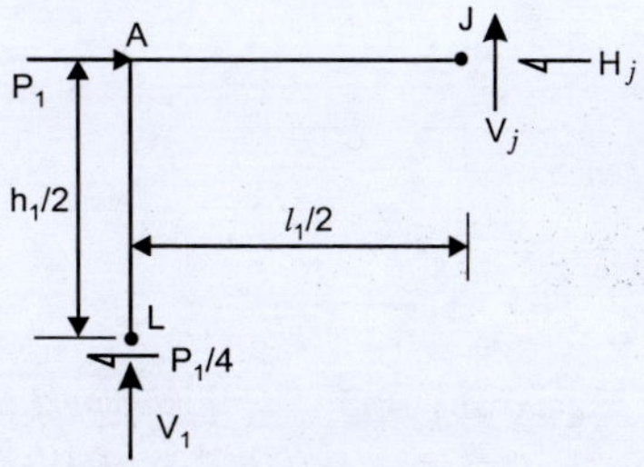

Fig. 10.17 (*a*)

From Fig. 10.17(a),

$$H_j = P_1 - P_1/4 = 3P_1/4$$

$$V_j = -V_1$$

$$M_{AJ} = V_j\, l_1/2$$

$$M_{AL} = \frac{P_1}{4} \times \frac{h}{2}$$

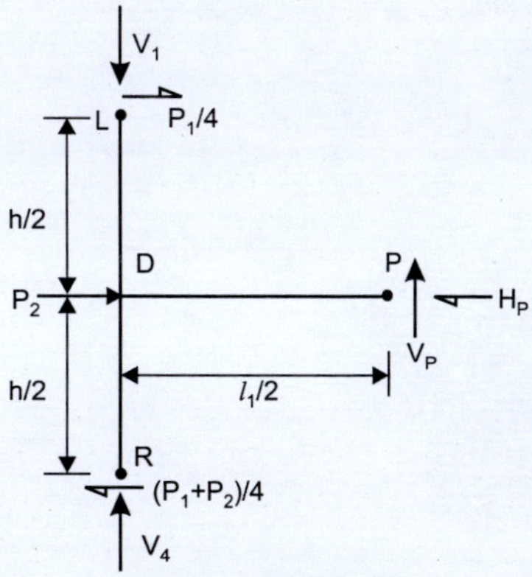

Fig. 10.17 (*b*)

From Fig. 10.17(b),

$$H_P = \frac{P_1}{4} - \frac{P_1 + P_2}{4} + P_2 = \frac{3P_2}{4}$$

$$V_P = V_1 - V_4$$

$$M_{DL} = \frac{P_1}{4} \times \frac{h}{2}$$

$$M_{DR} = \frac{(P_1 + P_2)}{4} \times \frac{h}{2}$$

$$M_{DP} = V_P \times l_1/2$$

From Fig. 10.17(*c*),

$$H_Q = H_P + H_M - \frac{(P_1 + P_2)}{2}$$

$$V_Q = V_P + V_M - V_S$$

$$M_{EM} = H_M \times h/2$$

$$M_{EP} = V_P \times l_1/2$$

$$M_{ES} = \frac{(P_1 + P_2)}{2} \times \frac{h}{2}$$

$$M_{EQ} = V_Q \times l_2/2$$

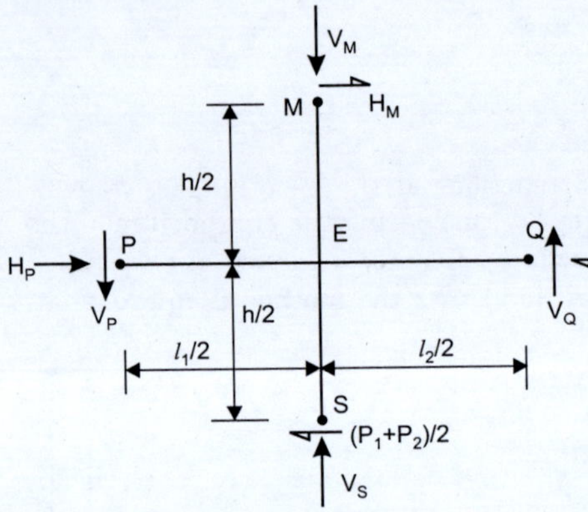

Fig. 10.17 (*c*)

Example 10.5 *Using the portal method, analyse the building frame subjected to horizontal forces (due to wind) as shown in Fig. 10.18. Sketch the bending moment diagram.*

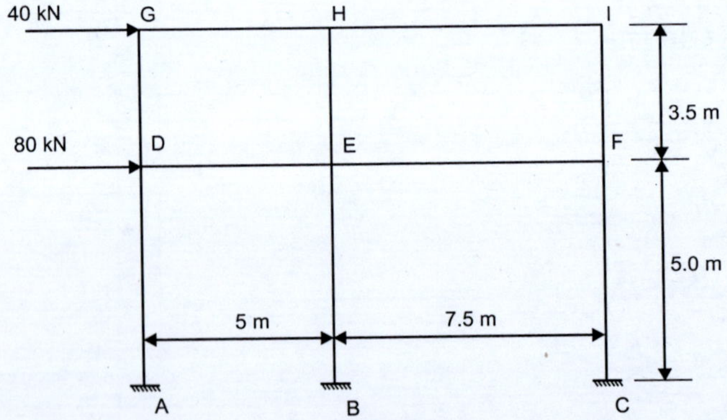

Fig. 10.18

Solution **Portal method:**

Points of contraflexure will be assumed to occur at midspan/mid-height of all the beams/columns.

Horizontal shear : Horizontal shear in interior column is assumed to be twice that in the outer columns.

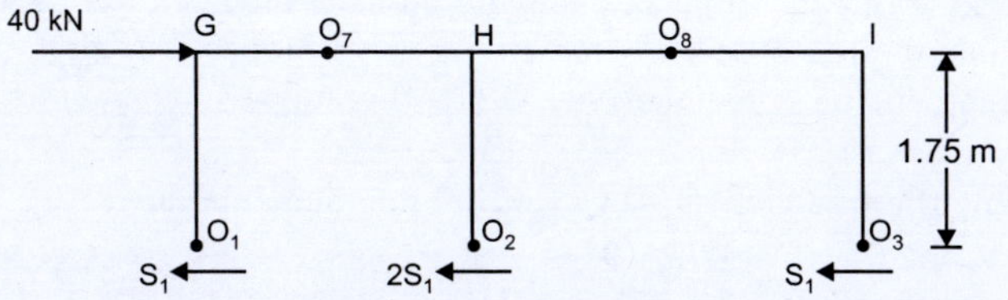

Fig. 10.19. Top storey.

Top storey

$$S_1 + 2S_1 + S_1 - 40 = 0$$

$$4S_1 = 40$$

$$S_1 = 10 \text{ kN}; \ 2S_1 = 20 \text{ kN}$$

Bottom storey

$$S_2 + 2S_2 + S_2 - 40 - 80 = 0$$

$$S_2 = 30 \text{ kN}; \ 2S_2 = 60 \text{ kN}$$

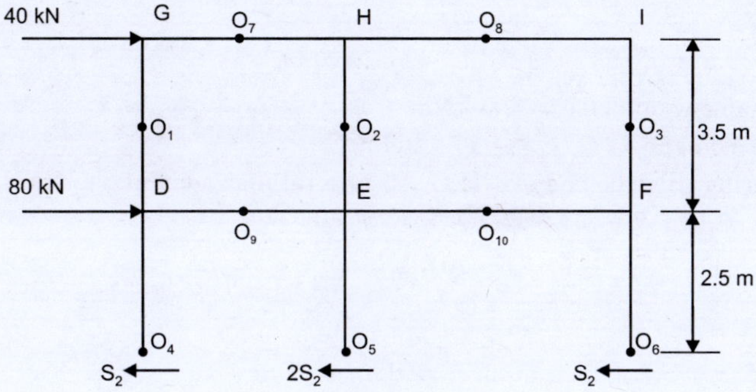

Fig. 10.20. Bottom storey.

Moments at the ends of columns

$$M_{GD} = M_{DG} = M_{IF} = M_{FI} = S_1 \times 1.75 = 10 \times 1.75 = 17.5 \text{ kNm } (\circlearrowright)$$

$$M_{HE} = M_{EH} = 20 \times 1.75 = 35 \text{ kNm } (\circlearrowright)$$

$$M_{DA} = M_{AD} = M_{FC} = M_{CF} = S_2 \times 2.5 = 30 \times 2.5 = 75 \text{ kNm } (\circlearrowright)$$

$$M_{EB} = M_{BE} = 60 \times 2.5 = 150 \text{ kNm } (\circlearrowright)$$

Beam shears :

Beam shears are evaluated by considering various free bodies bounded by hinges. We would find that when we consider the rotational equilibrium of each free body about the joint housed in the free body, the axial thrusts ***do not*** make any moment contribution. Hence in Figs. 10.20 (*a*), (*b*), (*c*), (*d*), (*e*) and (*f*) that follow, the values of the axial forces are not marked.

While working out the support moments on each member, we have to remember that the support moment plus external moment is zero. Hence if a shear force produces an anti-clockwise moment on a member say G O_7, the support moment is clockwise (marked positive).

Considering the free body O_1 G O_7 and taking moments about G

$$V_1 \times 2.5 - 10 \times 1.75 = 0$$

$$V_1 = \frac{10 \times 1.75}{2.5} = 7 \text{ kN}$$

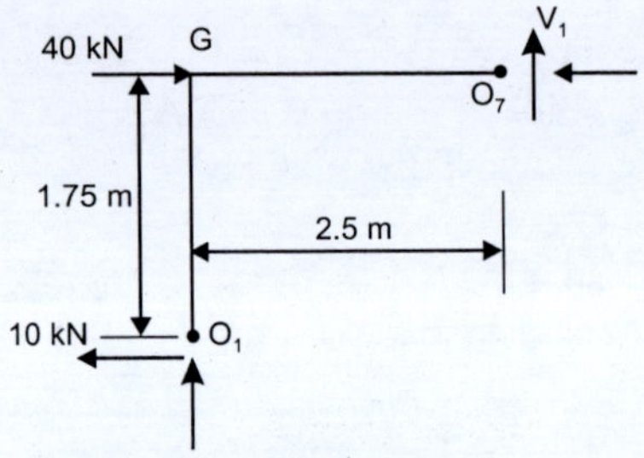

Fig. 10.20 (*a*)

Beam moment on G O_7 = 17.5 kNm

Column moment on G O_1 = – 17.5 kNm

Considering the free body O_7 H O_8 O_2 and taking moments about H.

$$V_2 \times 3.75 + 7 \times 2.5 - 20 \times 1.75 = 0$$

$$V_2 = \frac{-7 \times 2.5 + 20 \times 1.75}{3.75} = 4.67 \text{ kN}$$

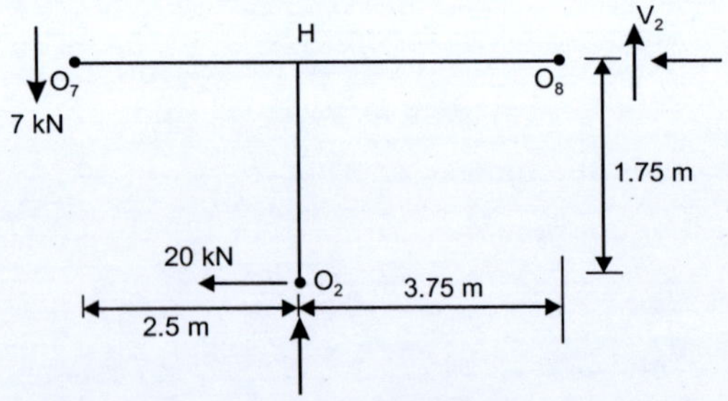

Fig. 10.20 (*b*)

Moment at joint H on H O_7 = 7 × 2.5 = 17.5 kNm

Moment at joint H on H O_8 = 4.67 × 3.75 = 17.5 kNm

Moment at joint H on H O_2 = – 20 × 1.75 = – 35 kNm

Considering the free body O_1 D O_4 O_9 and taking moments about D,

$$V_3 \times 2.5 - 10 \times 1.75 - 30 \times 2.5 = 0$$

$$V_3 = \frac{10 \times 1.75 + 30 \times 2.5}{2.5} = 37 \text{ kN}$$

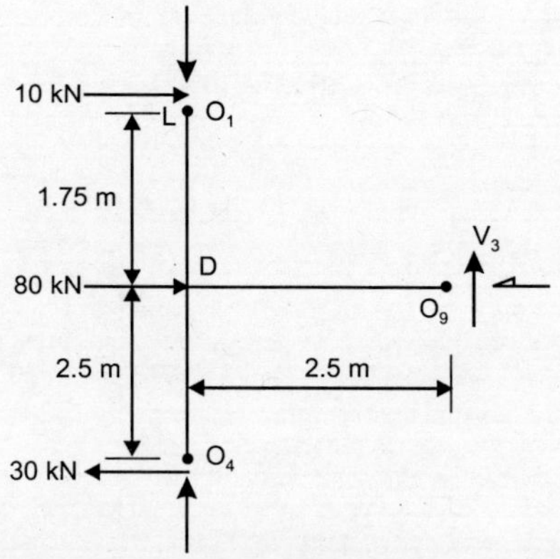

Fig. 10.20 (*c*)

Moment at D on D O_9 = 37 × 2.5 = 92.5 kNm

Moment D on D O_4 = – 30 × 2.5 = – 75 kNm

Moment D on D O_1 = – 10 × 1.75 = – 17.5 kNm

Considering the free body of O_9 E O_2 O_{10} O_5 and taking moments about E.

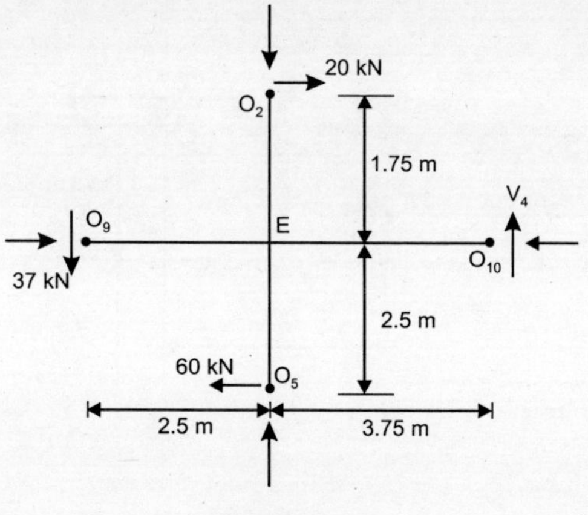

Fig. 10.20 (*d*)

$$V_4 \times 3.75 + 37 \times 2.5 - 20 \times 1.75 - 60 \times 2.5 = 0$$

$$V_4 = \frac{-37 \times 2.5 + 20 \times 1.75 + 60 \times 2.5}{3.75} = 24.67 \text{ kN}$$

Moment at E on E O_9 = 37 × 2.5 = 92.5 kNm

Moment at E on E O_2 = – 20 × 1.75 = – 35 kNm

Moment at E on E O_{10} = 24.67 × 3.75 = 92.5 kNm

Moment at E on E O_5 = – 60 × 2.5 = – 150 kNm

Joint I

Moment at joint I on I O_8 = 4.67 × 3.75 = 17.5 kNm

Moment at joint I on I O_3 = – 10 × 1.75 = – 17.5 kNm

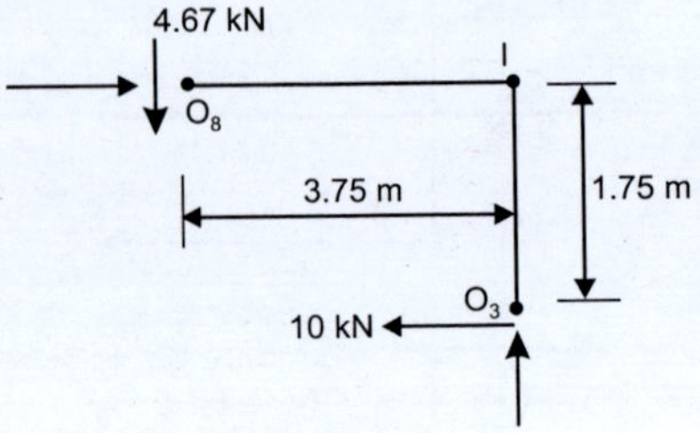

Fig. 10.20 (*e*)

Joint F

Moment at joint F on F O_3 = – 10 × 1.75 = – 17.5 kNm

Moment at joint F on F O_{10} = 24.67 × 3.75 = 92.5 kNm

Moment at joint F on F O_6 = – 30 × 2.5 = – 75 kNm

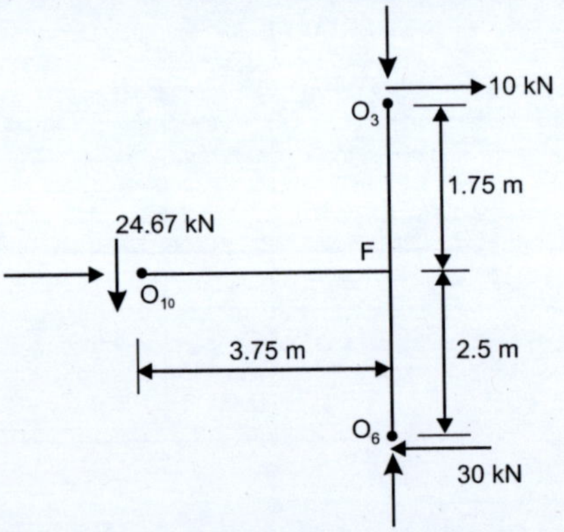

Fig. 10.20 (*f*)

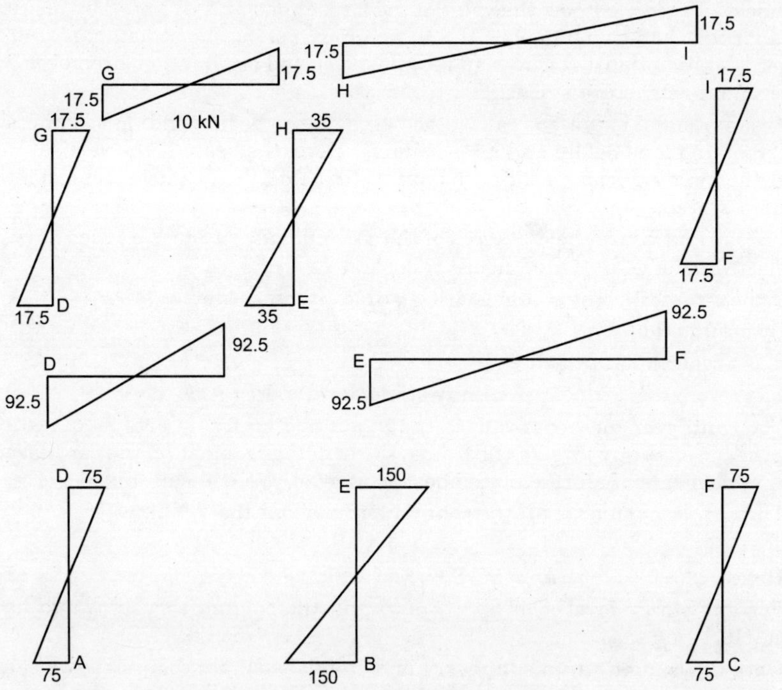

Fig. 10.21. Bending moment diagram.

Figure 10.21 shows the bending moment diagrams on the various members.

10.4.2 The Cantilever Method

This method assumes the building frame as a vertical cantilever fixed at the base and free at the top and subjected to lateral loads. Hence the columns on the windward side will be in tension and those on the leeward side will be in compression.

Assuming the wind to blow from left to right, the wind load will cause a clockwise overturning moment. For equilibrium an equal and opposite moment will have to be developed by the frame. This will be made available by the axial forces in the columns, which will be tensile in the windward columns and compressive in the leeward columns and will be of such magnitude as to create an anticlockwise moment required for the equilibrium.

In this method the following assumptions are made in the analysis:

1. Points of contra-flexure occur at midspan of beams and mid-height of columns and

2. Axial stresses in the columns are proportional to their distances from the centroidal vertical axis of the frame.

Although these assumptions are not very accurate, it makes the frame comfortable to analyse from the static considerations alone, just like in the case of the portal method.

Referring to Fig. 10.22, we find that the building frame shown is 12 degrees indeterminate. Introduction of the 10 hinges removes 10 constraints and we are left with 2 redundancies. When we know the vales of the 3 reactions at A, B and C by making use of only one equilibrium condition namely $\Sigma M = 0$, we are removing the remaining 2 redundancies. In other words assumption 1 removes 10 redundancies and assumption 2 removes 2 more. Hence the determinate structure status.

Assumption 2, it can be realized, is based on the beam bending assumption that plane sections remain plane before and after bending. Here, the beam is the vertical cantilever that we treat the building frame as. So we have to first find the (vertical) centroidal (Y-Y) axis of the column system.

If x_1, x_2 and x_3 are the distances of the 3 column axes from Y-Y, then $\bar{x}$ $(A_1 + A_2 + A_3) = \Sigma A_i x_i$. This would give us $\bar{x}$ and hence x_1, x_2 and x_3.

If the stresses in the columns are σ_1, σ_2, σ_3, then the forces are $P_i = \sigma_i \,.\, A_i$

By assumption 2,

σ_i is proportional to x_i;

$$\sigma_i = k_n\, x_i, \text{ at any storey level } n.\ P_i = A_i\, k_n\, x_i \qquad \text{...(10.1)}$$

The cantilever method assumes that the building frame acts as a vertical cantilever fixed at the base. At any level of the frame, the bending moment on the cantilever is obviously the sum of moments of all the forces above that level. We will call this M_1.

Then M_1 = moment of all the column forces about the Y-Y line

$$M_1 = \Sigma P_i x_i = k_n\ \Sigma A_i x_1^{\,2}$$

Hence, $$k_n = M_1 / \Sigma A_i x_1^{\,2} \qquad \text{...(10.2)}$$

For any storey level n, if k_n is known, all the column thrusts would be known from equation (10.1).

Once the column thrusts (at every mid-storey level) are computed, we can compute the shears and support moments at every joint by considering the free bodies bounded by hinges like in the portal method. There is freedom to vary the free bodies depending upon what forces/moments are to be evaluated?

Examples 10.6 and 10.7 would demonstrate this procedure.

Example 10.6 *Using the cantilever method, analyse the building frame subjected to horizontal forces as shown in Fig. 10.22. Sketch the bending moment diagrams.*

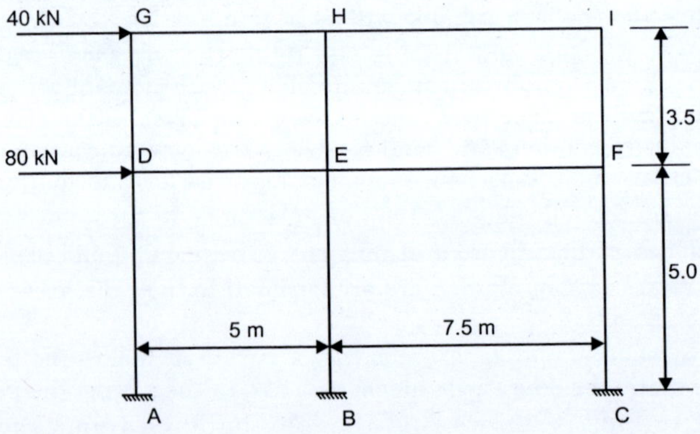

Fig. 10.22

Solution **Cantilever method:**

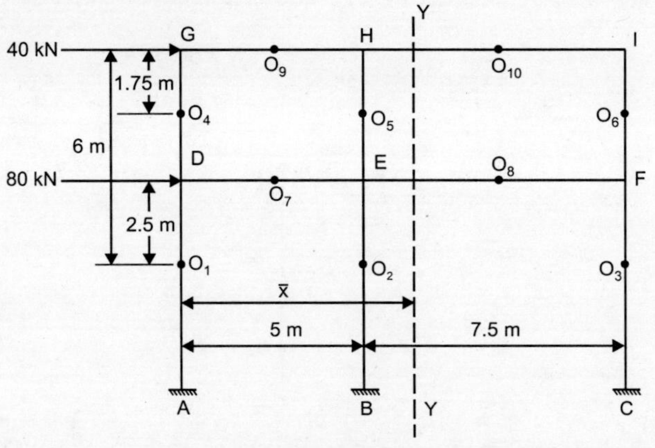

Fig. 10.23

Overturning moment at the mid-level of the columns of the first storey (O_1-O_2-O_3)

$$M_1 = 40 \times 6 + 80 \times 2.5 = 440 \text{ kNm}$$

Overturning moment at the mid-level of columns of the second storey (O_4-O_5-O_6)

$$M_2 = 40 \times 1.75 = 70 \text{ kNm}$$

The cross-sectional areas of all the columns are assumed to be the same (A). Taking moment of areas of the columns about GDA.

$$(A + A + A)\bar{x} = A \times 0 + A \times 5 + A \times (5 + 7.5)$$

$$\bar{x} = \frac{5A + 12.5A}{3A} = \frac{17.5A}{3A} = 5.83 \text{ m}$$

$$x_1 = 5.83 \text{ m}; \quad x_2 = 0.83 \text{ m and } x_3 = -6.67 \text{ m}$$

$$\Sigma x^2 = (5.83)^2 + (0.83)^2 + (6.67)^2 = 79.17$$

First storey:

Axial forces in the columns, P: ($P_i = k_n\, x_i$)

k_1 = constant of proportionality for the columns in the first storey

$$k_1 = \frac{M_1}{\Sigma x^2} = \frac{440}{79.17} = 5.558$$

$P_{AD} = 5.558 \times 5.83 = 32.40$ kN

$P_{BE} = 5.558 \times 0.83 = 4.61$ kN

$P_{CF} = 5.558 \times (-6.67) = -37.07$ kN (Leeward column and hence compressive force)

Second storey:

$$k_2 = \frac{M_2}{\Sigma x^2} = \frac{70}{79.17} = 0.884$$

$P_{GD} = 0.884 \times 5.83 = 5.15$ kN

$P_{HE} = 0.884 \times 0.83 = 0.73$ kN

$P_{IF} = 0.884 \times (-6.67) = -5.90$ kN (Leeward column and hence compressive force)

Shear forces in the beams :

These are obtained by considering the equilibrium of vertical forces acting on the joints. From Fig. 10.24.

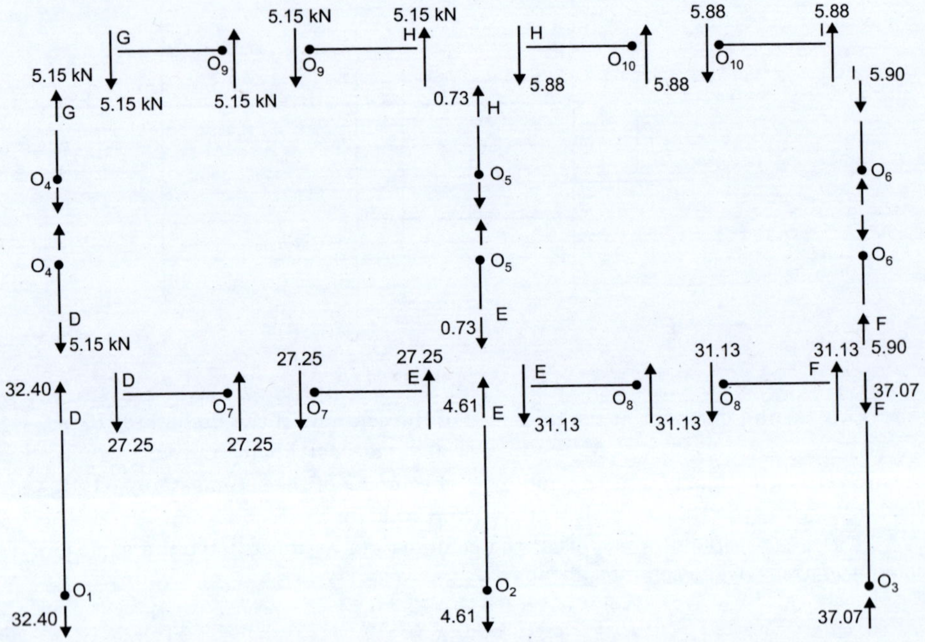

Fig. 10.24

Shear force in GH:

Shear force at G in GH = 5.15 kN (↑↓)

Shear force in DE:

Shear force at D in DE = 32.40 – 5.15 = 27.25 kN (↓)

(↑) (↓)

Similar to G and D,

The shear force at all joints are calculated and marked as shown in Fig. 10.24.

End moment calculations:

Moments on GH and HG = 5.15 × 5/2 = 12.875 kNm

Moments on HI and IH = 5.88 × 7.5/2 = 22.05 kNm

Moments on DE and ED = 27.25 × 5/2 = 68.125 kNm

Moments on EF and FE = 31.13 × 7.5/2 = 116.738 kNm

From the moment equilibrium of each joint, we can evaluate the column moments. They are plotted in Fig. 10.25.

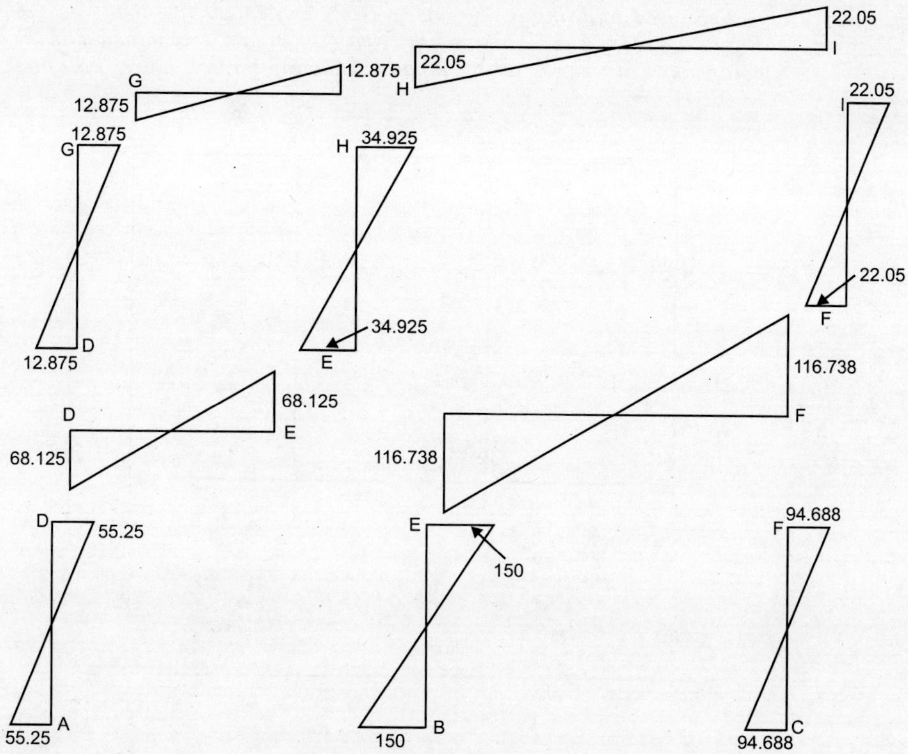

Fig. 10.25. Bending moment diagram.

End moments in the columns:

Joint G :

$$M_{GH} = 12.875 \text{ kNm};$$

$$M_{GD} = -12.875 \text{ kNm}$$

Because of the assumed hinge at the mid-height of GD,

$$M_{DG} = -12.875 \text{ kNm}$$

Joint D:

$$M_{DG} + M_{DE} + M_{DA} = 0$$

$$-12.875 + 68.125 + M_{DA} = 0$$

$$M_{DA} = -55.25 \text{ kNm}$$

$$M_{AD} = -55.25 \text{ kNm}$$

Likewise all other column moments are computed and plotted in Fig. 10.25.

Example 10.7 *A 3-storeyed building frame has to carry equivalent live loads as shown in Fig. 10.26. The inner columns have twice the area of cross-section as the outer columns. Find the bending moments on the bottom storey columns. Use cantilever method.*

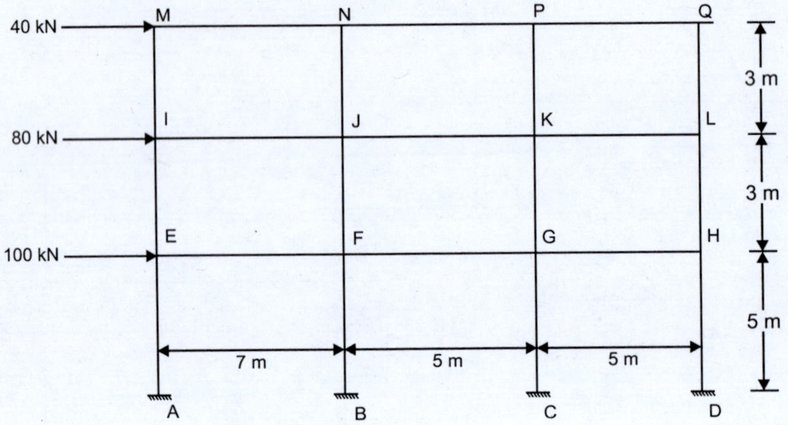

Fig. 10.26

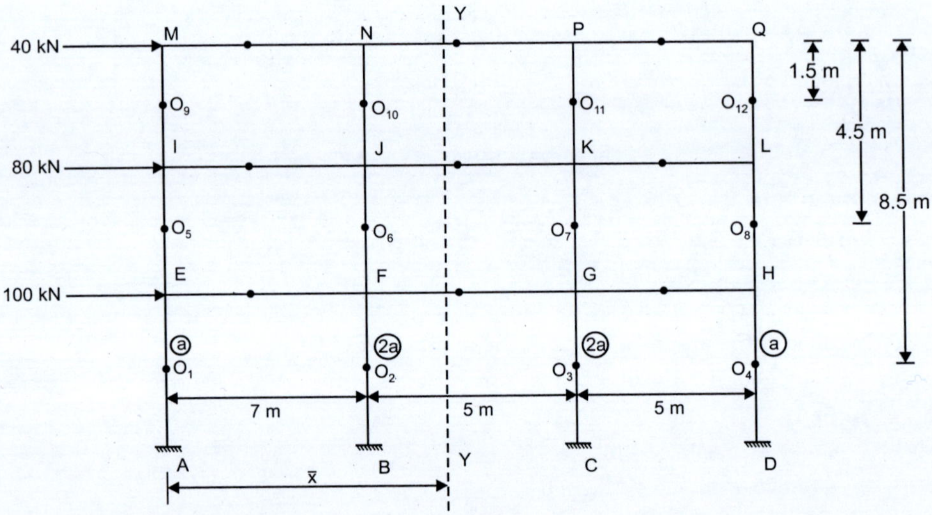

Fig. 10.27

Solution (Using the cantilever method)

To locate the centroidal axis Y-Y, let us take moments of areas of columns about A.

$$a.o + 2a \times 7 + 2a \times 12 + a \times 17 = (a + 2a + 2a + a)\ \bar{x}$$

$$\bar{x} = 55/6 = 9.167 \text{ m}$$

Hence, $\{x_i\}^T = [9.167 \quad 2.167 \quad -2.833 \quad -7.833]$ m

Let us assume hinges at the centre of all columns and beams.

The cantilever moment at the level of $O_1 - O_2 - O_3 - O_4$ is

$$M_1 = 100 \times 2.5 + 80 \times 5.5 + 40 \times 8.5$$

$$M_1 = 1030 \text{ kNm}$$

Assuming stresses in columns are proportional to x_i,

$$P_i = k_1\, x_i\, a_i$$

$$M_i = k_1 x_i^2\, a_i$$

$$\Sigma M_i = M_1 = k_1\, \Sigma x_i^2\, a_i$$

$$= k_1\, a[x_1^2 + x_2^2 + x_3^2 + x_4^2]$$

$$1030 = k_1\, a\ 158.11$$

$$k_1 a = 6.514$$

$$P_1 = 6.514 \times 9.167 = 59.71 \text{ kN}$$

$$P_2 = 6.514 \times 2.167 \times 2 = 28.23 \text{ kN}$$

$$P_3 = 6.514 \times (-2.83) \times 2 = 36.91 \text{ kN (Comp.)}$$

$$P_4 = 6.514 \times (-7.83) = 51.02 \text{ kN (Comp.)}$$

$\{P_i\}$ are the thrusts at O_1, O_2, O_3 and O_4

Likewise we will evaluate $\{P_j\}$ at level $O_5 - O_8$ and $\{P_k\}$ at level $O_9 - O_{12}$

Cantilever moment M_2 at the level $O_5 - O_8$ is

$$M_2 = 80 \times 1.5 + 40 \times 4.5 = 300 \text{ kNm}$$

$$k_2\, a = 300/158.11 = 1.8975$$

$$\{P_j\}^T = [17.4 \quad 8.22 \quad -10.75 \quad -14.86] \text{ kNm at second storey level.}$$

At the level of $O_9 - O_{12}$,

$$M_3 = 40 \times 1.5 = 60 \text{ kNm}$$

$$k_3\, a = 60/158.1 = 0.38$$

$$[P_k]^T = [3.48 \quad 1.64 \quad -2.15 \quad -2.97] \text{ kN at 3rd storey level.}$$

P_9 to P_{12} above will induce shears in the top level beams. These can be determined from the free bodies of the beams/columns (Fig. 10.28). We can also evaluate the beam moments and column moments.

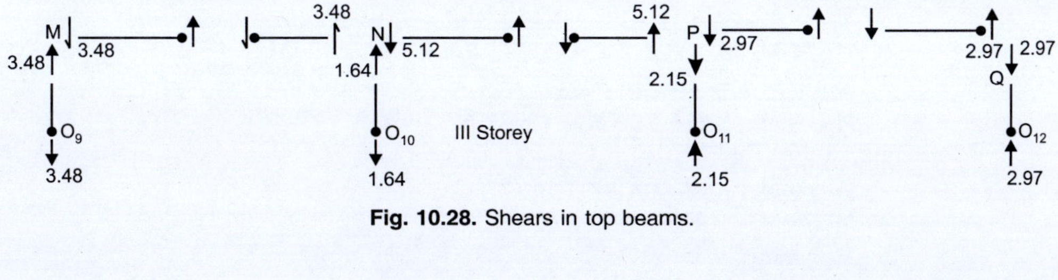

Fig. 10.28. Shears in top beams.

P_5 to P_8 and P_9 to P_{12} would influence the shears at the level EFGH as shown in Fig. 10.29.

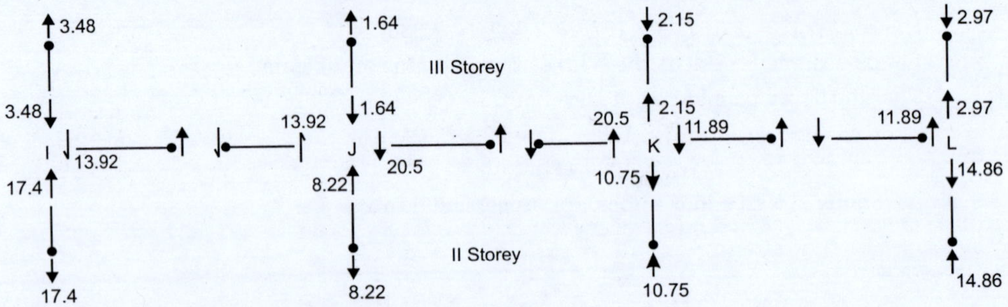

Fig. 10.29. Shears on beams I-J-K-L.

At the level EFGH, the shears would be influenced by the thrusts on the columns of I storey and II storey (Fig. 10.30).

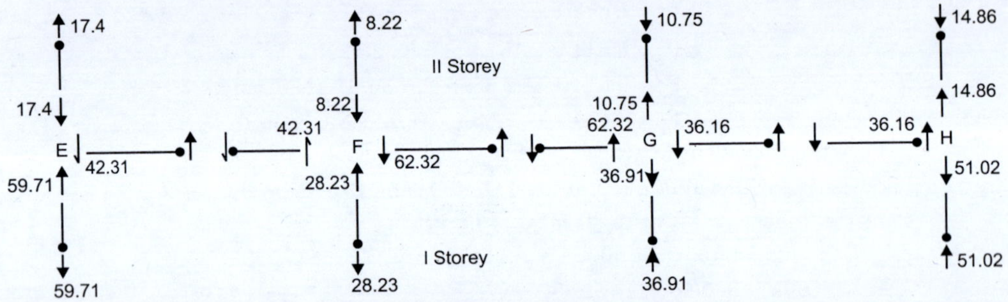

Fig. 10.30. Shears on beams E-F-G-H.

The beam shears at the ends of each beam are also the shears at the hinges at midspan. The shear at the hinge x half the span would be the bending moment (support moment) in each beam. These moments would balance the column moments (at the joints).

Because of the assumption of a hinge at midspan/mid height, the support moment will be the same at both supports of each member. Figure 10.31 shows all the support moments and the bending moment diagrams.

Top floor (see Fig. 10.28)

$$M_{MN} = M_{NM} = 3.48 \times 7/2 = 12.18 \text{ kNm} = -M_{MI} = -M_{IM}$$
$$M_{NP} = M_{PN} = 5.12 \times 5/2 = 12.8 \text{ kNm}$$
$$M_{PQ} = M_{QP} = 2.97 \times 5/2 = 7.425 \text{ kNm} = -M_{QL} = -M_{LQ}$$
$$M_{NJ} = M_{JN} = -M_{NP} - M_{NM} = -24.98$$
$$M_{PK} = M_{KP} = -M_{PN} - M_{PQ} = -12.8 - 7.425 = -20.225$$

Second floor (see Fig. 10.29)

$$M_{IJ} = M_{JI} = 13.92 \times 7/2 = 48.72$$
$$M_{JK} = M_{KJ} = 20.5 \times 5/2 = 51.25$$
$$M_{KL} = M_{LK} = 11.89 \times 5/2 = 29.725$$

$$M_{IE} = -M_{IM} - M_{IJ} = 12.18 - 48.72 = -36.54 \text{ kNm}$$

$$M_{FJ} = M_{JF} = -M_{JN} - M_{JI} - M_{JK} = 24.98 - 48.72 - 51.25 = -75 \text{ kNm}$$

$$M_{GK} = M_{KG} = -M_{KJ} - M_{KL} - M_{KP} = -51.25 - 29.725 + 20.225 = -60.75 \text{ kNm}$$

$$M_{LH} = -M_{LK} - M_{LQ} = -29.725 + 7.425 = -22.30 = M_{HL}$$

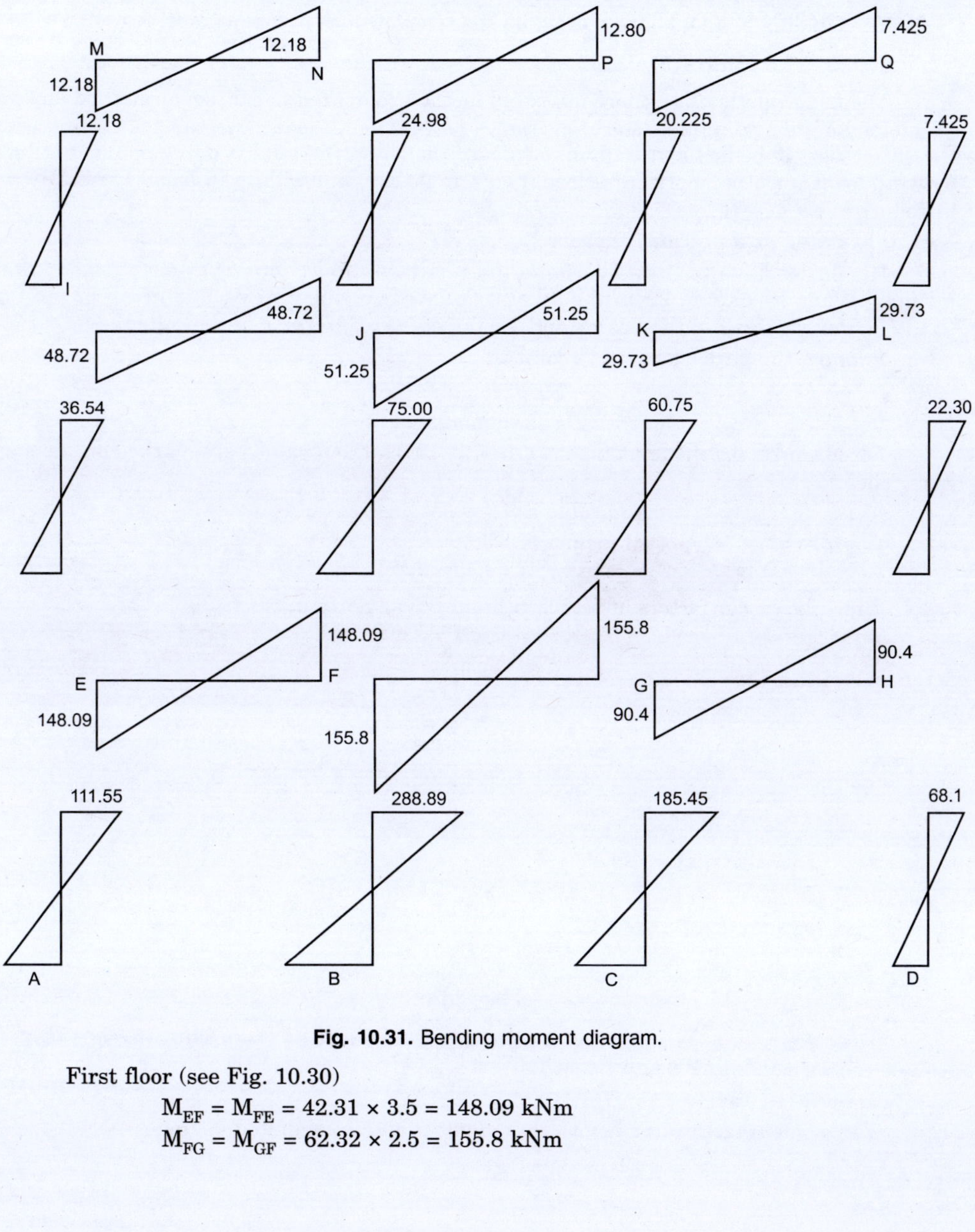

Fig. 10.31. Bending moment diagram.

First floor (see Fig. 10.30)

$$M_{EF} = M_{FE} = 42.31 \times 3.5 = 148.09 \text{ kNm}$$

$$M_{FG} = M_{GF} = 62.32 \times 2.5 = 155.8 \text{ kNm}$$

$M_{GH} = M_{HG} = 36.16 \times 2.5 = 90.4$ kNm

$M_{EA} = M_{AE} = -M_{EF} - M_{EI} = -148.09 + 36.54 = -111.55$ kNm

$M_{FB} = M_{BF} = -M_{FE} - M_{FJ} - M_{FG} = -148.09 + 75 - 155.8 = -288.89$ kNm

$M_{CG} = M_{GC} = -M_{GF} - M_{GK} - M_{GH} = -155.8 + 60.75 - 90.4 = -185.45$ kNm

$M_{HD} = M_{DH} = -M_{HL} - M_{HG} = 22.30 - 90.4 = -68.1$ kNm

The bending moments are plotted on the tension sides of members in Fig. 10.31.

10.4.3 The Factor Method

The factor method is a more improved method compared to cantilever method and portal method of analyzing multistoreyed frames for horizontal loads. We can take into account the stiffnessess of beams and columns while apportioning the total bending moment induced by wind loads (or other horizontal loads) on a multibay, multistoreyed frame. First, we will introduce a few new terms.

(*a*) **Girder and column factors**

Herein we have to first determine the relative stiffness $k(=I/l)$ at each joint of every member, beam or column. From the k values of every member at each joint we have to get (*i*) the ***girder factor*** **g** of the beams and (*ii*) the ***column factor*** **c** of the columns.

Funnily, the girder factor at a joint is

$$g = \frac{\Sigma\, k \text{ of columns}}{\Sigma\, k \text{ of all members}} \qquad ...(10.3)$$

For example, for the joint B shown in Fig. 10.32, the k values are marked on the members. Then

$$\Sigma k \text{ of column} = 4.5$$

$$\Sigma k \text{ of all members} = 7.5$$

$$g = 4.5/7.5 = 0.60$$

Thus, the girder factors of beams BC and BD at B are 0.6, 0.6.

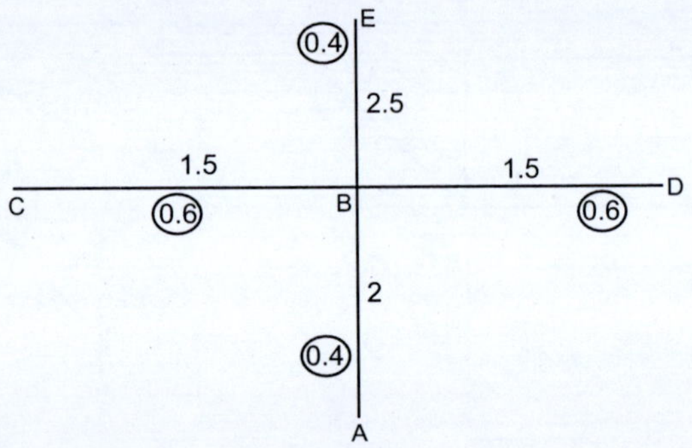

Fig. 10.32

Then the column factor for each column at the joint C is (1-g). Thus in Fig. 10.32, the column factor is 0.4 for BE and BA at B.

(*b*) **Moment factor**

The moment factor C or G for every member end is computed as under.

First we get the c (or) g for each member at each end. To these values of c (or) g we add half value of c (or) g from the opposite end and write down the modified values c_m or g_m.

Moment factors: C or G are got by multiplying c_m or g_m by the relative stiffness k. Thus

$$C = c_m\, k \text{ and } G = g_m\, k$$

(*c*) **Total column moments**

We must keep in mind that this analysis is for the action of horizontal loads on frames. For any given level we can determine the total moment caused by horizontal loads M_T. M_T will be taken by the frame in two ways.

1. by virtue of the axial thrusts (tension and compression in columns)
2. by the bending moments on the columns at any floor level.

If ΣC is the sum of column moment factors at any storey level,

Column moments $M_C = \dfrac{C}{\Sigma C} \cdot M_T$

(*d*) Beam moments at each joint are obtained by computing one more set of quantities called DF_B, (the familiar distribution factor but with a difference).

$DF_B = \left\{\dfrac{G}{\Sigma G \text{ at a joint}}\right\}$. This is computed for each beam at a joint and $M_B = DF_B \times$ Σ column moments at the joint.

10.4.4 Basis

The factor method is also an approximate method, but instead of making gross approximations like in the portal and cantilever methods, this takes into account the fact that

(*i*) At any joint the total moment is shared by all members in proportion to their stiffnesses.

(*ii*) That half the moment at each end gets carried over to the opposite end when we accommodate rotation at each joint $\left(\text{hence } \dfrac{c}{2} \text{ and } \dfrac{g}{2} \text{ from opposite end being added to obtain } c_m \text{ and } g_m\right)$.

(*iii*) A part of the overtuning moment on a frame is taken by the columns as bending moment and another part as thrust.

We will apply the procedure outlined in Art. 10.4.3 to solve a simple portal frame shown in Fig. 10.33 (example 10.8).

Example 10.8 *Analyse the portal frame in Fig. 10.33 by factor method.*

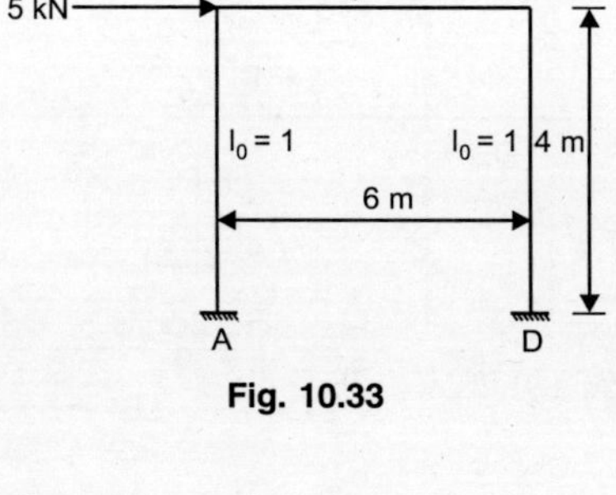

Fig. 10.33

Solution The real problem with the factor method is to organize the various quantities in a comprehensive table. First we will compute the preliminary quantities before tabulating.

$k_{BA} = 1/4$; $k_{BC} = k_{CB} = 1/6$; $k_{CD} = 1/4$, since $I = I_0$ for all members.

M_T = Total moment at level AD = 5 × 4 = 20 kNm

Table 10.1 (a)

①	②	③	④		⑤	⑥		⑦
			k = I/l			*Factor*		
No.	*Joint*	*Member*	*Column*	*Beam*	Σk	$c = \frac{\Sigma k_b}{\Sigma k}$	$g = \frac{\Sigma k_c}{\Sigma k}$	*c/2 or g/2 from far end*
1.	A	AB	0.25	xx*	0.25	1	0	0.2
2.	B	BA	0.25	—	0.417	0.4		0.5
3.		BC	—	0.167			0.6	0.3
4.	C	CB	—	0.167	0.417		0.6	0.3
5.		CD	0.25	—		0.4		0.5
6.	D	DC	0.25	xx	0.25	1	0	0.2

For columns fixed at base assume that the beam stiffness is a very large number xx.

Then $c \approx \frac{xx}{xx} = 1$ and $g \approx \frac{k}{xx} = 0$.

Table 10.1 (b)

①	②	③	⑥ + ⑦ = ⑧		⑨		⑩	⑪	⑫	⑬
No.	*Joint*	*Member*			*Moment factor (8 × 4)*		*Total col. moment* M_T	*Col. mom.* $M_c = C.M_T/\Sigma C$	*DFB* $= G/\Sigma G$	*Beam moment* $M_B = DF_B \cdot \Sigma M_C$
			c_m	g_m	*C*	*G*				
1.	A	AB	1.2		0.3		20	5.71		
2.	B	BA	0.9		0.225		20	4.29		
3.		BC		0.9		0.15			1	4.29
4.	C	CB		0.9		0.15			1	4.29
5.		CD	0.9		0.225		20	4.29		
6.	D	DC	1.2		0.3		20	5.71		

$\Sigma C = 0.3 + 0.3 + 0.225 + 0.225 = 1.05$

Bending moment diagram :

This method does not include any way of distinguishing the sign of the moments. However it is easy to judge the sense of the support moments for a set of horizontal forces. Figure 10.34 shows the bending moment diagrams.

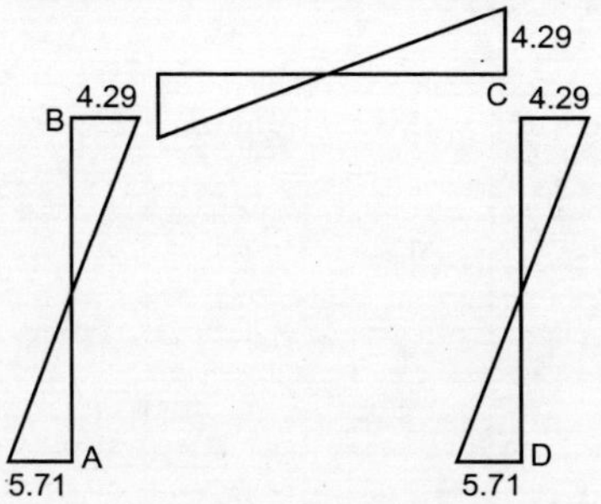

Fig. 10.34. Bending moment diagram.

If the problem is solved by exact method, the following would be the values

$M_{AB} = 6.0; M_{BC} = 4.0.$

Cantilever method would yield $M_{AB} = M_{BC} = 10.0.$

Portal method would result in $M_{AB} = M_{BC} = 5$ kNm.

Example 10.9 *Using the factor method, analyse the building frame subjected to horizontal forces, as shown in Fig. 10.35. Sketch the bending moment diagram. A, I are the same for all columns and beams.*

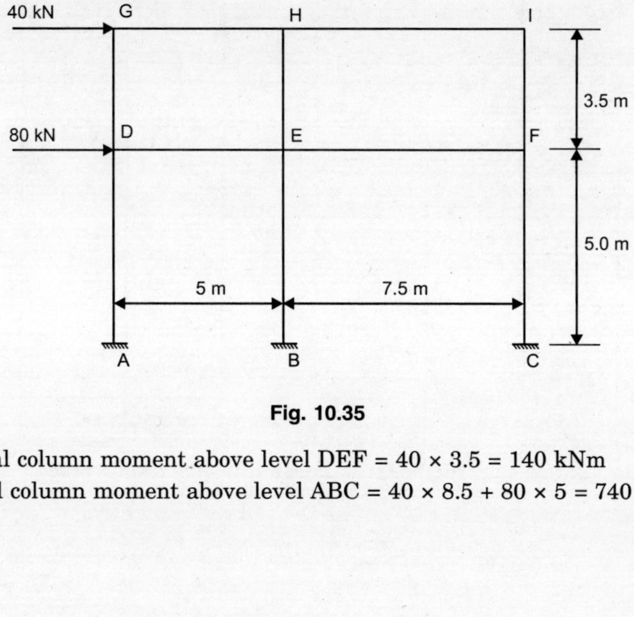

Fig. 10.35

Solution Total column moment above level DEF = 40 × 3.5 = 140 kNm

Total column moment above level ABC = 40 × 8.5 + 80 × 5 = 740 kNm

The computations are made in Table 10.2 (*a*) and (*b*).

Table 10.2 (a)

① No.	② Joint	③ Member		④ $k = I/l$		⑤ Σk	⑥ Factor		⑦ c/2, g/2 from far end
		Col	Beam	Col	Beam		$c = \Sigma k$ beam/Σk	$g = \Sigma k$ col/Σk	
1.	D	DA		0.2		0.686	0.29		0.5
2.			DE		0.2			0.71	0.3
3.		DG		0.286			0.29		0.3
4.	E		ED		0.2	0.819		0.59	0.36
5.		EH		0.286			0.41		0.27
6.			EF		0.133			0.59	0.40
7.		EB		0.2			0.41		0.50
8.	F		FE		0.133	0.772		0.79	0.3
9.		FI		0.286			0.21		0.16
10.		FC		0.2			0.21		0.50
11.	G	GD		0.286		0.486		0.59	0.15
12.			GH		0.2		0.41		0.23
13.	H		HG		0.2	0.772		0.46	0.21
14.		HE		0.286			0.54		0.21
15.			HI		0.133			0.46	0.34
16.	I		IH		0.133	0.419		0.68	0.23
17.		IF		0.286			0.32		0.11
18.	A	AD		0.2	xx	0.2	1	0	0.15
19.	B	BE		0.2	xx	0.2	1	0	0.21
20.	C	CF		0.2	xx	0.2	1	0	0.11

xx we can assume that the beam stiffness is a very large value.

Table 10.2 (b)

①	②	③		⑥ + ⑦ = ⑧		⑨		⑩	⑪	⑫	⑬
		Member				*Moment factor*					
No.	*Joint*	*Col*	*Beam*	*Col* (c_m)	*Beam* (g_m)	⑧ × ④ $C = C_m.K$	$G = g_m K$	*Total col. moment* M_T	*Col. moment* $M_C = C \times M_T/\Sigma C$	$DF_B = G/\Sigma G$	*Beam moment* $DF_B \cdot \Sigma M_C$
1.		DA		0.79		0.158		740	99.4		
2.	D		DE		1.01		0.202			1	122.6
3.		DG		0.59		0.169		140	23.2		
4.			ED		0.95		0.19			0.59	83.25
5.	E	EH		0.68		0.194		140	26.6		
6.			EF		0.99		0.132			0.41	57.85
7.		EB		0.91		0.182		740	114.5		
8.			FE		1.09		0.145			1	133.03
9.	F	FI		0.37		0.106		140	14.6		
10.		FC		0.71		0.142		740	89.3		
11.	G	GD		0.74		0.212		140	29.1		
12.			GH		0.64		0.128			1	29.1
13.			HG		0.67		0.134			0.559	16.5
14.	H	HE		0.75		0.215		140	29.5		
15.			HI		0.80		0.106			0.441	13.0
16.	I		IH		0.91		0.121			1	16.58
17.		IF		0.43		0.123		140	16.58		
18.	A	AD		1.15		0.230		740	144.7		
19.	B	BE		1.21		0.242		740	152.2		
20.	C	CF		1.11		0.222		740	139.6		

First storey: $\Sigma C = 0.158 + 0.182 + 0.142 + 0.23 + 0.242 + 0.222 = 1.176$

Second storey: $\Sigma C = 0.169 + 0.194 + 0.106 + 0.212 + 0.215 + 0.123 = 1.019$

The final moments are plotted in Fig. 10.36.

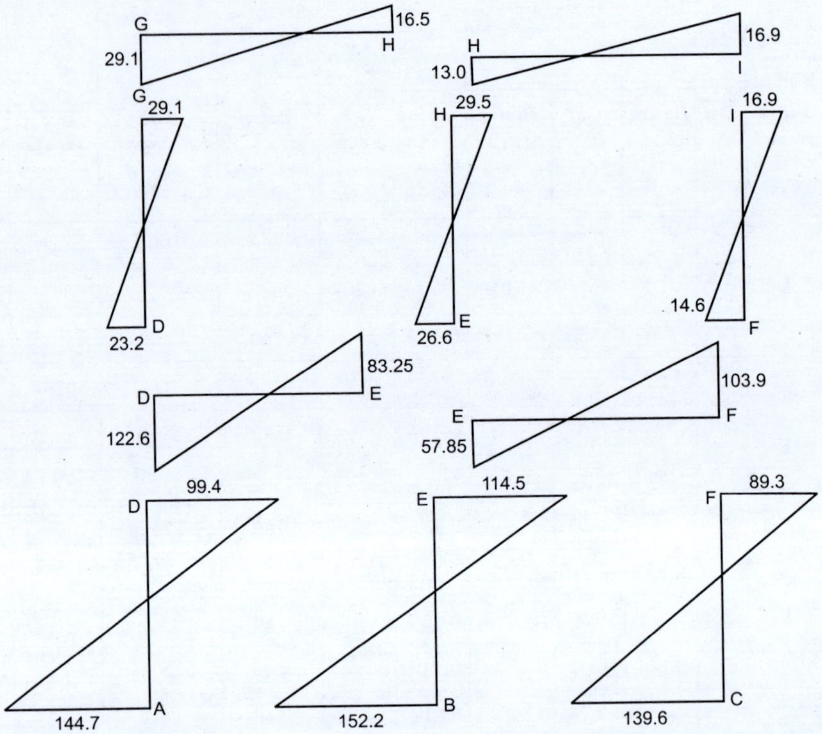

Fig. 10.36

PROBLEMS

1. Analyse the building frame shown in Fig. 10.37 by (*i*) Portal method, (*ii*) Cantilever method, and (*iii*) Factor method and tabulate the moments at AD, BE, CF, DE and EF by each method.

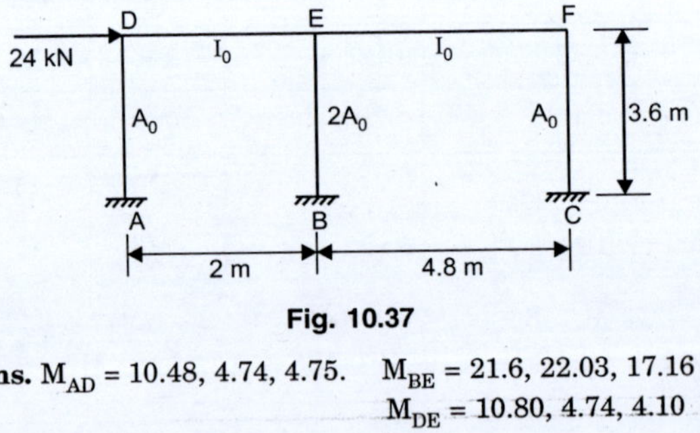

Fig. 10.37

[**Ans.** M_{AD} = 10.48, 4.74, 4.75. M_{BE} = 21.6, 22.03, 17.16 M_{CF} = 10.8, 17.29, 4.35
M_{DE} = 10.80, 4.74, 4.10 M_{EF} = 10.8, 17.29, 3.75]

2. Solve the frame in Fig. 10.38 by factor method and tabulate the moments on AD, BE, CF, DE, EF, GH and HI.

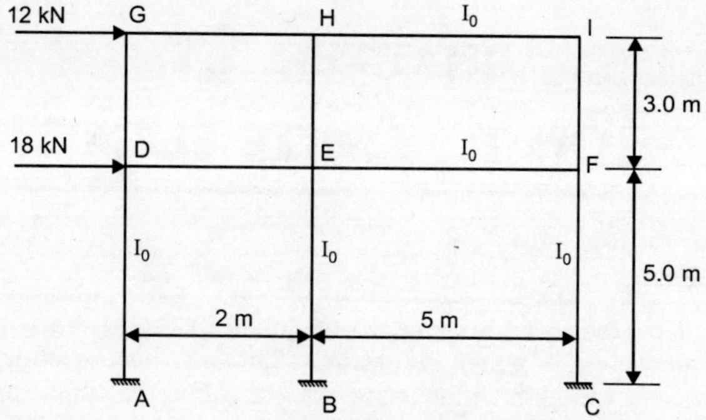

Fig. 10.38

3. Solve the frame in Fig. 10.38 above by the Cantilever method and the Portal method and compare all the end moments.

4. For the frame in problem 2 above, find the design moments in beam EF due to gravity loads. DL = 2.4 kN/m, LL = 3.6 kN/m. Draw the maximum bending moment diagram.

5. Figure 10.39 shows a 4 storeyed building frame. Compute the design bending moments on span BC due to a DL of 1.6 kN/m and LL of 2.2 kN/m on the beam, employing 2 cycle moment distribution on a substitute frame.

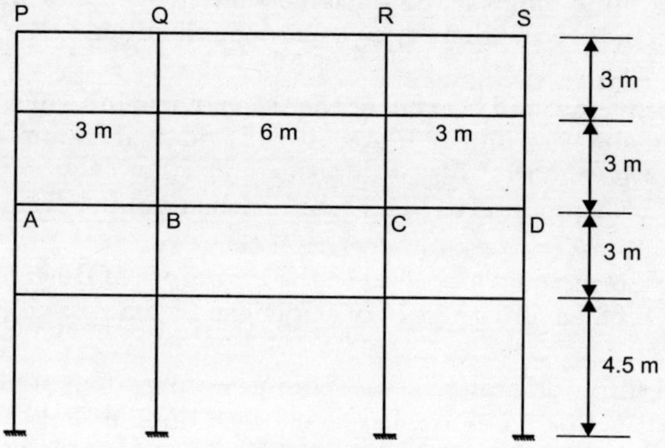

Fig. 10.39

6. For the frame in Fig. 10.39. Find the design moments on QR (Support and midspan).

CHAPTER 11
INTRODUCTION TO FINITE ELEMENT METHOD

11.1 PREAMBLE

Finite element method was originally developed in the aircraft industry to facilitate a refined analysis of complex air frame structures. Though the procedure was developed as a concept of structural analysis, the wider basis of the method makes it applicable to a variety of field problems such as heat conduction, fluid flow, hydro elasticity, diffusion-convection, mathematical physics, etc., in addition to structural analysis. In general, this method is applicable to almost all problems where a variational formulation of the physical phenomenon is feasible.

The basic concept involved in the finite element method is that every structure may be considered to be an assemblage of a finite number of individual structural components or elements. These elements can be put together in a number of ways and hence they can be used to represent exceedingly complex geometries. In the analysis of an elastic continuum by this method, the continuum is idealized as an assemblage of two or three dimensional elements of proper shape, with a finite number of interconnections. This approximation of discretisation is purely of a physical nature and there need be no approximation in the mathematical analysis of the substitute system.

Finite element method permits the incorporation of such complexities as non-homogeneous media, non linear material behaviour, dynamic loading, in situ stress conditions, arbitrary geometrics, discontinuities and other factors that cannot be readily handled by closed form procedures. This method has gained tremendous attention and popularity.

Most numerical techniques are based on the **Principle of Discretisation**— a procedure in which a complex problem of large extent is divided or discretised into smaller equivalent units or components.

Whereas in the finite difference method, the governing differential equation is discretised, in the finite element method, the physical body or continuum that constitutes the system is discretised. In boundary element method, only the boundary of the body is discretised into a number of segments or elements. Of the many ways to formulate the properties of individual elements, one is the variational approach which relies on the calculus of variations and involves the minimisation of a functional. For problems in solid mechanics, the functional turns out to be the potential energy, the complementary potential energy or some derivative of these. Accordingly, there are several variational approaches for deriving the element properties which may be broadly classified as

1. The displacement approach,
2. The force approach,
3. The hybrid approach, and
4. The mixed approach.

Out of the above, the displacement based finite element method is the most popular and more widely used because of its simplicity and easy applicability to different types of problems. Herein, the displacement approach (stiffness method) will be dealt with in some detail.

11.2 TERMINOLOGY

1. *Finite elements:* The "small" elements (1D, 2D or 3D) obtained by suitably subdividing the given domain to be analysed.
2. *Nodes or nodal points:* The intersections of the sides of the elements. The nodal points connecting the adjacent elements are called external nodes. The extra nodes used to increase the accuracy of solution but not connected to other elements, are called internal nodes.
3. *Nodal lines and nodal planes:* The interfaces between the elements.
4. *Linear elements:* Finite elements with straight sides.
5. *Higher order elements:* Finite elements with curved sides.
6. *Continuum:* This is defined as a domain in which matter exists at every point. We may think of a continuum as consisting of an infinite number of connected particles.
7. *Primary unknowns:* For displacement formulation based on potential energy, the main unknowns involved in the formulation are the nodal displacements $\{\delta\}$, which are called the primary unknowns. The corresponding forces would be denoted by $\{P\}$.
8. *Secondary unknowns:* Those unknowns derived from the primary unknowns, are known as secondary unknowns. In displacement based formulations, strains, stresses, moments, shear forces etc. are secondary unknowns.

11.3 PROCEDURE

The basic procedure in adopting FEM is outlined under.

1. *Discretise the continuum:* This is the process of subdividing the body into a number of elements. In a 2D problem, the elements would be 2D elements like triangle, rectangle, quadrilateral, sector etc.
2. *Select element displacement functions:* Simple mathematical functions are assumed to approximate the displacement field within each element. Since strains are functions of displacements, and stresses are functions of strains, this step amounts to defining the stress field within each element (in terms of some unknowns).
3. *Calculate element properties:* This step involves the use of element equilibrium equations to calculate element stiffness matrix $[k]$ (in terms of the unknowns in step 2), corresponding to a set of element co-ordinates choosen. We would end up with $[k]$ such that $\{\delta\} = [k]\{P\}$.
4. *Obtain the element load vector:* This is done, starting from the loading on the element. The loading can be concentrated, distributed or varying on the element area. This can be in the plane of the element or normal to the plane of the element. The load vector we want would be the equivalent set of loads at the element co-ordinates originally described in step 3.
5. *Assemble element properties:* Element load vectors are assembled to obtain structure load vector. This will accord with a set of chosen structure co-ordinates. [between the co-ordinates of all the elements making up the structure, and the co-ordinates of the structure itself, we can establish a unique number based relationship. This will depend upon the way the elements are connected or connectivity]. Likewise, element stiffness

matrices are assembled to generate the structure stiffness matrix. [In most 2D problems the sizes of element matrices are in the range 6 × 6 to 40 × 40. Structure stiffness matrices can be of sizes ranging from 40 × 40 to 2000 × 2000].

6. *Impose boundary conditions:* The structure for which we generated a "structure stiffness matrix" in step 5 is just an assemblage of elements, not yet attached to the ground. When we put it on the ground, it becomes a bridge or a dome or a frame—a real structure. This involves imposing the condition that displacements at some of the structure co-ordinates (introduced in step 5) touching the "ground" are either zero or restricted to a finite value. The structure stiffness matrix developed in step 5 shall therefore be modified to realize the condition that displacements at some of the co-ordinates need to be zero (or finite).

   When we modify the stiffness matrix thus, it will be non-singular, positive definite. Its diagonal values would be positive and dominant.

   If [K] is the stiffness matrix so modified then {F} = [K] {u}and {u}= $[K]^{-1}${F}. Here {F} is the vector of known forces at structure co-ordinates, and {u} the displacements.

7. From the structure displacement vector {u} obtained in step 6, we can *infer the displacements of all co-ordinates in* all *elements* using the connectivity information in step 5. In each element, if the displacements at the co-ordinates are given by {δ}, we can go on to *determine* the *displacement field, strains* and *stresses*.

8. Step 7 is almost the last step. But not quite. The assemblage of elements *models* (or approximates) the structural continuum. It is not the same as the continuum. Thus we may find some embarrassing results. For example, at a boundary between 2 elements we may have a stress of 1.2 N/mm^2 to the left and 2.8 N/mm^2 to the right. This is in fact not correct, nor possible. This inaccuracy due to discretisation gets smaller and smaller as we make our elements more in number and less in size. We can also adopt some stress smoothing techniques to achieve a continuous variation of stresses from the discontinuous output we get from step 7.

11.4 DISCRETISATION

Discretisation is the process of separating the length, area or volume we want to analyze into discrete (or separate) parts or elements.

11.4.1 1 D Elements

A beam can be approximated as a one dimensional structure. It can be split into one dimensional beam elements. So also, a continuous beam or a flexure frame (for example, a portal frame) can be discretised using 1 D beam elements.

A pinjointed truss is readily made up of discrete 1 D ties (or struts) which are duly assembled. This comes readily discretised. Even pinjoined space (3D) trusses are made up of 1 D elements (vide Fig. 11.1(*i*)).

11.4.2 2 D Elements

A plane wall, plate, diaphragm, slab, shell etc. can be approximated as an assemblage of 2 D elements (Fig. 11.1(*ii*)). Triangular elements are the most used ones. When our 2D domain has curved boundaries, it may be advantageous to choose elements that can have curved boundaries (Figs. 11.1(*ii*), (*aiii*) or (*biii*).

If the continuum has a curvature (shell, suspended membrane) we can choose plane elements with initial centroidal curvature (shell elements).

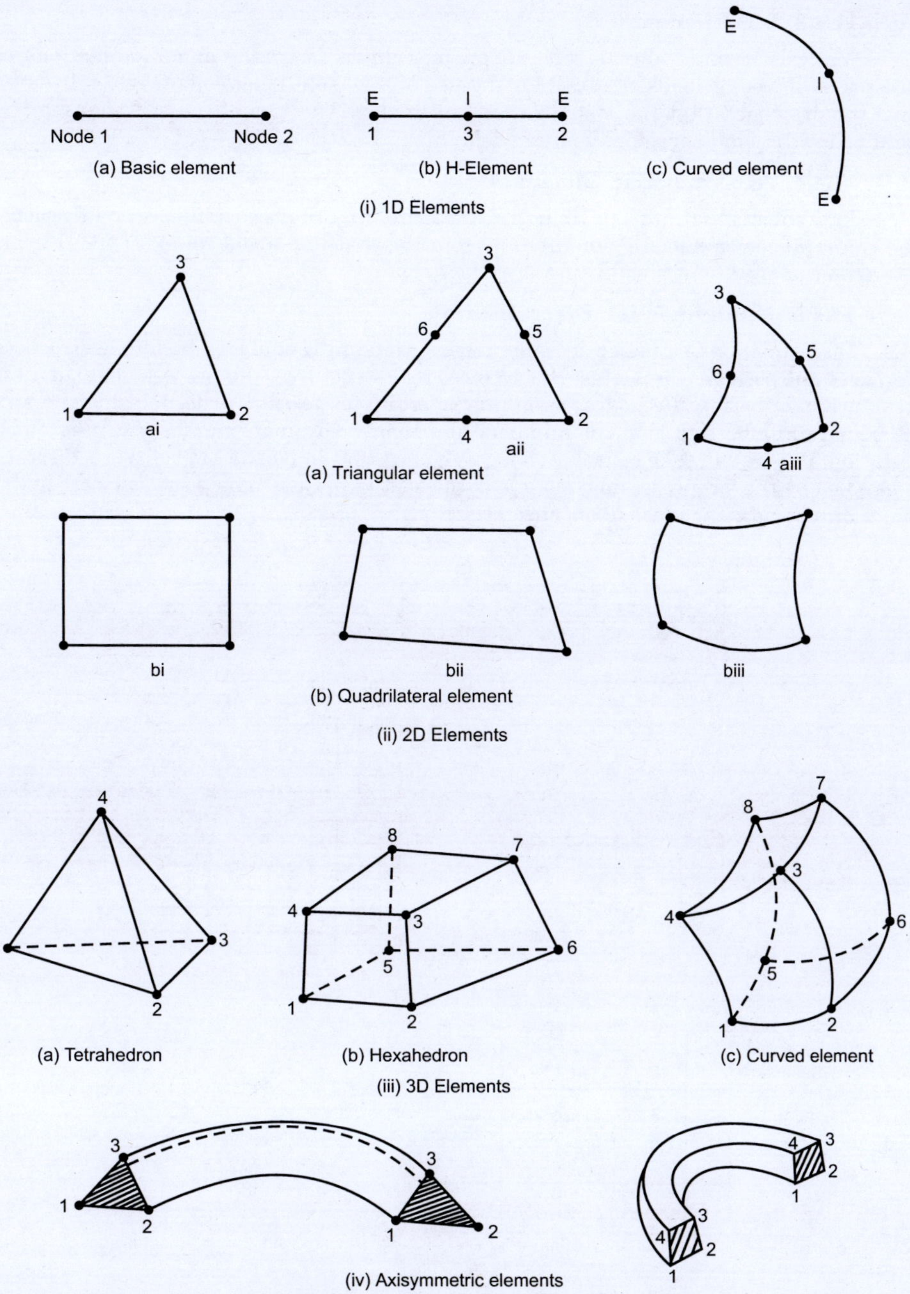

Fig. 11.1. Types of finite elements.

11.4.3 3 D Elements

Analysis of solid bodies (stresses in engine cylinders, stresses under foundations, contact stresses under point loads etc.) call for the use of 3 D (solid) elements (Fig. 11.1 (*iii*)). These have the drawback that the visualization is complex. The size of the stiffness matrix to be handled can become enormous and unwieldy.

11.4.4 Axisymmetric Elements

In problems that are axisymmetric in nature, we can make a huge simplification. A 1 D line can represent a surface generated by rotating that line about an axis. A 2 D triangle or rectangle can represent a solid ring (Fig. 11.1(*iv*))

11.4.5 Guidelines for Discretisation

Discretisation is a major decision making step in FE analysis. In simple structures this step does not pose any problems. But in most real structures, the process of subdividing the structure, numbering the nodes, assigning co-ordinates to each node, relating the structure co-ordinate numbers to element numbers and their co-ordinate numbers can become really daunting. But we can take consolation from the fact that in most FEM software, discretisation is handled by the 'preprocessing' module of the software. But, to help the software help us, we must form some clear ideas about discretisation.

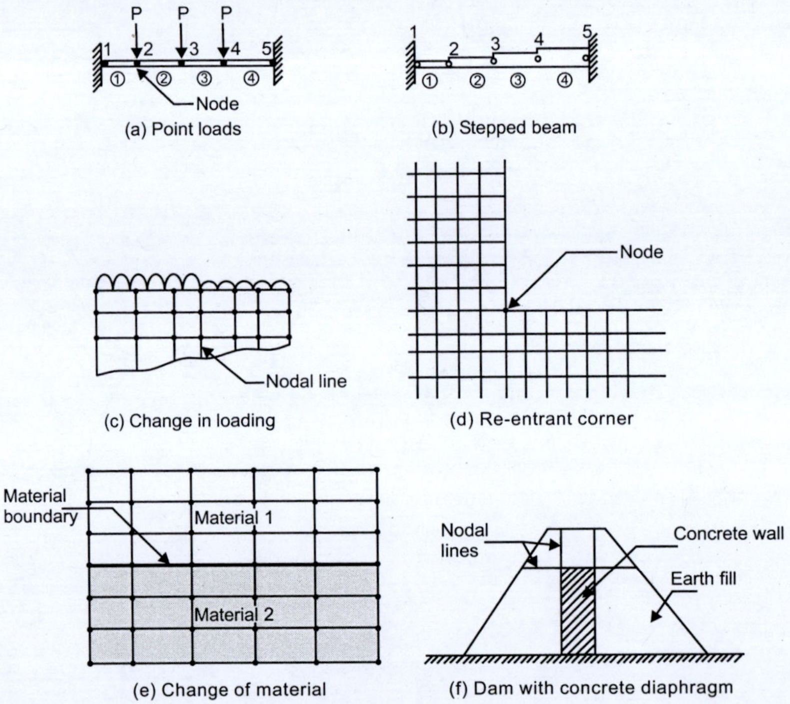

Fig. 11.2. Discretisation at discontinuities.

Figure 11.2(*a*) shows a beam with 3 concentrated loads. The analysis will be simple if we have a node at each support and at each load point. Thus we will have 4 beam elements. Figure 11.2(*b*) shows a stepped beam. It seems to be an obvious choice (and it is) that we have a node at every point at which the section changes. Thus we would have 4 elements and 5 nodes. Figure 11.2(*c*) shows a plate loaded in its own plane at the top boundary. When we divide the area into several rectangular elements, it is advisable to have a line of nodes along the point where the rate of loading changes.

In Fig. 11.2(*d*), we see a plane lamina with a re-entrant corner. This corner must have a node there. Likewise a line over which the material changes must be chosen as a nodal line (Figs. 11.2(*e*) and 11.2(*f*)). In Fig. 11.2(*f*) all the 3 boundaries between concrete and earth fill needs be taken as nodal lines. It is needless to add that all boundaries shall be nodal lines too.

Analysis of infinite and semi-infinite continua normally pose problems in FEA. We must be satisfied with finite and large elements to bound the so called infinite boundaries. Figure 11.3 shows a suggested discretisation pattern for an earthen dam. The elements adopted are both triangular and quadrilateral. Here, the values B_1, B_2 and H have to be decided upon from engineering judgement.

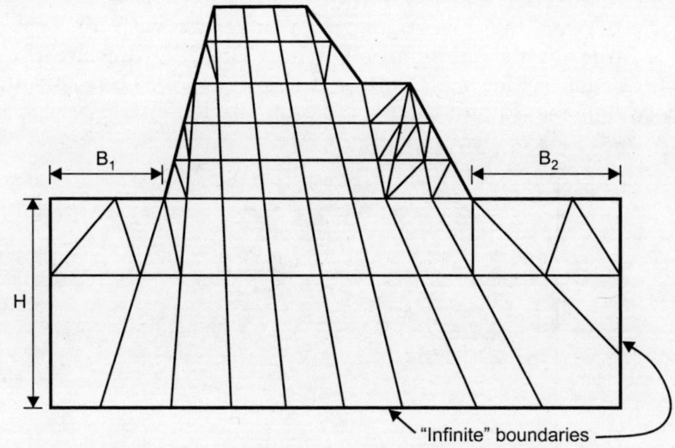

Fig. 11.3. Earth dam—Suggested discretisation.

When we combine fine and coarse meshes in a single configuration, we must take care that odd shaped elements are avoided. Elements with very acute or very obtuse angles (less than 30° or more than 150°), elongated elements (longest side ÷ shortest side > 2) are best avoided.

11.5 ELEMENT FUNCTIONS

There are probably more than 100 elements in 2 D to choose from. Most FE software give an element menu you can click on. Each one of these elements have been developed in the past 4 decades painstakingly by mathematically minded researchers to augment the wealth of FE bank. We will see herein some basis of developing a simple 2 D triangular constant strain element (CST element), truss element and beam element using elementary approach and using shape functions.

11.5.1 CST Element

When we develop an element, we start by stipulating a displacement model (or field or function) in the form of a polynomial (most common) or trigonometric function (less common). Functions like

$$u(x, y) = a_1 + a_2x + a_3y \quad \text{...(11.1)}$$

$$v(x, y) = a_4 + a_5x + a_6y \quad \text{...(11.2)}$$

lead to a constant strain field. This is because,

$$\varepsilon_x = \frac{\partial u}{\partial x} = a_2, \text{ constant}$$

$$\varepsilon_y = \frac{\partial v}{\partial y} = a_6, \text{ constant}$$

$$\gamma_{xy} = \frac{\partial u}{\partial y} + \frac{\partial v}{\partial x} = a_3 + a_5, \text{ constant.}$$

This will lead to a situation where σ_x, σ_y and τ_{xy} will take some constant values inside the element. If in reality these stresses vary inside the area of this element, our element would be incapable of revealing this. So, to get any precision with this type of element we will have to choose a large number of small elements. The following are the steps involved in deriving the stiffness matrix for a CST element shown in Fig. 11.4. This amounts to deriving a relation between 6 forces {P} and the 6 displacements {δ}.

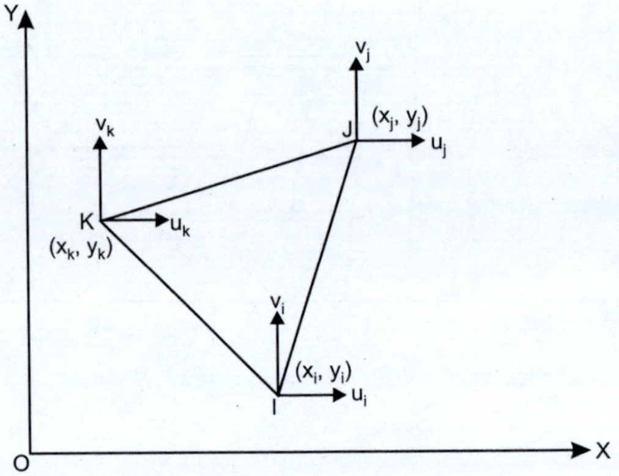

Fig. 11.4. CST element.

Consider a triangular element IJK with nodal co-ordinates (x_i, y_i), (x_j, y_j) and (x_k, y_k). Any node can be chosen as I but anti-clockwise naming is adopted. The displacements at the nodes, in the x and y directions are indicated as u and v respectively. Young's modulus E, thickness of the element t and Poisson's ratio ν are assumed as known values.

The displacement function within a triangle is assumed as in equations (11.1) and (11.2). Thus,

$$u = a_1 + a_2x + a_3y$$

$$v = a_4 + a_5x + a_6y$$

The x displacements at the nodes are then

$$\begin{bmatrix} u_i \\ u_j \\ u_k \end{bmatrix} = \begin{bmatrix} 1 & x_i & y_i \\ 1 & x_j & y_j \\ 1 & x_k & y_k \end{bmatrix} \begin{bmatrix} a_1 \\ a_2 \\ a_3 \end{bmatrix}$$

$$\{u\} = [A]\,\{a\}$$

$$\{a\} = [A]^{-1}\,\{u\} \quad ...(11.3)$$

This means that for finding a_1, a_2, a_3 we have to find inverse of the matrix [A]

$$[A]^{-1} = \begin{bmatrix} 1 & x_i & y_i \\ 1 & x_j & y_j \\ 1 & x_k & y_k \end{bmatrix}^{-1}$$

$$= \frac{1}{2\Delta} \begin{bmatrix} (x_j y_k - x_k y_j) & (x_k y_i - x_i y_k) & (x_i y_j - y_i x_j) \\ (y_j - y_k) & (y_k - y_i) & (y_i - y_j) \\ (x_k - x_j) & (x_i - x_k) & (x_j - x_i) \end{bmatrix}$$

where $$2\Delta = \begin{vmatrix} 1 & x_i & y_i \\ 1 & x_j & y_j \\ 1 & x_k & y_k \end{vmatrix} = 2 \times \text{area of triangle IJK}$$

a_2 and a_3 can be expressed as

$$a_2 = \frac{1}{2\Delta}(b_i u_i + b_j u_j + b_k u_k) \quad ...(11.4)$$

$$a_3 = \frac{1}{2\Delta}(c_i u_i + c_j u_j + c_k u_k) \quad ...(11.5)$$

where
$$b_i = (y_j - y_k) \qquad c_i = (x_k - x_j)$$
$$b_j = (y_k - y_i) \qquad c_j = (x_i - x_k)$$
$$b_k = (y_i - y_j) \qquad c_k = (x_j - x_i)$$

Similarly, we can write the nodal displacements in y direction as,

$$\begin{bmatrix} v_i \\ v_j \\ v_k \end{bmatrix} = \begin{bmatrix} 1 & x_i & y_i \\ 1 & x_j & y_j \\ 1 & x_k & y_k \end{bmatrix} \begin{bmatrix} a_4 \\ a_5 \\ a_6 \end{bmatrix}$$

Solving the above matrix equation, we get

$$a_5 = \frac{1}{2\Delta}(b_i v_i + b_j v_j + b_k v_k) \quad ...(11.6)$$

$$a_6 = \frac{1}{2\Delta}(c_i v_i + c_j v_j + c_k v_k)$$

The strain-displacement relationship for plane stress problems in 2-D is given by

$$\{\varepsilon\} = \begin{bmatrix} \varepsilon_x \\ \varepsilon_y \\ \gamma_{xy} \end{bmatrix} = \begin{bmatrix} \dfrac{\delta u}{\delta x} \\ \dfrac{\delta v}{\delta y} \\ \dfrac{\delta u}{\delta y} + \dfrac{\delta v}{\delta x} \end{bmatrix} = \begin{bmatrix} a_2 \\ a_6 \\ a_3 + a_5 \end{bmatrix} \quad ...(11.7)$$

Substituting the values of a_2, a_6, a_3 and a_5 obtained above,

$$\{\varepsilon\} = \frac{1}{2\Delta}\begin{bmatrix} b_i u_i + b_j u_j + b_k u_k \\ c_i v_i + c_j v_j + c_k v_k \\ c_i u_i + c_j u_j + c_k u_k + b_i v_i + b_j v_j + b_k v_k \end{bmatrix}$$

In matrix form {ε} can be written as

$$\begin{Bmatrix} \varepsilon_x \\ \varepsilon_y \\ \gamma_{xy} \end{Bmatrix} = \frac{1}{2\Delta}\begin{Bmatrix} b_i & 0 & b_j & 0 & b_k & 0 \\ 0 & c_i & 0 & c_j & 0 & c_k \\ c_i & b_i & c_j & b_j & c_k & b_k \end{Bmatrix}\begin{Bmatrix} u_i \\ v_i \\ u_j \\ v_j \\ u_k \\ v_k \end{Bmatrix}$$

This can be re-written as

$$\{\varepsilon\} = [B]\ \{\delta\} \qquad \text{...(11.8)}$$

where,
$$[B] = \frac{1}{2\Delta}\begin{Bmatrix} b_i & 0 & b_j & 0 & b_k & 0 \\ 0 & c_i & 0 & c_j & 0 & c_k \\ c_i & b_i & c_j & b_j & c_k & b_k \end{Bmatrix} \qquad \text{...(11.9)}$$

We also know that the stress strain relation is

$$\{\sigma\} = [D]\{\varepsilon\} \qquad \text{...(11.10)}$$

where [D] is the elasticity matrix given by

$$[D] = \frac{E}{(1-\nu^2)}\begin{bmatrix} 1 & \nu & 0 \\ \nu & 1 & 0 \\ 0 & 0 & \dfrac{1-\nu}{2} \end{bmatrix}$$

Let a set of forces {P} be assumed to be applied to the nodes of the element in Fig. 11.4. The stresses {σ} shall then be in equilibrium and consistent with these forces {P} at the nodes.

Then it is possible to form a relation between forces {P} and displacements {δ} in the form

$$\{P\} = [k]\ \{\delta\} \qquad \text{...(11.11)}$$

where [*k*] is known as the stiffness matrix and

$$\{\delta\}^T = [u_i \ v_i \ u_j \ v_j \ u_k \ v_k]$$

$$[k] = [B]^T\ [D]\ [B]\ V = [B]^T\ [D]\ [B]\ At \qquad \text{...(11.12)}$$

where **V** is the volume of the element, **A** is the area and **t** is the thickness. [B] is the strain displacement matrix in terms of the *x* and *y* co-ordinates of the corners of the triangular element.

CST element, as mentioned earlier, is the simplest form of 2D elements. We get the [*k*] matrix uniquely in terms of the geometry of the element (nodal co-ordinates and thickness) and the elasticity matrix [D].

For most other elements, this kind of simplicity is not available. In many cases we will have to perform several integrations (including numerical integrations) to generate the stiffness matrix of each element.

11.5.2 Stiffness Matrix for Plane Truss Element–Elementary Approach

Consider a truss element of length l as shown in Fig. 11.5.

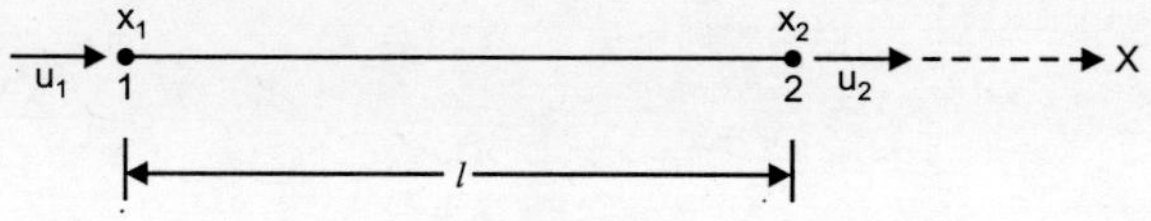

Fig. 11.5

Let u_1 and u_2 are the displacements at nodes. Force displacement relationship for this element is

$$\{F\} = [k]\,\{u\}$$

{F} = Nodal force vector

{u} = Nodal displacement vector

[k] = Element stiffness matrix

As the element has two nodal degrees of freedom, the size of the stiffness matrix will be [2 × 2].

$$[k] = \begin{bmatrix} k_{11} & k_{12} \\ k_{21} & k_{22} \end{bmatrix}$$

where k_{ij} = Force required at co-ordinate i to produce unit deformation at co-ordinate j with the condition that no other displacement occurs except at co-ordinate j.

Apply a unit deformation in the axial direction of the truss element at node '1' only.

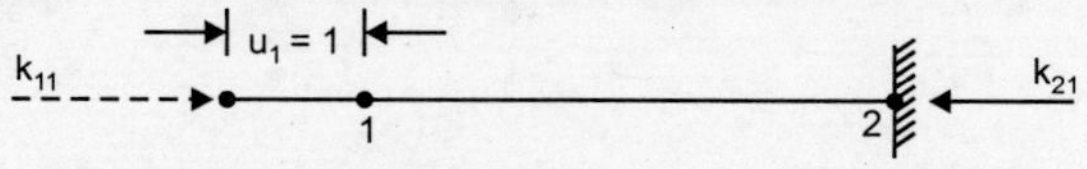

$$\text{Deformation applied at node '1'} = \begin{pmatrix}\text{Force required}\\ \text{at node '1'}\end{pmatrix} \times \frac{l}{AE}$$

$$1 = k_{11} \times \frac{l}{AE}$$

$$k_{11} = \frac{AE}{l}\,(1)$$

Force at co-ordinate 2,

$$k_{21} = \frac{-AE}{l}\,(1)$$

$$k_{21} = \frac{AE}{l}\,(-1)$$

Apply a unit deformation along the axial direction of the truss element at node '2' only.

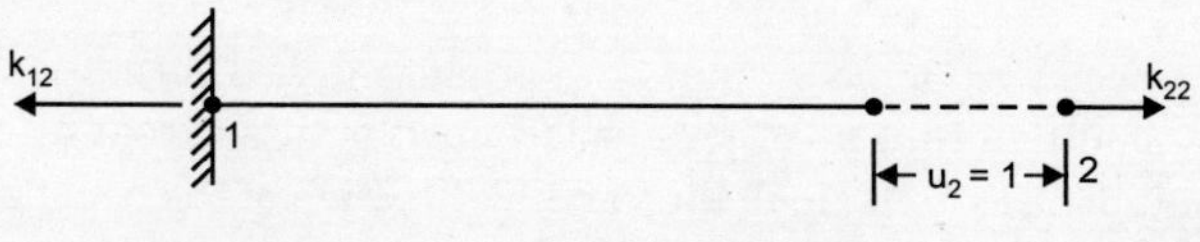

$$\left.\begin{array}{l}\text{Force required at '1'}\\ \text{when a unit deformat-}\\ \text{ion is applied at '2'}\end{array}\right\} = k_{12} = \frac{-AE}{l}(1)$$

$$k_{12} = \frac{AE}{l}(-1)$$

Force required at node '2' = $k_{22} = \frac{AE}{l}(1)$

The stiffness matrix

$$[k] = \begin{bmatrix} k_{11} & k_{12} \\ k_{21} & k_{22} \end{bmatrix}, \ [k] = \begin{bmatrix} \frac{AE}{l}(1) & \frac{AE}{l}(-1) \\ \frac{AE}{l}(-1) & \frac{AE}{l}(1) \end{bmatrix}$$

$$[k] = \frac{AE}{l}\begin{bmatrix} 1 & -1 \\ -1 & 1 \end{bmatrix}$$ is the stiffness matrix for a truss element.

11.5.3 Stiffness Matrix for Plane Beam Element–Elementary Approach

The Fig. 11.6 below shows a beam element of length 'l'.

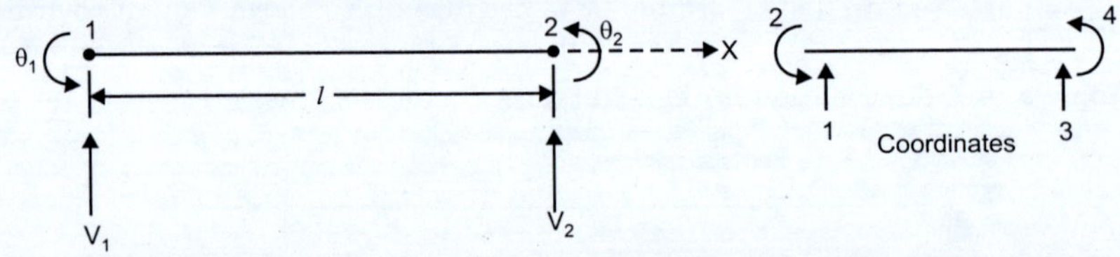

Fig. 11.6. Beam element.

Let V_1, V_2, θ_1, θ_2 are the displacements (translations and rotations) at nodes 1 and 2 respectively.

We know that the force displacement relationship is given by

$$\{F\} = [k]\{u\}$$

where $[k]$ = Stiffness matrix of beam element.

As the element has four degrees of freedom, the size of the stiffness matrix will be (4 × 4).

$$[k] = \begin{bmatrix} k_{11} & k_{12} & k_{13} & k_{14} \\ k_{21} & k_{22} & k_{23} & k_{24} \\ k_{31} & k_{32} & k_{33} & k_{34} \\ k_{41} & k_{42} & k_{43} & k_{44} \end{bmatrix}$$

where k_{ij} = Force required at co-ordinate i to produce unit deformation at co-ordinate j with the condition that no other displacement occurs except at co-ordinate j.

Apply a unit deformation along the vertical direction of the beam element at co-ordinate 'V_1' only.

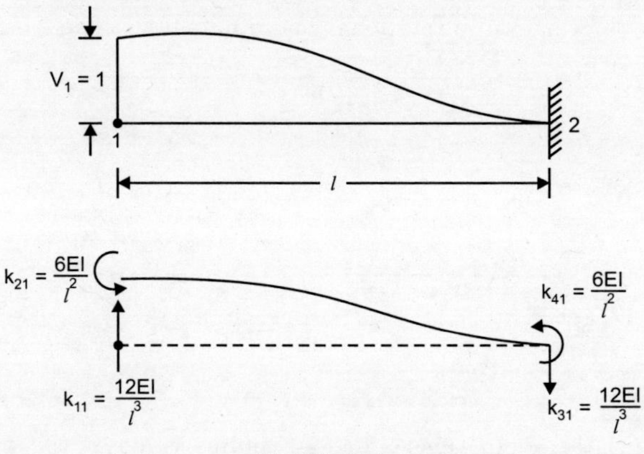

Fig. 11.7 *(a)*

Apply a unit deformation along the rotational co-ordinate θ_1 at node 1 only.

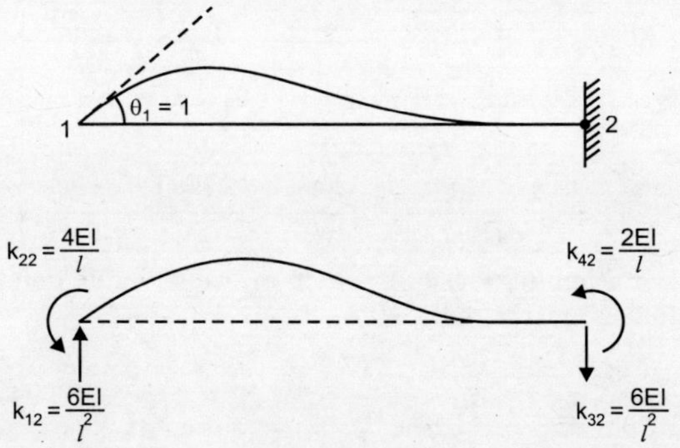

Fig. 11.7 *(b)*

Apply a unit deformation in vertical co-ordinate 'V_2' at node 2 only.

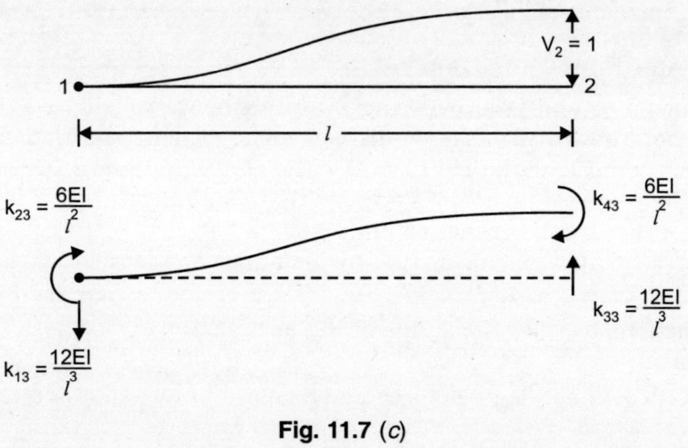

Fig. 11.7 *(c)*

Apply a unit deformation in rotational co-ordinate θ_2 at node 2 only.

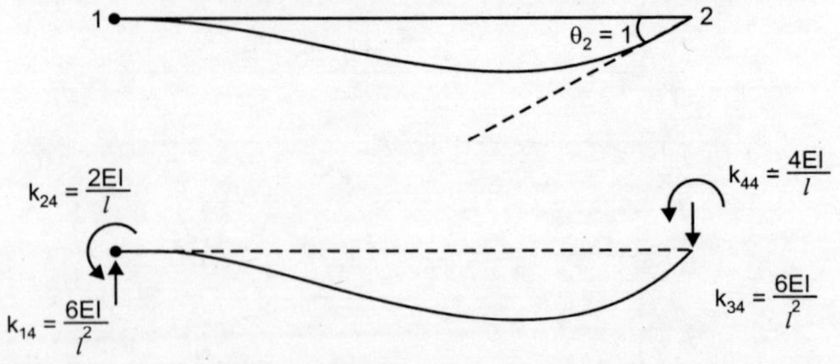

Fig. 11.7 (*d*)

Figure 11.7 (*a*), (*b*), (*c*), (*d*) shows the four column values of the stiffness matrix.

$$[k] = \begin{bmatrix} \frac{12\,EI}{l^3} & \frac{6\,EI}{l^2} & \frac{-12\,EI}{l^3} & \frac{6\,EI}{l^2} \\ \frac{6\,EI}{l^2} & \frac{4\,EI}{l} & \frac{-6\,EI}{l^2} & \frac{2\,EI}{l} \\ \frac{-12\,EI}{l^3} & \frac{-6\,EI}{l^2} & \frac{12\,EI}{l^3} & \frac{-6\,EI}{l^2} \\ \frac{6\,EI}{l^2} & \frac{2\,EI}{l} & \frac{-6\,EI}{l^2} & \frac{4\,EI}{l} \end{bmatrix}$$

When only the rotation degrees of freedom are considered, deleting first and third rows and columns, the stiffness matrix reduces to

$$[k] = \begin{bmatrix} \frac{4\,EI}{l} & \frac{2\,EI}{l} \\ \frac{2\,EI}{l} & \frac{4\,EI}{l} \end{bmatrix}$$

$$[k] = \frac{EI}{l}\begin{bmatrix} 4 & 2 \\ 2 & 4 \end{bmatrix}.$$

11.5.4 Shape Functions-General

In finite element analysis, over the finite elements considered the true variation of displacement is not known. We assume displacement models. Here, many times, we have to deal with functions whose analytical form is either totally unknown or the function may not be subjected to the required operations. Under these circumstances, it is desirable to replace the given function by another function that can be handled more easily. This operation of replacing or representing a given function by a simpler one is referred as "shape function" in finite element literature.

Shape functions express the shape of the assumed displacement function over the domain of the element.

When the *i*th element degree of freedom has unit value, all other degrees of freedom are zero.

If an element with two d.o.f. is considered, if N_1 and N_2 are linear functions,

When $N_1 = 1$ at node 1

$N_1 = 0$ at node 2

When $N_2 = 1$ at node 2

$N_2 = 0$ at node 1.

Also, $N_1 + N_2 = 1$ for any axial co-ordinate along the bar.

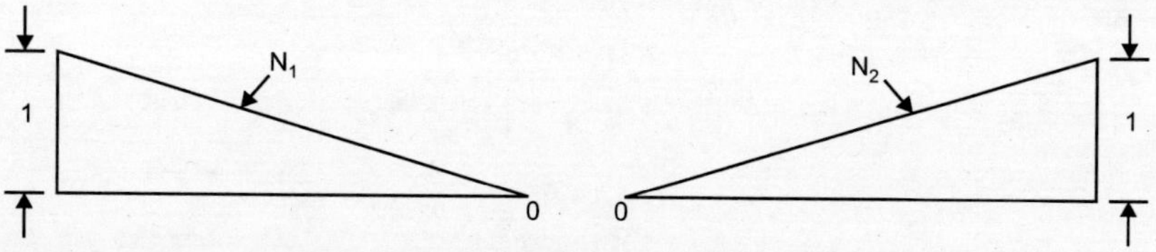

Fig. 11.8

11.5.5 Shape Function for Plane Truss Element

Consider a truss element of length 'l' as shown in Fig. 11.9. u_1, u_2 are generalized co-ordinate displacements in axial direction.1, 2 are nodes.

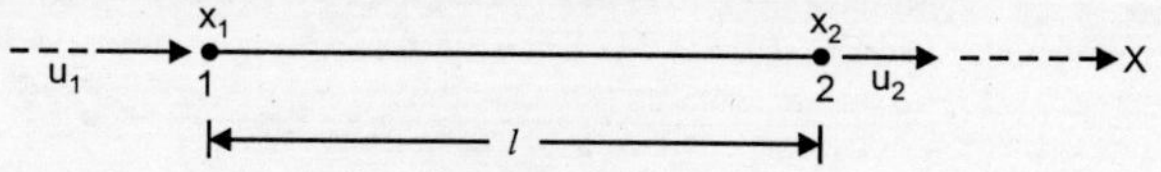

Fig. 11.9

The polynomial for this two-noded element is,

$$u = a_1 + a_2 x$$

Here, the total number of co-efficients 'a' is equal to the total number of degrees of freedom. Here one d.o.f. *i.e.*, axial displacement at each node is considered.

$$\{u\} = \begin{bmatrix} 1 & x \end{bmatrix} \begin{Bmatrix} a_1 \\ a_2 \end{Bmatrix} \quad \text{...(1)}$$

The above polynomial function should satisfy the compatibility requirements:

At node 1 $\quad u = u_1 = a_1 + a_2 x_1$

At node 2 $\quad u = u_2 = a_1 + a_2 x_2$

$$\{\delta\} = \begin{Bmatrix} u_1 \\ u_2 \end{Bmatrix} = \begin{bmatrix} 1 & x_1 \\ 1 & x_2 \end{bmatrix} \begin{Bmatrix} a_1 \\ a_2 \end{Bmatrix}$$

$$\begin{Bmatrix} a_1 \\ a_2 \end{Bmatrix} = \begin{bmatrix} 1 & x_1 \\ 1 & x_2 \end{bmatrix}^{-1} \begin{Bmatrix} u_1 \\ u_2 \end{Bmatrix}$$

$$\begin{Bmatrix} a_1 \\ a_2 \end{Bmatrix} = \frac{1}{x_2 - x_1} \begin{bmatrix} x_2 & -x_1 \\ -1 & 1 \end{bmatrix} \begin{Bmatrix} u_1 \\ u_2 \end{Bmatrix}$$

$$\begin{Bmatrix} a_1 \\ a_2 \end{Bmatrix} = \frac{1}{l} \begin{bmatrix} x_2 & -x_1 \\ -1 & 1 \end{bmatrix} \begin{Bmatrix} u_1 \\ u_2 \end{Bmatrix} \quad \text{...(2)} \quad (\because\ x_2 - x_1 = l)$$

Substitute equation (2) in (1),

$$\begin{Bmatrix} u_1 \\ u_2 \end{Bmatrix} = \begin{bmatrix} 1 & x \end{bmatrix} \frac{1}{l} \begin{bmatrix} x_2 & -x_1 \\ -1 & 1 \end{bmatrix} \begin{Bmatrix} u_1 \\ u_2 \end{Bmatrix}$$

$$\{u\} = \begin{Bmatrix} u_1 \\ u_2 \end{Bmatrix} = \frac{1}{l} \begin{bmatrix} x_2 - x & -x_1 + x \end{bmatrix} \begin{Bmatrix} u_1 \\ u_2 \end{Bmatrix}$$

$$\{u\} = \begin{bmatrix} \frac{x_2 - x}{l} & \frac{x - x_1}{l} \end{bmatrix} \begin{Bmatrix} u_1 \\ u_2 \end{Bmatrix}$$

$$\{u\} = \begin{bmatrix} N_1 & N_2 \end{bmatrix} \begin{Bmatrix} u_1 \\ u_2 \end{Bmatrix} = N_1 u_1 + N_2 u_2$$

Here N_1, N_2 are the shape functions corresponding to nodes 1, 2.

$$N_1 = \frac{x_2 - x}{l} \qquad N_2 = \frac{x - x_1}{l} \qquad (x_2 = l, x_1 = 0)$$

$$N_1 = \frac{l - x}{l} = \frac{1 - x}{l} \qquad N_2 = \frac{x}{l}$$

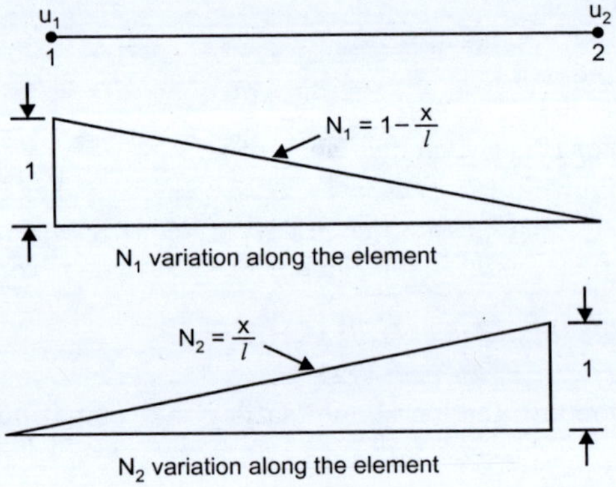

Fig. 11.10

11.5.6 Stiffness Matrix for Plane Truss Element–Using Shape Functions

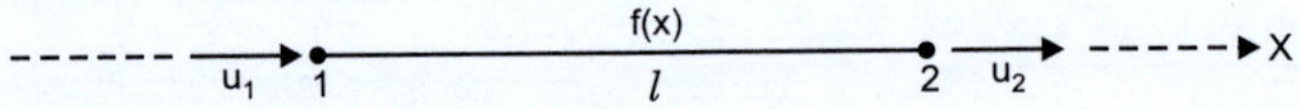

Fig. 11.11

For the truss element shown in Fig. 11.11. u_1, u_2 are the displacements at nodes 1, 2

$$u(x) = N_1(x)u_1 + N_2(x)u_2$$

where $N_1(x)$, $N_2(x)$ are the shape functions.

$$N_1 = \frac{l - x}{l} \quad N_2 = \frac{x}{l}$$

$$u(x) = \begin{bmatrix} \frac{l - x}{l} & \frac{x}{l} \end{bmatrix} \begin{Bmatrix} u_1 \\ u_2 \end{Bmatrix}$$

$u(x) = [N]\ \{u\}$

$\{u\}$ = Nodal displacement vector

[N] = Shape function matrix.

Strain:

$$\varepsilon(x) = \frac{d}{dx} u(x) = \left[\frac{d}{dx}[N]\right]\{u\} \qquad ...(1)$$

$$\varepsilon(x) = \begin{bmatrix} \frac{-1}{l} & \frac{1}{l} \end{bmatrix} \begin{Bmatrix} u_1 \\ u_2 \end{Bmatrix}$$

$$\varepsilon(x) = [B]\ \{u\} = \{U\}^T\ [B]^T \qquad ...(2)$$

where $[B] = \frac{d}{dx}[N]$ is called strain-displacement matrix.

By the principle of virtual work,

Internal work done = External work done.

Internal work done, $W_I = \int \frac{1}{2}\ \varepsilon\ .\ \sigma\ dV$

where $\sigma = E\varepsilon,\quad dV = A\ .\ dx$

σ = Stress, ε = Strain, A = Area

E = Modulus of elasticity, V = Volume

$$W_I = \int \frac{1}{2}\ \varepsilon\ .\ \sigma\ dV = \int \frac{1}{2}\ \varepsilon.\ E\ .\ \varepsilon\ .\ A\ .\ dx$$

$$= \int \frac{1}{2}(\varepsilon E\ \delta\varepsilon + \delta\varepsilon E\varepsilon)\ A\ .\ dx$$

External work done, $W_E = \int f(x)\ .\ \delta u(x)\ dx$

$$W_I = W_E$$

$$\int_0^l \varepsilon\ .\ E\ .\ \delta\varepsilon\ .\ A\ .\ dx = \int_0^l f(x)\ .\ \delta u(x)\ dx \qquad ...(3)$$

Substitute equations (1) and (2) in left and right side of equation (3)

$$\delta U \int_0^l AE\,[B]^T\ .\ [B]\ .\ \{u\}\ dx = [\delta U] \int_0^l f(x)\ .\ \{N\}\ dx$$

$$\{U\}^T \int_0^l AE\,[B]^T\ .\ [B]\ dx = \int_0^l f(x)\ .\ \{N\}\ dx$$

$$\{U\}^T\ [k] = \{F\}$$

where $[k] = \int_0^l AE\ .\ [B]^T\ .\ [B]\ dx$

$$= \int_0^l AE \begin{Bmatrix} -1/l \\ 1/l \end{Bmatrix} \begin{Bmatrix} \frac{-1}{l} & \frac{1}{l} \end{Bmatrix} dx$$

$$[k] = \int_0^l AE \begin{bmatrix} \frac{1}{l^2} & \frac{-1}{l^2} \\ \frac{-1}{l^2} & \frac{1}{l^2} \end{bmatrix} dx$$

$$[k] = \frac{AE}{l^2}\begin{bmatrix} 1 & -1 \\ -1 & 1 \end{bmatrix} \int_0^l dx$$

$$[k] = \frac{AE}{l}\begin{bmatrix} 1 & -1 \\ -1 & 1 \end{bmatrix}$$

11.5.7 Shape Function for Plane Beam Element

Consider a 2D-beam element shown in Fig. 11.12.

1, 2 are nodes

V_1, V_2, θ_1, θ_2 are generalized co-ordinates displacements.

V_1, V_2 are translations in 'Y' direction.

θ_1, θ_2 are rotations about 'Z' direction.

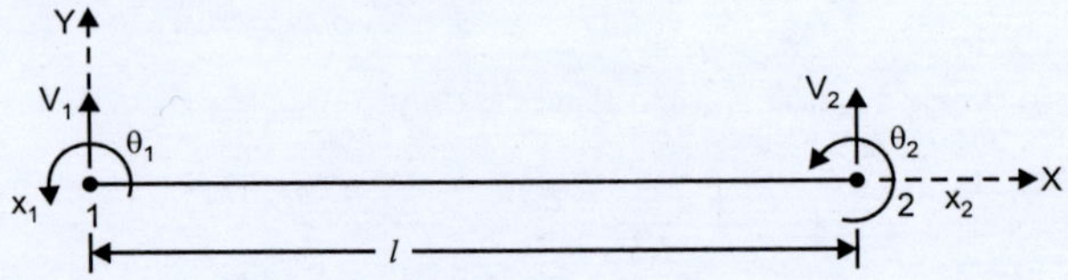

Fig. 11.12. Beam element.

The displacement vector for the above element,

$$\{\delta\} = \begin{Bmatrix} V_1 \\ \theta_1 \\ V_2 \\ \theta_2 \end{Bmatrix}$$

where $\theta_1 = \frac{\partial V_1}{\partial x}$ $\theta_2 = \frac{\partial V_2}{\partial x}$.

$$V = \{\delta\} = a_1 + a_2 x + a_3 x^2 + a_4 x^3$$

The complete cubic displacement function is appropriate because there are totally four degrees of freedom (a transverse displacement and a small rotation at each node).

The above equation should satisfy the shape function requirements such as compatibility and displacement completeness, etc.

$$\theta = \frac{\partial V}{\partial x} = a_2 + 2a_3\, x + 3a_4 x^2$$

Using the local co-ordinates,

At node 1	At node 2
$x_1 = 0$	$x_2 = l$

At node (1),

$$V_1 = a_1 + a_2 x_1 + a_3\, x_1^{\,2} + a_4 x_1^{\,3}, \quad \text{and}$$

$$\theta_1 = \frac{\partial V_1}{\partial x_1} = a_2 + 2a_3 x_1 + 3a_4 x_1^{\,2}$$

Put $x_1 = 0$ in the above equations, we get

$$V_1 = a_1, \ \theta_1 = a_2$$

At node (2),

$$V_2 = a_1 + a_2x_2 + a_3x_2^{\,2} + a_4x_2^{\,3}, \quad \text{and}$$

$$\theta_2 = \frac{\partial V_2}{\partial x_2} = a_2 + 2a_3\,x_2 + 3a_4x_2^{\,2}$$

Put $x_2 = l$ in the above equations, we get

$$V_2 = a_1 + a_2l + a_3l^2 + a_4l^3$$
$$\theta_2 = a_2 + 2a_3l + 3a_4l^2$$

$$\{\delta\} = \begin{Bmatrix} V_1 \\ \theta_1 \\ V_2 \\ \theta_2 \end{Bmatrix} = \begin{bmatrix} 1 & 0 & 0 & 0 \\ 0 & 1 & 0 & 0 \\ 1 & l & l^2 & l^3 \\ 0 & 1 & 2l & 3l^2 \end{bmatrix} \begin{Bmatrix} a_1 \\ a_2 \\ a_3 \\ a_4 \end{Bmatrix}$$

$$\begin{Bmatrix} a_1 \\ a_2 \\ a_3 \\ a_4 \end{Bmatrix} = \begin{bmatrix} 1 & 0 & 0 & 0 \\ 0 & 1 & 0 & 0 \\ 1 & l & l^2 & l^3 \\ 0 & 1 & 2l & 3l^2 \end{bmatrix}^{-1} \begin{Bmatrix} V_1 \\ \theta_1 \\ V_2 \\ \theta_2 \end{Bmatrix}$$

$$\begin{Bmatrix} a_1 \\ a_2 \\ a_3 \\ a_4 \end{Bmatrix} = \frac{1}{l^4} \begin{bmatrix} l^4 & 0 & 0 & 0 \\ 0 & l^4 & 0 & 0 \\ -3l^2 & -2l^3 & 3l^2 & -l^3 \\ 2l & l^2 & -2l & l^2 \end{bmatrix} \begin{Bmatrix} V_1 \\ \theta_1 \\ V_2 \\ \theta_2 \end{Bmatrix}$$

$$\begin{Bmatrix} a_1 \\ a_2 \\ a_3 \\ a_4 \end{Bmatrix} = \begin{bmatrix} 1 & 0 & 0 & 0 \\ 0 & 1 & 0 & 0 \\ -3/l^2 & -2/l & 3/l^2 & -1/l \\ 2/l^3 & 1/l^2 & -2/l^3 & 1/l^2 \end{bmatrix} \begin{Bmatrix} V_1 \\ \theta_1 \\ V_2 \\ \theta_2 \end{Bmatrix} \qquad ...(1)$$

We know,

$$V = a_1 + a_2x + a_3x^2 + a_4x^3$$

$$V = \begin{bmatrix} 1 & x & x^2 & x^3 \end{bmatrix} \begin{Bmatrix} a_1 \\ a_2 \\ a_3 \\ a_4 \end{Bmatrix} \qquad ...(2)$$

Substitute (1) in (2)

$$V = [1 \quad x \quad x^2 \quad x^3] \begin{bmatrix} 1 & 0 & 0 & 0 \\ 0 & 1 & 0 & 0 \\ -3/l^2 & -2/l & 3/l^2 & -1/l \\ 2/l^3 & 1/l^2 & -2/l^3 & 1/l^2 \end{bmatrix} \begin{Bmatrix} V_1 \\ \theta_1 \\ V_2 \\ \theta_2 \end{Bmatrix}$$

$$V = \left[\left(1 - \frac{3x^2}{l^2} + \frac{2x^3}{l^3}\right)\left(x - \frac{2x^2}{l} + \frac{x^3}{l^2}\right)\left(\frac{3x^2}{l^2} - \frac{2x^3}{l^3}\right)\left(\frac{-x^2}{l} + \frac{x^3}{l^2}\right)\right]\begin{Bmatrix} V_1 \\ \theta_1 \\ V_2 \\ \theta_2 \end{Bmatrix}$$

$$V = [N_1 \quad N_2 \quad N_3 \quad N_4]\ \{\delta\}$$

$$V = [N]\ \{\delta\}$$

$$[N] = [N_1 \quad N_2 \quad N_3 \quad N_4]$$

$$N_1 = 1 - \frac{3x^2}{l^2} + \frac{2x^3}{l^3} \qquad N_2 = x - \frac{2x^2}{l} + \frac{x^3}{l^2}$$

$$N_3 = \frac{3x^2}{l^2} - \frac{2x^3}{l^3} \qquad N_4 = \frac{-x^2}{l} + \frac{x^3}{l^2}$$

$$N_1 + N_2 + N_3 + N_4 = 1$$

when $x = 0$ or $x = l$

At node 1, $N_1 = 1$ and $N_2 = N_3 = N_4 = 0$

Similarly,

At node 2, $N_2 = 1$ $\quad \dfrac{\partial N_1}{\partial x} = \theta_1 = 1$ and $N_1 = N_3 = N_4 = 0$

At node 3, $N_3 = 1$ and $N_1 = N_2 = N_4 = 0$

At node 4, $N_4 = 1$ $\quad \dfrac{\partial N_2}{\partial x} = \theta_2 = 1$ and $N_1 = N_2 = N_3 = 0$

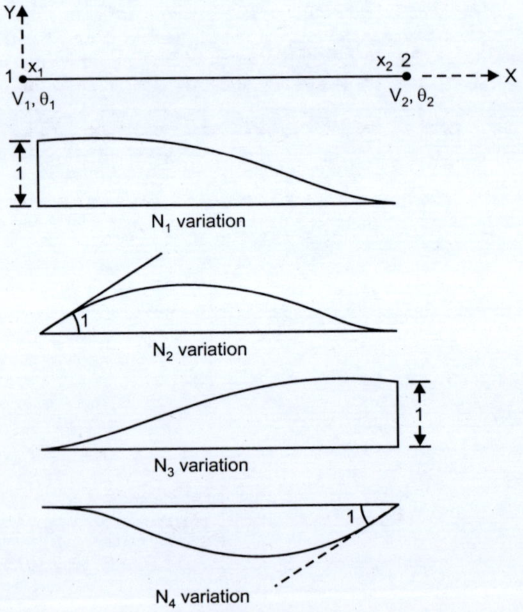

Fig. 11.13

11.5.8 Stiffness Matrix for Plane Beam Element–Using Shape Functions

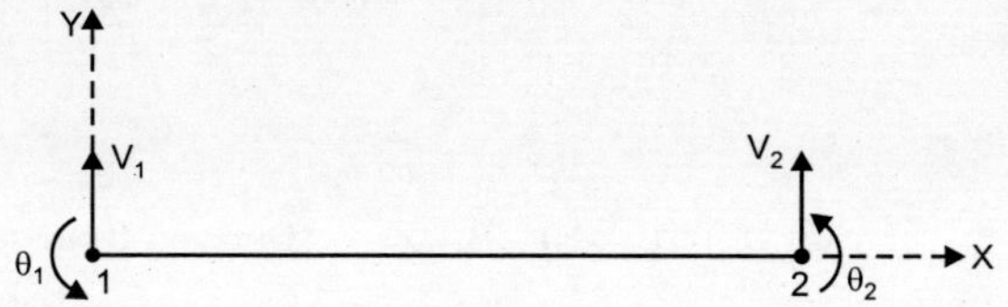

Fig. 11.14

For the beam element shown in Fig. 11.14.

V_1, θ_1, V_2, θ_2 are the nodal displacements

$$V(x) = [N_1 \quad N_2 \quad N_3 \quad N_4] \begin{Bmatrix} V_1 \\ \theta_1 \\ V_2 \\ \theta_2 \end{Bmatrix} = [N]\{\delta\} \qquad ...(1)$$

The shape functions for this beam element;

$$N_1(x) = 1 - \frac{3x^2}{l^2} + \frac{2x^3}{l^3}$$

$$N_2(x) = x - \frac{2x^2}{l} + \frac{x^3}{l^2}$$

$$N_3(x) = \frac{3x^2}{l^2} - \frac{2x^3}{l^3}$$

$$N_4(x) = \frac{-x^2}{l} + \frac{x^3}{l^2}$$

The curvature, $$\frac{d^2V}{dx^2} = \left[\frac{d^2}{dx^2}(N)\right]\delta = B.\ \delta \qquad (\because \quad V = N\delta)$$

where B = Strain displacement matrix.

$$[N] = [N_1 \quad N_2 \quad N_3 \quad N_4]$$

$$[N] = \left[\left(1 - \frac{3x^2}{l^2} + \frac{2x^3}{l^3}\right)\left(x - \frac{2x^2}{l} + \frac{x^3}{l^2}\right)\left(\frac{3x^2}{l^2} - \frac{2x^3}{l^3}\right)\left(\frac{-x^2}{l} + \frac{x^3}{l^2}\right)\right]$$

$$[B] = \frac{d^2}{dx^2}(N)$$

$$[B] = \left[\left(\frac{-6}{l^2}+\frac{12x}{l^3}\right)\left(\frac{-4}{l}+\frac{6x}{l^2}\right)\left(\frac{6}{l^2}-\frac{12x}{l^3}\right)\left(\frac{-2}{l}+\frac{6x}{l^2}\right)\right]$$

Stiffness matrix

$$[k] = \int_0^l [B]^T \, EI \,.\, [B]\, dx$$ (Derived from strain energy of system. Refer previous derivation)

$$[k] = EI \int_0^l \begin{bmatrix} \left(\frac{-6}{l^2}+\frac{12x}{l^3}\right) \\ \left(\frac{-4}{l}+\frac{6x}{l^2}\right) \\ \left(\frac{6}{l^2}+\frac{12x}{l^3}\right) \\ \left(\frac{-2}{l}+\frac{6x}{l^2}\right) \end{bmatrix} \left[\left(\frac{-6}{l^2}+\frac{12x}{l^3}\right)\left(\frac{-4}{l}+\frac{6x}{l^2}\right)\left(\frac{6}{l^2}-\frac{12x}{l^3}\right)\left(\frac{-2}{l}+\frac{6x}{l^2}\right)\right] dx$$

By integrating, we get

$$[k] = EI \begin{bmatrix} \frac{12}{l^3} & \frac{6}{l^2} & \frac{-12}{l^3} & \frac{6}{l^2} \\ \frac{6}{l^2} & \frac{4}{l} & \frac{-6}{l^2} & \frac{2}{l} \\ \frac{-12}{l^3} & \frac{-6}{l^2} & \frac{12}{l^3} & \frac{-6}{l^2} \\ \frac{6}{l^2} & \frac{2}{l} & \frac{-6}{l^2} & \frac{4}{l} \end{bmatrix}$$

11.6 SELECTING ELEMENT FIELDS

To approximate the variation of displacements within an element, it is customary to choose simple mathematical functions (polynomial, trigonometric etc.). These functions are called *displacement fields* or *displacement models*.

Polynomial functions are more generally selected as displacement models. This is because they are easy to integrate and differentiate. A polynomial of an arbitrary degree permits a recognizable approximation to the true function. It is also easy to vary the degree of approximation. A polynomial of infinite order corresponds to an exact solution.

Thus a one dimensional displacement function can be

$$f(x) = c_1 + c_2x + c_3x^2 + c_4x^3 + \ldots\ldots + c_{n+1}\, x^n \qquad \ldots(11.13)$$

where c_1, c_2, c_3, ... are known as generalized co-ordinates or generalized displacement amplitudes.

The greater the number of terms in the approximation, the more closely the exact solution is represented.

Two dimensional displacement functions:

If we use an nth order polynomial, the components of displacements u and v in the x and y directions respectively can be written as

$$u = c_1 + c_2x + c_3y + c_4x^2 + c_5xy + c_6y^2 + \ldots + c_m y^n$$

$$v = c_{m+1} + c_{m+2}\,x + c_{m+3}\,y + c_{m+4}x^2 + \ldots + c_{2m}y^n \qquad \ldots(11.14)$$

A complete two dimensional polynomial is conveniently illustrated by means a Pascal triangle.

Pascal triangle	Name	No. of terms
1	Constant	1
$x \quad y$	Linear	3
$x^2 \quad xy \quad y^2$	Quadratic	6
$x^3 \quad x^2y \quad xy^2 \quad y^3$	Cubic	10
$x^4 \quad x^3y \quad x^2y^2 \quad xy^3 \quad y^4$	Quadric	15
$x^5 \quad x^4y \quad x^3y^2 \quad x^2y^3 \quad xy^4 \quad y^5$	Quintic	21

PASCAL TRIANGLE

Each of the above polynomial forms of displacement model can be truncated at any desired degree to give a constant, linear, quadratic or higher terms. For example, a linear model is

$$u = c_1 + c_2x + c_3\,y$$

$$v = c_4 + c_5x + c_6y$$

and has 6 unknowns.

These unknowns will have to be expressed in terms of the 6 displacements u_i, u_j, u_k, v_i, v_j and v_k of the nodes i, j and k.

This leads to a situation wherein we have to specify more nodal displacements if we choose to involve more constants in the displacement fields. For example, if we take

$$u = c_1 + c_2x + c_3y + c_4x^2 + c_5xy + c_6y^2 \quad \text{and}$$

$$v = c_7 + c_8x + c_9y + c_{10}x^2 + c_{11}xy + c_{12}\,y^2$$

we have to think of 6 + 6 = 12 nodal displacements. This can be the u and v displacements of 6 nodes shown in Fig. 11.1 (*aii*). It is also possible to specify more than 2 displacements per node (even in plane stress cases) by specifying the derivatives of u and v as displacements [we are already familiar, in matrix methods, with calling slopes, $\frac{\partial w}{\partial x}$, as displacements]. In 2D cases we can classify $\frac{\partial u}{\partial x}, \frac{\partial v}{\partial y}$ etc. as (generalized) displacements in order to meet the demand for more nodal displacements. Such elements employing derivatives of u and v as displacements are called *higher order elements*.

The advantage of having more terms in the displacement polynomial is that we can accommodate changes in strains (and stresses) within an element.

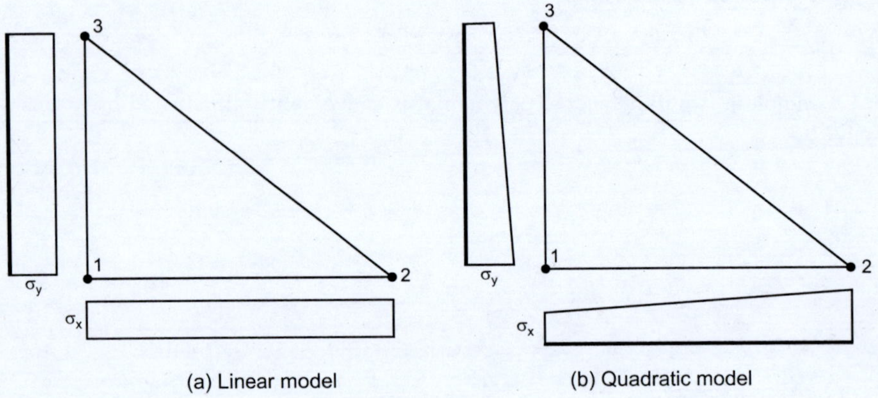

Fig. 11.15. Linear and quadratic models.

While developing a new element, care should be taken to employ polynomials with due regard to the Pascal triangle. We have to make sure we have included all lower degree terms before using a higher degree term. In addition, for isotropic materials, we have to maintain symmetry between x and y terms. For example, if there is an x^2y term, we must include a xy^2 term also. And the number of constants in the polynomials shall match with the degrees of freedom of the element.

Example 11.1 *For element No.18 of a CST formulation, finite element analysis yields the following nodal displacements of the nodes A, B and C.*

Node	*Co-ordinates (mm)*		*Displacements (mm)*	
	x	y	u	v
A	0	0	0.11	0.72
B	100	0	0.23	0.91
C	0	80	0.47	0.60

Compute σ_x, σ_y and τ_{xy}, assuming $E = 3 \times 10^4$ MPa, $v = 0.28$ and $t = 3$ mm.

Solution The variations of u and v within a CST element are linear.

Hence, $$\varepsilon_x = \frac{\partial u}{\partial x} = \frac{\Delta_u}{x_2 - x_1} = \frac{0.23 - 0.11}{100} = 0.0012$$

$$\varepsilon_y = \frac{\partial v}{\partial y} = \frac{\Delta_v}{y_3 - y_1} = \frac{0.6 - 0.72}{80} = -0.0015$$

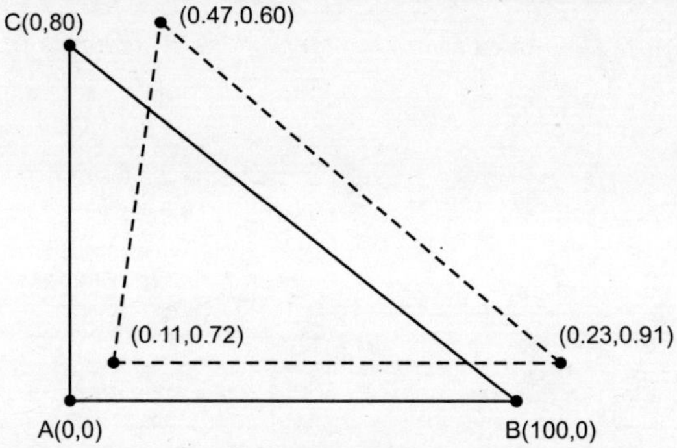

Fig. 11.16

$$\gamma_{xy} = \frac{\partial u}{\partial y} + \frac{\partial v}{\partial x} = \frac{0.47 - 0.11}{80} + \frac{0.91 - 0.72}{100} = 0.0064$$

$$\{\sigma\} = [D]\ \{\varepsilon\}$$

$$\begin{Bmatrix} \sigma_x \\ \sigma_y \\ \tau_{xy} \end{Bmatrix} = \frac{E}{(1-\nu^2)} \begin{bmatrix} 1 & \nu & 0 \\ \nu & 1 & 0 \\ 0 & 0 & \frac{1-\nu}{2} \end{bmatrix} \begin{Bmatrix} \varepsilon_x \\ \varepsilon_y \\ \gamma_{xy} \end{Bmatrix}$$

$$= 3.255 \times 10^4 \begin{bmatrix} 1 & 0.28 & 0 \\ 0.28 & 1 & 0 \\ 0 & 0 & \frac{1-0.28}{2} \end{bmatrix} \begin{bmatrix} 0.0012 \\ -0.0015 \\ 0.0064 \end{bmatrix}$$

$$\{\sigma\} = \begin{Bmatrix} 25.4 \\ -37.9 \\ 75.0 \end{Bmatrix} \text{ MPa}$$

Each of these stresses is constant all over the triangle.

The very idea of any finite element analysis is that the entire stress picture is known, once the nodal displacements are known. Example 11.1 can also be solved by directly using equation (11.8) (The student is advised to try this).

11.7 DEVELOPING NODAL LOAD VECTORS

Finite element method deals with nodal forces and nodal displacements. Hence, in finite element method, whatever be the nature of actual loading on the continuum, they have to be converted into concentrated loads at the assigned co-ordinates. There are 2 methods of expressing distributed loads as nodal loads. They are

1. Direct method or lumped load method, and
2. Variational method–Principle of virtual work

11.7.1 Direct Method

(*a*) Method of Inspection

In this method, we inspect the distributed loading on any element and assign appropriate lumped loads to the nodes.

Example 11.2 *Compute the nodal loads on each of the 3 elements in Fig. 11.17.*

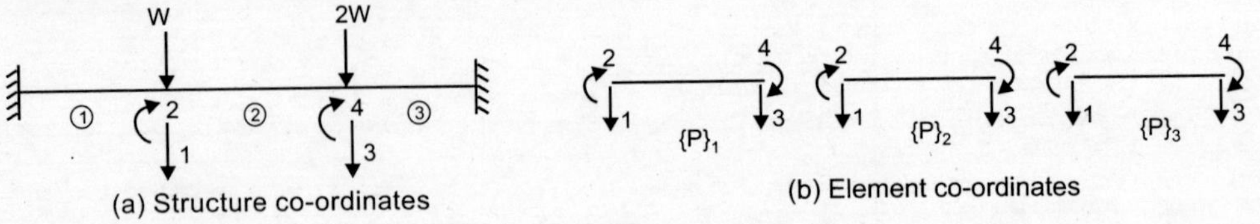

Fig. 11.17

Solution

$$\{P\}_1^T = \begin{bmatrix} 0 & 0 & \frac{W}{2} & 0 \end{bmatrix},$$

$$\{P\}_2^T = \begin{bmatrix} \frac{W}{2} & 0 & W & 0 \end{bmatrix},$$

$$\{P\}_3^T = [W \quad 0 \quad 0 \quad 0],$$

$$\{F\}^T = [W \quad 0 \quad 2W \quad 0].$$

Example 11.3 *Determine the element load vectors and global load vector for the system in Fig. 11.18.*

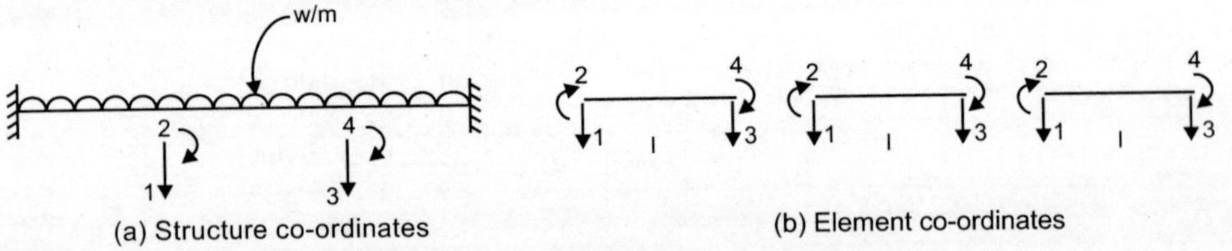

Fig. 11.18

Solution

$$\{P\}_1^T = \begin{bmatrix} \frac{wl}{2} & 0 & \frac{wl}{2} & 0 \end{bmatrix} = \{P\}_2^T = \{P\}_3^T$$

$$\{F\}^T = [wl \quad 0 \quad wl \quad 0]$$

Method of inspection is not comprehensive. This does not take into account the ability of the vertical loads to cause angular displacement (or rotation). Hence all moment components in the load vectors are (erroneously) assumed as zero.

(*b*) **Equivalent Loads**

For 1D elements (*e.g.*, A beam element) the ability of transverse loads (Fig. 11.18 (*b*)) on elements to create rotations at the co-ordinates positions can be easily taken into account as below.

Let the loading produce rotations of θ_A and θ_B at A and B. We want moments F_1 and F_2 that would produce the same rotations at A and B. Suppose (Fixed end) moments M_1 and M_2 nullify the slopes θ_A and θ_B. Then $F_1 = -M_1$ and $F_2 = -M_2$. F_1 and F_2 are the fixed end moments with their signs reversed.

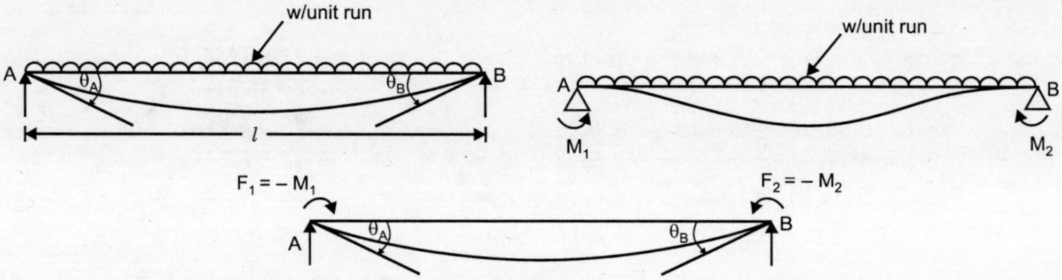

Fig. 11.19

Thus for a w/metre run on a span l,

$$\{P_1 \;\; P_2 \;\; P_3 \;\; P_4\} = \left[-\frac{wl}{2} \quad \frac{wl^2}{12} \quad -\frac{wl}{2} \quad \frac{-wl^2}{12}\right]$$

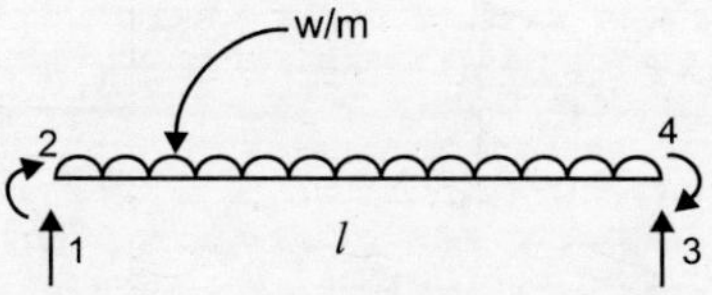

Fig. 11.20

(Here P_1 and P_3 are evidently the support reactions with their signs reversed)

Example 11.4 *For the beam in Fig. 11.18, determine the {P} vectors and the {F} vector by equivalent loads method.*

Solution

$$\{P\}_1 = \{P\}_2 = \{P\}_3 = \left[-\frac{wl}{2} \quad \frac{wl^2}{12} \quad -\frac{wl}{2} \quad \frac{-wl^2}{12}\right]^T$$

$$\{F\}^T = [wl \quad 0 \quad wl \quad 0]$$

Example 11.5 *A long side wall ABCD of a rectangular water tank 5 m × 2 m × 2.5 m is acted upon by water pressure (Fig. 11.21). It is divided into elements 0.2 m long and 0.12 m high. Find the 12 nodal forces on an element near the base given that*

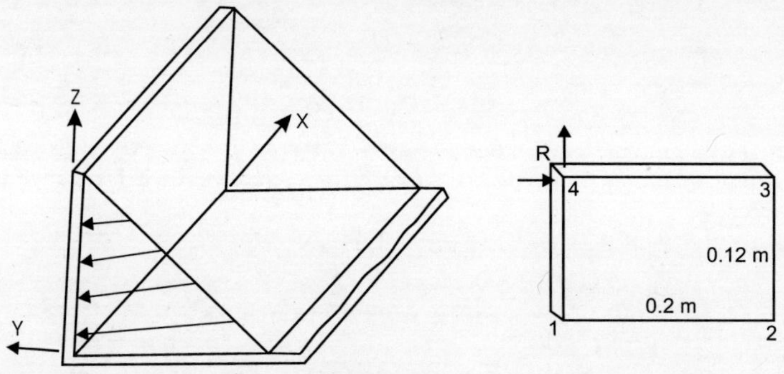

Fig. 11.21

Water pressure at bottom	$= 25\ kN/m^2$
Water pressure at 0.12 m from bottom	$= 23.8\ kN/m^2$
Vertical load on top of element	$= 19\ kN/m$
Vertical load at bottom of element	$= 19.9\ kN/m$
Tension from short walls at the level 2.38 m from top	$= 23.8\ kN/m$
Tension from short walls at 2.5 m from top	$= 25\ kN/m$

Solution Method of inspection would be adopted to get the nodal forces

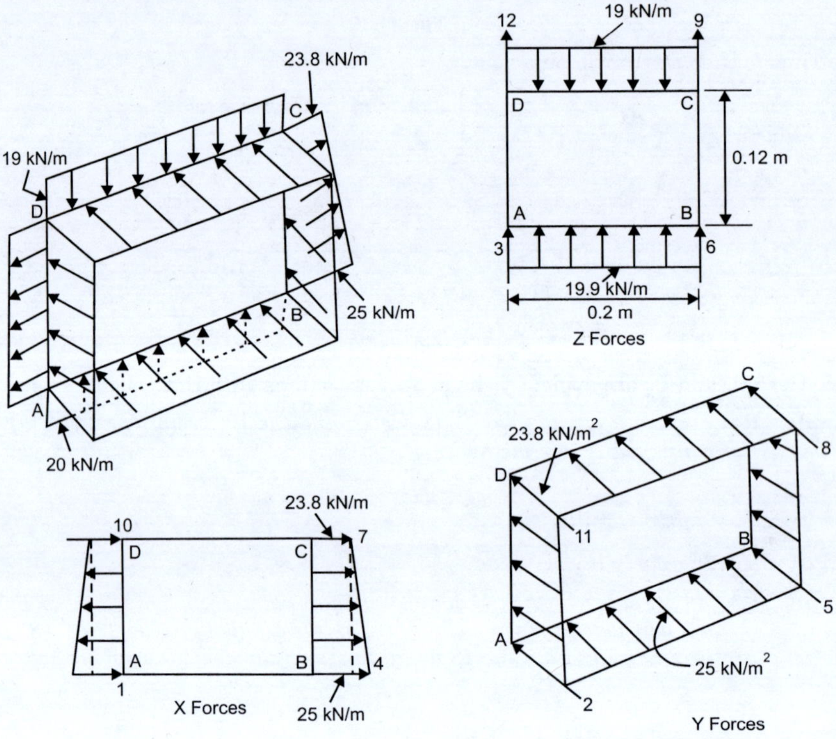

Fig. 11.22

Figure 11.22 shows the X,Y and Z forces. Z forces are, strictly speaking, body forces. X forces are boundary forces Y forces are surface forces. However Z forces would be treated as boundary forces.

Z forces $\quad P_3 = P_6 = 19.9 \times 0.2/2 = 1.99$ kN

$P_9 = P_{12} = 19 \times 0.2/2 = 1.9$ kN

X forces $$-P_1 = P_4 = \frac{23.8 \times 0.12}{2} + \frac{(25 - 23.8) \times 0.12}{2} \times \frac{2}{3}$$

$= 1.428 + 0.048 = 1.476$ kN

$-P_{10} = P_7 = 1.428 + 0.048/2 = 1.452$ kN

Y forces first the forces along lines AB and CD would be computed

Forces on $$\text{CD} = \frac{23.8 \times 0.12 \times 0.2}{2} + 1.2 \times \frac{0.12}{2} \times 0.2 \times \frac{1}{3}$$

$$= 0.2856 + 0.0048$$

$$= 0.2904 \text{ kN}$$

$$P_8 = P_{11} = 0.2904/2 = 0.1452 \text{ kN}$$

Force on AB = 0.2952; $P_2 = P_5 = 0.1476$ kN

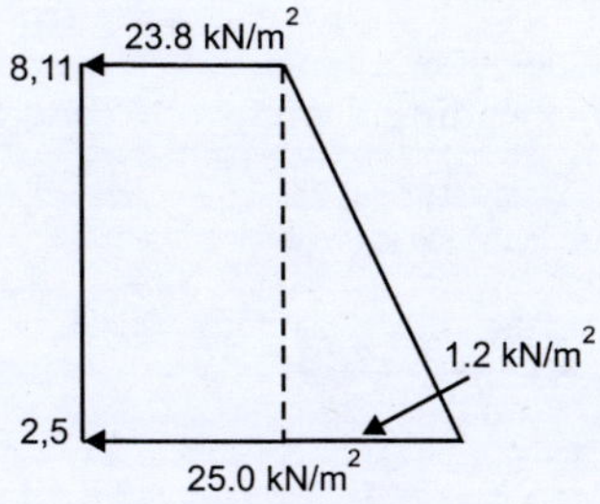

Fig. 11.23

$\{P\}^T$ = [– 1.476, 0.1476, 1.99, 1.476, 0.1476, 1.99, 1.452, 0.1452, 1.90, – 1.452, 0.1452, 1.90]

11.7.2 Variational Method

Principle of virtual work: Consider an element subjected to distributed loading {F} and the equivalent nodal loads be $\{P^e\}$. Let {U}, {δ} are the displacements corresponding to the force vectors {F} and $\{P^e\}$ respectively.

According to the principle of virtual work,

$$\{\delta\}^T \{P^e\} = \int_l \{U\}^T \{F\}\, dx \qquad \text{...(1)}$$

The displacements {U} in terms of nodal displacements {δ} is given by

$$\{U\} = [N]\ \{\delta\}$$

$$\{U\}^T = [N]^T\ \{\delta\}^T \qquad \text{...(2)}$$

Substituting equation (2) in (1),

$$0 = \int_l [N]^T \{\delta\}^T \{F\}\, dx - \{\delta\}^T \{P^e\}$$

$$\{\delta\}^T \left\{ \int_l [N]^T \{F\}\, dx - \{P^e\} \right\} = 0$$

$$\{P^e\} = \int_l [N]^T \{F\}\, dx$$

Example 11.6 *A plane beam element of length 'l' is subjected to a uniformly distributed load of intensity w per metre run over the entire span. Using the shape functions, derive the equivalent joint loads.*

Solution

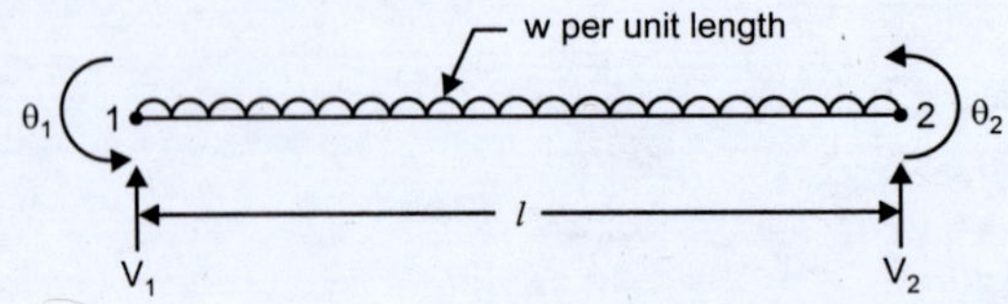

1, 2 → Nodes
V_1, V_2, θ_1, θ_2
→ displacement coordinates

Fig. 11.24. Beam element with nodal displacements.

Upward forces and anticlockwise moments are assumed positive. The shape functions of this beam element with corresponding degrees of freedom are given by,

$$N_1 = 1 - \frac{3x^2}{l^2} + \frac{2x^3}{l^3}$$

$$N_2 = x - \frac{2x^2}{l} + \frac{x^3}{l^2}$$

$$N_3 = \frac{3x^2}{l^2} - \frac{2x^3}{l^3}$$

$$N_4 = \frac{-x^2}{l} + \frac{x^3}{l^2}$$

We know that

$$\{U\} = [N]\{\delta\}$$

$$\{U\} = [N_1 \quad N_2 \quad N_3 \quad N_4]\begin{Bmatrix} \delta_1 \\ \delta_2 \\ \delta_3 \\ \delta_4 \end{Bmatrix}$$

$$\{U\} = \left[\left(1 - \frac{3x^2}{l^2} + \frac{2x^3}{l^3}\right)\left(x - \frac{2x^2}{l} + \frac{x^3}{l^2}\right)\left(\frac{3x^2}{l^2} - \frac{2x^3}{l^3}\right)\left(\frac{-x^2}{l} + \frac{x^3}{l^2}\right)\right]\begin{Bmatrix} \delta_1 \\ \delta_2 \\ \delta_3 \\ \delta_4 \end{Bmatrix}$$

We have $$\{P^e\} = \int_0^l [N]^T \{F\}\, dx$$

The force acting on the beam

$$\{F\}\, dx = \{-w\}\, dx$$

(Negative sign indicates load direction downward)

Transform of shape function matrix

$$[N]^T = \begin{bmatrix} \left(1 - \dfrac{3x^2}{l^2} + \dfrac{2x^3}{l^3}\right) \\ \left(x - \dfrac{2x^2}{l} + \dfrac{x^3}{l^2}\right) \\ \left(\dfrac{3x^2}{l^2} - \dfrac{2x^3}{l^3}\right) \\ \left(\dfrac{-x^2}{l} + \dfrac{x^3}{l^2}\right) \end{bmatrix}$$

Equivalent joint loads

$$\{P^e\} = \int_0^l \begin{bmatrix} \left(1 - \dfrac{3x^2}{l^2} + \dfrac{2x^3}{l^3}\right) \\ \left(x - \dfrac{2x^2}{l} + \dfrac{x^3}{l^2}\right) \\ \left(\dfrac{3x^2}{l^2} - \dfrac{2x^3}{l^3}\right) \\ \left(\dfrac{-x^2}{l} + \dfrac{x^3}{l^2}\right) \end{bmatrix} \{-w\}\, dx$$

$$= -w \begin{bmatrix} \left(x - \dfrac{3x^3}{3l^2} + \dfrac{2x^4}{4l^3}\right) \\ \left(\dfrac{x^2}{2} - \dfrac{2x^3}{3l} + \dfrac{x^4}{4l^2}\right) \\ \left(\dfrac{3x^3}{3l^2} - \dfrac{2x^4}{4l^3}\right) \\ \left(\dfrac{-x^3}{3l} + \dfrac{x^4}{4l^2}\right) \end{bmatrix}_0^l = \begin{bmatrix} \left(l - l + \dfrac{l}{2}\right) - 0 \\ \left(\dfrac{l^2}{2} - \dfrac{2l^2}{3} + \dfrac{l^2}{4}\right) - 0 \\ \left(l - \dfrac{l}{2}\right) - 0 \\ \left(\dfrac{-l^2}{3} + \dfrac{l^2}{4}\right) - 0 \end{bmatrix}$$

$$= \{P^e\} = -w \begin{bmatrix} +l/2 \\ +l^2/12 \\ +l/2 \\ -l^2/12 \end{bmatrix} = \begin{bmatrix} -wl/2 \\ -wl^2/12 \\ -wl/2 \\ +wl^2/12 \end{bmatrix} \begin{bmatrix} \text{-ve sign indicates downward force} \\ \text{-ve sign indicates clockwise moment} \\ \text{-ve sign indicates downward force.} \\ \text{+ve sign indicates anticlockwise moment} \end{bmatrix}$$

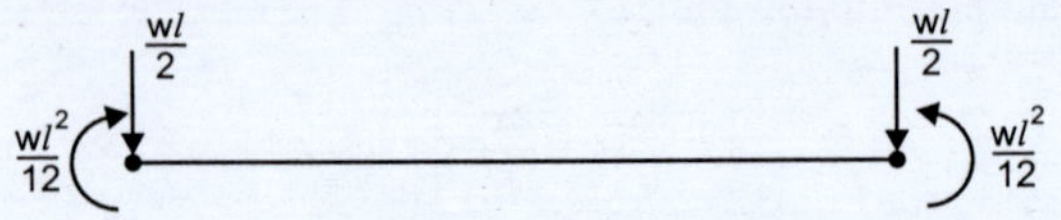

Fig. 11.24. (a) Equivalent joint loads.

Example 11.7 *For the beam with loading and co-ordinates shown in Fig. 11.25, find out the equivalent joint loads using shape functions.*

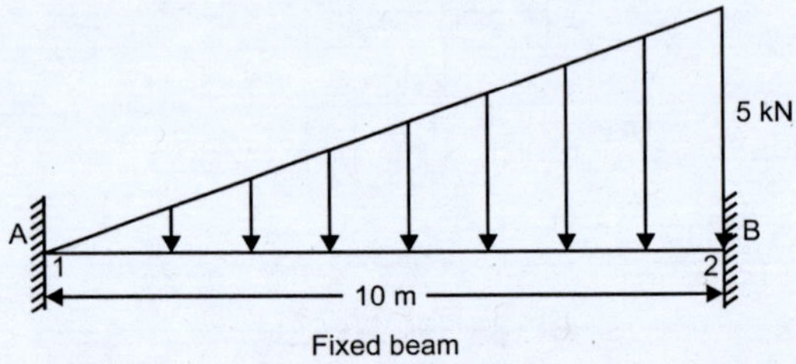

Fig. 11.25

Solution

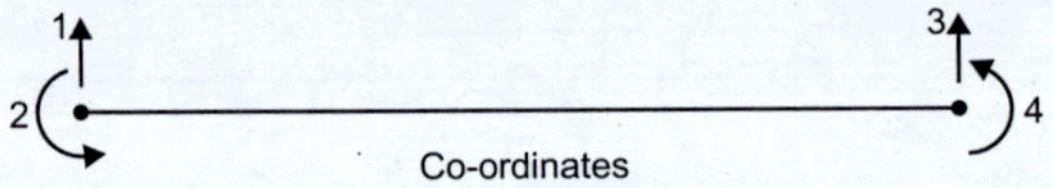

For the beam and co-ordinates given, the shape functions are

$$N_1 = 1 - \frac{3x^2}{l^2} + \frac{2x^3}{l^3}$$

$$N_2 = x - \frac{2x^2}{l} + \frac{x^3}{l^2}$$

$$N_3 = \frac{3x^2}{l^2} - \frac{2x^3}{l^3}$$

$$N_4 = \frac{-x^2}{l} + \frac{x^3}{l^2}$$

$[N] = [N_1 \quad N_2 \quad N_3 \quad N_4]$

= Shape function matrix.

$$[N]^T = \begin{bmatrix} N_1 \\ N_2 \\ N_3 \\ N_4 \end{bmatrix} = \begin{bmatrix} \left(1 - \frac{3x^2}{l^2} + \frac{2x^3}{l^3}\right) \\ \left(x - \frac{2x^2}{l} + \frac{x^3}{l^2}\right) \\ \left(\frac{3x^2}{l^2} - \frac{2x^3}{l^3}\right) \\ \left(\frac{-x^2}{l} + \frac{x^3}{l^2}\right) \end{bmatrix}$$

For the loading given in the beam.

$$\left.\begin{array}{r}\text{Force at any section XX at}\\ \text{a distance '}x\text{' from left end}\end{array}\right\} = \frac{Wx}{l}$$

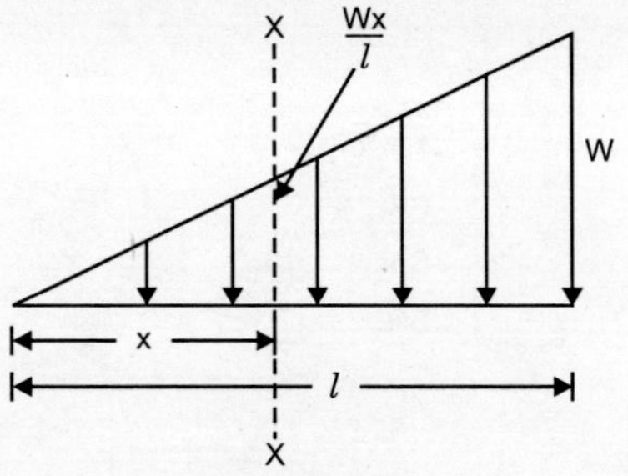

Fig. 11.26

$$= \frac{5x}{10} = \frac{x}{2}$$

$$\{F\} = \left\{\frac{-x}{2}\right\} dx \qquad \left(\begin{array}{l}\text{Negative sign indi-}\\ \text{cates downward load}\end{array}\right)$$

Equivalent joint loads

$$\{P^e\} = \int_0^l [N]^T \{F\}\, dx$$

$$= \int_0^{10} \begin{bmatrix} \left(1 - \dfrac{3x^2}{l^2} + \dfrac{2x^3}{l^3}\right) \\ \left(x - \dfrac{2x^2}{l} + \dfrac{x^3}{l^2}\right) \\ \left(\dfrac{3x^2}{l^2} - \dfrac{2x^3}{l^3}\right) \\ \left(\dfrac{-x^2}{l} + \dfrac{x^3}{l^2}\right) \end{bmatrix} \left\{-\frac{x}{2}\right\} dx = \frac{1}{2}\int_0^{10} \begin{bmatrix} \left(-x + \dfrac{3x^3}{l^2} - \dfrac{2x^4}{l^3}\right) \\ \left(-x^2 + \dfrac{2x^3}{l} + \dfrac{(-x^4)}{l^2}\right) \\ \left(\dfrac{-3x^3}{l^2} + \dfrac{2x^4}{l^3}\right) \\ \left(\dfrac{x^3}{l} - \dfrac{x^4}{l^2}\right) \end{bmatrix} dx$$

$$= \frac{1}{2}\begin{bmatrix} \left(\dfrac{-x}{2} + \dfrac{3x^4}{4l^2} - \dfrac{2x^5}{5l^3}\right) \\ \left(\dfrac{-x^3}{3} + \dfrac{2x^4}{4l} - \dfrac{x^5}{5l^2}\right) \\ \left(\dfrac{-3x^4}{4l^2} + \dfrac{2x^5}{5l^3}\right) \\ \left(\dfrac{x^4}{4l} - \dfrac{x^5}{5l^2}\right) \end{bmatrix}_0^{10} = \frac{1}{2}\begin{bmatrix} \left(\dfrac{-10^2}{2} + \dfrac{3}{4}\times\dfrac{10^4}{10^2} - \dfrac{2\times10^5}{5\times10^3}\right) \\ \left(\dfrac{-10^3}{3}\right) + \dfrac{2\times10^4}{4\times10} - \dfrac{10^5}{5\times10^2} \\ \left(\dfrac{-3\times10^4}{4\times10^2} + \dfrac{2\times10^5}{5\times10^3}\right) \\ \left(\dfrac{10^4}{4\times10} - \dfrac{10^5}{5\times10^2}\right) \end{bmatrix}$$

$$\{P^e\} = \frac{1}{2}\begin{bmatrix} (-50+75-40) \\ (-333.33+500-200) \\ (-75+40) \\ (250-200) \end{bmatrix}$$

$$= \frac{1}{2}\begin{bmatrix} -15 \\ -33.33 \\ -35 \\ 50 \end{bmatrix}$$

$$= \begin{Bmatrix} -7.5 \\ -16.67 \\ -17.5 \\ 25 \end{Bmatrix} \quad \begin{bmatrix} -\text{ve sign} \longrightarrow \text{downward force} \\ -\text{ve sign} \longrightarrow \text{clockwise moment} \\ -\text{ve sign} \longrightarrow \text{downward force} \\ +\text{ve sign} \longrightarrow \text{anticlockwise moment} \end{bmatrix}$$

11.8 NUMBERING SYSTEMS

In a F.E. formulation, the main thing that decides efficiency is the method of discretisation. The next in importance is the method of numbering the nodes in the assemblage, or 'labeling' the assemblage. Figure 11.27 is an assemblage of 10 rectangular elements. Each element has 8 d.o.f. Hence the force vector and displacement vector have a size of 8 × 1. The stiffness matrix will have a size of 8 × 8.

11.8.1 Banded Structure

Most stiffness matrices have a banded structure. This means that on either side of the leading diagonal there are non zero values. But far away from the diagonal, to the right or to the left, the matrix has zero elements. For example, the following matrix is banded.

$$[M] = \begin{bmatrix} 10 & 7 & 2 & 0 & 0 & 0 \\ 7 & 12 & 3 & 7 & 0 & 0 \\ 2 & 3 & 12 & 8 & 6 & 0 \\ 0 & 7 & 8 & 6 & 4 & 2 \\ 0 & 0 & 6 & 4 & 6 & 5 \\ 0 & 0 & 0 & 2 & 5 & 9 \end{bmatrix} \quad \begin{matrix} \text{Band width, } B = 5 \\ \text{Semi-band width, } b = 3 \end{matrix}$$

It's band width is the maximum number of contiguous non zero elements in a row. For the matrix in question, band width B = 5.

Because of banding, the matrix [M] has 26 non zero elements out of a possible 36 elements. When we deal with matrices of 1000 × 1000, a banded structure, say with a semi-band width of 80, considerably saves the demand on storage. Instead of storing 10^6 numbers, it is enough if we store 80,000 numbers. The saving is about 92%.

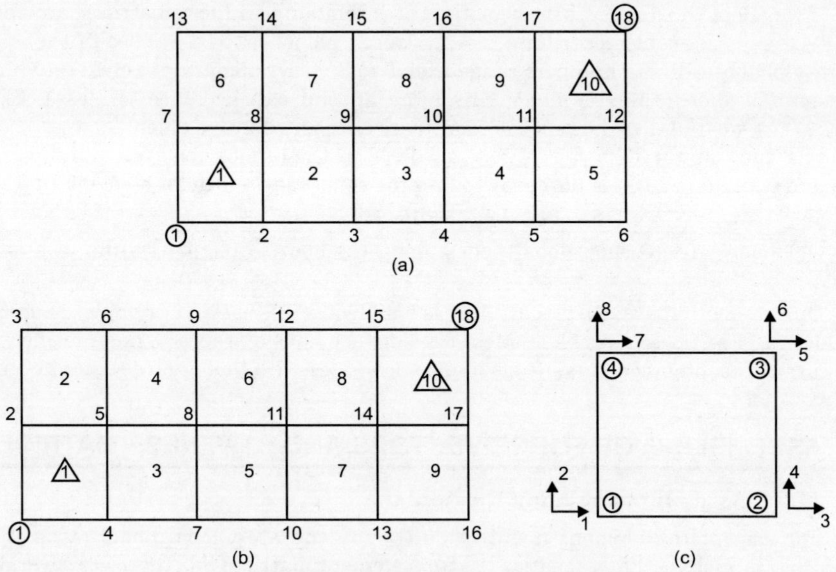

Fig. 11.27. Labeling of elements and nodes.

With this in mind, let us examine the two numbering systems in Fig. 11.27 (*a*) and (*b*). Both have 18 nodes and 18 × 2 = 36 d.o.f. The stiffness matrix (that will have to be inverted) will be of size 36 × 36. This matrix would be assembled by considering the [*k*] matrices of each element (from 1 to 10) and sending the numbers to the appropriate boxes in the 36 × 36 frame work of the global matrix [K].

Let us say we are trying to deal with element 3 (Fig. 11.28).

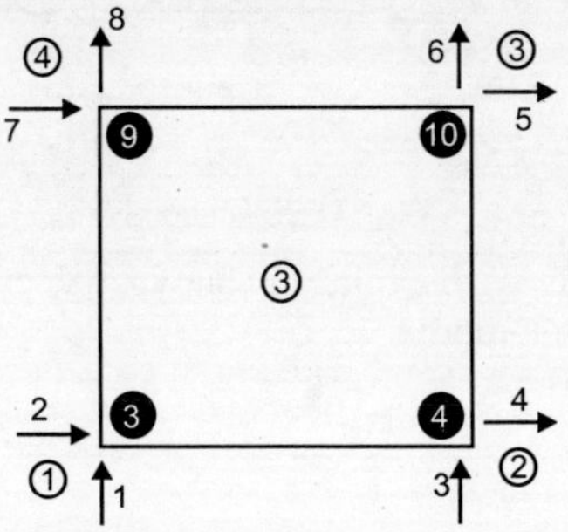

Fig. 11.28

It has 4 nodes ①, ②, ③ and ④. They correspond to nodes ❸, ❹, ❿ and ❾ of the system. The table below shows the correspondence between local and global coordinates.

Local coordinates	1	2	3	4	**5**	6	**7**	8
Global coordinates	5	6	7	8	**19**	20	**17**	18

If we adopt the direct stiffness method of assembling stiffness matrices, any term in the [k] matrix, say $k_{5,7}$ would contribute to $k_{19,17}$ of the global matrix. We see in the above table that the global coordinate numbers range from 5 to 20 (covering a width of 16 columns). When we assemble some other element, this band spread can be more (or less). Examining Fig. 11.27 (a), we find that the range of coordinate numbers due to element 10 would be between column 21 and column 36 (b = 36 – 21 + 1 = 16). For the numbering system in Fig. 11.16, the semi-band width b is 16. It is now easy to find the semi-band width for element 10 (Fig. 11.16).

$$b = (18 - 14 + 1) \times 2 = 10$$

Obviously, numbering global nodes along the short direction results in a lower band width.

Optimum numbering of global nodes is not always this easy. For complex shapes labeling the nodes can itself pose a problem. Most FE software have a preprocessing routine which also takes care of labeling elements, nodes and co-ordinates with due regard to minimum possible band width of [K].

11.9 ASSEMBLING OF FORCE VECTORS AND STIFFNESS MATRICES

11.9.1 Assembling Similar Elements

For assembling element nodal loads {P} to generate Global nodal loads {F} and for generating the global stiffness matrix [K] from element matrices $[k]_i$, there are several methods available. We will describe here the direct stiffness method due to Rubinstein. We will consider a very small assembly of 3 triangular elements (Fig. 11.29).

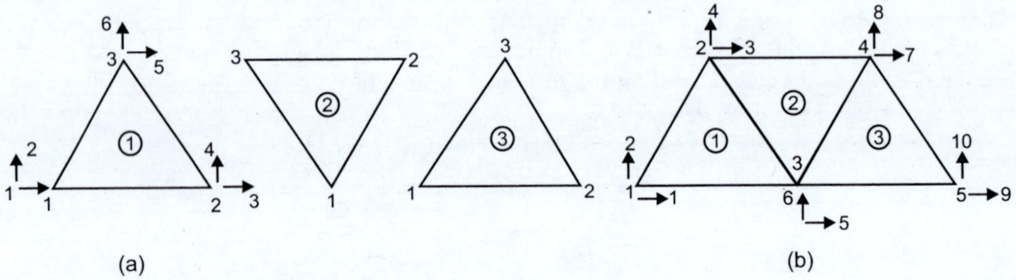

Fig. 11.29. Assembling.

Table 11.1(a) below gives the relation between local coordinate numbers and global coordinates for each of the 3 elements.

Table 11.1 (a)

Element No.	***Coordinates of the elements***	***1***	***2***	***3***	***4***	***5***	***6***
1.	Global coordinates	1	2	5	6	3	4
2.	Global coordinates	5	6	7	8	3	4
3.	Global coordinates	5	6	9	10	7	8

Table 11.1 (b)

				5	6	1	2	3	4	*el*2	
		1	2	3	4	5	6	7	8	9	10
el2	1	element 1 (1 to 6)									
	2										
5	3			k_{55}	k_{56}	k_{51}	k_{52}	k_{53}	k_{54}		
6	4			k_{65}	k_{66}	k_{61}	k_{62}	k_{63}	k_{64}		
1	5			k_{15}	k_{16}	k_{11}	k_{12}	k_{13}	k_{14}		
2	6			k_{25}	k_{26}	k_{21}	k_{22}	k_{23}	k_{24}		
3	7			k_{35}	k_{36}	k_{31}	k_{32}	k_{33}	k_{34}		
4	8			k_{45}	k_{46}	k_{41}	k_{42}	k_{43}	k_{44}		
	9										
	10									Element 3 (5 to 10)	

Assembling members of element 2 into the global matrix [K] (3 to 8)

Table 11.1 (*b*) shows the regions into which the stiffness terms of the 3 elements contribute to the global stiffness matrix. This scheme can be effectively adopted for programming for assemblage of element matrices. Having done that, it should be easy to assemble element nodal forces, say of elements 1, 2 and 3 in Fig. 11.29 to form Global nodal forces.

The global load vector would be of size 10. In to this vector, contributions would be made by the 3 elements as below (Table 11.2).

Table 11.2

Element 1	1	2	5	6	3	4	
Element 2	5	6	7	8	3	4	← Global coord. nos.
Element 3	5	6	9	10	7	8	
	1	2	3	4	5	6	← Element coord. nos.

Table 11.2 each element contributes loads to the global coordinate number listed. This can also be relisted as below :

Table 11.3. Contributions to {F} vector.

	P_1	P_2	P_5	P_6	P_3	P_4					el1
Global coord. number			P_7	P_8	P_1	P_2	P_5	P_6			el2
					P_1	P_2	P_5	P_6	P_3	P_4	el3
	1	2	3	4	5	6	7	8	9	10	

11.9.2 Assembling Different Type of Elements

It is common for an assemblage to be made up of several kinds of elements. For examples a floor can be made up of beams and slabs. Its finite element model would then contain beam elements (1D) and plate elements (2D). It is also possible for plane continua to be made up of triangular and quadrilateral elements.

Figure 11.30 shows a combination of linear, triangular and quadrilateral elements.

Example 11.8 *Assemble the elements 1,2 and 3 in Fig. 11.30 to develop the global load vector and global stiffness matrix, given that*

$$\{P_1\}^T = [8\ 0\ 6\ 0\ 2\ 0]$$
$$\{P_2\}^T = [5\ 1\ 3\ 0\ 6\ 0\ 9\ 2]$$
$$\{P_3\}^T = [0\ 2\ 0\ 2]$$

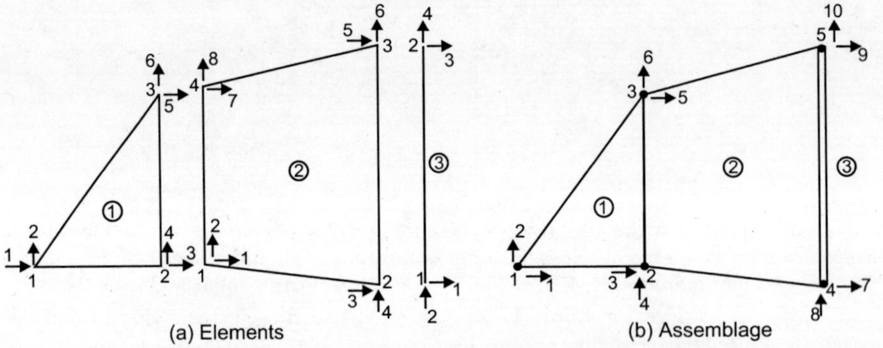

Fig. 11.30

and
$$[k_1] = \begin{bmatrix} 16 & 6 & 3 & 1 & 0.5 & 0.2 \\ 6 & 9 & 4 & 2 & 3 & 0.5 \\ 3 & 4 & 4 & 3 & 2 & 1 \\ 1 & 2 & 3 & 16 & 6 & 3 \\ 0.5 & 3 & 2 & 6 & 9 & 1 \\ 0.2 & 0.5 & 1 & 3 & 1 & 4 \end{bmatrix};$$

$$[k_2] = \begin{bmatrix} 25 & 8 & 6 & 3 & 1.5 & 1 & 0.6 & 0.2 \\ 8 & 9 & 7 & 4 & 3 & 1.2 & 0.8 & 0.6 \\ 6 & 7 & 6 & 4 & 7 & 2 & 1.2 & 1 \\ 3 & 4 & 4 & 3 & 9 & 6 & 3 & 1.5 \\ 1.5 & 3 & 7 & 9 & 25 & 9 & 6 & 2 \\ 1 & 1.2 & 2 & 6 & 9 & 9 & 3 & 3 \\ 0.6 & 0.8 & 1.2 & 3 & 6 & 3 & 6 & 1 \\ 0.2 & 0.6 & 1 & 1.5 & 2 & 3 & 1 & 3 \end{bmatrix}; \ [k_3] = \begin{bmatrix} 18 & 0 & 4 & 0 \\ 0 & 12 & 0 & 0 \\ 4 & 0 & 18 & 0 \\ 0 & 0 & 0 & 12 \end{bmatrix}$$

Solution The element coordinate – global coordinate relation is tabulated under:

Element coords	*1*	*2*	*3*	*4*	*5*	*6*	*7*	*8*	
el1	1	2	3	4	5	6	–	–	Global
el2	3	4	7	8	9	10	5	6	coord.
el3	7	8	9	10	–	–	–	–	nos.

Now we get the global vector F by importing from element vectors into the appropriate squares.

	el1	8	0	6	0	2	0	–	–	–	–
$[F]^T =$	el2	–	–	5	1	9	2	3	0	6	0
	el3	–	–	–	–	–	–	0	2	0	2
		1	2	3	4	5	6	7	8	9	10

$[F]^T = [8\ \ 0\ \ 11\ \ 1\ \ 11\ \ 2\ \ 3\ \ 2\ \ 6\ \ 2]$

The global stiffness matrix [K] is likewise assembled by importing the appropriate *k* terms of the 3 elements into the appropriate squares of the [K] matrix.

		1	2	3	4	5	6	7	8	9	10			
	1	16	6	3	1	0.5	0.2					1		
	2	6	9	4	5	3	0.5					2		
	3	3	4	4,25	3,8	2,0.6	1,0.2	6	3	1.5	1	3	1	
	4	1	2	3,8	16,9	6,0.8	3,0.6	7	4	3	1.2	4	2	
	5	0.5	3	2,0.6	6,0.8	9,6	1,1	1.2	3	6	3	5	7	
[K] =	6	0.2	0.5	1,0.2	3,0.6	1,1	4,3	1	1.5	2	3	6	8	
	7			6	7	1.2	1	6,18	4,0	7,4	2,0		3	1
	8			3	4	3	1.5	4,0	3,12	9,0	6,0		4	2
	9			1.5	3	6	2	7,4	9,0	25,18	9,0		5	3
	10			1	1.2	3	3	2,0	6,0	9,0	9,12		6	4
	el1	1	2	3	4	5	6							
	el2			1	2	7	8	3	4	5	6			
	el3							1	2	3	4			

$$
[K] = \begin{array}{c} \\ \end{array}
\begin{array}{cccccccccc|c}
1 & 2 & 3 & 4 & 5 & 6 & 7 & 8 & 9 & 10 & \\
16 & 6 & 3 & 1 & 0.5 & 0.2 & 0 & 0 & 0 & 0 & 1 \\
6 & 9 & 4 & 2 & 3 & 0.5 & 0 & 0 & 0 & 0 & 2 \\
3 & 4 & 29 & 11 & 2.6 & 1.2 & 6 & 3 & 1.5 & 1 & 3 \\
1 & 2 & 11 & 25 & 6.8 & 3.6 & 7 & 4 & 3 & 1.2 & 4 \\
0.5 & 3 & 2.6 & 6.8 & 15 & 2 & 1.2 & 3 & 6 & 3 & 5 \\
0.2 & 0.5 & 1.2 & 3.6 & 2 & 7 & 1 & 1.5 & 2 & 3 & 6 \\
0 & 0 & 6 & 7 & 1.2 & 1 & 24 & 4 & 11 & 2 & 7 \\
0 & 0 & 3 & 4 & 3 & 1.5 & 4 & 15 & 9 & 6 & 8 \\
0 & 0 & 1.5 & 3 & 6 & 2 & 11 & 9 & 43 & 9 & 9 \\
0 & 0 & 1 & 1.2 & 3 & 3 & 2 & 6 & 9 & 21 & 10
\end{array}
$$

11.10 COMPUTATION OF NODAL DISPLACEMENTS

This is the step that calls for the use of computers desperately. If the size of the system stiffness matrix exceed 5 × 5, (which it will), it is almost impossible to invert the matrix manually. Of course FEM is a totally computer based method.

Once we generate [K] and {F}, we go to the next step of computing $\{u\}$, the displacements at the global coordinates. Since

$$\{F\} = [K]\{u\} \qquad \text{...(11.15)}$$

$$\{u\} = [K]^{-1}\{F\}$$

But [K] as developed in step 11.9 can not be inverted. [K] has to be modified by taking into account the fact that the structure for which [K] is generated has to be secured to the ground. In other words we have to arrest displacements at sufficient number of coordinates of the system so that the structure is stable. When this is not done the [K] matrix would be singular and cannot be inverted. (Its determinant would be zero)

Tampering with the [K] matrix would be a cumbersome affair and would involve a lot of labour.

For example, suppose we want to modify [K] in equation (11.14) so as to arrest displacements u_1, u_2 and u_8 of the structure in Fig. 11.30(*b*), we can take any one of the following courses.

11.10.1 Enforcing Zero Displacement Boundary Conditions

(*a*) **Unit Diagonal Method**

In this method, K matrix would be modified as follows.

K_{11}, K_{22} and K_{88} would be redefined as unity. All other terms in rows 1, 2 and 8 and columns 1, 2 and 8 would be made zero. In addition, F_1, F_2 and F_8 are also replaced with zeros.

This trick would ensure that

$u_1 = u_2 = u_8 = 0$. The altered matrix can be inverted, or we can solve the set of equations to obtain the displacement vector $\{u\}$.

Example 11.9 *Solve the matrix equation* $\{F\} = [K]\{u\}$

where $\{F\}^T = [100, 120, -10]$ *and* [K] *is* $\begin{bmatrix} 12 & 6 & 2 \\ 6 & 48 & 4 \\ 2 & 4 & 24 \end{bmatrix}$. *Make sure that* $u_1 = 0$.

Solution

$$\begin{bmatrix} F_1 \\ F_2 \\ F_3 \end{bmatrix} = [K] \begin{bmatrix} u_1 \\ u_2 \\ u_3 \end{bmatrix}$$

$$\begin{bmatrix} 100 \\ 120 \\ -10 \end{bmatrix} = \begin{bmatrix} 12 & 6 & 2 \\ 6 & 48 & 4 \\ 2 & 4 & 24 \end{bmatrix} \begin{bmatrix} u_1 \\ u_2 \\ u_3 \end{bmatrix} \text{ or } \begin{bmatrix} 0 \\ 120 \\ -10 \end{bmatrix} = \begin{bmatrix} 1 & 0 & 0 \\ 0 & 48 & 4 \\ 0 & 4 & 24 \end{bmatrix} \begin{bmatrix} u_1 \\ u_2 \\ u_3 \end{bmatrix}$$

$$u_1 = 0$$

$$48u_2 + 4u_3 = 120$$

$$4u_2 + 24u_3 = -10$$

$$u_2 = 2.43; \quad u_3 = -0.845$$

$$[u]^T = [0 \quad 2.43 \quad -0.845]$$

(*b*) **Large Diagonal Method**

In this method, the [K] and {F} terms of the coordinates for which displacements is to be made zero are modified as under:

For the [K] matrix the appropriate diagonal term is multiplied by a large number, say 10^{20}. The corresponding force term is modified as zero.

Example 11.10 *For the [K] and {F} values in example 11.9, use the large diagonal method to arrest the displacement at coordinate 3.*

Solution {F} = [K] {*u*}

After modification on coordinate 3

$$\begin{bmatrix} 100 \\ 120 \\ 0 \end{bmatrix} = \begin{bmatrix} 12 & 6 & 2 \\ 6 & 48 & 4 \\ 2 & 4 & 24 \times 10^{20} \end{bmatrix} = \begin{bmatrix} u_1 \\ u_2 \\ u_3 \end{bmatrix}$$

$$100 = 12u_1 + 6u_2 + 2u_3 \quad \text{...}(i)$$

$$120 = 6u_1 + 48u_2 + 4u_3 \quad \text{...}(ii)$$

$$0 = 2u_1 + 4u_2 + 24 \times 10^{20} u_3 \quad \text{...}(iii)$$

From (*iii*), $u_3 \approx 0$

$$100 = 12u_1 + 6u_2$$

$$120 = 6u_1 + 48u_2$$

$$u_1 = 7.56; \ u_2 = 1.56$$

$$\{u\}^T = [7.56 \quad 1.56 \quad 0]$$

(*c*) **Row Column Delete Method**

This method is essentially the same as unit diagonal method. Here we completely eliminate the rows and columns in the stiffness matrix corresponding to coordinates with zero displacements. We would therefore be dealing with a smaller number of equations.

But this is normally not done in computer practice since this will involve renumbering all members of the [*k*] matrix.

For large matrices this can become a really time consuming operation.

In some structures the ground points may sink or move. If the amount of displacement is known, such known displacement will have to be preassigned to the right u_i.

Example 11.11 *For the structure in example 11.9, assign a displacement of 0.02 to u_1.*

Solution Coordinate 1 is to have a pre assigned displacement. Hence make $K_{11} = 1$ and make other members of row 1 and column 1 equal to zero. In addition change F_1 to u_0, in this case 0.02. We get

$$\begin{bmatrix} 1 & 0 & 0 \\ 0 & 48 & 4 \\ 0 & 4 & 24 \end{bmatrix} \begin{Bmatrix} u_1 \\ u_2 \\ u_3 \end{Bmatrix} = \begin{bmatrix} 0.02 \\ 120 - k_{12}u_0 \\ -10 - k_{13}u_0 \end{bmatrix}$$

$$u_1 = 0.02$$

$$48u_2 + 4u_3 = 120 - 6 \times 0.02 = 119.88$$

$$4u_2 + 24u_3 = -10 - 2 \times 0.02 = -10.04$$

$$u_2 = 2.56;\ u_3 = -0.846$$

$$[u]^T = [0.02\ 2.56 = -0.846]$$

11.11 CALCULATION OF PRIMARY UNKNOWNS

Displacements $\{u\}$ are the primary unknowns. They are obtained from equation (11.15).

11.11.1 Features of the Assembled Stiffness Matrix

(*i*) The system stiffness matrix has its non zeros terms clustered about its main diagonal, while locations distant from the diagonal are zero terms. The non zero terms are contained in a band centred on the main diagonal.

(*ii*) The stiffness matrices are symmetric. This characteristic can be used to advantage in storing the matrices.

Half band width

$$[k] = \begin{bmatrix} 5 & 3 & 4 & 0 & 0 & 0 \\ 3 & 6 & 7 & 2 & 0 & 0 \\ 4 & 7 & 4 & 3 & 5 & 0 \\ 0 & 2 & 3 & 9 & 6 & 4 \\ 0 & 0 & 5 & 6 & 7 & 3 \\ 0 & 0 & 0 & 7 & 3 & 5 \end{bmatrix}$$

Since the matrix shown above is symmetric, it is enough if we store the elements above the leading diagonal. Moreover, beyond the band the values are zero. Hence, techniques available to store and use the values above the leading diagonal and within the band can be used. That is, with the leading diagonal as the first column, the values to be stored will be

$$[k] = \begin{bmatrix} 5 & 3 & 4 \\ 6 & 7 & 2 \\ 4 & 3 & 5 \\ 9 & 6 & 4 \\ 7 & 3 & 0 \\ 5 & 0 & 0 \end{bmatrix}$$

The reduction in the band width and the reduction in the number of elements reduces the required storage memory and the solution time.

(*iii*) The diagonal terms are always positive.

Solutions of equations

Here the simultaneous equation (11.15) are solved to obtain the primary unknowns. Equation solving schemes can be generally classified as direct and iterative. Direct solution techniques are generally used for linear problems whereas iterative techniques are required for the solution of non linear problems.

Gaussian elimination and a number of its modifications are examples of direct methods and Jacobi, Gauss Seidel and successive over relaxation (SOR) are examples of iterative techniques.

The choice of technique to be adopted depends upon the size of the problem envisaged and upon the type of computer available.

The following methods can be used for the solution of positive definite, symmetric and banded systems.

(*i*) Cholesky's square root method,
(*ii*) Halfband Gauss elimination solution technique,
(*iii*) Half block technique,
(*iv*) Skyline technique, and
(*v*) Frontal solution technique.

11.12 CALCULATION OF SECONDARY UNKNOWNS

Using the nodal displacements $\{u\}$ obtained in the last step, the stresses within the element are calculated in this step. Here the nodal displacements of a particular element are extracted from the complete set of nodal displacements.

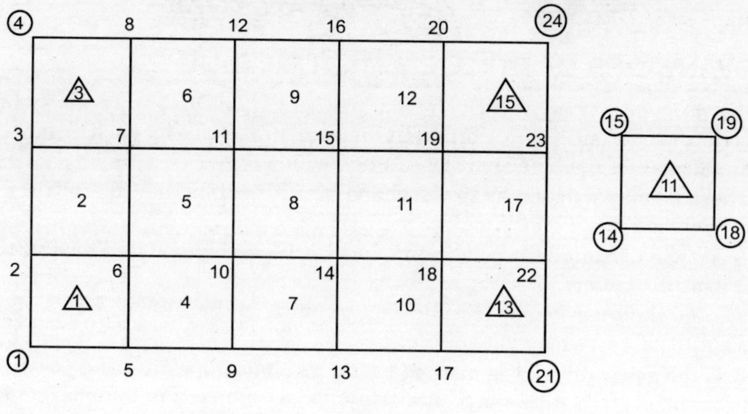

Fig. 11.31

For the configuration in Fig. 11.31 if only one displacement is assumed for each node, the complete set will consist of displacements u_1 to u_{24}. For the calculation of stresses in elements say, 11 the displacements δ_{15}, δ_{14}, δ_{18} and δ_{19} are to be extracted from the complete set.

The strains can be calculated, using

$$\{\varepsilon\} = [B]\ \{\delta\}$$

We already know that [B] is a matrix of local coordinates (x, y) for the element. Hence the strains at different points can be calculated by substituting the coordinates of those point.

However, for a simple triangular element (CST element) the [B] matrix is independent of the coordinates and therefore the strain is constant throughout the element [see eqn. (11.7)].

Using the strains, the stresses can be calculated by using the relation

$$\{\sigma\} = [D]\{\varepsilon\} = [D]\{B\}\{\delta\} \quad \text{...(11.16)}$$

Generally we calculate the stresses and strains at the centroid of the element. After knowing the normal and shear stresses, the principal stresses can be calculated from

$$\sigma_1 \text{ (or) } \sigma_2 = \frac{\sigma_x + \sigma_y}{2} \pm \sqrt{\left(\frac{\sigma_x - \sigma_y}{2}\right)^2 + \tau_{xy}^{\;2}}$$

The maximum shearing stress

$$\tau_{max} = \pm\sqrt{\left(\frac{\sigma_x - \sigma_y}{2}\right)^2 + \tau_{xy}^{\;2}} \text{ or } \tau_{max} = \left(\frac{\sigma_1 - \sigma_2}{2}\right)$$

and the angle defining the principal plane can be obtained using

$$\tan 2\alpha = -\frac{2\tau_{xy}}{\sigma_x - \sigma_y}$$

Standard computer routines give the output in a tabular form or graphical form.

In F.E. software, the presentation of results is done by a module known as post processor. This module includes several refinements of colour graphics. Direct stresses, σ_x, σ_y, shear stress τ_{xy}, principal stresses σ_1, σ_2, maximum shear, von mises stresses, deviatrix stresses etc. are contoured and each zone assigned a colour coding. This enables the user to identify highly stressed areas. This would normally help in interactive redesigning of structures and structural components.

11.13 ADVANTAGES OF FINITE ELEMENT METHOD

1. Since the properties of each element are evaluated separately, an obvious advantage is that we can incorporate different material properties for each element. Thus almost any degree of nonhomogeneity can be included.
2. There is no restriction as to the shape of the medium; hence arbitrary and irregular geometries cause no difficulty.
3. Any type of boundary condition can also be accommodated easily; this can be an important factor.
4. The method can easily handle factors such as non-linearities, arbitrary loading conditions and time dependence.
5. It is the generality of the method that make it so appealing and powerful for solution of problems in a wide range of disciplines in engineering and mathematical physics.

11.14 PITFALLS OF FINITE ELEMENT METHOD

1. It is of dubious utility when applied to problems without a proper physical comprehension of the problem.
2. There are many reasons which make the [K] matrix singular and the solution of the equation

$$\{F\} = [K]\{u\}$$

impossible. Some of the reasons are:

(*i*) Ill conditioned configurations,

(*ii*) Violent variations in element sizes used,

(*iii*) Inadequate arresting of boundary displacements, and

(*iv*) Improper choice of units – resulting in very large or very small numbers in the computational procedures.

3. Proper cross checks are needed by applications to known/solved problems. Otherwise we might end up with gross errors. Hence, however sophisticated the software, it is necessary to validate the same by applying static checks and from solved cases. The validation is more for our understanding of the use of software, especially in making inputs.

11.15 CONVERGENCE

We saw that a CST element is incapable of showing a change in strain (hence stress) within the element. Thus a stress variation like in curve A is apt to be shown as in B (Fig. 11.32).

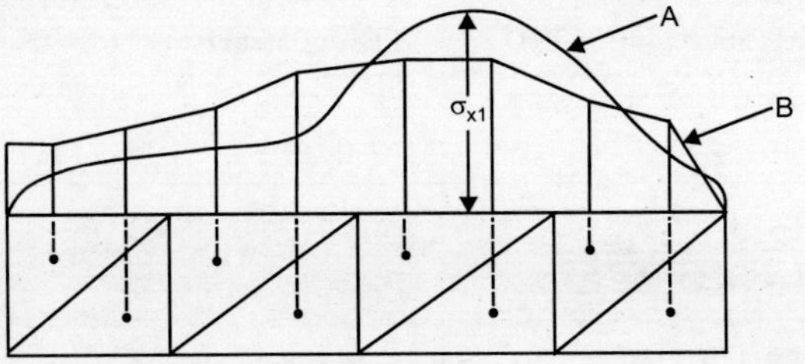

Fig. 11.32

Convergence means the process of successively moving towards the exact or correct solution.

The approximation would be better if we have more elements. That does not mean we can have 20,000 elements instead of 8. For one thing, it will result in enormous computing time. For another, it will end up with considerable round off errors. So there is an optimum number of elements for best results. Suppose we want to find σ_{x1} in Fig. 11.32.

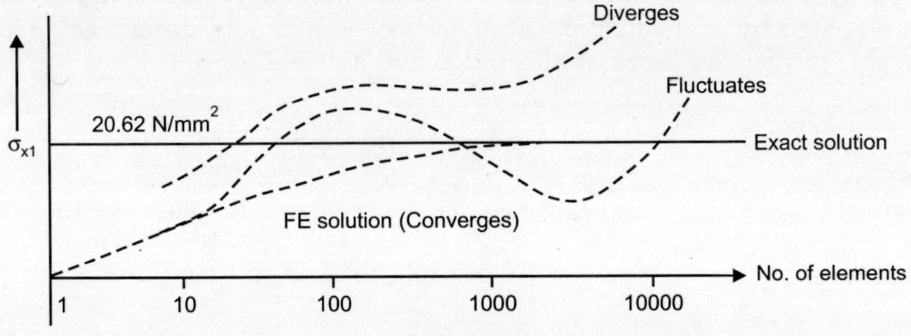

Fig. 11.33

Figure 11.33 shows that, as we make the mesh finer (and elements more in number), our FEA shows a better and better estimate of σ_{x1}. This shows our analysis converges to the exact value. Wrong formulations and improper choice of elements may hamper convergence and result in divergence or fluctuations.

When in the process, the successive approximate solutions approach the exact solution without changing sign or direction, it is termed as monotonic convergence. On the other hand if the successive solutions shows fluctuation in results, it is termed as non-monotonic convergence.

11.16 COMMONLY USED F.E. SOFTWARE

Since its introduction in the sixties, FEM has grown in sophistication and versatility. They have also become so very user friendly, that we can (and many do) use them for analyzing and designing very complex systems even without a knowledge of FEM. This of course is highly objectionable and risky and shall be scrupulously avoided. Table 11.4 lists some of the more common FE softwares and their capabilities.

Table 11.4 Finite element softwares.

S.No.	*Name of software*	*Capabilities*
1.	ANSYS	Elastic plastic thermal, Non linear, Vibration
2.	SAP 2000 and SOLID SAP, NONSAP	FEA of structures 3D modeling, Non-linear analysis
3.	LUCAS	Analysis and design of bridges, buildings
4.	STAAD PRO	Analysis and design of structures
5.	STRAP	Bridges, R.C.
6.	STRUDL	Comprehensive FEA
7.	CADS	Analysis and design
8.	NISA-CIVIL	Analysis and design of buildings, civil structures

Generally, the more general a program, the more inputs it would need. It is better to procure job specific design software.

India, it must be realized, is rich in software industry. There is always more value for software prepared by a team experienced in the analysis and design; this is because they would be able to visualize more contingencies and possibilities in the process.

PROBLEMS

1. Given that $u = 0.012 + 0.2x + 0.13x^2 + 0.05y + 0.005y^2 + 0.11xy$ determine the expression for ε_x.

[**Ans.** $0.2 + 0.26x + 0.11y$]

2. In problem 1, determine ε_x at the 3 nodes of a triangle, A (0, 0), B(1, 0) and C(0, 1).

[**Ans.** 0.2, 0.46, 0.31]

3. 2 triangular 6 d.o.f. elements join to make an 8 d.o.f. assemblage. The nodal load vectors of the elements are

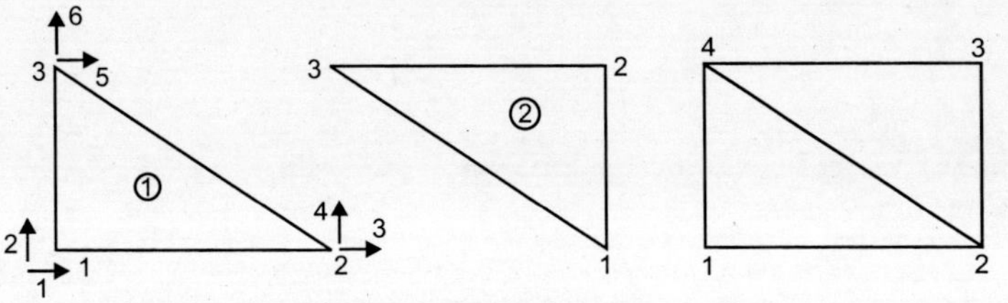

Fig. 11.34

$$\text{element 1 } \{P\}^T = [0 \quad 1 \quad 0 \quad 2 \quad 0 \quad 3]$$
$$\text{element 2 } \{P\}^T = [1 \quad 0 \quad 2 \quad 2 \quad 3 \quad 0]$$

determine the global load vector.

[**Ans.** $[F]^T = [0 \quad 1 \quad 1 \quad 2 \quad 2 \quad 2 \quad 3 \quad 3]$

4. In an efficient FE program how would the following system matrix be stored?

$$k = \begin{bmatrix} 5 & 2 & 0 & 0 \\ 2 & 3 & 4 & 0 \\ 0 & 4 & 6 & 7 \\ 0 & 0 & 7 & 8 \end{bmatrix}$$

$$\left[\textbf{Ans.} \begin{bmatrix} 5 & 2 \\ 3 & 4 \\ 6 & 7 \\ 8 & 0 \end{bmatrix} \right]$$

5. For a 3 nodal 6 d.o.f. element the following displacement model is proposed

$$u = ax^2 + bx + c$$
$$v = dy^2 + ey + f$$

Comment on its validity. [**Ans.** No good. Skips row 2 of Pascal triangle]

6. In a 3 noded CST element the displacement model for u is $u = ax + by + c$. After processing, nodal displacements are found to be $u_1 = 0$; $u_2 = 0.0172$ and $u_3 = 0.010$. The nodes are A(0,0), B(0,4) and C(4,4). Compute ε_x. [**Ans.** $\varepsilon_x = -0.0068$]

7. A rectangular plate analysed with 4 elements is found to have nodal displacements (in microns) as shown in Fig. 11.35. Find the strains in element 3, if the displacement model is $u = a_1x + b_1y + c_1$ and $v = a_2x + b_2y + c_2$.

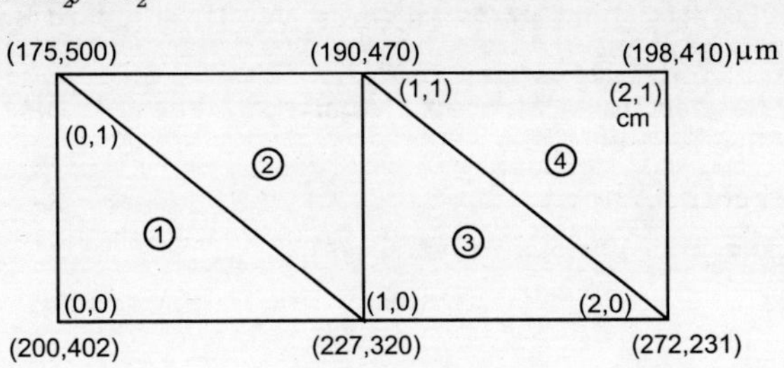

Fig. 11.35

[**Ans.** $\{\varepsilon_x\}^T = [45 \quad 150 \quad -126] \times 10^{-4}$]

Bibliography

1. Punmia, B.C., Ashok Jain and Arun Jain, Theory of Structures, Laxmi Publications, New Delhi, 1992.
2. Gupta, S.P., Pandit, G.S., Gupta, R., Theory of Structures, Vol. I & II, Tata McGraw-Hill Publishing Company Limited, New Delhi, 1999.
3. Prakash Rao, D.S., Structural Analysis, University Press, 1996.
4. Ramamrutham, S., Theory of Structures. Dhanpat Rai & Sons, Delhi, 1986.
5. Reddy, C.S., Basic Structural Analysis, Tata McGraw-Hill Publishing Company Limited, New Delhi, 1997.
6. Khurmi, R.S., Theory of Structures, S. Chand & Company Ltd., New Delhi, 1998.
7. Junnarkar, S.B. and Shah, H.J., Mechanics of Structures—Volume-II, Charotar Publishing House, Anand, 1996.
8. Negi, L.S. and Jangid, R.S., Structural Analysis, Tata McGraw-Hill Publishing Company Limited, New Delhi, 1997.
9. Anand, A.S., Theory of Structures, Satya Prakashan, New Delhi, 1989.
10. Vazirani, V.N., Ratwani, M.M., Analysis of Structures, Khanna Publishers, Delhi, 1976.
11. Ramchandra. Dr., Design of steel structures, Vol. II, Standard Book House, New Delhi, 1991.
12. Kinney, J.S., Indeterminate Structural Analysis, Addison-Wesley Publishing Company, Inc., Massachusetts, 1962.
13. Lecture Notes of the Authors.
14. Question Papers from Various Universities.

INDEX

R

S

T

U

V

W